SOLAR ENERGY

Solar Energy

Edited by

Radu D. Rugescu

Solar Energy

Edited by Radu D. Rugescu

CBS, Edition 2016

Published by InTech

Janeza Trdine 9, 51000 Rijeka, Croatia

Technical Editor Teodora Smiljanic

Cover Designer Dino Smrekar

Additional hard copies can be obtained from orders@intechopen.com

Solar Energy, Edited by Radu D. Rugescu
p. cm.
ISBN 978-953-307-052-0

Preface

The present book on "Solar Energy" is entitled to reveal some of the latest concepts and research in the direct exploitation of solar energy and incorporates eighteen chapters, authored by fifty-two selected, international contributors. All these contributions are developed in the two of the areas that we believe to be the most promising and efficient in solar energy: the new generation of PV cells and the solar array-gravity draught accelerators. From new advances in the Material Science to provide high efficiency photocells, up to the airborne solar "chimney" and the closed-circuit chimney gravitational accelerator for use on Mars and Moon, the original authors explain the new concepts in a high-level, first-hand presentation, which characterizes the entire book. The chapters are designed to gradually attract the interest of the reader by means of their content. Despite the small space available, or rather surely due to this constraint, the stories of the new technologies are presented in a very synthetic and easy-to-read manner, a necessary quality for time saving today.

The present "Solar Energy" science book hopefully opens a series of other firsthand texts in new technologies with practical impact and subsequent interest. They might include the ecological combustion of fossil fuels, space technology in the benefit of local and remote communities, new trends in the development of secure Internet Communications on an interplanetary scale, new breakthroughs in the propulsion technology and others. The editors will be pleased to see that the present book is open to debate and they will wait for the readers' reaction with great interest. Critics and proposals will be equally welcomed.

The editor addresses special thanks to the contributors for their high quality and courageous initiative, and to the Technical Corp. of editors that conveyed the text into a pleasant and cozy presentation.

Bucharest, January 18, 2010

Editor

Prof. Dr. Eng. Radu D. Rugescu
University Politehnica of Bucharest
Romania E.U.

Contents

1

Potential of the Solar Energy on Mars

Dragos Ronald Rugescu[1] and Radu Dan Rugescu[2]
[1]University of California at Davis,
[2]University Politehnica of Bucharest
[1]U.S.A.,
[2]Romania E.U.

1. Introduction

The problem of energy accessibility and production on Mars is one of the three main challenges for the upcoming colonisation of the red planet. The energetic potential on its turn is mainly dependent on the astrophysical characteristics of the planet. A short insight into the Mars environment is thus the compulsory introduction to the problem of energy on Mars. The present knowledge of the Martian environment is the result of more than two centuries of attentive observation on its astronomical appearance and, more recently, on its on-site astrophysical features. Recent surface measurements of Martian geology, meteorology and climate had fixed the sometime-unexpected image of a completely desert planet. Mars is one of the most visible of the seven planets within the solar system and thusfor its discovery cannot be dated, still the interest for Mars is old. It was easily observed from the ancient times by necked eye and the peculiar reddish glance of the planet had induced the common connection of the poor planet with the concept of war. The god of war and the planet that inherited his name had provoked, still from antiquity, curiosity and disputes on the most diverse themes. These disputes are at a maximum right now regarding the habitability of Mars. The red planet owes his color to still unexplained causes, where a yet undisclosed chemistry of iron oxides seems to be the main actor. The visit card of Mars is fast increasing in the quantity of data and is now quite well known (Bizony, 1998), as we observe from the description that follows.

1.1 Mars as seen before the space age

As far as the knowledge of the solar system has gradually extended, from optical, ground-based observations to the present astrophysical research on site, Mars appears as the fourth planet as starting from the Sun. The reddish planet of the skies, nicely visible by necked eyes, has attracted the most numerous comments during the time regarding the presence of life on an extraterrestrial planet. With all other eight planets, except for Pluto-Charon doublet, Mars aligns to a strange rule by orbiting the Sun at a distance that approximates a multiple of $\sqrt{2}$ from that of the Earth. This means that the rough 149.6 mil km of the Earthsemi-major axis is followed by a rough 212 mil km for Mars. In fact there are 227.92 mil km at mean from the center of Sun. The power rule of Titius-Bode[1], modified several times, but originally described as $a = (4 + 3 \times \operatorname{sgn} n \times 2^{n-1}) / 10 \,|\, n = \overline{0,9}$ gives a better distribution,

Planet	n	Titius-Bode rule	Actual semi-major axis
Mercury	0	0.4	0.39
Venus	1	0.7	0.72
Earth	2	1.0	1.00
Mars	3	1.6	1.52
Asteroids	4	2.8	2.80
Jupiter	5	5.2	5.20
Saturn	6	10.0	9.54
Uranus	7	19.6	19.20
Neptune/Pluto	8	38.8	30.10/39.20
Sedna	9	77.2	75.00

Table 1. Mars within Titius-Bode's rule (astronomical units)

It is immediately seen that the primary solar radiation flux is roughly two times smaller for Mars than it is for Earth. More precisely, this ratio is equal to 2.32. This observation for long has suggested that the climate on Mars is much colder than the one on Earth. This has not removed however the belief that the red planet could be inhabited by a superior civilization. Nevertheless, beginning with some over-optimistic allegations of Nicolas Camille Flammarion (*Flamarion*, 1862) and other disciples of the 19-th century, the planet Mars was for a century considered as presenting a sort of life, at least microbial if not superior at all. The rumor of Mars channels is still impressing human imagination. When estimates begun to appear regarding the Martian atmosphere and figures like 50 mbar or 20 mbar for the air pressure on Martian ground were advanced (Jones 2008), a reluctant wave of disapproval has been produced. It was like everybody was hoping that Mars is a habitable planet, that we have brothers on other celestial bodies and the humankind is no more alone in the Universe. As more data were accumulating from spectroscopic observations, any line of emission or absorption on Mars surface was immediately related to possible existence of biological effects. Even during the middle 20-th century the same manner was still preserving. In their book on "Life in the Universe" Oparin and Fesenkov are describing Mars in 1956 as still a potential place for biological manifestations (*Oparin & Fesenkov*, 1956). The following two excerpts from that book are relevant, regarding the claimed channels and biological life on Mars: "...up to present no unanimous opinion about their nature is formed, although nobody questions that they represent real formations on the planet (Mars)..." and at the end of the book "On Mars, the necessary conditions for the appearance and the development of life were always harsher than on Earth. It is out of question that on this planet no type of superior form of vegetal or animal life could exist. However, it is possible for life, in inferior forms, to exist there, although it does not manifest at a cosmic scale."

---

[1] In 1768, Johann Elert Bode (1747-1826), director of Berlin Astronomical Observatory, published his popular book, "Anleitung zur Kenntnis des gestirnten Himmels" (*Instruction for the Knowledge of the Starry Heavens*), printed in a number of editions. He stressed an empirical law on planetary distances, originally found by J.D. Titius (1729-96), now called "Titius-Bode Law".

The era of great *Mars Expectations*, regarding extraterrestrial life, took in fact its apogee in 1938, when the radio broadcast of Howard Koch, pretending to imaginary fly the coverage of Martian invasion of Earth, had produced a well-known shock around US, with cases of extreme desperation among ordinary people. Still soon thereafter, this sufficed to induce a reversed tendency, towards a gradual diminution of the belief into extraterrestrial intelligence and into a Martian one in particular. This tendency was powered by the fact that no proofs were added in support of any biological evidence for Mars, despite the continuous progress of distant investigations. Still every of the 36 great oppositions of Mars, since the "canali" were considered by Giovanni Schiaparelli in 1877, prior to the space age in 1957, like the series in 1901, 1911, 1941, 1956 was only adding subjective dissemination of channels reports, with no other support for the idea that on Mars any form of biological life exists. After Schiaparelli and Flammarion, the subjective belief in a Martian life is successively claimed by the well known astronomers Antoniadi, Percival Lowel, A. Dollfus, G. A. Tihov and others.

Any documentation of a hostile environment on Mars was received with adversity. Despite later spectroscopic and radiometric measurements from Earth, which were revealing a very thin atmosphere and extreme low temperatures, still in the immediate down of the space age the pressure on the Martian soil was yet evaluated at 87 mbar (*Oparin & Fesenkov*, 1956), overrating by more than ten times the actual value, irrefutably found after 1964. It is a pregnant evidence of how subjective the world could be in administrating even the most credible scientific data in this delicate subject. A piece of this perception is surviving even today.

1.2 Mars during the space age

With the availability of a huge space carrier, in fact a German concept of Görtrupp, the Soviets started to built a Mars spacecraft during 1959, along to the manned spacecraft Vostok. The launch took place as early as in October 1960, mere 3 years after Sputnik-1, but proved that only mechanical support is of no much use. The restart of the accelerator stage from orbit was a failure that repeated a few weeks later with similar non-results. The boost towards Mars commenced again with Mars-1 and a companion during the 1962 window, ending in failure again. It followed this way that a much smaller but smarter US device, called Mariner-4, despite of the nose-hood miss-separation of its companion Mariner-3, had marked the history of Mars in 1964 with a shaking fly-by of the planet and several crucial observations. These stroke like a thunder: the radio occultation experiment was suddenly revealing an unexpectedly low atmospheric pressure on Mars of approximately 5 mbar, much far below the most of the previous expectations. The life on Mars was bluntly threatened to became a childish story. Primitive but breathtaking images from Mariner-4 were also showing a surface more similar to the Moon's one than any previous expectation could predict. A large number of craters were mixed with dunes and shallow crevasses to form a desert landscape and this appearance produced a change in the perception of the formation of the solar system at large. No channel was revealed and none of the previously mentioned geometrical marks on the Martian surface. The wave of disappointment grew rapidly into a much productive bolster for deepening these investigations, so incredible those results were. Mars's exploration proceeded with Mariner-6 and 7 that performed additional Martian fly-byes in 1967 only to confirm the portrait of a fully deserted planet.

The present hostile environment on Mars and the absence of life seem for now entirely proven, but it still remains to understand when this transform took place, whether there was sometime a more favorable environment on Mars and all issues regarding the past of Mars.
We say *for now* because our human nature bolsters us towards an ideal which we only dream of, but which, as we shall prove, is almost accomplishable to the level of present technologies. Anyhow, the nephews of our nephews will perhaps take it up to the end. We are speaking of Mars's colonization, process that could only take place after the complete transformation of the surface and of the atmosphere to closely resemble those on Earth, process we call terraforming. Whether planet Mars worth more than its terraforming, then it deserves all the money spent with this process.
For the moment however, the Martian environment is far of being earth-like, as the compared data from the next table reveal (*Almanac* 2010):

Parameter	Mars	Earth	units.
Orbital eccentricity	ε = 0,09341233	0.01669518	-
Semi-major axis	a = 227,920,000	149,597,871	km
Focal semi-chord	p = 225,931,201	149,556,174	km
Perihelion	r_P= 206,629,462	147,100,308	km
Aphelion	r_A= 249,210,538	152,095,434	km
Inclination to ecliptic	i = 1.84964°	0°	°
Ascending node	Ω = 49.6296364°	0°	°
Perihelion argument	ω = 336.04084°	102.94719°	°
Sideral revolution	T = 59355072	31558118	s
Equatorial radius	R_e= 3,397,515.0*	6,378,136.6	m
Flattening factor	f = 1/154.321	1/298.25642	-
Rotational velocity	w =7.088244 $\cdot 10^{-05}$	7. 292700 $\cdot 10^{-05}$	rad/s
Obliquity	v = 25.19°	23.45°	°
Gravitational field**	K = 4.28283149 $\cdot 10^4$	3.986004391 $\cdot 10^5$	km^3/s^2
Solar irradiance	q = 589.2	1367.6	W/m^2
Ground pressure***	p_0 = 6.36	1013.25	mbar
Ground temperature	T_0= 210.0	283.0	°K .

* (JPL 1998)
** Earth without Moon
*** Mean for Martian latitude and year

Table 2. Comparative Mars and Earth astrophysical data

The data in table 2 are based on the mean eccentricities of the year 2009 as given in the reference. Accordingly, a value of g=6.67428 $\cdot 10^{-11}$ is used for the universal constant of gravitation. Second is the time unit used to define the sidereal periods of revolution around the Sun and is derived, on its turn, from the solar conventional day of 24 hours in January 1, 1900. The atmosphere of Mars is presently very well known and consists, in order, of Carbon dioxide (CO_2) 95.32%, Nitrogen (N_2) 2.7%, Argon (Ar) 1.6%, Oxygen (O_2) 0.13%, Carbon monoxide (CO) 0.07%, Water vapor (H_2O) 0.03%, Nitric oxide (NO) 0.013%, Neon (Ne) 2.5 ppm, Krypton (Kr) 300 ppb, Formaldehyde (H_2CO) 130 ppb [1], Xenon (Xe) 80 ppb, Ozone (O_3) 30 ppb, Methane (CH_4) 10.5 ppb and other negligible components. This composition is further used to assess the effect of solar radiation upon the dissociation and los of the upper Mars atmosphere and upon potential greenhouse gases.

The present atmosphere of Mars is extremely tenuous, below 1% of the Earth one and seemingly unstable. Seasonal worming and cooling around the poles produce variations of up to 50% in atmospheric pressure due to carbon dioxide condensation in winter. The values in table 18.2 are rough means along an entire year, as measured by Viking-1 and Viking-2 lenders, Pathfinder and Phoenix station recently. The greenhouse effect of carbon dioxide is considered responsible for 5° increment of atmospheric temperature, very low however, with only occasional and local worming above 0°C on privileged equatorial slopes. The chart of the present Martian atmosphere (*Allison & McEwen*, 2000) is given in figure 18.1. The exponential constant for pressure on Mars is H=11,000 m.

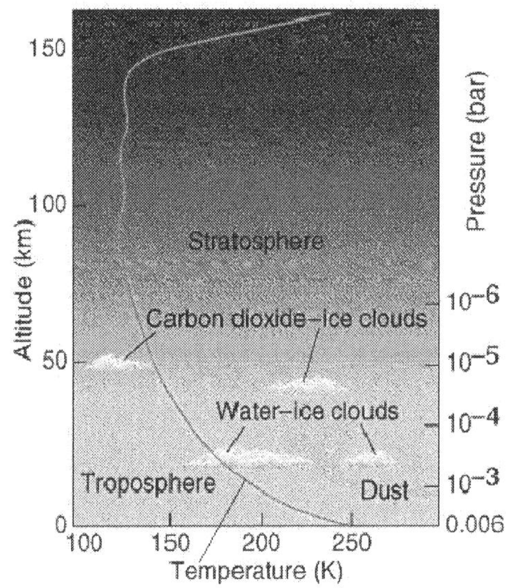

Fig. 1. Mars's atmosphere profile.

These atmospheric characteristics stand as the database in evaluating the efficiency of the solar-gravitational draught, closed-circuit powerplant, which we propose to be employed as an energy source on Mars.

1.3 Mars presumptive past

It is generally considered that the planetary atmospheres, including that of Mars, went through major transformations at the beginning of formation of the solar system (Brain 2008). Present models show a gradual and fast depletion of Martian atmosphere, with essential implications for designing the Mars's terraforming. Even today, the depletion of Mars's atmosphere continues on a visible scale. The observations, recently made by Mars Express and Venus Express, have allowed scientists to draw similar conclusions for both Mars and Venus, through direct comparisons between the two planets. The phenomenon proves this way as nonsingular and systematic (*Brain,* 2008). The results have shown that both planets release beams of electrically charged particles to flow out of their atmospheres. The particles are being accelerated away by interactions with the solar wind released by the Sun. This phenomenon was observed by the two spacecrafts while probing directly into the magnetic regions behind the planets, which are the predominant channels through which electrically-charged particles escape. The findings show that the rate of escape rose by ten times on Mars when a solar storm struck in December 2006. Despite the differences among the two atmospheres, the magnetometer instruments have discovered that the structure of the magnetic fields of both planets is alike. This is due to the fact that the density of the ionosphere at 250 km altitude is surprisingly similar, as shows the Venus Express magnetometer instrument. By observing the current rates of loss of the two atmospheres, planetary scientists try to turn back the clock and understand what they were like in the past. To build a model of Mars's atmosphere past we are only basing our presumptions on the in-situ observed tracks impressed by that past evolution. A very detailed computer modeling of the charged particles erosion and escape may be found in a recent paper from Astrobiology (*Terada et al.*, 2009), with the tools of magneto-fluid-dynamics and today's knowledge of the Martian upper atmosphere and of the solar radiation intensity.

First, recent orbital observations of networks of valleys in high lands and cratered areas of the Southern Hemisphere, especially those by Mars Express and Mars suggest that Mars exhibited a significant hydrologic activity during the first Gyr (Giga-year) of the planet lifetime. Some suggest that an ancient water ocean equivalent to a global layer with the depth of about 150 m is needed for the explanation of the observed surface features (*Bullock & Moore*, 2007). These authors suppose that the evolution of the Martian atmosphere and its water inventory since the end of the magnetic dynamo (*Schubert et al.*, 2000) at the late Noachian period about 3.5-3.7 Gyr ago was dominated by non-thermal atmospheric loss processes. This includes solar wind erosion and sputtering of oxygen ions and thermal escape of hydrogen (*Zhang et al.*, 1993). Recent studies (*Wood at al.*, 2005) show that these processes could completely remove a global Martian ocean with a thickness of about 10–20 m, indicating that the planet should have lost the majority of its water during the first 500 Myr. As far as Mars was exposed before the Noachian period to asteroid impacts, as the entire solar planets did, their effect on atmospheric erosion was simulated. Furthermore, the study uses multi-wavelength observations by the ASCA, ROSAT, EUVE, FUSE and IUE satellites of Sun-like stars at various ages for the investigation of how high X-ray and EUV fluxes of the young Sun have influenced the evolution of the early Martian atmosphere. Terada et al. apply for the first time a diffusive-photochemical model and investigate the heating of the Martian thermosphere by photo-dissociation and ionization processes and by exothermic chemical reactions, as well as by cooling due to CO_2 IR-radiation loss. The model used yields high exospheric temperatures during the first 100–500 Myr, which result in blow off for hydrogen and even high loss rates for atomic oxygen and carbon. By applying a hydrodynamical model for the estimation of the atmospheric loss rates, results were obtained, which indicate that the early Martian atmosphere was strongly evaporated by the young Sun and lost most of its water during the first 100 - 500 Myrs after the planets origin. The efficiency of the impact erosion and hydrodynamic blow off are compared, with the conclusion that both processes present the same rating.

It is a common believe now that during the early, very active Sun's lifetime, the solar wind velocity was faster then the one recorded today and the solar wind mass flux was higher during this early active solar period (Wood *et al.*, 2002, 2005; Lundin *et al.*, 2007).

As the solar past is not directly accessible, comparisons to neighboring stars of the same G and K main-sequence, as observed e. g. By the Hubble Space Telescope's high-resolution spectroscopic camera of the H Lyman-α feature, revealed neutral hydrogen absorption associated with the interaction between the stars' fully ionized coronal winds and the partially ionized interstellar medium (Wood *et al.* 2002, 2005). These observations concluded in finding that stars with ages younger than that of the Sun present mass loss rates that increase with the activity of the star. A power law relationship was drawn for the correlation between mass loss and X-ray surface flux, which suggests an average solar wind density that was up to 1000 times higher than it is today (see Fig. 2 in Lammer *et al.*, 2003a).

This statement is considered as valid especially for the first 100 million years after the Sun reached the ZAMS (Kulikov *et al.* 2006, 2007), but not all observations agree with this model. For example, observations by Wood *et al.* (2005) of the absorption characteristic of the solar-type G star τ-Boo, estimated as being 500-million-year-old, indicate that its mass loss is about 20 times less than the loss rate given by the mass loss power law, for a star with a similar age. Young stars of similar G- and K-type, with surface fluxes of more than 10^6 erg/cm^2/s require more observations to ascertain exactly what is happening to stellar winds during high coronal activity periods. Terada *et al.* (2009) are using a lower value for the mass loss, which is about 300 times larger than the average proton density at the present-day Martian orbit. These high uncertainties regarding the solar wind density during the Sun's youth prevent from attracting a confident determination of how fast the process of atmospheric depletion took place. To overcome this uncertainty, Terada *et al.* (2009) has applied a 3-D multispecies MHD model based on the Total Variation Diminishing scheme of Tanaka (1998), which can self-consistently calculate a planetary obstacle in front of the oncoming solar wind and the related ion loss rates from the upper atmosphere of a planet like Mars.

Recently, Kulikov *et al.* (2006, 2007) applied a similar numerical model to the early atmospheres of Venus and Mars. Their loss estimates of ion pickup through numerical modeling strongly depended on the chosen altitude of the planetary obstacle, as for a closer planet to the star more neutral gas from the outer atmosphere of the planet can be picked up by the solar plasma flow. We shall give here a more developed model of the rarefied atmosphere sweeping by including the combined effect of electrically charged particles and the heterogeneous mixture of gas and fine dust powder as encountered in the Martian atmosphere along the entire altitude scale.

Data were sought from the geology of some relevant features of the Mars surface like the Olympus Mons (Phillips *et al.*, 2001; Bibring *et al.*, 2006), the most prominent mountain in the solar system, to derive an understanding of the past hydrology on the red planet and the timing of geo-hydrological transformations of the surface. Olympus Mons is an unusual structure based on stratification of powdered soils. To explain the features, computer models with different frictional properties were built (McGovern and Morgan, 2009). In general, models with low friction coefficient reproduce two properties of Olympus Mons' flanks: slopes well below the angle of repose and exposure of old, stratigraphically low materials at the basal scarp (Morris and Tanaka, 1994) and distal flanks (Bleacher et al., 2007) of the edifice. The authors show that such a model with a 0.6° basal slope produces asymmetries in averaged flank slope and width (shallower and wider downslope), as actually seen at Olympus Mons. However, the distal margins are notably steep, generally creating planar to convex-upward topography.

This is in contrast to the concave shape of the northwest and southeast flanks and the low-slope distal benches (next figure) of Olympus Mons. Incremental deformation is concentrated in the deposition zone in the center of the edifice.

Outside of this zone, deformation occurs as relatively uniform outward spreading, with little internal deformation, as indicated by the lack of discrete slip surfaces. This finding contradicts the evidence for extension and localized faulting in the northwest flank of Olympus Mons. The wedgelike deformation and convex morphology seen in the figure appear to be characteristic of models with constant basal friction. McGovern and Morgan (2009) found that basal slopes alone are insufficient to produce the observed concave-upward slopes and asymmetries in flank extent and deformation style that are observed at Olympus Mons; instead, lateral variations in basal friction are required.

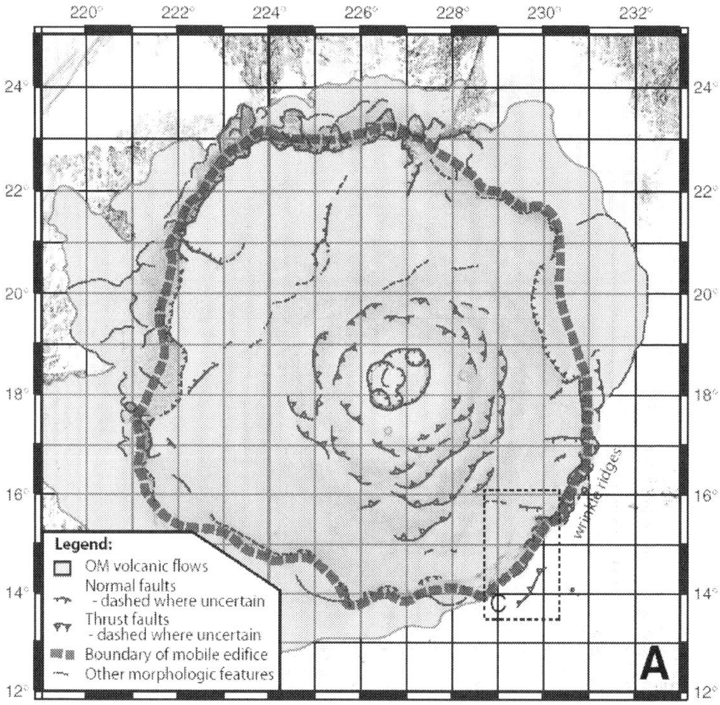

Fig. 2. Geology and interpretation of the Olympus Mons.

The conclusion is that these variations are most likely related to the presence of sediments, transported and preferentially accumulated downslope from the Tharsis rise. Such sediments likely correspond to ancient phyllosilicates (clays) recently discovered by the Mars Express mission. This could only suggest that liquid water was involved into the formation of ancient Olympus Mons, but neither of the conclusions of this work is strong enough. Up to the present, those findings are some of the hardest arguments for ancient water on Mars for a geologically long period.

The importance of this finding is in the promise that a thick atmosphere and wet environment could have been preserving for a long period and thus the Earth-like climate could withstand the depletion influences long enough. This is the type of encouraging arguments that the community of terraforming enthusiasts are waiting for. Problems are still huge, starting with the enormous amounts of energy required for the journey to Mars and back.

2. Energy requirements for the Earth-Mars- Earth journey

The problem of transportation from Earth to Mars and vice versa are restricted by the high amount of energy that must be spent for one kg of payload to climb into the gravitational field of the Sun up to the Mars level orbit and back to the Earth's orbit, where the difference in the orbital speed must also be zeroed. Astrophysical data from the first paragraph allow computing exactly this amount of energy. With the inertial, reactive propulsion of today, two strategies of transferring from Earth to Mars are possible and they belong to the *high thrust navigation* and to *low thrust navigation*. The energy requirements we consider are those for transfer from a low altitude, circular orbit around the starting planet to the low altitude, circular orbit around the destination planet. The very energy of injection into the corresponding low orbits should be added as a compulsory and given extra quantity, as far as the fragmentation of flight for pausing into a transfer orbit neither adds nor diminishes the energetic consumption of the total flight. It remains to determine the energy consumption for the transfer itself.

The motion into a field of central forces always remains planar and is easily describer in a polar, rotating referential by the equation of Binet (Synge and Griffith, 1949), based on the polar angle θ, named the *true anomaly*. For a gravitational field this equation reads

$$\frac{d^2u}{d\theta^2} + u = \frac{K}{h^2u^2} \tag{1}$$

The solution is a conical curve $r(\theta)$ that may be an ellipse, parabola or hyperbola, depending on the initial position and speed of the particle. The expression of the orbit equation may be found while resuming the Cartesian, planar orthogonal coordinates $\{xOy\}$, where the position vector and the velocity are given by (Mortari, 2001)

$$\mathbf{r} = \frac{p}{1+\varepsilon\cos\theta}\begin{bmatrix}\cos\theta\\ \sin\theta\end{bmatrix}, \quad \mathbf{v} = \sqrt{\frac{K_\oplus}{p}}\begin{bmatrix}-\sin\theta\\ e+\cos\theta\end{bmatrix}. \tag{2}$$

Between the local position on the conical orbit and the corresponding orbital velocity the relation exists, that derives from the conservation of energy or simply by calculating the velocity from the integral of the equation of motion, in this case with the Earth gravitational parameter $K_\oplus$,

$$\mathbf{v} = \sqrt{\frac{K_\oplus}{p}} \cdot \frac{1+e\cos\varphi}{p}\begin{bmatrix}0 & -1\\ 1 & \dfrac{e}{\sin\varphi}\end{bmatrix}\cdot\mathbf{r}\,. \tag{3}$$

Multiplying the matrices and considering the case when the orbital eccentricity lies between 0 and 1, that means elliptical orbits, the usual energy equation of thc orbital speed appears (Synge and Griffith, 1949)

$$v^2 = K_\oplus\left(\frac{2}{r} - \frac{1}{a}\right). \tag{4}$$

These formulae are only needed to access the energy requirements for an impulsive transfer towards Mars and thus a so-called high thrust interplanetary transfer is resulting. The results supplied this way are bound to some approximations when considering the Earth-Mars actual transfer. They mainly come from the fact that the two planets move along slightly non-coplanar orbits, the angle between the two orbital planes is that given in Table 18.2, namely the inclination of Mars's orbit to the ecliptic is

$$i = 1.84964°.$$

This circumstance induces a very little difference in the transfer energy however and all previous results from the coplanar approximation preserve quite correct.

2.1 Energy for Earth-Mars high thrust transfers

With moderately high thrust propulsion the acceleration for the transfer acts quick and the time of flight gets its actual minimum. The amount of energy is directly given by the equation of motion on the quasi-best, quasi-Hohmann transfer orbit that starts from around the Earth and ends around Mars (figure below).

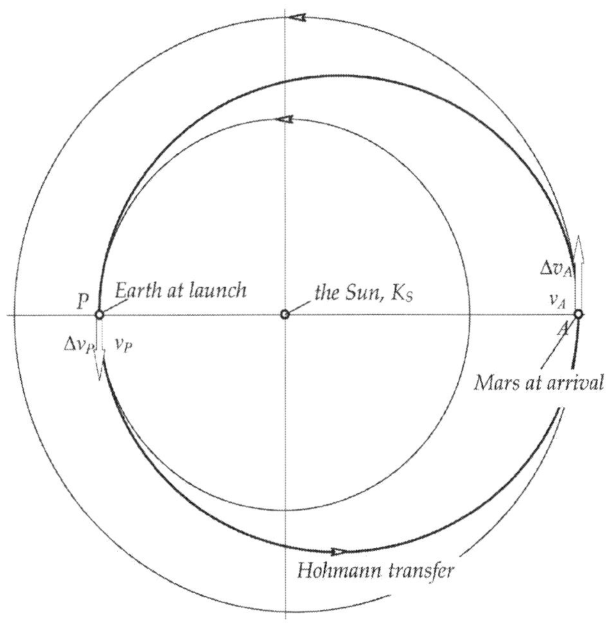

Fig. 3. Minimum energy Earth-Mars transfer

To enter the higher altitude orbit as referred to the Sun, the transfer spacecraft must be provided with extra kinetic energy at the Earth orbit level, exactly equal to the perihelion energy of the transfer Hohmann ellipse, v_P^2. This difference is easily computed by using equation (4).

When we introduce the gravitational parameter of the Sun $K_{\odot}$ and the location of the perihelion at Earth orbit level r_P, the extra speed required to the spacecraft is $\Delta v_P = v_P - v_E = 32.746 - 29{,}7944 = 2.952$ km/s. With the same procedure we observe that at the arrival on Mars orbit level, the spacecraft requires additional energy and speed to remain on the Mars

orbit level around the Sun, namely $\Delta v_A = v_M - v_A$=24,126-21,472=2.654 km/s. The total velocity requirement for the spacecraft equals the considerable amount of Δv_Σ=5.606 km/s.
To this requirement the exit energy from the Earth gravity and the braking energy to avoid fall on Mars must be added. In order to remain with the required extra velocity when the spacecraft leaves the Earth field of gravity, the semi-axis of the departure hyperbola must comply with the value given by

$$v_\infty^2 = K_\oplus \left(\frac{1}{a_{\oplus e}} \right) = 2.952^2 \tag{5}$$

from where the value results,

$$a_{\oplus e} = K_\oplus / 8.714304 = 45759.478 \text{ km} \tag{6}$$

with the corresponding hyperbolic escape velocity at the altitude of a LEO (*Low Earth Orbit*, h=200 km) of

$$v_e^2 = K_\oplus \left(\frac{2}{R_0 + h} + \frac{1}{a_{\oplus e}} \right) = 130.14 \tag{7}$$

from where the value results

$$v_e = 11.408 \text{ km/s}$$

that must be actually supplied to the spacecraft to properly initiate the transfer to Mars. The local parabolic velocity were v_2=11.019 km/s only.
At the Mars end of the flight, the spacecraft is challenged with a continuous fall towards Mars, as it enters the gravitational field of Mars with a hyperbolic velocity at infinity of 2.654 km/s. Similar computations show that at Mars arrival, namely at an altitude of merely 50 km above the Martian surface, where the aerodynamic breaking begins to manifest, the velocity relative to Mars raises to 5.649 km/s and must be completely slowed down. These transfer velocities are ending in 17.057 km/s.
The high amount of energy at approach and landing on Mars can be diminished by air breaking, at least in part. It only remains then to assure the launching velocity from Earth, with its considerable amount. In terms of energy, this requirement equals 65 MJ for each kilogram of end-mass in hyperbolic orbit, to which the energy for lifting that kilogram from the ground to the altitude of the orbit must be considered. This extra energy of less than 2 MJ looks negligible however, as compared to the kinetic energy requirement from above.
It must be considered that a value of 16 MJ only is required for the return travel to Earth from MLO (Mars low orbit). It is thus hard to understand the reasoning for the so-called *one-way trips to Mars*, or to be open *"no return trips to Mars"*, warmly proposed by some experts in Mars colonization from Earth.

3. A thick atmosphere facing erosion

Long term events in the outer atmosphere of the planet are related to the interaction of the solar radiation and particles flow (solar wind) with the very rarefied, heterogeneous fluid envelope of the planet. A model of an electrically charged gas was recently used by Tanaka et al. (2009) to approach the erosion rate of the Martian atmosphere under high UV solar radiation. We add now the double phase fluid model that covers the dust dispersion proved even in 1975 by the Viking spacecraft to flow at high altitudes into the tinny atmosphere. First the equations of conservation and motion for the heterogeneous and rarefied fluid should be introduced.

3.1 Equations governing the exo-atmospheric erosion

While writing the equations of motion of a material particle of the rarefied gas and its condensed, particle content the relative referential related to the moving planet is considered. The equations of motion, referenced to a non-inertial coordinate system, depend on the relative acceleration of the particles and thus on the *transport* and *complementary* acceleration terms, given by Rivals and Coriolis formulae (Rugescu, 2003)

$$\begin{aligned} \mathbf{a}_t &= \mathbf{a}_M + \left(\mathbf{\Omega}_t^2 + \mathbf{\Omega}_t'\right)\cdot \mathbf{x}, \\ \mathbf{a}_C &= 2\mathbf{\Omega}_t \wedge \mathbf{v}, \end{aligned} \tag{8}$$

where $\mathbf{\Omega}_t$ is the anti-symmetric matrix of the transport velocity (rotation). Four distinct types of time derivatives exist when the relative fluid motion is described (Rugescu, R. D., 2000). With an apostrophe for the absolute, total (material) time derivative and $d\,/dt$ for the local total time derivative the vectors and associated matrices of relative motion are

$$\begin{aligned} &\mathbf{r} = \begin{pmatrix} x \\ y \\ z \end{pmatrix}, \quad \mathbf{v} \equiv \begin{pmatrix} u \\ v \\ w \end{pmatrix} = \frac{d\mathbf{r}}{dt}, \quad \mathbf{a} \equiv \begin{pmatrix} a \\ b \\ c \end{pmatrix} = \frac{d^2\mathbf{r}}{dt^2}, \\ &\boldsymbol{\omega}_t = \begin{pmatrix} \omega_x \\ \omega_y \\ \omega_z \end{pmatrix}, \quad \mathbf{\Omega}_t = \begin{bmatrix} 0 & -\omega_z & \omega_y \\ \omega_z & 0 & -\omega_x \\ -\omega_y & \omega_x & 0 \end{bmatrix}. \end{aligned} \tag{9}$$

Only the centripetal component $\mathbf{a}_c = -\mathbf{\Omega}_t^2 \cdot \mathbf{r}$ preserves the direction of the position vector **r**, the other components presenting ortho-normal properties.

Consequently, the global effect of the relative, non-inertial motion upon the acoustical behavior of the fluid can only be described in a computational 3-Dimensional scheme, although partial effects can be evaluated under simpler assumptions.

The twin viscosity coefficient model observes the reology of the fluid with coefficients of viscosity introduced. The following NS-type, laminar, unsteady hyperbolic matrix equation system is written for both the gas fraction and the condensed part of the fluid mixture, and stands for example as the background of the *Eagle* solver (Tulita et al., 2002), used in many previous aerodynamic investigations at QUB and UPB. A single type of condensed particles is considered, in other words the chemical interactions are seen as minimal and not interfering with the flow process.

To express more precisely the conservation of the gas and particles that fill the same elemental volume dV we observe that a fraction only α of the lateral area of the frontier is occupied by the gas, while the remaider fraction (1-α) is traversed by the condensed particles. Consequently, the area where the pressure acts upon the gas is αA_i for each facet of the cubic volume element (Fig. 4).

In orthogonal and 3D non-inertial coordinates this equations write out for the gas fraction and for the condensate as:

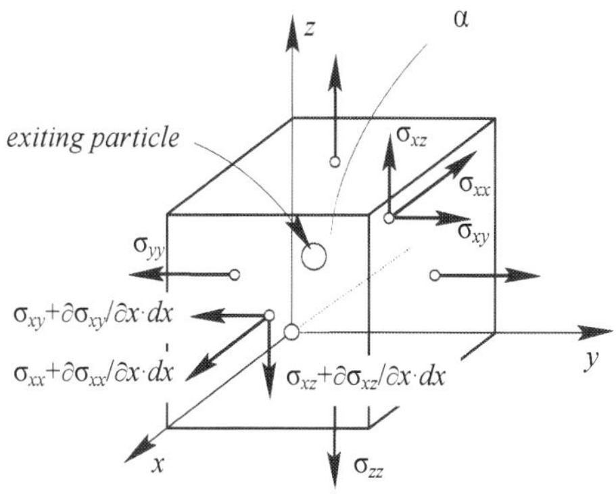

Fig. 4. The heterogeneous fluid

$$\frac{\partial \mathbf{E}}{\partial t}+\frac{\partial \mathbf{F}}{\partial x}+\frac{\partial \mathbf{G}}{\partial y}+\frac{\partial \mathbf{H}}{\partial z}+\mathbf{K}=\mathbf{0}\,, \tag{10}$$

$$\iiint_{ATR}\left[\left(\frac{\partial E}{\partial t}+\frac{\partial F}{\partial x}+\frac{\partial G}{\partial y}+\frac{\partial H}{\partial z}\right)M_p f+K\right]dRdT\,dA=0\,, \tag{11}$$

where the vectors are used as defined by the following expressions, in connection with the above system (3) and (4):

$$\mathbf{E}=\begin{bmatrix}\rho\\ \rho u\\ \rho v\\ \rho w\\ \rho h-p\\ \rho_c\end{bmatrix}, \tag{12}$$

$$\mathbf{F}=\begin{bmatrix}\rho u\\ \rho u^2+\alpha p-2\mu\alpha\frac{\partial u}{\partial x}-\alpha\left(\frac{\mu}{3}+\mu'\right)\nabla\cdot\mathbf{V}\\ -\alpha\mu\left(\frac{\partial v}{\partial z}+\frac{\partial w}{\partial y}\right)\\ -\alpha\mu\left(\frac{\partial w}{\partial x}+\frac{\partial u}{\partial z}\right)\\ \rho hu-\lambda\frac{\partial T}{\partial x}\\ \iiint_{\mathbf{v}_pTR}M_p(R)f\,dR\,dT\,d\mathbf{v}_p\end{bmatrix}, \tag{13}$$

$$\mathbf{G}=\begin{bmatrix} \rho v \\ -\mu\alpha\left(\dfrac{\partial u}{\partial y}+\dfrac{\partial v}{\partial x}\right) \\ \rho v^2+\alpha p-2\alpha\mu\dfrac{\partial v}{\partial y}-\alpha\left(\dfrac{\mu}{3}+\mu'\right)\nabla\cdot\mathbf{V} \\ -\alpha\mu\left(\dfrac{\partial w}{\partial y}+\dfrac{\partial v}{\partial z}\right) \\ \rho h v-\lambda\dfrac{\partial T}{\partial y} \\ \iiint\limits_{\mathbf{v}_p T R} M_p(R) f\, dR\, dT\, d\mathbf{v}_p \end{bmatrix}, \tag{14}$$

$$\mathbf{H}=\begin{bmatrix} \rho w \\ -\mu\alpha\left(\dfrac{\partial u}{\partial z}+\dfrac{\partial w}{\partial x}\right) \\ -\alpha\mu\left(\dfrac{\partial v}{\partial z}+\dfrac{\partial w}{\partial y}\right) \\ \rho w^2+\alpha p-2\alpha\mu\dfrac{\partial w}{\partial z}-\alpha\left(\dfrac{\mu}{3}+\mu'\right)\nabla\cdot\mathbf{V} \\ \rho h w-\lambda\dfrac{\partial T}{\partial z} \\ \iiint\limits_{\mathbf{v}_p T R} M_p(R) f\, dR\, dT\, d\mathbf{v}_p \end{bmatrix}, \tag{15}$$

$$\mathbf{K}=\begin{bmatrix} 0 \\ \gamma_{Mx}+z\omega'_y-y\omega'_z+\omega_x(y\omega_y+z\omega_z)-x(\omega_y^2+\omega_z^2)+ \\ \qquad +2(w\omega_y-v\omega_z)-\rho f_x+S_x \\ \gamma_{My}+x\omega'_z-z\omega'_x+\omega_y(z\omega_z+x\omega_x)-y(\omega_z^2+\omega_x^2)+ \\ \qquad +2(u\omega_z-w\omega_x)-\rho f_y+S_y \\ \gamma_{Mz}+y\omega'_x-x\omega'_y+\omega_z(x\omega_x+y\omega_y)-z(\omega_x^2+\omega_y^2)+ \\ \qquad +2(v\omega_x-u\omega_y)-\rho f_z+S_z \\ -\mathbf{V}\cdot\nabla p-\mathbf{\Pi}:\nabla\circ\mathbf{V}+S \end{bmatrix} \tag{16}$$

The following notations were used in the expression of the vector **K**, referring to the distribution of condensed particles:

$$S_x \equiv \iiint\limits_{u_p TR}\left(\frac{9}{2}\frac{\mu_g c}{\rho_p R^2}-3\frac{d\ln R}{dt}-\frac{d\ln f}{dt}\right)(u-u_p)M_p f(x,r,T,u_p,t)\,dR\,dT\,du_p\,,$$

$$S_y \equiv \iiint_{v_p\, TR} \left(\frac{9}{2} \frac{\mu_g c}{\rho_p R^2} - 3\frac{d\ln R}{dt} - \frac{d\ln f}{dt} \right) (v - v_p) M_p f(x,r,T,u_p,t) dR\, dT\, dv_p \, ,$$

$$S_z \equiv \iiint_{w_p\, TR} \left(\frac{9}{2} \frac{\mu_g c}{\rho_p R^2} - 3\frac{d\ln R}{dt} - \frac{d\ln f}{dt} \right) (w - w_p) M_p f(x,r,T,u_p,t) dR dT\, dw_p \, ,$$

$$S \equiv \iiint_{\mathbf{v}_p\, TR} \left\{ \left(e_p + v_p^2 / 2 \right) D_p - \left[A_p (\mathbf{v} - \mathbf{v}_p)^2 + B_p \right] M_p f(x,r,T,u_p,t) \right\} dR dT\, d\mathbf{v}_p \, .$$

The equations involving the solid phase (dust particles) depend upon the distribution of the particles along their size, temperature, velocity and their distribution in space as a variable of time. Consequently some extra terms in the four expressions from above are to be written for those functions as follows:

$$A = \begin{bmatrix} u_p \\ v_p \\ w_p \\ \mathbf{v}_p \end{bmatrix}, \; E = \begin{bmatrix} u_p \\ v_p \\ w_p \\ M_p f(h_p + \mathbf{v}_p^2) \end{bmatrix}, \; F = \begin{bmatrix} u_p^2 \\ u_p v_p \\ u_p w_p \\ M_p f(h_p + \mathbf{v}_p^2) u_p \end{bmatrix} \tag{17}$$

$$G = \begin{bmatrix} u_p v_p \\ v_p^2 \\ v_p w_p \\ M_p f(h_p + \mathbf{v}_p^2) v_p \end{bmatrix}, H = \begin{bmatrix} u_p w_p \\ v_p w_p \\ w_p^2 \\ M_p f(h_p + \mathbf{v}_p^2) w_p \end{bmatrix}, \tag{18}$$

$$K = \begin{bmatrix} A_p (u - u_p) - f_x \\ A_p (v - v_p) - f_y \\ A_p (w - w_p) - f_z \\ \left[(h_p + \mathbf{v}_p^2) \nabla \cdot \mathbf{v}_p + B_p - 2\mathbf{v}_p \cdot \mathbf{f} \right] M_p f \end{bmatrix}. \tag{19}$$

Other three simplifying notations were used to define the factors from the vector (18.19) as follows:

$$A_p \equiv \frac{9}{2} \frac{\mu_g c}{\rho_p R^2} - 3\frac{d\ln R}{dt} - \frac{d\ln f}{dt} \tag{20}$$

$$B_p \equiv \frac{3}{\rho_p R_p} \left[\alpha_c (T_p - T) + \sigma_0 \left(\varepsilon_p T_p^4 - \varepsilon T^4 \right) \right] \tag{21}$$

$$D_p \equiv \frac{d\left(M_p f_p \right)}{dt} + M_p f_p \nabla \cdot \mathbf{v}_p \tag{22}$$

While ionization is acting on the rarefied fluid the terms including the external, field forces include the Lorenz forces that appear in the fluid. Coefficients and properties of the field parameters are taken into account within the model developed by Terada (Terada et al. 2009),

Solar wind parameters	*Present*	*4.5 Ga*
Solar wind proton density	3 cm^{-3}	1000 cm^{-3}
Solar wind velocity	450 km s^{-1}	2000 km s^{-1}
B_{IMF} (Parker spiral)	2 nT	60 nT
XUV flux (normalized to martian orbit)	1 × (present-day low solar activity)	100 × (present-day moderate solar activity)

Table 3. Solar flux parameters after Terada

A 3-D computational code is under development for the simulation of the exosphere motion and erosion into an inertial reference frame, bound to the planet surface. To image the interaction of the solar flux with the upper atmosphere of Mars, this interaction as given by the Terada simulations is presented in figure 5.

Fig. 5. Ion of O+ erosion simulated by Terada for a Mars atmosphere cca. 4.5 Ga ago.
A density of $N(O^+)=100\ cm^{-3}$ is indicated with a blue transparent isosurface contour.

The authors of the simulation show that if they use the same assumption as Pérez-de-Tejada (1992) or Lammer *et al.* (2003b) that the cold ions are lost through the entire circular ring area around the planet's terminator, a maximum O+ loss rate of about $1.2\times10^{29}\ s^{-1}$ is obtained. Integrating this loss rate over a period of Δt=150 million years, a maximum water loss equivalent to a global Martian ocean with a depth of 70 m is obtained, which is quite impressive and shows that the erosion of the atmosphere is extremely severe.

4. Solar-gravitational draught on Mars

The innovative model of the solar-gravitational air accelerator for use on Mars and other celestial bodies with thin atmosphere or without atmosphere is the closed circuit tower in figure 6.

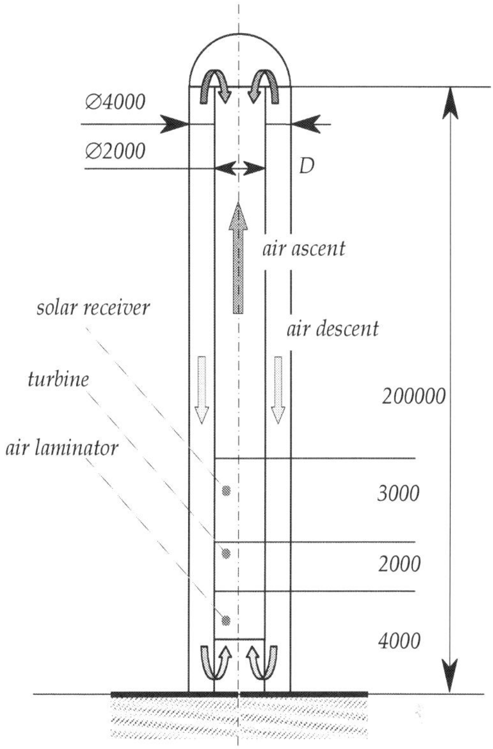

Fig. 6. Closed-circuit thermal-gravity tower for celestial bodies without atmosphere.

The air ascent tunnel is used as a heater with solar radiation collected through the mirror array, while the descent tunnels are used as air coolers to close the gravity draught.
According to the design in Fig. 6, a turbine is introduced in the facility next to the solar receiver, with the role to extract at least a part of the energy recovered from the solar radiation and to transmit it to the electric generator, for converting to electricity. The heat from the flowing air is thus transformed into mechanical energy with the payoff of a supplementary air rarefaction and cooling in the turbine.
The best energy extraction will take place when the air recovers entirely the ambient temperature before the solar heating, although this desire remains for the moment rather hypothetical. To search for the possible amount of energy extraction, the quotient ω is introduced, as further defined. Some differences appear in the theoretical model of the turbine system as compared to the simple gravity draught wind tunnel previously described.
The process of air acceleration at tower inlet is governed by the same energy (Bernoulli) incompressible (constant density ρ_0 through the process) equation as in the previous case,

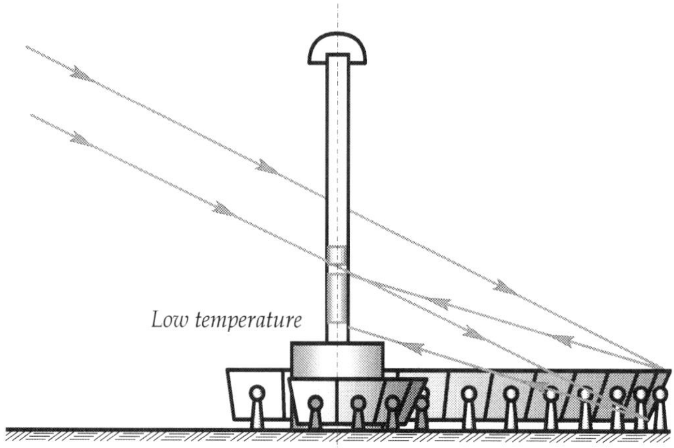

Fig. 7. Solar array concept for the closed-circuit thermal-gravity tower.

$$p_1 = p_0 - \frac{\dot{m}^2}{2\rho_0 A^2} \,. \tag{23}$$

The air is heated in the solar receiver with the amount of heat q, into a process with dilatation and acceleration of the airflow, accompanied by the usual pressure loss, called sometimes "*dilatation drag*" [8]. Considering a constant area cross-section in the heating solar receiver zone of the tube and adopting the variable γ for the amount of heating rather then the heat quantity itself,

$$\gamma = \frac{\rho_0 - \rho_2}{\rho_0} = 1 - \beta \,, \tag{24}$$

with a given value for

$$\beta = \frac{T_1}{T_2} < 1 \,, \tag{25}$$

the continuity condition shows that the variation of the speed is given by

$$c_2 = c_1 / \beta \,. \tag{26}$$

No global impulse conservation appears in the tower in this case, as long as the turbine is a source of impulse extraction from the airflow. Consequently the impulse equation will be written for the heating zone only, where the loss of pressure due to the air dilatation occurs, in the form of eq. 27,

$$p_2 + \frac{\dot{m}^2}{\rho_2 A^2} = p_1 + \frac{\dot{m}^2}{\rho_0 A^2} - \Delta p_R \,. \tag{27}$$

A possible pressure loss due to friction into the lamellar solar receiver is considered through Δp_R. The dilatation drag is thus perfectly identified and the total pressure loss Δp_Σ from outside up to the exit from the solar heater is present in the expression

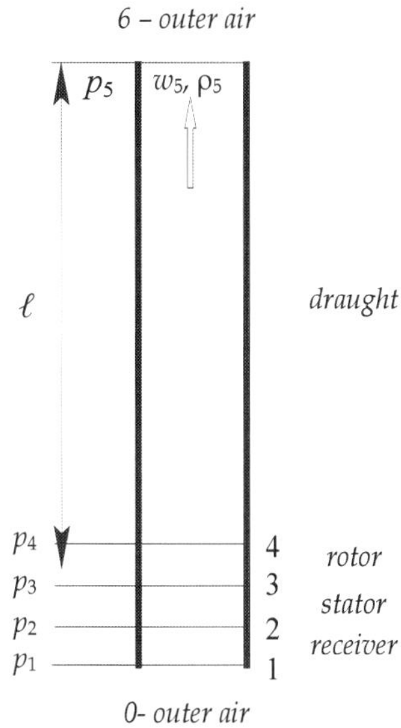

Fig. 8. Main stations in the turbine cold-air draught tower.

$$p_2 = p_0 - \frac{\dot{m}^2}{2\rho_0 A^2} - \frac{\dot{m}^2}{\rho_2 A^2} + \frac{\dot{m}^2}{\rho_0 A^2} - \Delta p_R \equiv p_0 - \Delta p_\Sigma \,. \tag{28}$$

Observing the definition of the rarefaction factor in (24) and using some arrangements the equation (28) gets the simpler form

$$p_2 = p_0 - \frac{\dot{m}^2}{\rho_0 A^2} \cdot \frac{\gamma + 1}{2(1-\gamma)} - \Delta p_R \,. \tag{29}$$

The thermal transform further into the turbine stator grid is considered as isentropic, where the amount of enthalpy of the warm air is given by

$$q = \frac{p_1}{\rho_1} \cdot \frac{1-\beta}{\beta} + \frac{\dot{m}^2}{\rho_1^2 A^2} \cdot \left[\frac{1 - \frac{\Gamma}{2}\beta}{\beta} + \frac{\frac{\Gamma}{2} - 1}{\beta^2} \right] - \frac{\Delta p_R}{\rho_1} \cdot \frac{1}{\beta} \,.$$

If the simplifying assumption is accepted that, under this aspect only, the heating progresses at constant pressure, then a far much simpler expression for the enthalpy fall in the stator appears,

$$\Delta h_{23} = \omega q = \omega c_p T_2 \gamma \,. \tag{30}$$

To better describe this process a choice between a new rarefaction ratio of densities ρ_3/ρ_2 or the energy quota ω must be engaged and the choice is here made for the later. Into the isentropic stator the known variation of thermal parameters occurs,

$$\frac{T_3}{T_2} = 1 - \omega\gamma\ , \tag{31}$$

$$\frac{p_3}{p_2} = (1 - \omega\gamma)^{\frac{\kappa}{\kappa-1}}\ , \tag{32}$$

$$\frac{\rho_3}{\rho_2} = (1 - \omega\gamma)^{\frac{1}{\kappa-1}}\ . \tag{33}$$

The air pressure at stator exit follows from combining (32) and (29) to render

$$p_3 = \left[p_0 - \frac{\dot{m}^2}{\rho_0 A^2} \cdot \frac{\gamma + 1}{2(1-\gamma)} - \Delta p_R \right] (1 - \omega\gamma)^{\frac{\kappa}{\kappa-1}}\ . \tag{34}$$

Considering a Zölly-type turbine the rotor wheel is thermally neutral and no variation in pressure, temperature and density appears. The only variation is in the air kinetic energy, when the absolute velocity of the airflow decreases from c_3 to $c_3 \sin\alpha_1$ and this kinetic energy variation is converted to mechanical work outside. Consequently $\rho_4 = \rho_3$, $p_4 = p_3$, $T_4 = T_3$ and

$$c_4 = \frac{c_1}{(1-\gamma)(1-\omega\gamma)^{\frac{1}{\kappa-1}}}\ . \tag{35}$$

The air ascent in the tube is only accompanied by the gravity up-draught effect due to its reduced density, although the temperature could drop to the ambient value. We call this quite strange phenomenon the *cold-air draught*. It is governed by the simple gravity form of Bernoulli's equation of energy,

$$p_5 = p_3 - g\rho_3\ell\ . \tag{36}$$

The simplification was assumed again that the air density varies insignificantly during the tower ascent. The value for p_3 is here the one in (35). At air exit above the tower a sensible braking of the air occurs in compressible conditions, although the air density suffers insignificant variations during this process.
The Bernoulli equation is used to retrieve the stagnation pressure of the escaping air above the tower, under incompressible conditions

$$p_6{}^* = p_5 - \frac{\Gamma}{2}\rho_5 c_5^2 = p_5 + \frac{\Gamma}{2} \cdot \frac{\dot{m}^2}{\rho_3 A^2} = p_5 + \frac{\Gamma}{2} \cdot \frac{\dot{m}^2}{\rho_0 A^2} \cdot \frac{\rho_0}{\rho_3}\ . \tag{37}$$

Value for p_5 from (36) and for the density ratio from (24) and (33) are now used to write the full expression of the stagnation pressure as

$$p_6{}^* = (p_0 - \Delta p_R)(1-\omega\gamma)^{\frac{\kappa}{\kappa-1}} - \frac{\dot{m}^2}{\rho_0 A^2} \cdot \frac{\gamma+1}{2(1-\gamma)} \cdot (1-\omega\gamma)^{\frac{\kappa}{\kappa-1}} + \frac{\dot{m}^2}{\rho_0 A^2} \cdot \frac{\Gamma}{2} \cdot \frac{1}{(1-\gamma)\cdot(1-\omega\gamma)^{\frac{1}{\kappa-1}}} - g\rho_4\ell \tag{38}$$

It is observed again that up to this point the entire motion into the tower hangs on the value of the mass flow-rate, yet unknown. The mass flow-rate itself will manifest the value that fulfils now the condition of outside pressure equilibrium, or

$$p_6{}^* = p_0 - g\rho_0\ell \,. \tag{39}$$

This way the local altitude air pressure of the outside atmosphere equals the stagnation pressure of the escaping airflow from the inner tower. Introducing (38) in (39), after some re-arrangements the dependence of the global mass flow-rate along the tower, when a turbine is inserted after the heater, is given by the final developed formula:

$$R^2(\gamma) \equiv \frac{\dot{m}^2}{2g\ell\rho_0^2A^2} = \frac{1-\gamma}{(\gamma+1)(1-\omega\gamma)^{\frac{\kappa+1}{\kappa-1}} - \Gamma}(1-\omega\gamma)^{\frac{1}{\kappa-1}}\left\{1-(1-\gamma)(1-\omega\gamma)^{\frac{1}{\kappa-1}} + \right.$$
$$\left. + \frac{p_0}{g\rho_0\ell}\left[(1-\omega\gamma)^{\frac{\kappa}{\kappa-1}} - 1\right] - \frac{\Delta p_R}{g\rho_0\ell}(1-\omega\gamma)^{\frac{\kappa}{\kappa-1}}\right\} \tag{40}$$

where the notations are again recollected

$\gamma = \dfrac{\rho_0 - \rho_2}{\rho_0}$, the dilatation by heating in the heat exchanger, previously denoted by r;

ω = the part of the received solar energy which could be extracted in the turbine;

Δp_R = pressure loss into the heater and along the entire tube either.

All other variables are already specified in the previous chapters. It is clearly noticed that by zeroing the turbine effect ($\omega = 0$) the formula (40) reduces to the previous form, or by neglecting the friction, which stays as a validity check for the above computations. For different and given values of the efficiency ω the variation of the mass flow-rate through the tube depends parabolically of the rarefaction factor γ.

Notice must be made that the result in (40) is based on the convention (30). The exact expression of the energy q introduced by solar heating yet does not change this result significantly. Regarding the squared mass flow-rate itself in (40), it is obvious that the right hand term of its expression must be positive to allow for real values of R^2. This only happens when the governing terms present the same sign, namely

$$\left\{(\gamma+1)(1-\omega\gamma)^{\frac{\kappa+1}{\kappa-1}} - \Gamma\right\} \cdot \left\{1-(1-\gamma)(1-\omega\gamma)^{\frac{1}{\kappa-1}} + \frac{p_0}{g\rho_0\ell}\left[(1-\omega\gamma)^{\frac{\kappa}{\kappa-1}} - 1\right] - \frac{\Delta p_R}{g\rho_0\ell}(1-\omega\gamma)^{\frac{\kappa}{\kappa-1}}\right\} : 0 \,. \tag{41}$$

The larger term here is the ratio $p_0/(g\rho_0\ell)$, which always assumes a negative sign, while not vanishing. The conclusion results that the tower should surpass a minimal height for a real R^2 and this minimal height were quite huge. Very reduced values of the efficiency ω should be permitted for acceptably tall solar towers. This behavior is nevertheless altered by the first factor in (41) which is the denominator of (30) and which may vanish in the usual range of rarefaction values γ. A sort of thermal resonance appears at those points and the turbine tower works properly well.

5. Reasons and costs for terraforming Mars

Thicken Mars' atmosphere, and make it more like Earth's. Earth's atmosphere is about 78% Nitrogen and 21% Oxygen, and is about 140 times thicker than Mars' atmosphere. Since Mars is so much smaller than Earth (about 53% of the Earth's radius), all we'd have to do is bring about 20% of the Earth's atmosphere over to Mars. If we did that, not only would Earth be relatively unaffected, but the Martian atmosphere, although it would be thin (since the force of gravity on Mars is only about 40% of what it is on Earth), would be breathable, and about the equivalent consistency of breathing the air in Santa Fe, NM. So that's nice; breathing is good.

Mars needs to be heated up, by a lot, to support Earth-like life. Mars is cold. Mars is damned cold. At night, in the winter, temperatures on Mars get down to about -160 degrees! (If you ask, "Celcius or Fahrenheit?", the answer is first one, then the other.) But there's an easy fix for this: add greenhouse gases. This has the effect of letting sunlight in, but preventing the heat from escaping. In order to keep Mars at about the same temperature as Earth, all we'd have to do is add enough Carbon Dioxide, Methane, and Water Vapor to Mars' atmosphere. Want to know something neat? If we're going to move 20% of our atmosphere over there, we may want to move 50% of our greenhouse gases with it, solving some of our environmental problems in the process.

These greenhouse gases would keep temperatures stable on Mars and would warm the planet enough to melt the icecaps, covering Mars with oceans. All we'd have to do then is bring some lifeforms over and, very quickly, they'd multiply and cover the Martian planet in life. As we see on Earth, if you give life a suitable environment and the seeds for growth/regrowth, it fills it up very quickly.

So the prospects for life on a planet with an Earth-like atmosphere, temperature ranges, and oceans are excellent. With oceans and an atmosphere, Mars wouldn't be a red planet any longer. It would turn blue like Earth! This would also be good for when the Sun heated up in several hundred million years, since Mars will still be habitable when the oceans on Earth boil. But there's one problem Mars has that Earth doesn't, that could cause Mars to lose its atmosphere very quickly and go back to being the desert wasteland that it is right now: Mars doesn't have a magnetic field to protect it from the Solar Wind. The Earth's magnetic field, sustained in our molten core, protects us from the Solar Wind.

Mars needs to be given a magnetic field to shield it from the Solar Wind. This can be accomplished by either permanently magnetizing Mars, the same way you'd magnetize a block of iron to make a magnet, or by re-heating the core of Mars sufficiently to make the center of the planet molten. In either case, this allows Mars to have its own magnetic field, shielding it from the Solar Wind (the same way Earth gets shielded by our magnetic field) and allowing it to keep its atmosphere, oceans, and any life we've placed there.

But this doesn't tell us how to accomplish these three things. The third one seems to us to be especially difficult, since it would take a tremendous amount of energy to do. Still, if you wanted to terraform Mars, simply these three steps would give you a habitable planet.

The hypothetical process of making another planet more Earth-like has been called terraforming, and terraforming Mars is a frequently mentioned possibility in terraforming discussions. To make Mars habitable to humans and earthly life, three major modifications are necessary. First, the pressure of the atmosphere must be increased, as the pressure on the surface of Mars is only about 1/100th that of the Earth. The atmosphere would also need the addition of oxygen. Second, the atmosphere must be kept warm. A warm atmosphere would melt the large quantities of water ice on Mars, solving the third problem, the absence of water.

Terraforming Mars by building up its atmosphere could be initiated by raising the temperature, which would cause the planet's vast CO_2 ice reserves to sublime and become atmospheric gas. The current average temperature on Mars is −46 °C (-51 °F), with lows of −87 °C (-125 °F), meaning that all water (and much carbon dioxide) is permanently frozen. The easiest way to raise the temperature seems to be by introducing large quantities of CFCs (chlorofluorocarbons, a highly effective greenhouse gas) into the atmosphere, which could be done by sending rockets filled with compressed CFCs on a collision course with Mars. After impact, the CFCs would drift throughout Mars' atmosphere, causing a greenhouse effect, which would raise the temperature, leading CO_2 to sublimate and further continuing the warming and atmospheric buildup. The sublimation of gas would generate massive winds, which would kick up large quantities of dust particles, which would further heat the planet through direct absorption of the Sun's rays. After a few years, the largest dust storms would subside, and the planet could become habitable to certain types of algae and bacteria, which would serve as the forerunners of all other life. In an environment without competitors and abundant in CO_2, they would thrive. This would be the biggest step in terraforming Mars.

6. Conclusion

The problem of creating a sound source of energy on Mars is of main importance and related to the capacity of transportation from Earth to Mars, very limited in the early stages of Mars colonization, and to the capacity of producing the rough materials in situ. Consequently the most important parameter that will govern the choice for one or another means of producing energy will be the specific weight of the powerplant. Besides the nuclear sources, that most probably will face major opposition for a large scale use, the only applicable source that remains valid is the solar one. As far as the solar flux is almost four times fainter on Mars than on Earth, the efficiency of PVC remains very doubtfull, although it stands as a primary candidate. This is why the construction of the gravity assisted air accelerators looks like a potential solution, especially when rough materials will be available on Mars surface itself. The thermal efficiency of the accelerator for producing a high power draught and the propulsion of a cold air turbine remains very high and attractive. The large area of the solar reflector array is still one of the basic drawbacks of the system, that only could be managed by creating very lightweight solar mirrors, but still very stiff to withstand the winds on Mars surface.

7. References

*** (1977), Scientific Results of the Viking Project, *Journal of Geophysical Research*, vol. 82, no. 28, A.G.U., Washington, D.C.

*** *The Astronomical Almanac*, 2010, U.S. Naval Observatory and H.M. Nautical Almanac Office.

Michael Allison and Megan McEwen, 2000. A post-Pathfinder evaluation of aerocentric solar coordinates with improved timing recipes for Mars seasonal/diurnal climate studies. Planetary and Space Science, 48, 215-235.

Asimov, Isaac (1979), *Civilisations extraterrestres*, Ed. L'Etincelle, Montreal, Quebec, Canada.

Andre L. Berger, 1978. Long Term Variations of Daily Insolation and Quaternary Climatic Changes. Journal of the Atmospheric Sciences, volume 35(12), 2362-2367.

Berger A and Loutre MF, 1992. Astronomical solutions for paleoclimate studies over the last 3 million years. Earth and Planetary Science Letters, 111, 369-382.

Bibring, J.P., and 42 others (2006), Global mineralogical and aqueous Mars history derived from OMEGA/Mars Express data, *Science*, vol. 312, pp. 400–404.

P. Bizony (1998), *The Exploration of Mars-Searching for the Cosmic Origins of Life*, Aurum Press, London.

David Brain (2008), ESA observations indicate Mars and Venus are surprisingly similar, *Thaindian News*, March 6th <esa-observations-indicate-mars-and-venus-are-surprisingly-similar_10024474.html>.

Bullock, M. A., and J. M. Moore (2007), Atmospheric conditions on early Mars and the missing layered carbonates, *Geophys. Res. Lett.*, 34, L19201, doi:10.1029/2007GL030688.

Sylvio Ferraz-Mello (1992), Chaos, resonance, and collective dynamical phenomena in the solar system, *Proceedings of the 152nd Symposium of the IAU*, Angra dos Reis, Brazil, July 15-19, 1991, Springer, ISBN 0792317823, 9780792317821, 416 pp.

Flammarion, Nicolas Camille (1862), *La pluralité des mondes habités*, Didier, Paris.

Jones, Barrie W. (2008), Mars before the space age, *International Journal of Astrobiology*, Volume 7, Number 2, 143-155.

McGovern, Patrick J., and Morgan, Julia K. (2009), Mars Volcanic spreading and lateral variations in the structure of Olympus Mons, *Geology*, 37, pp 139-142.

D. Mortari, "*On the Rigid Rotation Concept in n-Dimensional Spaces*" Journal of the Astronautical Sciences, vol. 49, no. 3, July-September 2001.

Oparin, A. I. and Fesenkov, V. G. (1956), *Jizni vo vselennoi* (*Life in Universe*-Russ.), Ed. Academy of Science of USSR.

Phillips, R.J., Zuber, M.T., Solomon, S.C., Golombek , M.P., Jakosky, B.M., Banerdt, W.B., Smith, D.E., Williams, R.M.E., Hynek, B.M., Aharonson , O., and Hauck, S.A. (2001), Ancient geodynamics and global-scale hydrology on Mars: *Science*, vol. 291, pp. 2587–2591.

Rugescu, R. D. (2003), Sound Pressure Behavior in Relative Fluid Mechanics, *Proceedings of the 10th International Congress on Sound and Vibration (ICSV10)*, Stockholm, Sweden, July 7-10, pp. 3169-3176.

Rugescu, R. D. (2000), *On the Principles of Relative Motion of Continua*, Scientific Bulletin of U.P.B., series A (Applied Mathematics and Physics), 62, 2/2000, pp. 97-108;

Schubert, G., Russell, C.T., and Moore, W.B. (2000), Timing of the martian dynamo, *Nature* 408, pp. 666–667.

Sheehan, W. (1996), *The Planet Mars: A History of Observation & Discovery*, University of Arizona Press.

Smith, D. E., Lerch, F. J., Nerem, R. S., Zuber, M. T., Patel, G. B., Fricke, S. K., Lemoine, F. G. (1993), An Improved Gravity Model for Mars: Goddard Mars Model 1, Journal of Geophysical Research (ISSN 0148-0227), vol. 98, no. E11, p. 20, 871-889.

Standish, E. M. (1998), JPL IOM 312.F-98-048, (DE405/LE405 Ephemeris).

Terada, N., Kulikov, Y. N., Lammer, H., Lichtenegger, H. I. M., Tanaka, T., Shinagawa, H., & Zhang, T. (2009), Atmosphere and Water Loss from Early Mars Under Extreme Solar Wind and Extreme Ultraviolet Conditions, *Astrobiology*, Vol. 9, No. 1, 2009, © Mary Ann Liebert, Inc., DOI: 10.1089/ast.2008.0250

C. Tulita, S. Raghunathan, E. Benard, *Control of Steady Transonic Periodic Flow on NACA-0012 Aerofoil by Contour Bumps*, Department of Aeronautical Engineering, The Queen's University of Belfast, Belfast, Northern Ireland, United Kingdom, 2002;

Synge, J. L., and Griffith, B. A., (1949), Principles of Mechanics, second ed., McGraw-Hill Book Company, Inc., New York, Toronto, London.

Wood, B.E., Müller, H.-R., Zank, G.P., Linsky, J.L., and Redfield, S. (2005) New mass-loss measurements from astrospheric Ly-α absorption, *Astrophys. J.*, 628:L143–L146.

Zhang, M.H.G., Luhmann, J.G., Nagy, A.F., Spreiter, J.R., and Stahara, S.S. (1993), Oxygen ionization rates at Mars and Venus: relative contributions of impact ionization and charge exchange. *J. Geophys. Res.*, 98, pp. 3311–3318.

2

Surface-Barrier Solar Cells Based On Monocrystalline Cadmium Telluride with the Modified Boundary

P.M. Gorley[1], V.P. Makhniy[1], P.P. Horley[1,2], Yu.V. Vorobiev[3] and J. González-Hernández[2]

[1]*Science and Education Center "Semiconductor Material Science and Energy-Efficient Technology" at Yuri Fedkovych Chernivtsi National University, 58012 Chernivtsi,*

[2]*Centro de Investigación en Materiales Avanzados S.C., Chihuahua / Monterrey, 31109 Chihuahua,*

[3]*Centro de Investigación y de Estudios Avanzados del IPN, Unidad Querétaro, 76230 Querétaro,*

[1]*Ukraine*

[2,3]*México*

1. Introduction

Cadmium telluride is one of the most promising materials for solar cell (SC) applications due to its unique physical and chemical parameters. In the first place, it has the band gap $E_g \approx 1.5$ eV (300 K) close to the optimal value for photovoltaic conversion (Fahrenbruch & Bube, 1983; Donnet, 2001). The highest temperature and radiation stability of CdTe in comparison with Si and GaAs (Ryzhikov, 1989; Korbutyak et al., 2000) permits to use SCs based on cadmium telluride under elevated temperatures and a considerable flux of ionizing radiation. The possible alternative to CdTe with the similar band gap - gallium arsenide and its solid solutions - are far more difficult to obtain and expensive due to the rarity of Ga (Mizetskaya et al., 1986; Kesamanly & Nasledov, 1973; Andreev et al., 1975).

Solar cells may use diode structures of different kind: p-n-junction, heterojunction (HJ) or surface barrier contact. Despite CdTe has a bipolar conductivity, creation of p-n junction cell based on it is impractical due to high resistivity of p-CdTe and technological difficulty to make ohmic contacts to this material. Therefore, heterojunctions offer more versatile solution by allowing larger parameter variation of junction components than those acceptable for p-n junctions (Alferov, 1998). Additionally, direct band gap of cadmium telluride allows to use this material in thin film form, which was confirmed experimentally for thin film junction nCdS/pCdTe (Sites & Pan, 2007). Alas, the crystalline parameters and coefficients of thermal expansion for CdS (as well as the other semiconductors with wider band gap) significantly differ from those of CdTe (Milnes & Feucht, 1972; Sharma & Purohit, 1979; Simashkevich, 1980), so that the resulting HJ would inherit a significant concentration of the defects at the junction boundary, which will decrease the performance of the solar cell. Moreover, despite the low cost of thin film cells comparing to those based on the bulk material, the technology for the thin-film CdTe has a significant perspectives to be improved (Chopra & Das, 1983; Britt & Ferekides, 1993). Therefore, it seems more appropriate to use monocrystalline cadmium telluride, which has well established and reliable technology (Ryzhikov, 1989; Korbutyak et al., 2000; Mizetskaya et al., 1986). It is preferable to design a technological method for manufacturing photovoltaic devices that could be easily adapted for creation of similar structures based on thin film form of CdTe after minute correction of the corresponding technological regimes.

For this type of applications, surface-barrier diode (SBD) is definitely a good candidate due to its advantages over other diode types (Strikha & Kil'chitskaya, 1992) - simple single-cycle technology necessary to create mono- and multi-element photodiodes of arbitrary area and topology over mono- or poly-crystalline substrates, as well as anomalously low temperatures of barrier contact deposition so that the parameters of the base substrates do not undergo any significant changes in the process. Additionally, substrates of any conductivity type are suitable for formation of SBDs, including those with pre-deposited ohmic contacts. The presence of a strong sub-surface electric field favors efficient separation of non-equilibrium carriers generated by the high-energy phonons. Finally, SBDs can have much lower values of series resistance R_0 comparing with the p-n junctions and hetero-junctions, as they have one semiconductor region in place of two.

However, many of the recent papers (Amanullah, 2003; Mason et al., 2004; Kim et al., 2009, 2009; Gnatyuk et al., 2005; Higa et al., 2007) are rather dedicated to ionizing radiation detectors based on high-resistive CdTe. In contrast with SC, here the surface effects are of far lower importance because ionizing particles penetrate deeper into the material. Independently on the application area of the SBD devices, they should have the largest possible height of the potential barrier. Analysis of literature sources points the impossibility to obtain SBDs with high barrier, minimal series resistance and surface recombination rate using the traditional technological methods (Milnes & Feucht, 1972; Strikha & Kil'chitskaya, 1992; Rhoderick, 1978; Valiev et al., 1981; Sze & Kwok, 2007) - thus, the new methodology should be involved.

One of the perspective ways to solve this problem involves technologies that modify the sub-surface properties of the base substrates, at the same time keeping parameters of bulk material free from significant changes. Here we analyze the experimental results concerning electrical, optical and photoelectric properties of n-CdTe substrates with the modified substrate and surface-barrier solar cells based on them.

2. Objects and methodology of investigations

The base substrates with dimensions 4×4×1 mm^3 were cut from the bulk CdTe monocrystal, grown by Bridgeman method. The base material featured intrinsic defect electron conductivity 0.1 - 0.05 $\Omega^{-1}\cdot cm^{-1}$ at 300 K as was not doped during the growth process. The substrates were polished mechanically and chemically in the solution of $K_2Cr_2O_7$:H_2O:HNO_3 in proportion 4:20:10 with further rinsing in de-ionized water. As a result, the surface of the substrates gained a mirror-reflective look, and the samples featured a weak photo-luminescence (PL) at 300 K. The similar luminescence is also observed for the cleft surfaces, but it is completely absent in the mechanically-polished plates.

One of the largest sides of the plates with the mirror-reflective surface was deposited with indium ohmic contacts by soldering. Before the creation of a rectifying contact, which was formed with a semitransparent layer of gold deposited by vacuum sputtering, the contact-bearing side of the substrate was subjected to a different additional treatment. The first group of the samples was annealed in the air; further on, they will be referred to as $CdTe:O_2$ (Makhniy et al., 2009). The second group was processed in boiling aquatic suspension of base metal salts (Li_2CO_3, K_2CO_3 and Na_2CO_3), further addressed CdTe:AS (Makhniy & Skrypnyk, 2008). The third group of the samples was formed by chemically-etched substrates not subjected to any additional treatment; these will be further referred to as CdTe. The SBDs based on them served as reference material for comparative studies of modified surface diodes made of the samples belonging to $CdTe:O_2$ and CdTe:AS groups.

The dark current-voltage and capacitance-voltage curves (CVC and CpVC) were measured using the common methodology (Vorobiev et al., 1988; Batavin et al., 1985). The luminescence of the samples was excited with He-Ne and N_2-lasers (wave lengths 0.63 and 0.337 μm, respectively). The radiation, reflection and transmission spectra (N_ω, R_ω and T_ω) were obtained with a universal setup, allowing measurements in standard and differential modes (Makhniy et al., 2004). The spectra were registered automatically with a recording equipment KS-2, also allowing to obtain relaxation curves for the photoluminescence intensity. The light source for measurements of reflection and transmission spectra was a xenon lamp with a smooth spectrum in the investigated energy ranges. All the obtained spectra were corrected for non-linearity of the measuring system. PL spectra were plotted as a number of photons per unitary energy interval N_ω versus photon energy $\hbar\omega$.

The light source used for measuring of photoelectric characteristics was an incandescent lamp with a tungsten filament and a deuterium lamp. Fine tuning of the illumination level in the ranges of 4-5 orders of magnitude was done using a set of calibrated filters. The integral light and loading characteristics of the solar cell were measured with a common methodology (Koltun, 1985; Koltun, 1987). To study the spectral distribution of photosensitivity S_ω we used monochromator DMR-4 with energy dispersion 0.5 - 6.0 eV and precision 0.025 eV/mm. The Si and ZnSe photodiodes with known absolute current sensitivity were used as reference detectors.

The temperature measurements were performed in the ranges 300 - 450 K. The sample was deposited into a specially designed thermal chamber allowing appropriate illumination, quick variation and steady temperature maintenance with the precision of ±1 K. The photovoltaic efficiency of the solar cells was studied by comparing their photoelectric parameters with the reference ITO-Si cell, which under 300 K had the efficiency of 10% for AM2 illumination.

3. Optical properties

As it was shown by the previous studies (Makhniy et al., 2004), the surface modification of the n-CdTe substrates changes their optical properties in a different way depending on the annealing conditions. This approach also works for CdTe:AS and $CdTe:O_2$ samples featuring a sharp efficiency increase for the edge A-band of luminescence η (Fig. 1), which at 300 K can reach several percents. At the same time, for the substrates with mirror-reflecting surface this parameter does not exceed 0.01%. As the effective length of laser radiation in cadmium telluride is $\ell_{PL} \leq 10^{-5}$ cm, the photoluminescence takes place in a narrow sub-surface layer. Therefore, the luminescence intensity I_{PL} in the first approximation can be considered inversely proportional to the concentration of surface defects N_S. As surface modification decreases this concentration for more than two orders of magnitude, it makes a good motivation to use the substrates with modified surface for photodiode applications sensitive in short-wave region.

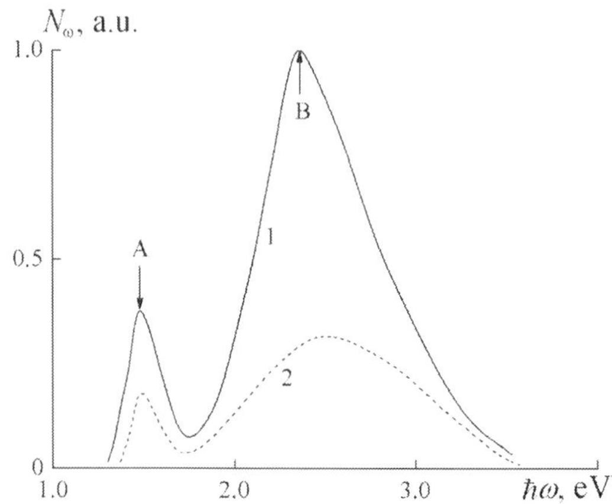

Fig. 1. PL spectra for the substrates CdTe:O_2 with 1) free surface and 2) covered with a golden film

To the contrast to CdTe:AS, the PL spectra of CdTe:O_2 samples feature a wide B-band situated in the intrinsic absorption area of CdTe at $\hbar\omega > E_g$, Fig. 1. The half-width $\Delta\hbar\omega_{1/2}$ of this band at 300 K is about 0.7 - 0.8 eV. It was found that the intensity of this high-energy band changes with time but that of A-band remains stable, also the peak of A-band $\hbar\omega_m$ shifts towards the low-energy region. It is important to mention that the stationary values of intensity I_{PL} and energy $\hbar\omega_m$ can be achieved several minutes after enabling the laser excitation (Fig. 2). Photoluminescence spectrum shown in Fig. 1, curve 1, was measured namely under such conditions. The experimental dependence of time constant for B-band $t_{1/2}(T)$ can be described by the Arrhenius law (inset to Fig. 2) with activation energy ~ 0.2 eV.

Our investigations revealed that the deposition of a semitransparent golden film onto the surface of CdTe:O_2 stabilizes I_{PL} (Fig. 2, curve 2). Despite the figure shows only the initial part of the I_{PL}(t) curve, it remains unchanged not only after several hours, but also after switching laser illumination on and off for several times. This result is very important from the applied point of view, because semitransparent golden film works as an efficient barrier contact required for the optimal performance of SCs based on CdTe:O_2 substrates (Ciach et al., 1999).

The existence of high-energy B-band in the PL spectra of CdTe:O_2 samples can be explained by quantization of carrier energy caused by the presence of nano-scale structure formations, which is confirmed with the images obtained by the AFM Nanoscope-III in the periodic contact mode (Fig. 3). As one can see, the surface of annealed CdTe:O_2 samples is composed with granules some 10 - 50 nm in size (Fig. 3a), which may eventually join into a larger (100 - 500 nm) formations (Fig. 3b).

Annealing the samples under the optimal conditions (e.g., temperature and time) will result in optimization of A-band intensity corresponding to the edge luminescence (Makhniy et al., 2009). Using the relation $I_{PL} \sim N_S^{-1}$, one may come to the conclusion that such high-luminescent samples should have the minimal surface recombination rate. The AFM image of such substrates CdTe:O_2 reveals formation of nano-grains of various sizes (Fig. 3b). It is

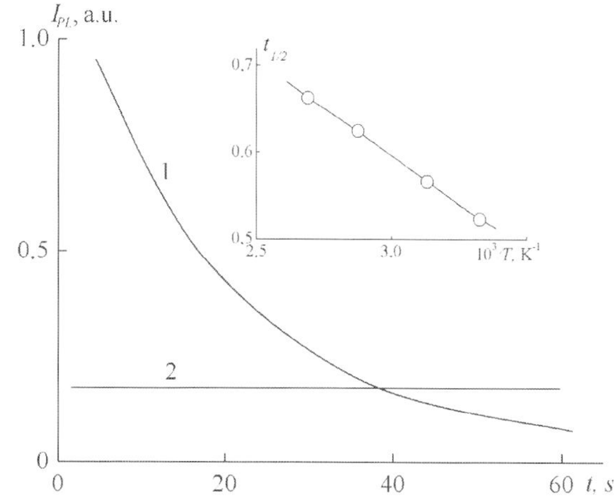

Fig. 2. Time dependence of B-band intensity at 300 K for $CdTe:O_2$ samples 1) with free surface and 2) covered with golden film. The inset shows the temperature dependence of time variable $t_{1/2}$

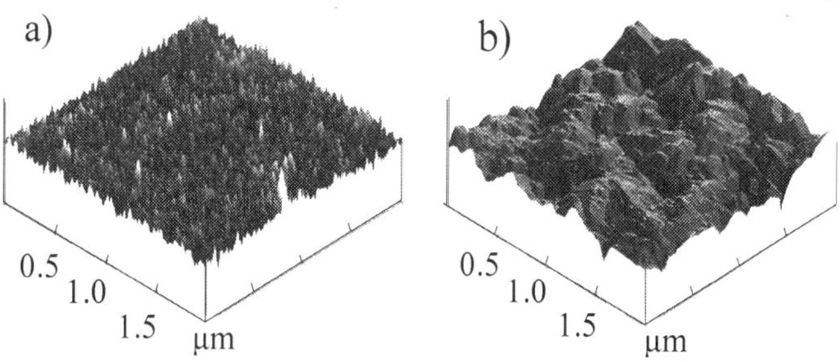

Fig. 3. AFM images of a) the original and b) air-annealed n-CdTe substrates

worth mentioning that large and small grains play a different role for the properties of surface-modified $CdTe:O_2$. The transition energy is defined by the quantization rule (Zayachuk, 2006; Pool & Owens, 2006)

$$\Delta E = \frac{n^2\hbar^2}{2d^2}\left(\frac{1}{m_n^*}+\frac{1}{m_p^*}\right), \tag{1}$$

where m_n^* and m_p^* are the effective masses of electron and a hole; d is lateral dimension of nano-object responsible for the energy peak $\hbar\omega_m$ in the PL spectrum. The depth of the quantum well for such objects is

$$\Delta E = \hbar\omega_m - E_g\,. \tag{2}$$

Using the values m_n^* = 0.11 m_0; m_p^* = 0.35m_0; E_g=1.5 eV and $\hbar\omega_m$≈2.5 eV defined from the experiment, we applied expressions (1) and (2) to estimate nano-particle dimension d to be

about 5 nm, which is two times smaller than the minimum observed size of the small grains ($d \approx 10$ nm) appearing in the AFM images. This controversy can be removed taking into account that the majority of the grains have a pyramidal shape (Fig. 3b), so that the B-band may be formed by contribution from their top parts, which are narrower than their base. The secondary proof of such possibility consists in presence of the photons with energy $\hbar\omega > \hbar\omega_m$, most probably caused by nano-objects with dimensions smaller than 5 nm.
It is important to emphasize that B-band can not be caused by the luminescence of CdO film that may be eventually formed during the annealing process. In the first place, we did not observe any visible radiation for the samples with CdO film created by photo-thermal oxidation over the substrates with mirror-smooth surface. Moreover, the differential reflection spectrum R_ω' of such samples has a peak corresponding to E_g of cadmium oxide, which is absent in the spectra of the modified substrates, Fig. 4.

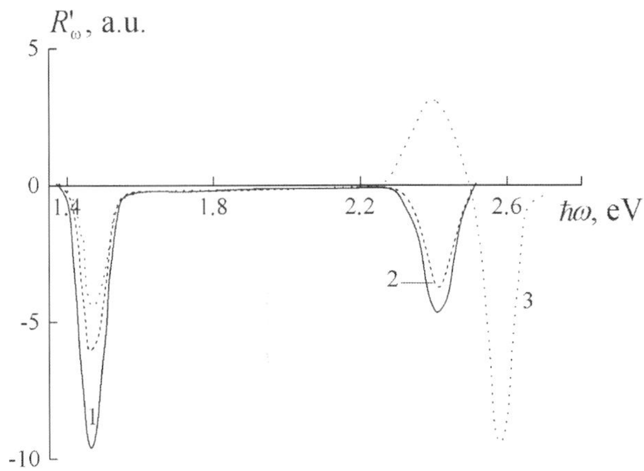

Fig. 4. Differential reflection spectra for various substrates: 1) CdTe, 2) CdTe:O_2 and 3) heterostructures CdO/CdTe at T = 300 K

The peaks located at 1.5 and 2.4 eV will correspond to the energy distance between the conduction band E_C and the edges of the valence band: the main sub-band E_{VA} and spin-orbital split band E_{VB}. The peak at $\hbar_{\omega m} \approx 2.6$ eV correlates with the band gap of CdO at 300 K (Madelung, 2004). The high-energy "tail" of B-band continues much further than it should if being solely defined by E_g(CdTe). It is worth noting that optical transmission / absorption of the samples is defined with the sample group. The "smooth" transmission curve T_ω for CdTe and CdTe:AS substrates has a sharp high-energy edge at $\hbar_{\omega m} \approx 1.5$ eV corresponding to the band gap of CdTe, Fig. 5.
In contrast, the samples annealed in the air feature a significant decrease of T_ω with red-shifted high-energy transmission edge intercepting the abscissa axis at the energy 1.3 eV, which is significantly lower than E_g of cadmium telluride. The detected peculiarities of transmission spectrum can be explained by the presence of super-grains (100 - 500 nm in size) at the surface of the samples. Such grains may cause light scattering and multiple reflections decreasing the absolute value of T_ω. As these processes intensify for the larger $\hbar\omega$, it may be the cause of the observed red-shift of T_ω curve. It is worth mentioning that in the

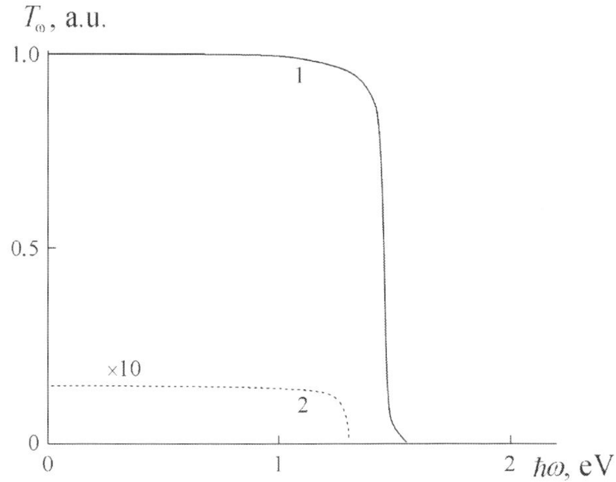

Fig. 5. Optical transmission spectra for different substrates: 1) CdTe and CdTe:AS; 2) $CdTe:O_2$ at 300 K.

photosensitivity spectra S_ω of Au- $CdTe:O_2$ diodes the sensitivity increases drastically for the energies exceeding the band gap of CdTe. It seems feasible to suggest that the observed quantum-scale surface texture is most probably formed by some kind of self-organization phenomena. This hypothesis is confirmed by a considerably narrow temperature and time intervals of the annealing process (Makhniy et al., 2003) that enable formation of the modified structure. Moreover, reproducing the same parameters under the different conditions (noble gas atmosphere or vacuum) does not lead to the desired effect, suggesting that one of the atmosphere components (in particular oxygen) plays an important role in the formation of a nano-crystalline surface structure. The definition of such dependence requires additional experimental and theoretical studies.

4. Electrical properties and carrier transport mechanisms

4.1 Potential barrier height

The height of the potential barrier φ_0 is one of the most important parameters of SBD: it limits open-circuit voltage U_{OC} of the photovoltaic device and enters the exponent in the expression describing over-barrier and tunneling currents, which with increase of φ_0 will yield a lower dark current. Also, higher barrier will favor a larger working temperature of a solar cell.

It was found out that the SBDs based on CdTe:AS and $CdTe:O_2$ feature significantly higher barrier than that achievable for the devices based on the non-modified CdTe. This point can be clearly seen from Fig. 6, where we plot forward-biased branches of current-voltage curves (CVCs) of the studied structures in their linearity region. It is important to highlight that the difference between the slopes of the segments for all three groups of diodes studied is insignificant, proving the close similarity of series resistance in the system diode base - deposited contacts. On the other hand, it also suggests that the contribution of the modified layer into the value of R_0 is negligibly small.

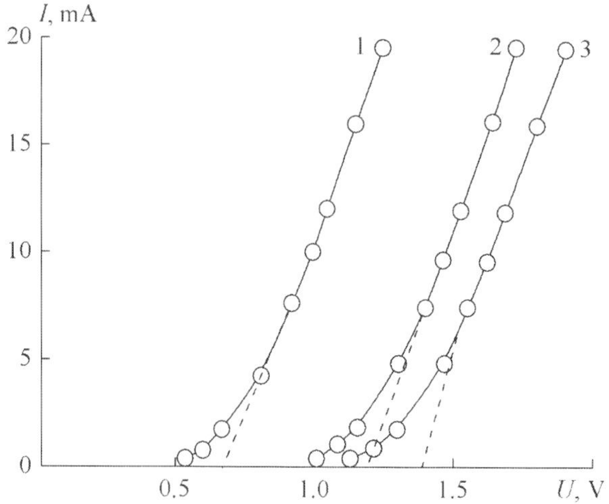

Fig. 6. Direct branches of CVC for the SBDs based on different substrates: 1) CdTe, 2) CdTe:AS, 3) CdTe:O_2 at 300 K.

The latter can be confirmed by the equal $I(U)$ dependence for the different diode types (Fig. 6). The same φ_0 and R_0 observable for the SBDs that differ by processing of their substrates (treatment in the solutions of Li, Na and K salts) suggest that various impurities have the same behavior in the resulting device. It is worth noting that the values of φ_0 determined from the direct CVC branches agree with the results obtained from the capacity measurements. The further analysis of the electrical properties revealed that the surface state apart from influencing φ_0 also change the character of physical processes in the SBDs, which is reflected in the dependences of forward / reverse currents as a function of voltage.

4.2 Direct current formation mechanisms in the surface barrier diodes

Figure 7 displays the initial segments of direct CVC branches measured for the modified surface diodes. As one can see from the figure, for $eU \geq 3kT$ the curves can be successfully approximated with the following expression (Rhoderick, 1978; Fahrenbruch & Bube, 1983)

$$I = I_0 \exp\left(eU / nkT \right) \tag{3}$$

where I_0 is a cut-off current at $U = 0$, and n is the non-ideality coefficient, which can be easily found as a slope of the straight CVC segments plotted in semi-logarithmic coordinates.

For Au-CdTe diodes the value of n is equal to the unity, suggesting the dominating over-barrier carrier transport for the SBDs made on the base of moderately-doped substrates (Rhoderick, 1978; Makhniy, 1992). It also means that the dielectric layer between metal and semiconductor is tunneling-transparent with the barrier height lower than $E_g/2$. The dependence $I_0(T)$ is mainly defined by the factor (Sze & Kwok, 2007)

$$I_0 \approx \left(- \varphi_0 / kT \right), \tag{4}$$

which gets an excellent confirmation in the experiment. Plotting $I_0(T)$ in $\ln I_0 - 1/T$ coordinates, one can achieve a good linear fit with the slope of 0.9 eV corresponding to φ_0 at 0 K.

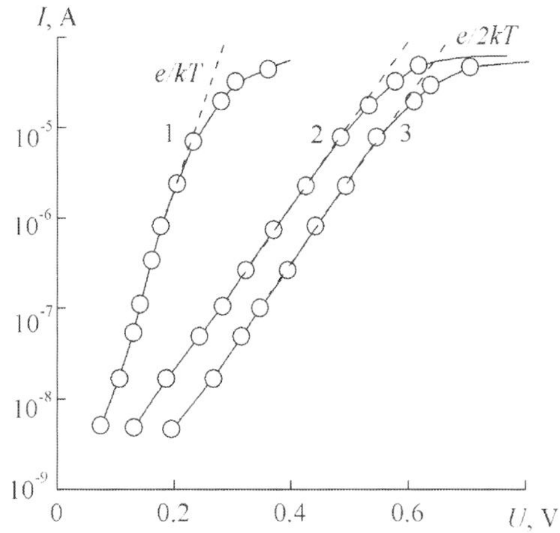

Fig. 7. The direct CVC branches of the diodes based on: 1) CdTe, 2) CdTe:AS and 3) CdTe:O_2 at 300 K.

In contrast to Au-CdTe contacts, the potential barrier height in the SBD with a modified surface will significantly exceed the value of E_g(CdTe)/2 = 0.75 eV at 300 K (Fig. 6). For such diodes the dominating carrier transport mechanism (Makhniy, 1992) is a recombination current in the space charge region (SCR) involving deep impurity levels. It can be also described by formula (3) with non-ideality coefficient $n = 2$. As one can see from Fig. 7, the initial direct CVC branches for Au-CdTe:AS and Au-CdTe:O_2 diodes fit excellently with the expression (3) for the temperature T = 300 K. However, it is also possible to approximate the experimental data with another exponential factor (Makhniy 1992)

$$I_0 \approx (-\varphi_0 / kT), \tag{5}$$

For this case, the measured $I_0(T)$ plotted in ln I_0 - $1/T$ coordinates will represent a straight line with a slope 1.6 eV corresponding to the band gap of cadmium telluride at 0 K.
Deviation of the experimental points from a straight line under the higher bias is caused by the voltage drop over series resistance of the diode, reducing the applied voltage U to the voltage at the barrier U_0 as

$$U_0 = U - IR. \tag{6}$$

Taking into account (6), one can re-write the expression (3) in the form (Makhniy 1992)

$$\ln I - \frac{eU}{kT} = \ln I_0 - \frac{eR_0}{nkT} I\,. \tag{7}$$

When the over-barrier current is dominating (which is correct for the voltages about φ_0) the non-ideality coefficient is equal to the unity and the direct branches of CVC can be efficiently plotted in the coordinates ln I_0 - eU/kT. These experimental data can be nicely fit with expression (7) in the case of the diodes with modified surface (Fig. 8) with ordinate interception yielding the ln I_0 for the given temperature.

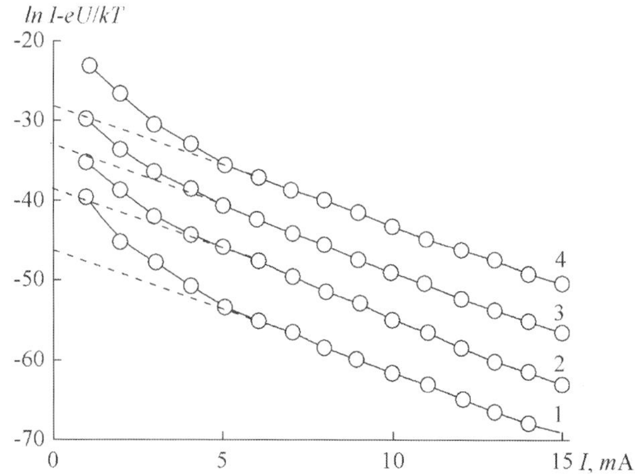

Fig. 8. Comparison of experimental data for direct CVC branches (Au-CdTe:AS diode) with theoretical formula (7) for the temperatures 1) 300 K, 2) 320 K, 3) 340 K and 4) 360 K

The dependence $I_0(T)$ obeys the expression (4) with $\varphi_0 = 1.6$ eV, which corresponds to the barrier height at 0 K. Taking into account the linear dependence $\varphi_0(T) = \varphi_0(0) - \gamma_\varphi T$ and using $\gamma_\varphi = 1.4 \cdot 10^{-3}$ eV/K one can estimate $\varphi_0 = 1.18$ eV at the room temperature, which correlates well with the experimental result 1.2 eV, Fig. 6.

While Fig. 8 presents the data for Au-CdTe:AS diode only, the similar results were obtained for all the structures studied. Therefore, the direct current transport in surface-barrier diodes based on CdTe has a predominant over-barrier character. In the diodes based on the substrates with a modified surface, the current is defined by carrier recombination in the space charge region (low direct bias) or over-barrier emission (high bias).

4.3 Reverse current in surface barrier diodes

It is important to emphasize that over-barrier and recombination cut-off currents I_0 at 300 K are always below 10^{-10} A (Fig. 7). Moreover, these currents in a theoretical model feature a weak dependence on the voltage, such as $U^{1/2}$ (Rhoderick, 1978; Sze & Kwok, 2007). However, in the experiments we obtained steeper curve of inverse current $I_{inv}(U)$, suggesting other current transport mechanisms.

It is logical to assume that in the SBDs based on wide-band semiconductors (which holds for CdTe) subjected to inverse bias tunneling of the carriers will become a dominating mechanism, including both inter-band tunneling and tunneling *via* the local levels (Makhniy, 1992). The validity of this hypothesis is supported by Fig. 9, where the inverse CVC exhibit a good fitting with the formula for tunneling current in an abrupt junction:

$$I_{inv} = a \exp\left(-\frac{b}{\sqrt{\varphi_0 - eU}}\right). \tag{8}$$

Here a and b are the coefficients that can be found from the parameter substrate and the diode structure. The negative sign under the square root reflects the negative bias, and the voltage in the expression (8) should be substituted with the negative sign. The high slope of

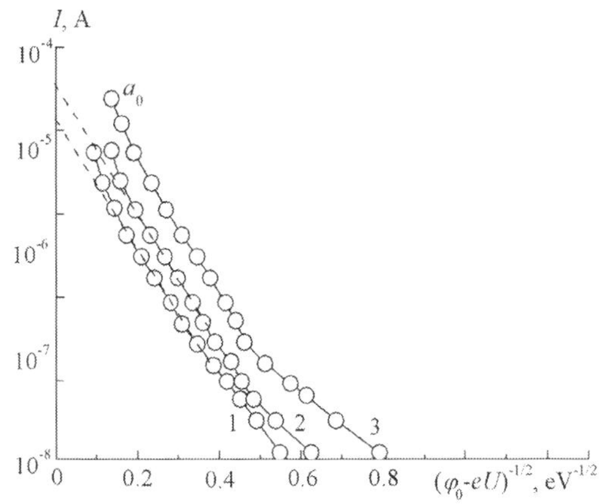

Fig. 9. Comparison of theoretical calculations according to (8) with the inverse CVC branches for the diodes based on: 1) CdTe, 2) CdTe:AS, and 3) CdTe:O_2 at 300 K.

the initial branches of inverse CVCs signals the parameter difference for the diode structures but not the substrates themselves. It is worth noting that the coefficient b depends on the height and width of the barrier, the depth of the local centers in the SCR and some other parameters (Makhniy, 1992), which requires more detailed study. The same can be commented concerning the dependence $I_{inv}(U)$ under high bias (Fig. 9), which in addition to tunneling may have also include the avalanche processes (Mizetskaya et al., 1986).

While the solution of the aforementioned tasks is important for the photodiodes, it is not that much crucial for solar cell applications that require a special attention to the parameters φ_0, n, I_0, and E_g appearing in the expression for the direct current.

5. Photoelectric properties of surface barrier diodes

5.1 Integral illuminated current voltage curves

The CVC of illuminated diode can be described as (Fahrenbruch & Bube, 1983; Koltun, 1987)

$$I = I_0\left[\exp\left(\frac{eU}{nkT}\right) - 1\right] - I_p \tag{9}$$

with photocurrent I_p, dark current $I_0(U = 0)$ and non-ideality coefficient n, which is determined by the current transport mechanism. To determine the exact mechanisms involved in formation of CVCs for illuminated solar cells, we need introduce open circuit voltage (at $I = 0$) and short circuit current I_{SC} (for $U = 0$) that upon being substituted into (9) for a particular case $eU \geq 3kT$ would yield the expression

$$I_p = I_{SC} = I_{SC}^0 \exp(eU_L / nkT). \tag{10}$$

Here I_{SC}^0 denotes the cut-off current for illuminated cell voltage $U_L = 0$, which may coincide with I_0 only in the case if the formation mechanisms for the light and dark currents are the same. Therefore, investigation of the integral light CVCs allow to determine the corresponding current formation mechanism, as well as to reveal both common and different traits between the electric and photoelectric properties of the materials studied. Analysis of $I_{SC}(U_{OC})$ curves shows that they are qualitatively similar, differing only in non-ideality value for the various SBD groups. Thus, for Au-CdTe diodes the value of n = 1 suggests over-barrier transport of photo-generated carriers, which can be also illustrated by fitting data presented in Fig. 10 (curve 1). It is worth mentioning that I_{SC}^{0} is close to the cut-off current I_0, obtained for the dark CVC for the same diode at 300 K.

The energy slope of the segment $\ln I_{SC}^{0} - 1/T$ yields the value of 0.9 eV that coincide with the potential barrier height φ_0 at 0 K, determined from the temperature dependence of the dark cut-off current I_0. In this way, one can conclude that dark and light currents of Au-CdTe contacts are formed by over-barrier emission of the carriers. To the contrary, the measured CVC of the illuminated SBDs with a modified surface will require non-ideality coefficient to be n=2, equation (10), to achieve the appropriate fitting illustrated in Fig. 10 (curve 2) for Au-CdTe:AS diode. As it was mentioned before, such dependence is characteristic for carrier recombination in the space charge region involving local centers, while the photosensitivity data for Au-CdTe suggests the dominating inter-band generation of the carriers.

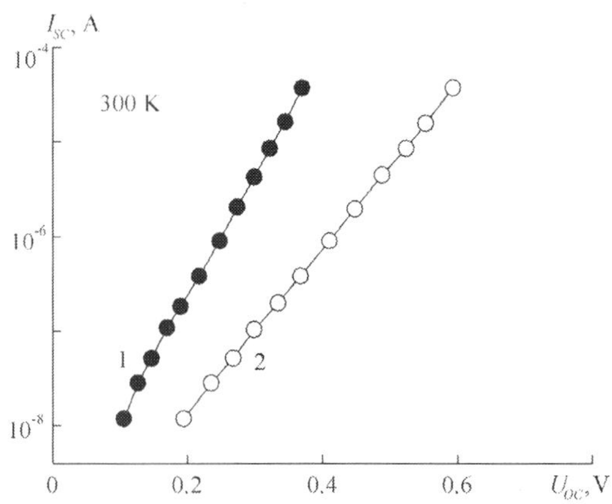

Fig. 10. Dependence of short circuit current on the open circuit voltage for Au-CdTe (1) and Au-CdTe:AS (2) diodes.

This controversy can be eliminated (Ryzhikov, 1989): under the stationary conditions the carrier generation rate G should be equal to the recombination rate R. However, if these phenomena are caused by distinct mechanisms, they will be described by the different analytical expressions. It is worth mentioning that the measurement of the dependences $I_{SC}(U_{OC})$ is performed in so-called compensation mode, when the photocurrent is equalized with dark current under the presence of direct bias. Due to this, one actually monitors recombination of photo-carriers, which yields the expression similar to that of the dark current. It is important that for the SBDs based on the substrates with a modified surface the dark current under the low direct bias is controlled by the recombination processes in space charge region (see Section 4). As carrier generation in the surface barrier diodes takes place in the space charge region due to the fundamental absorption of high-energy photons with $\hbar\omega > E_g$, it is more probable that they will recombine at the same device region taking advantage of the local impurity centers rather than *via* the inter-band recombination.

5.2 Dependence of I_{SC} and U_{OC} on illumination level and the temperature

For all the SBDs studied, the short circuit current depends linearly on the power of the incident light flux L, which is a consequence of linearity of photo-carrier generation as

$$I_{SC} = \beta L, \tag{11}$$

where β is a proportionality coefficient independent on L. As one can see from Fig. 11, the equation (11) holds well for illumination power spanning over several orders of magnitude. Using the expressions (10) and (11) it is easy to show that

$$U_{OC} = nkT\left(\ln\left(I_{SC} / I_{SC}^0\right)\right) = nkT\ln\left(\beta L / I_{SC}^0\right). \tag{12}$$

As one can see from Fig. 12, expression (12) describes well the experimental dependences of $U_{OC}(L)$ for the low illumination conditions. The tendency of $U_{OC}(L)$ curves to saturate under high L is caused by the potential barrier compensation. The higher value of U_{OC} for the SBDs with a modified surface in comparison with that of Au-CdTe diodes is explained by the significant difference of potential barrier heights for these rectifying structures (see Sec. 2).

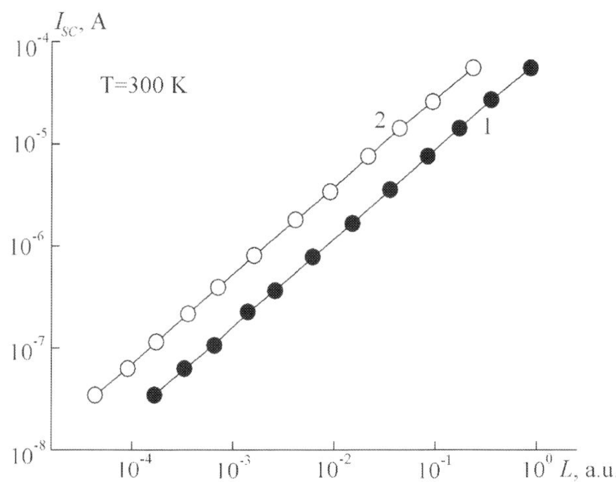

Fig. 11. Dependence of short circuit current on radiation power for Au-CdTe (1) and Au-CdTe:AS (2) diodes

Our investigations revealed that the temperature increase leads to a slight growth of short circuit current, which is caused by an insignificant decrease of series resistance R_0 characterizing the device. Under high illumination open circuit voltage tends to saturate, so that in the first approximation $U_{OC} \approx \varphi_0$ or E_g depending on formation mechanism of the light CVCs (see above). Therefore, under otherwise equal conditions, the open circuit voltage should diminish with increase of temperature, which was confirmed experimentally for all SBDs studied. It is important to emphasize that $U_{OC}(T)$ curve for SBDs with a modified surface features much weaker slope than that observable for the reference diodes. This can be explained by the different temperature dependence of variation coefficient for the band gap E_g and barrier height γ_φ (see Sec. 4).

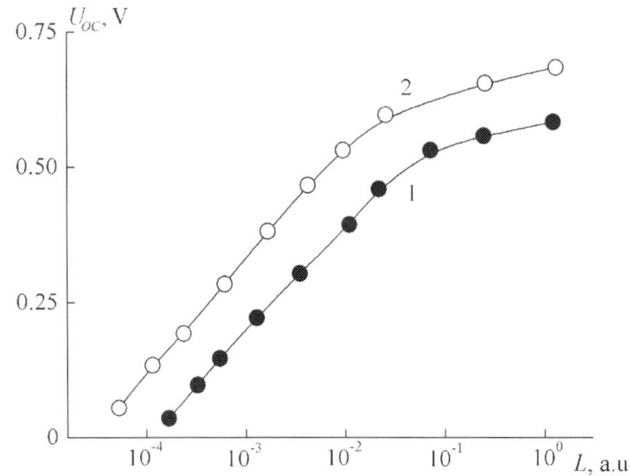

Fig. 12. Dependence of open circuit voltage on the illumination level for Au-CdTe (1) and Au-CdTe:AS (2) diodes

5.3 Spectral characteristics

The previous investigations of optical, electrical and photoelectrical properties of the SBDs predict several peculiar and promising characteristics of these structures. In the first place, the low energy limit $\hbar\omega_{min}$ of the photosensitivity spectrum S_ω is determined with high-energy edge of the transmission spectrum. According to the data presented in Fig. 5, the low-energy edge of S_ω curves for Au-CdTe and Au-CdTe:AS diodes will be about 1.5 eV, while for Au-CdTe:O_2 contacts it is somewhat lower, reaching 1.3 eV (Fig. 13).

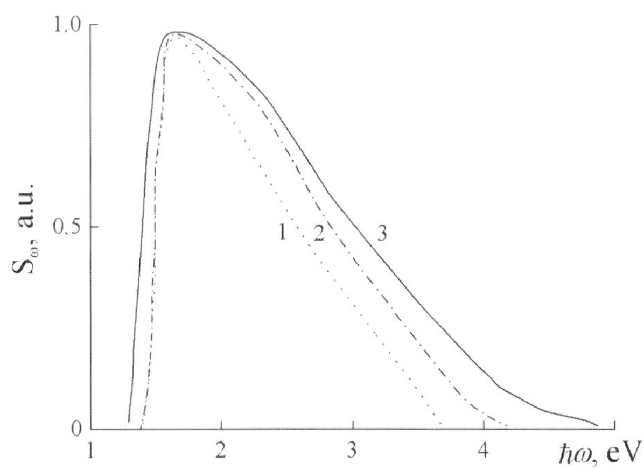

Fig. 13. Photosensitivity spectra for the SBD based on the substrates of CdTe (1), CdTe:AS (2) and CdTe:O_2 (3) at 300 K

High-energy photosensitivity, as it was discussed above, is defined by the surface recombination rate, which in turn depends on concentration of the defects N_S at the junction

boundary. Thus, high short-wave sensitivity should be expected in SBDs based on the substrates with a maximum efficiency of edge luminescence band. This conclusion is confirmed by the data presented in Fig. 13, featuring wider photo-sensitivity peaks for Au-CdTe:O_2 diodes, which are based on the substrates with the most intensive A-band (see Section 3) reaching 10% for 300 K. At the same time, for case of CdTe:AS and CdTe substrates, the corresponding A-band values would be 1-2% and 0.01%, respectively, strongly influencing the behavior of S_ω spectra for photon energies $\hbar\omega > E_g$ (Fig. 13).
Investigation of the temperature dependence of the SBD spectra reveals that higher values of T mainly affect their low-energy "tail". The edge of S_ω under the temperature varying in the ranges 300-450 K almost replicates the dependence $E_g(T)$. In the contrast to this, the temperature dependence of the surface defects (*i.e.*, their concentration, energy distribution and scattering cross-sections) is negligibly weak, as one may conclude from the photosensitivity spectra for $\hbar\omega > E_g$.

Type of the surface barrier diode		Au-CdTe	Au-CdTe:AS	Au-CdTe:O_2
Parameters	φ_0, eV	0.7	1.2	1.35
	R_0, Ω	30	30	32
	U_{OC}, V	0.4	0.6	0.7
	I_{SC}, mA/cm^2	10	15	22
	Efficiency, %	5	9	13

Table 1. Main parameters of the solar cells

5.4 Main parameters of surface barrier solar cells

The diodes with the effective photosensitive area of $2 \cdot 10^{-2}$ cm^2 were illuminated from the side of a semi-transparent golden contact by light equivalent to AM2 solar illumination conditions. All the measurements were performed at 300 K and the best results obtained were summarized in Table 1.
The values characterize photovoltaic devices in "as-obtained" form without anti-reflection coatings or any special optimization of solar cell design. Due to this, the values of fill factor *ff* are under 0.7 because of considerable series resistance (see Table 1). The reduction of the latter would allow increase of both I_{SC} and *ff*, improving the efficiency of the SCs.

6. Conclusions

The authors developed a special technological approach consisting in a special surface treatment of monocrystalline plates of cadmium telluride before deposition of the barrier contact, which significantly improves the electrical and photoelectric properties of the surface-barrier devices of metal-semiconductor junction type created on their base. The annealing of n-CdTe substrates ($\rho \sim$ 10-20 Ω cm) in the air or aquatic suspense of alkaline metals increases the potential barrier height φ_0 up to 1.2-1.4 eV in comparison with $\varphi_0 \approx 0.7$ eV for non-annealed substrates. The SBDs with modified surface features much better open circuit voltage and short circuit current. The efficiency of solar cells based on the diodes studied under AM2 illumination was within the ranges 8-13% at the room temperature. The developed technology for surface-barrier solar cells is simple, cheap and ecologically clean. In addition, the proposed technological principles can be used for producing photovoltaic devices based on thin films of cadmium telluride.

7. Acknowledgements

This research was supported in part by the Ministry for Education and Science (Ukraine) through the financing provided for the scientific projects carried out at the Department of Optoelectronics, Department of Energy Engineering and Electronics, as well as at the Science and Education Center "Semiconductor Material Science and Energy-Efficient Technology" at the Chernivtsi National University, Research Project #SU/447-2009 for the years 2009-2010. We also acknowledge the financial support of CONACYT (México) for the research performed in CINVESTAV-Querétaro, Research Project CB-2005-01-48792 for the years 2007-2010.

8. References

Alferov, Zh.I. (1998). The history and the future of semiconductor structures. *Fiz. Tekhn. Poluprov.*, Vol. 32, No. 1, 3-19

Amanullah, F.M. (2003). Effect of isochronal annealing on CdTe and the study of electrical properties of Au–CdTe Schottky devices. *Canadian Journal of Physics*, Vol. 81, No. 3, 617-624

Andreev, V.M.; Dolginov, L.M. & Tret'yakov. D.N. (1975). *Liquid epitaxy in technology of semiconductor devices*, Soviet radio, Moscow

Batavin, V.V.; Kontsevoj, Yu.A. & Fedorovich. Yu.V. (1985). *Measurement of the parameters of semiconductor materials and structures*, Radio i svyaz', Moscow

Britt, J. & Ferekides, C. (1993). Thin film CdS/CdTe solar cell with 15.8% efficiency. *Applied Physics Letters*, Vol. 62, 2851-2852

Chopra, K.L. & Das, S.R. (1983). *Thin film solar cells*, Springer, New York

Ciach, R.; Demich, M.V.; Gorley, P.M.; Kuznicki, Z.; Makhniy, V.P.; Malimon, I.V. & Swiatek, Z. (1999). Photo and X-ray sensitive heterostructures based on cadmium telluride, *J. Cryst. Growth*, Vol. 197, No. 3, 675-679

Donnet, D. (2001). Cadmium telluride solar sells. In: *Crean Electricity from Photovoltaic.* Archer, M.D. & Hill, R. (Eds.), 245-276, Imperial College Press

Fahrenbruch, A.L. & Bube, R.H. (1983). *Fundamentals of solar cells: photovoltaic solar energy conversion*. Academic Press, New York

Gnatyuk, V.A.; Aoki, T.; Hatanaka, Y. & Vlasenko, O.I. (2005). Metal–semiconductor interfaces in CdTe crystals and modification of their properties by laser pulses, *Applied Surface Science*, Vol. 244, 528-532

Higa, A.; Owan, I.; Toyama, H.; Yamazato, M.; Ohno, R. & Toguchi, M. (2007). Properties of Al Schottky Contacts on CdTe(111)Cd Surface Treated by He and H_2 Plasmas *Japanese Journal of Applied Physics*, Vol. 46, 2869-2872

Kesamanly, F.P. & Nasledov, D.N. (Eds.) (1973). *Gallium arsenide: production, properties and applications*, Nauka, Moscow

Kim, K.; Cho, S.; Suh, J. Won, J. Hong, J. & Kim. S. (2009). Schottky-type polycrystalline CdZnTe X-ray detectors. *Current Applied Physics*, Vol. 9, No. 2, 306-310

Koltun, M.M. (1985). *Optics and metrology of the solar cells*, Nauka, Moscow

Koltun, M.M. (1987). *Solar cells*, Nauka, Moscow

Korbutyak, D.V.; Mel'nychuk, S.V.; Korbut, Ye.V. & Borysyuk, M.M. (2000). *Cadmium telluride: impurity defect states and detector properties*, Ivan Fedoriv, Kyiv

Madelung, O. (2004). *Semiconductors: Data Handbook*, Springer

Makhniy, V.P. & Skrypnyk, M.V. (2008). Patent for the useful model UA №31891 published 25.04.2008.

Makhniy, V.P. (1992). *Physical processes in the diode structures based on wide-band A^2B^6 semiconductors*. Dissertation of the Doctor in Physics and Mathematics, Chernivtsi

Makhniy, V.P.; Demych, M.V. & Slyotov, M.M. (2003). Declaration patent UA №5010A published 22.04.2003.

Makhniy, V.P.; Skrypnyk, M.V. & Demych, M.V. (2009a). Patent for a useful model UA №40056 published 23.05.2009.

Makhniy, V.P.; Slyotov, M.M. & Skrypnyk, N.V. (2009b). Peculiar optical properties of modified surface of monocrystalline cadmium telluride, *Ukr. J. Phys. Opt.*, Vol. 10, No. 1, 54-60

Makhniy, V.P.; Slyotov, M.M.; Stets, E.V.; Tkachenko, I.V.; Gorley, V.V. & Horley, P.P. (2004). Application of modulation spectroscopy for determination of recombination center parameters. *Thin Solid Films*, Vol. 450, 222-225

Mason, W.; Almeida, L.A.; Kaleczyc, A.W. & Dinan, J.H. (2004). Electrical characterization of Cd/CdTe Schottky barrier diodes. *Applied Physics Letters*, Vol. 85, No. 10, 1730-1732

Milnes, A. & Feucht, D. (1972). *Heterojucntions and metal semiconductor junctions*, Academic Press, New York

Mizetskaya, I.B.; Oleynik, G.S.; Budennaya, L.D.; Tomashek, V.N. & Olejnik. N.D. (1986). *Physico-chemical bases for the synthesis of monocrystals of semiconductor solid solutions $A^{II}B^{VI}$*, Naukova dumka, Kiev

Pool, Ch. Jr. & Owens, F. (2006). *Nanotechnologies*, Tekhnosfera, Moscow

Rhoderick, E.H. (1982). *Metal-semiconductor contacts*, Clarendon Press, Oxford

Ryzhikov, V.D. (1989). *Scintillating crystals of the semiconductor compounds A^2B^6: Obtaining, properties, applications*, NIITEKhIM, Moscow

Sharma, B.L. & Purohit, R.K. (1979). *Semiconductor heterojunctions*, Soviet radio, Moscow

Simashkevich, A.V. (1980). *Heterojunctions based on semiconductor $A^{II}B^{VI}$ compounds*, Sthtiintsa, Kishinev

Sites, J. & Pan, J. (2007). Strategies to increase CdTe solar-cell voltage. *Thin Solid Films*, Vol. 515, No. 15, 6099-6102.

Strikha, V.I. & Kil'chitskaya, S.S. (1992). *Solar cells based on the contact metal-semiconductor*, Energoatomizdat, St. Petersburg

Sze, S.M. & Kwok, K.Ng. (2007). *Physics of semiconductor devices*, J. Willey & Sons, New Jersey

Valiev, K.A.; Pashintsev, Yu.I. & Petrov, G.V. (1981). *Using contacts metal-semiconductor in electronics*, Soviet radio, Moscow

Vorobiev, Yu.V.; Dobrovol'skiy, V.N. & Strikha, V.I. (1988). *Methods in semiconductor studies*, Vyshcha shkola, Kiev

Zayachuk, D.M. (2006). *Low-scale structures and super-lattices*, Lviv Polytechnic, Lviv

3

Control of a 3KW Polar-Axis Solar Power Platform with Nonlinear Measurements

John T. Agee and Adisa A. Jimoh
Tshwane University of Technology, Pretoria, South Africa

1. Introduction

Environmental concerns and the finiteness of fossil fuels have engendered a global embrace for alternative energy systems. Botswana is a country blessed with abundant solar energy resources: having a mean solar day of 8.8 hours and 320 days of clear sunshine in a year (Anderson & Abkenari, 1999; Botswana Energy Report, 2003). It also experiences an excellent mean solar radiation intensity of 5.8/KW.m^2 (Anderson & Abkenari, 1999). Given that the country currently meets about 70% of her electricity needs through imports from the Southern African Power Pool (SAPP) (Botswana Power Corporation, 2004; SADC, 2004; Matenge & Masilo, 2004), the country is well motivated to integrate solar power into its energy generation base.

Earlier attempts at integrating solar power generation into the national energy mix in Botswana in the eighties, and, in fact, up to the nineties, advocated the use of solar power installations of a few Watts' capacity for lighting in small rural communities. Moreover, such solar power projects employed static installations. Static solar power installations generally have lower daily and seasonal efficiencies than sun-tracking installations. Compared to tracking systems, the lower efficiencies of static solar installations often mean that additional photovoltaic panels must be mounted to meet the required output capacity, thus raising the over-all cost of the facility. Consequently, the above-mentioned solar power philosophy imploded by reason of two shortcomings: customers' perception that initial installation costs were unduly high; and the sentiments of financial institutions that the business value of such small capacity installations was insignificant. Experiences around the SADC countries generally show that such integration philosophies have always not been sustainable (Geche & Irvine, 1996; Mogotsi, 2002; Lasschuit *et al*, 2009). For Botswana, as well as for several other countries in the SADC (Southern African Development Council) region, the high initial costs of solar power installations have been a major hindrance to the massive adoption of solar energy for rural communities (BPC, 2005; Solarie 2005).

The current development of solar power equipment for use in Botswana, and the possible subsequent extension to other SADC countries benefited from the findings reported above. First of all, the current efforts concentrate on the development of solar power equipment that could support rural entrepreneurial activities, in addition to basic lighting needs. This approach is rooted in the understanding that sustainability could be enhanced with the stimulation of economic activities. This approach is also thought to be supported by the fact that rural individuals could be motivated to acquire solar power systems that might enhance their economic wellbeing; and also that, such solar power installations with added economic value would attract the support of local financial institutions. Tables 1 to 3 show examples of typical rural enterprises around Botswana, and the example power requirements (SADC Report, 2004; de Lazzer, 2005). Additional scenarios that could be considered include cold storage facilities for anti-retroviral drugs, rural guest houses for tourism, battery charging and welding businesses. In all studied cases, the power requirements could be considered to be in the range 1-3KW; this forming the basis for the 3KW rating of the solar power platform studies presented in the chapter. Secondly, the current approach for equipment development uses tracking solar power systems, as opposed to the earlier approach that utilized static solar power systems. An extensive discussion comparing tracking solar power systems is presented elsewhere (de Lazzer, 2005; Agee et al 2006a). Suffice it to state that, polar-axis tracking systems present an option capable of producing 97.5% the output power of two-axis solar power systems; and this at acquisition and maintenance costs similar to those of the cheaper single-axis installations. This comparative economic advantage informed our choice of the polar-axis tracking solar power systems for the study reported in this chapter.

For the rest of the chapter, the physical structure and the data of the solar platform system is presented in section two. Dynamic modelling of the platform and model studies is presented in section three. Sensor characteristics modelling and validation form the contents of section four. Two controllers are investigated for the enhancement of the dynamic performance of the polar-axis solar power platform. The design and comparative analysis and discussions of these controllers is presented in section five of the chapter. Section six contains the conclusions and recommendations for further research. A list of references is included in section seven, to conclude the chapter.

Equipment	Power Consumption (W)
Refrigerator with freezer	550
Lighting bulb 60W (x2)	120
Television 51cm color	80
Radio portable	6
Fan	250
TOTAL	**1006**

Table 1. Energy requirements of a rural bar

Equipment	Power Consumption (W)
4 computers	1200
light	60
Radio-cassette	6
TV color	80
TOTAL	**1346**

Table 2. Energy requirements of a rural internet café.

2. Description of the 3KW polar-axis solar tracker hardware

Ten, 300KW solar panels were required to realize the 3KW design power level. Thus, the platform carries ten Shott 300W solar panels. In addition, two smaller Shell SQ 80W solar panels are provided to compensate for the energy looses and the power required in the electrical drive system (Alternative Energy Store, 2005; Shell.com, 2005). The detailed design of the 3KW solar power platform is presented in (de Lazzer, 2005). The weight of each of the 300W solar panels is 46.6Kg. The weight of the 80W solar panel is 7.5Kg /panel. The total weight is ≈500 Kg. The platform is 7 meters long by 3.8 meters wide. Therefore, the area of the platform is 26.4m². For the 300W solar panels, the weight is distributed on all the frame perimeter. The concentrated force value is 0.1 N/mm. For the 80W solar panels, the weight is distributed on the frame length only and its value is 0.03 N/mm. The solar panels are fixed symmetrically with respect to the beams of the platform. It is assumed that the structure experiences no dynamic effects. The load can therefore be classified as static. The arrangement of solar panels is shown in Figure 1. The standing 3KW platform is shown in Figure 2. The drive system consist of a d.c motor linked to the platform through a gear train having a gear ratio of 800. Additional provision was made for manual control for the purposes of field experimentation in the Botswana environment. This manual provision for the seasonal adjustment of the longitudinal inclination of the platform is visible from Figure 2, where it appears as a knob on the stem supporting the platform.

Equipment	Power (W)
4 refrigerators with freeze	2100
3 lights	120
Hi Fi	180
2 TV 51cm color	160
TOTAL	**2560**

Table 3. Energy requirements for a rural clinic

Parameters and their values		
R_a=5Ω	L_a=0.003H	B=3.95.10 $^{-6}$ Kg.ms^{-1}
K_b=0.0636V/ rad/s	K_m=0.00711 Kgm/A	K=0.01Kg m²/ s²
J_M=7.72.10^{-6} Kg m²	J_L=970Kgm²	N=1/n=1 /800

Table 4. System parameters

The platform is a sensor-based tracking solar power platform. As suggested by the name, this type of tracking solar power system employs two photosensitive detectors to determine the position of the sun. Usually, two sensors are positioned on an imaginary line parallel to the east-west axis passing through the centre of the array of solar panels. They are arranged so that they produce a differential output whenever the active surface of the solar panels is not aligned perpendicular to the direction of sunrays. When the incident solar radiation is perpendicular to the plane of the array of PV cells, both sensors generate equal amount of current. If however, the incident solar radiation is not perpendicular to the array, then one of the sensors produces an output current greater than that of the second sensor. The differential output current from the sensor arrangement is a current whose magnitude depends of the angle of misalignment of the panels. The sign of the resultant current indicates the direction of the sun. The control system utilises the output of the sensor arrangement to control the motor that rotates the platform of solar panels until the electrical signal in each of the light detectors becomes equal.

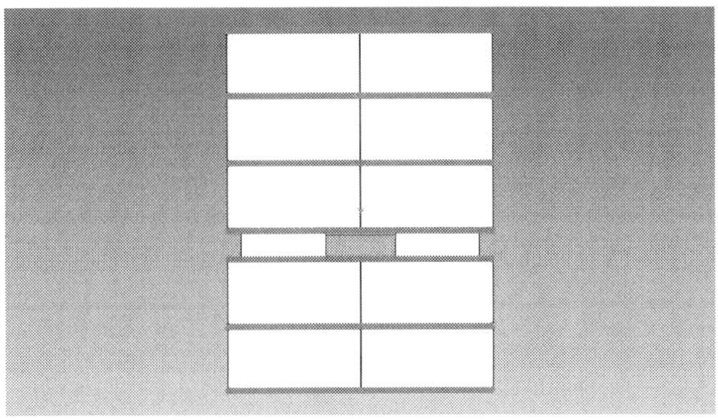

Fig. 1. Arrangement of solar panels on the platform.

For the purpose of investigating the dynamic performance of the platform, as well as for controller design , a model of the platform is developed, based on the application of relevant physical laws. The platform modeling is presented in the next section, section three of the chapter.

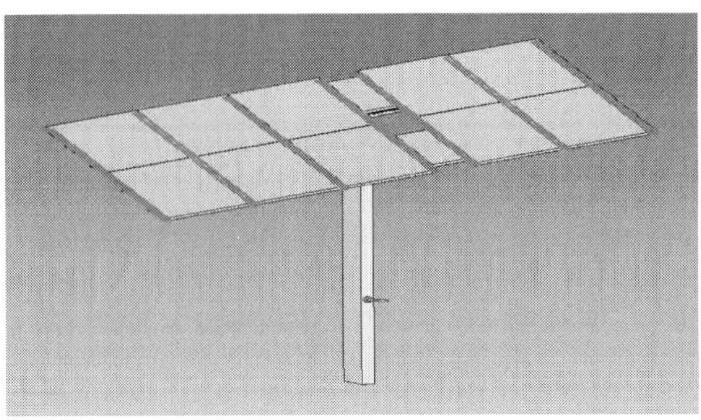

Fig. 2. The 3KW platform viewed from above

3. Mathematical modelling and analysis of the dynamics of the basic platform

The subsequent modeling presented in this section concerns the east-west motion of the platform. The block diagram representation of the platform in the east-west direction is shown in Figure 3.

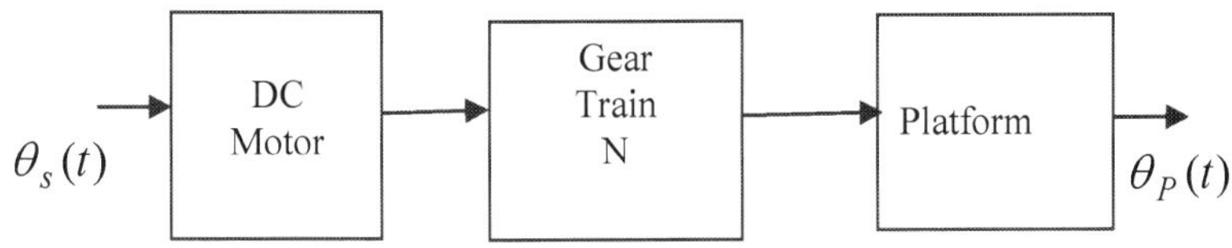

Fig. 3. Block diagram of 3KW solar power platform

Where: $\theta_s(t)$ is the instantaneous direction of sunlight and $\theta_p(t)$ the instantaneous position of the platform.
Hence, the model of the solar tracker is the system of dynamic equations linking the separately excited dc motor to the platform through the gear train as derived in the sequel.

3.1 Modelling of the separately excited DC motor

The typical equivalent circuit arrangement for a separately excited DC motor is shown in Figure 4. An applied armature voltage e_a creates an armature current i_a given by (Kuo & Golnaraghi, 2003):

$$e_a = R_a i_a + L_a \frac{di_a}{dt} + K_b \frac{d\theta_m}{dt} \qquad (1)$$

where $e_a(t)$: armature voltage (V); $i_a(t)$: armature

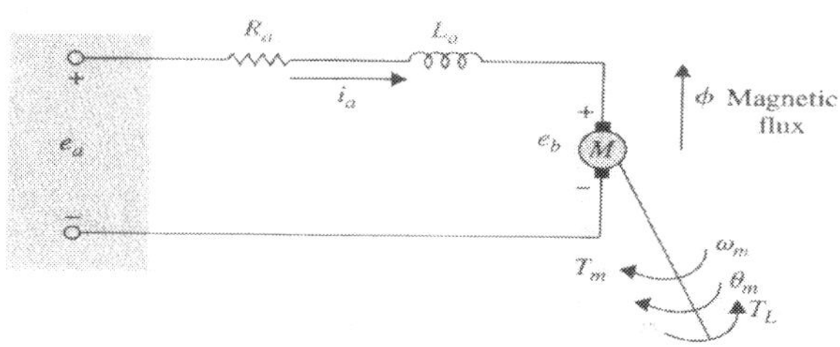

Fig. 4. Schematic diagram of a separately excited DC motor.

current (A); R_a: armature resistance (Ω); L_a: armature inductance (H); K_b: back-emf constant (V/rad/s) and $\theta_m(t)$: rotor displacement (rad.). This current causes a torque

$$T_m = K_m i_a \qquad (2)$$

where $T_m(t)$ is torque(N.m.) and K_m the torque constant (N.m/A).
The torque of the DC motor is coupled to drive the platform through the motor shaft and a gear train. The torque causes an angular displacement of the rotor θ_m, given by (Kuo & Golnaraghi, 2003):

$$T_m = J_t \frac{d^2\theta_m}{dt^2} + B\frac{d\theta_m}{dt} + K\theta_m \qquad (3)$$

where $J_t = J_m + N^2 J_l$ and J_m: moment of inertia of the motor ($kg.m^2$); J_l: moment of inertia of the solar power platform ($kg.m^2$); N: gear-train ratio between motor and platform; B: viscous-friction coefficient of the motor ($kg.m.s^{-1}$); K: spring constant ($kg.m^2.s^{-2}$). After taking the Laplace transform of equations (1)-(3), it is straightforward to obtain the open-loop transfer function for the system $G(s)=\theta_m/V_a$ as shown in equation (4).

$$G(s) = \frac{K_m / L_a J}{s^3 + \left\{\frac{R_a J_t + L_a B}{L_a J_t}\right\} s^2 + \left\{\frac{R_a B + K L_a + K_b K_m}{L_a J_t}\right\} s + \frac{R_a K}{L_a J_t}} \tag{4}$$

Note further that, the angular position of the platform θ_p is related to the motor angular position θ_m through the gear ratio:

$$\theta_p / \theta_m = N = 1/800 \tag{5}$$

The gear ratio was decided by comparing similar applications (Kuo & Golnaraghi, 2003).
The parameters of the open-loop platform system are given in the Table 4. A substitution of these parameters in equation (4) results in the open-loop transfer function:

$$G(s) = \frac{1559.2}{s^3 + 1666.7s^2 + 109.87s + 10965} = \frac{b}{s^3 + a_3 s^2 + a_2 s + a_1} \tag{6}$$

In subsection 3.2, the dynamic behaviour of this open loop system was investigated with the view to determining what controller would be most suitable for improving the platform dynamic performance.

3.2 Simulation of the open-loop platform system

The analysis of the dynamic performance of the platform was simulated using MATLAB. The following simulations were carried out.

3.2.1 Time-domain characterisation of the open-loop platform system

The system was simulated for a unit step increase in the input voltage. The results are shown in Figure 5. The performance of the system, from Figure 5, could be summarized as in Table 5. ***Settling Time*:** It is evident from Figure 5 and the summary in Table 5 that the settling time of 105 seconds for the platform well exceeds one minute. This is too long and would not be suitable for the successful tracking of sunlight, since the direction of the sun rays is likely to change significantly before the platform settles down to the last command. Improvements in the settling time would be required. The peak overshoot is 96% of the final value. This is much higher than the maximum 17% overshoot acceptable in literature (Kuo & Golnaragh, 2003). The maximum overshoot must be reduced. For a third order system, the damping ratio is not strictly defined (Kuo & Golnaraghi, 2003). However, the contribution of the root s_3=-1670 in the transient response is negligible. The complex pole-pair s_1, s_2=-0.031±j2.56 are the significant poles of the system. Their damping factor ξ is equal to 0.0121. This is much less than the damping factor of ≈ 0.707 required for optimum plant performance (Norman, 2004). An appropriate control strategy should enhance the damping of the system.

Settling time	105 Sec.
Peak overshoot	96%
Steady state error	0.875%
Eigenvalues	-1670, -0.031±j2.56
Damping factor	0.0121

Table 5. Summary of the dynamic performance of the platform

Steady State Error: The steady state error of 0.875% is well less than the typical tolerance band values of either 2% or 5% and does not need any further improvement.

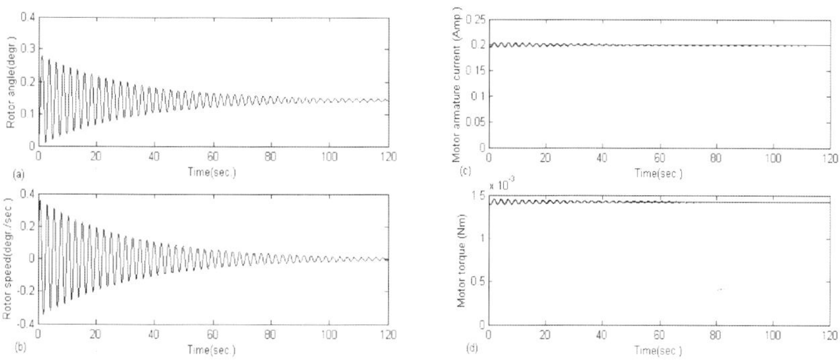

Fig. 5. Dynamic response of solar platform without feedback control

3.2.2 Frequency analysis of the open loop system

The frequency response of the open system is shown in Figure 6. From the Bode plot, the Gain and Phase margins are read. The Gain margin is 40.86dB and the phase margin is 10.4 deg. While the gain margin seems adequate, the phase margin is too low; and is hence due

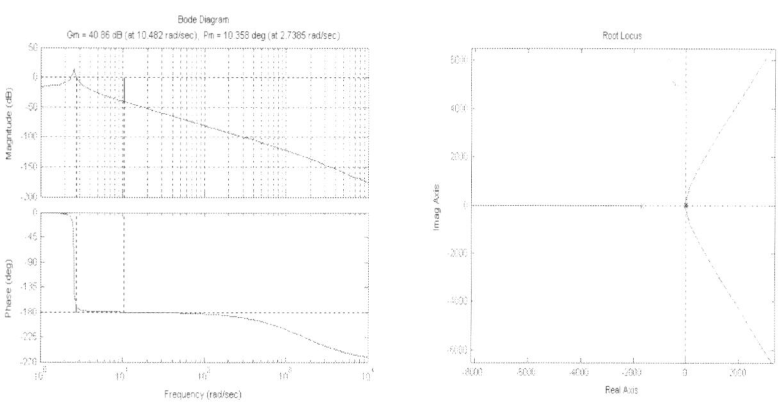

(a) (b)

Fig. 6. (a) Bode plot of open- loop platform (b) The root locus of open-loop platform for further improvements. More over, the roots-locus plots (Figure 6.b) show that even for very small increases in the forward gain of the system, the dominant pole pair crosses the imaginary axis into the right half of the s-plane. It could be concluded that a control strategy that only increases the forward gain of the system, such as proportional control, would aid instability. In the next section of the paper, we present the design of a controller that avoids increases in the forward gain of the system; and which replaces the existing dominant pole-pair of the plant with ones that have the optimum damping factor.

3.3 Summary results from dynamic simulations of the basic platform system

The open-loop simulations of the platform model show that the steady state error of the open-loop dynamics is satisfactory, hence needs no further improvements. The settling time of 105 seconds is well in excess of practical requirements for such systems; and needs to be substantially reduced. The controller design shall specify a target settling time of 2 seconds. The settling time of the open-loop platform system is, in fact, due to the poor damping factor ξ= 0.0121. Alternatively, the poor settling time could be interpreted as a consequence of the dominant pole-pair s_1, s_2=-0.031±j2.56 existing close to the origin of the s-plane. It therefore seems plausible to explore the use of a controller or controller-types that will cancel the existing dominant pole-pair and replace them with those that optimise the damping of the platform. Linear systems are considered to be optimally damped if the damping factor $\xi \approx 1/\sqrt{2}$ (Kuo & Golnaragh, 2003).

4. Sensor modelling and characterisation

For the purpose of effecting feedback control of the platform, a sensor required to measure the relative orientation of the platform axis $\theta_p(t)$ for feedback was required. As indicated in section two of the chapter, a photovoltaic position sensor was employed for the measurement of the misalignment $\alpha(t)$ between the orientation of the platform axis and the direction of sun rays. Hence the sensor measurements are a function of the variable $\alpha(t) = \theta_s - \theta_p(t)$. The sensor consists of two photocells. The derivation of the sensor output relationship begins with a recollection of the theory of the photovoltaic cell.

4.1 The photovoltaic cell

The photovoltaic cell is a two terminal device which consists of a photodiode. The photodiode may be a p-n junction or p-i-n structure. When the cell absorbs light, mobile electrons and positively charged holes are created. If the absorption occurs within the junction's depletion region, or one diffusion length away from it, these carriers are swept from the junction by the built-in field of the depletion region, producing a photocurrent (Nelson, 2003). A detailed representation of the current phenomenon in a photocell, accounting for internal diode current is given as:

$$i_0 = I_{PH} - I_D = I_{PH} - I_S \cdot \{\exp(V / mV_T) - 1\}. \tag{7}$$

Where: I_{PH}: Photo current; I_D: Diode current; I_S: Diode reverse saturation current; m: Diode "ideally factor" m= 1-5V_T; Thermal voltage: $V_T = kT / e$; k: constant of Boltzmann; T: absolute temperature; *e:* charge of an electron. Internal voltage drop in practical photocells is accounted for by the addition of a series resistor R_S. Also leakage currents could be observed in photocells, which may be described as due to a parallel resistor R_p; from these we could write the photocurrent of the photocell as:

$$0 = I_{PH} - I_D - I_P - i_0 \tag{8}$$

And

$$I_P = \frac{V_D}{R_P} = \frac{V + i_0 R_S}{R_P} \tag{9}$$

Hence,

$$i_0 = I_{PH} - I_S \left\{ \exp\left(\frac{V + i_0 R_S}{mV_T} \right) - 1 \right\} - \frac{V + i_0 R_S}{R_P} \tag{10}$$

4.2 The photovoltaic position sensor

Two photovoltaic cells were required to measure the position of the solar power platform, relative to the axis of sunlight. A photocell each was installed on either side of a line running perpendicular to the direction of rotation of the platform. Let these cells be "A" and "B" respectively, as shown in Fig. 7. The opening of the enclosure has a width of *W*. Each cell has the breadth *D* and height *C*. The cells are located a distance *L* behind the opening. Each cell generates a current, I_A and I_B, respectively, proportional to the intensity of its incident radiation. The difference between I_A and I_B is conditioned to generate an error voltage which could then be used to control the motion of the platform, as explained in the next subsection (Kuo & Golnaragh, 2003). It shall be shown below, the relationship between the cell currents I_A, I_B, and angle of misalignment α, of the platform.

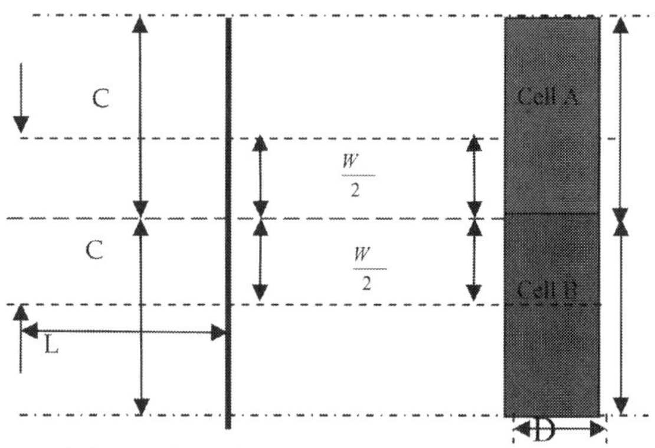

Fig. 7. Arrangement of photocells in the photovoltaic position sensor

4.3 Relating the light-intensity, exposed area of the cell and cell photocurrent

Consider the ray geometry of the photovoltaic position sensor with two identical photocells as shown in Fig. 8. With light incident perpendicular to both cells (angle of misalignment α=0), the area of each cell, exposed, is given by

$$A_A = A_B = (W/2) * D \tag{11}$$

Both cells intercept the same amount of incident radiation, producing the same amount of current. In such a case, the difference between the two current values is zero and no error signal is produced. Now, consider the situation in which an angular misalignment of $\alpha \neq 0^0$, causes the incident light to enter the enclosure of the sensor at an angle of inclination α to the axis of the two cells. Then, from Fig. 8, the exposed areas of cells A and B are respectively given by:

$$A_A = D(\frac{W}{2} + h); A_B = D(\frac{W}{2} - h) \tag{12}$$

Also, from the geometry in Fig. 8,

$$h = L\tan\alpha \therefore A_A = D(\frac{W}{2} + L\tan\alpha); A_B = D(\frac{W}{2} - L\tan\alpha) \tag{13}$$

With the intercepted light energy being proportional to the light-sensing surface area of the intercepting cell,

$$P_{RA} = I_R D(\frac{W}{2} + L\tan\alpha); P_{RB} = I_R D(\frac{W}{2} - L\tan\alpha) \tag{14}$$

The photocurrent I_{PH} of a cell is known to be proportional to the light energy, therefore:

$$I_0 \propto P_R \tag{15}$$

Therefore, from Eq. (14), and Eq. (15), one can write:

$$\begin{aligned} i_A &= K_1 P_{RA} = K_1 I_R D(\frac{W}{2} + L\tan\alpha); \\ i_B &= K_1 P_{RB} = K_1 I_R D(\frac{W}{2} - L\tan\alpha) \end{aligned} \tag{16}$$

Substituting Eq. (16) into Eq. (10) yields:

$$\begin{aligned} i_A &= K_1 I_R D\frac{W}{2} + K_1 I_R DL\tan\alpha - I_S(e^{\frac{V_O \log(1+I_R)}{mV_T}} - 1) \\ &- \frac{V_O \log(1 + I_R) + I_A R_S}{R_P} \end{aligned} \tag{17}$$

and,

$$\begin{aligned} i_B &= K_1 I_R D\frac{W}{2} - K_1 I_R DL\tan\alpha - I_S(e^{\frac{V_O \log(1+I_R)}{mV_T}} - 1) \\ &- \frac{V_O \log(1 + I_R) + I_B R_S}{R_P} \end{aligned} \tag{18}$$

Now, the differential output current for the cells would now be approximated as:

$$i_A - i_B = 2K_1 I_R DL\tan(\theta_s - \theta_p) \tag{19}$$

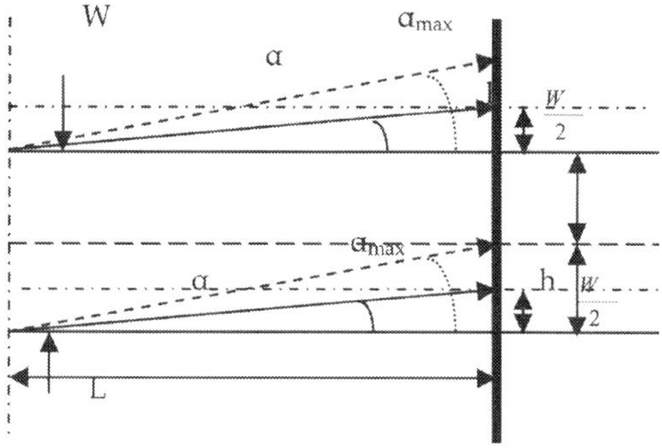

Fig. 8. Ray geometry in the sensor

4.4 The feedback signal

The circuit for feedback signal generation is shown in Figure 9. As derived in (Agee *et al*, 2009), the resultant output current, representing the difference between the direction of the sun rays θ_s and the axis of the rotor θ_m is given by (20):

$$i_A - i_B = 2K_1 I_R DLan(\theta_S - \theta_P) \tag{20}$$

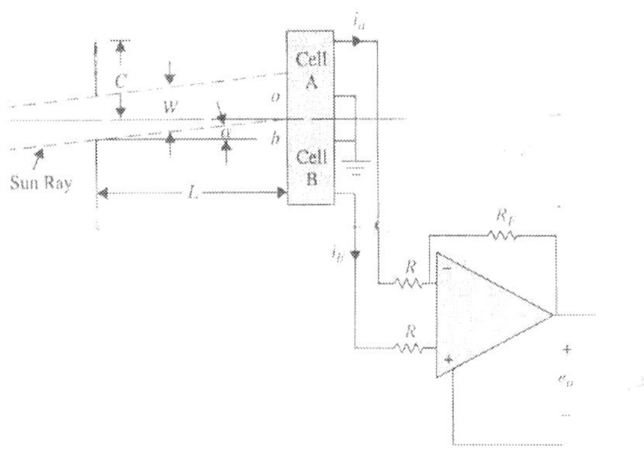

Fig. 9. Sensor signal conditioning for feedback

where i_A, i_B are cell currents from cells A and B respectively. The error voltage signal e_o is straight away derived from Figure 9, to be:

$$e_o = -R_F(i_A - i_B) = -2K_1 I_R DLR_F \tan(\theta_S - \theta_P) \tag{21}$$

However, with the introduction of a unity-gain inverting amplifier in the follow-up stage, we have

$$e_{o1} = -e_o = R_F(i_A - i_B) = 2K_1 I_R DLR_F \tan(\theta_S - \theta_P) \tag{22}$$

Thus with the measurements included for feedback control, the armature voltage V_a could now be expressed as

$$V_a = K_F(e_{o1}) = 2K_F K_1 I_R DLR_F \tan\alpha; \qquad \alpha = \theta_S - \theta_P \tag{23}$$

where, K_F is a constant of proportionality. For linear controller design, a linearised representation of sensor characteristics is required. By approximating $\tan(\theta_S - \theta_P) \cong \theta_S - \theta_P$, the linear sensor feedback model takes the form:

$$V_a = K_F(e_{o1}) = 2K_F K_1 I_R DLR_F \alpha \tag{24}$$

4.5 Validation of sensor model

The data in Table 6, together with equations (17) and (18) were used for the numerical validation of sensor model. Simulations were done in MATLAB. Fig. 10 and Fig. 11, respectively, confirm the nonlinear relationships between I_A, I_B and α. Figure 12 show the nonlinear relationship between α and the differential output cutrrent of the sensor.

PARAMETER	VALUE	PARAMETER	VALUE
Light intensity I_R	12 [mW/cm^2]	R_S	5 [Ω]
Sensitivity of sensor K_1	0.55 [AW^{-1}]	D	5 [cm]
m, the ideal diodes factor	1	L	0.8 [cm]
Temperature voltage V_T	25.7 [mV] at T = 25°C	W	3.5 [cm]
Reverse saturation current	$I_S = 2 \times 10^{-6}$[A]	C	5 [cm]
Photo-voltage V_O	0.5 [V]	Area	12.5 [cm^2]
Parallel resistance R_P	1 [KΩ]		

Table 6. Data used for the simulation of the photovoltaic position sensor

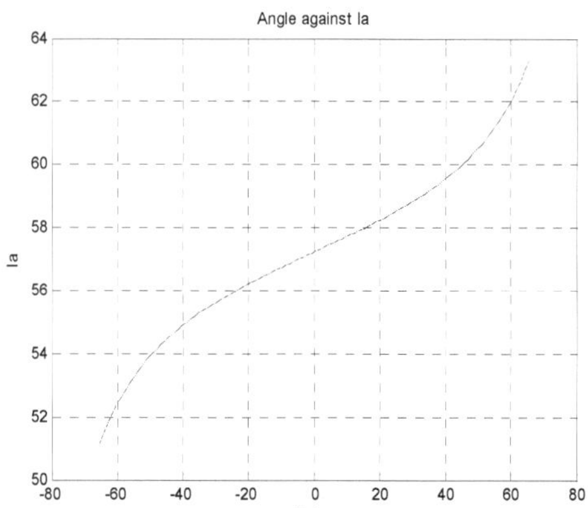

Fig. 10. Current I_A versus angle α

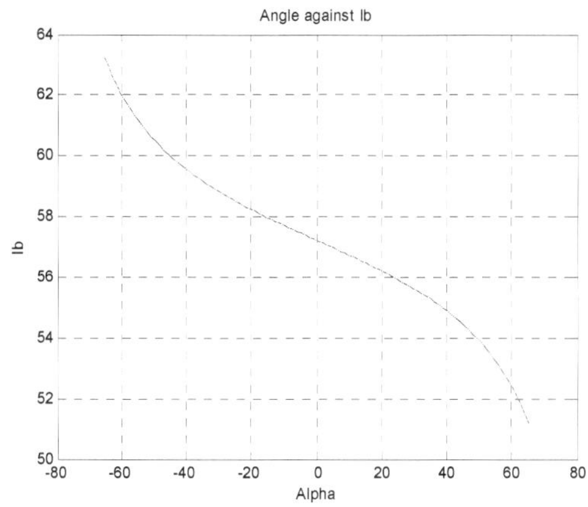

Fig. 11. Current I_B versus angle α

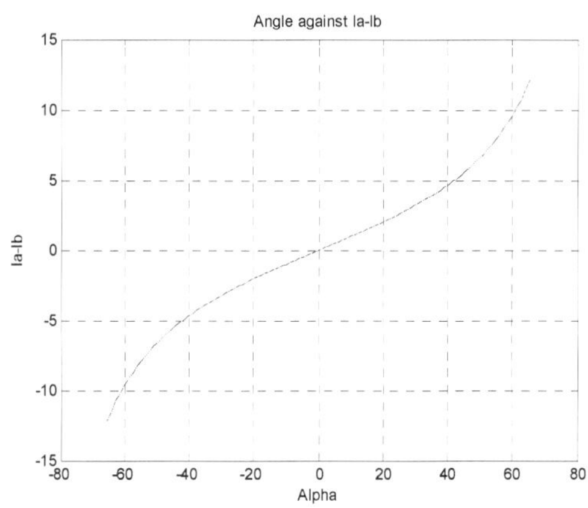

Fig. 12. I_A - I_B related with angle α.

5. Controller design

The structure for the platform system with controller is shown in Figure 13. This shows the embedding of the photosensor into the feedback network.

5.1 Performance specification

The controller was required to modify the response of the system in such a manner as to achieve the following performance specification:

- Settling time for 2% tolerance band = 2 seconds.
- Damping factor = 0.71

First, the pole-cancellation controller is designed and validated. A critical appraisal of the performance of the controller in the presence parameter variations and nonlinearities is presented. Finally, the design of the nonlinear controller, and a comparative analysis of its performance with respect to the pole-cancellation controller are presented.

5.2 Design of the pole cancellation controller for the platform

The structure for the system with controller is shown in Figure 13. The Figure shows the position of the sensor presented in section (4) of the chapter. It has been explained earlier that, both the settling time and the damping of the platform system need improvement. Our solution was to use a controller whose transfer function zeros cancelled the undesirable poles of the platform transfer function, G(s). Then, the poles of the controller were placed so as to achieve the desired closed-loop dynamic performance. Accordingly, the structure of the controller is specified to be the notch filter with the following transfer function:

$$G_C(s) = \frac{K_1 s^2 + \alpha_1 s + \beta_1}{K_2 s^2 + \alpha_2 s + \beta_2} = \frac{N_1(s)}{D_1(s)} \tag{25}$$

as shown in Figure 14 (Agee *et al*, 2006).

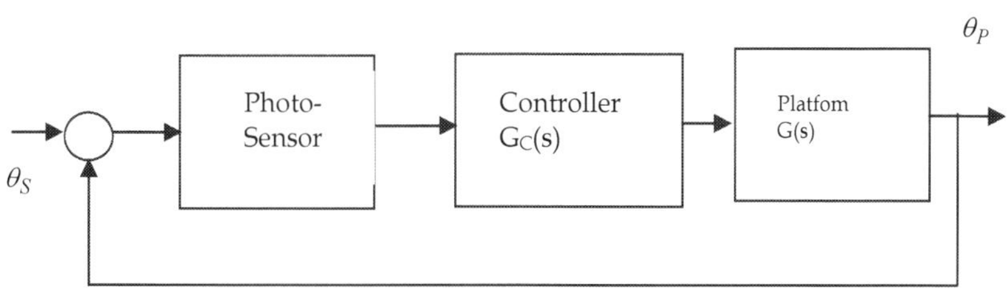

Fig. 13. Block diagram of controlled platform

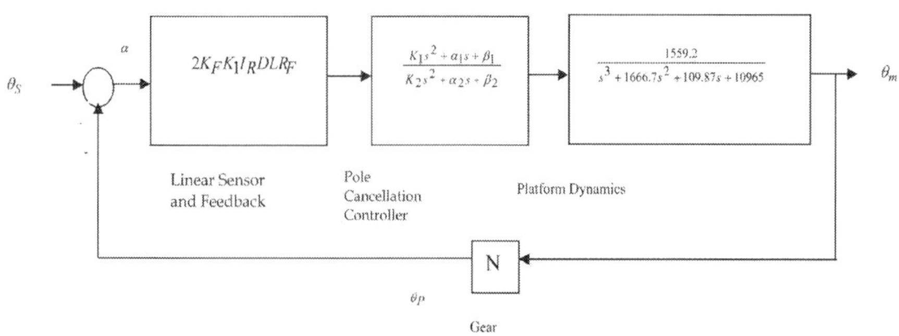

Fig. 14. Closed-loop block diagram of platform with a linear sensor

5.2.1 Tuning of pole-cancellation controller

The controller design entails the determination of the six coefficients K_1, α_1, β_1, K_2, α_2 and β_2. Note that the zeros of the controller are required to cancel the dominant pole-pair of the open-loop transfer function of the platform system. With the pole-cancellation controller included as in Figure 14, the closed-loop transfer function of the platform becomes equation (25). To determine K_1, α_1 and β_1 for the controller, consider the closed loop transfer function for the controlled system with linear sensor characteristics. Then, using the fact that the dominant pole-pair of the open loop transfer function is : -0.031±j2.56;

$$C(s) = \frac{(2K_F K_1 I_R DLNR_F G_C(s)G(s)}{1 + (2K_F K_1 I_R DLNR_F)G_C(s)G(s)} \tag{26}$$

An equivalent second-order polynomial for the zeros of the notch filter is obtained. Hence, $K_1 = 1$; α_1 =0.062 and β_1 =6.5546. Now, let $2K_F K_1 I_R DL = 1$, such that,

$$G_C(s)G(s) = \frac{(s+0.031-2.56j)(s+0.031+2.56j).1559.2}{D_1(s)(s+1670)(s+0.031-2.56j)(s+0.031+2.56j)}$$

or,

$$G_C(s)G(s) = \frac{1559.2}{(K_2 s^2 + \alpha_2 s + \beta_2).(s+1670)}$$

Then, for the closed-loop platform with pole-cancellation,

$$C(s) = \frac{1559.2NR_F}{(K_2 s^2 + \alpha_2 s + \beta_2).(s+1670) + 1559.2NR_F} \tag{27}$$

$$C(s) = \frac{1559.2NR_F / K_2}{s^3 + (\frac{\alpha_2}{K_2} + 1670).s^2 + (\frac{\beta_2}{K_2} + 1670\frac{\alpha_2}{\beta_2})s + (1670\frac{\beta_2}{K_2} + 1559.2\frac{NR_F}{K_2})} \tag{28}$$

Also, for a damping factor of 0.71, the settling time is given by: $t_{s_{1\%}} = 4.6 / \xi\omega_n$ (Kuo & Golnaraghi, 2003). Thus, for a settling time of 2 seconds, we deduce ω_n = 3.24 rad/s. The new dominant pole-pair is given by $s_{1,2} = -\xi\omega_n \pm \omega_n\sqrt{1-\xi^2}$, yielding the dominant second-order factor $s^2 + 2\xi\omega_n + \omega_n{}^2$. With the third-pole of the closed-loop system situated at s= -d, we have the characteristic equation of the controlled system to be

$$s^3 + (d + 2\xi\omega_n)s^2 + (\omega_n^2 + 2\xi\omega_n d)s + d\omega_n^2 = 0 \tag{29}$$

Compare equations (28) and (29) to obtain the following relations:

$$\begin{aligned} 160y + z &= d\omega_n^2 \\ \omega_n^2 + 2\xi\omega_n d &= y + 1670x \\ d + 2\xi\omega_n &= 1670 + x \end{aligned} \tag{30}$$

Where,

$$x = \frac{\alpha_2}{K_2}, z = \frac{1.949 R_F}{K_2}; y = \frac{\beta_2}{K_2} \tag{31}$$

The simultaneous solution of equation (30) yields:

$$\begin{aligned} \omega_n &= 1670\xi \pm \sqrt{697225\xi^2 - 2788900 + z / \Delta} \\ \Delta &= d - 1670 \end{aligned} \tag{32}$$

It becomes a matter of substitutions to verify that for the given settling time of 2 seconds and damping factor $\xi = 0.71$, equation (32) is satisfied for

$$z = 2893.664, \Delta = 0.0001 \tag{33}$$

and further that the closed loop pole d=-1670.0001, $\alpha_2 / K_2 = 4.6008; \beta_2 / K_2 = 8.7648$. By setting K_2=1, obtain the complete transfer function of the controller as:

$$G_C(s) = \frac{s^2 + 0.0620s + 6.5546}{s^2 + 4.6008s + 8.7648} \tag{34}$$

and $R_F = z / 1.949 = 1.485 K\Omega$

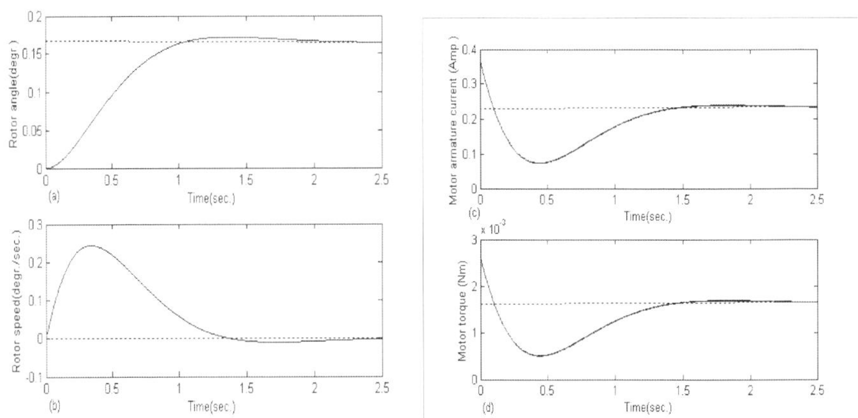

Fig. 15. Analysis of platform system controlled by pole-cancellation

5.3 Performance of the platform equipped with pole-cancellation controller

The transfer function of the closed-loop system with controller becomes:

$$C(s) = \frac{1665855}{s^3 + 1674s^2 + 7688s + 17505} \tag{35}$$

Both the transient response and the frequency response of the controlled system were simulated. The results are shown in Figure 15. It can be seen that, with the controller included, the system does not oscillate like it did before the introduction of the controller.

The overshoot is 4.3%. Overshoot peak has been reduced by 95%, as compared to the case of the uncontrolled system. The damping factor is improved to 0.71. The settling time is now reduced to 2 seconds. The steady state error is only 0.01 rads. This is much better than for the uncontrolled system. In fact, the steady state error has been reduced by 99%. The system is stable. The gain margin is improved by 75% to about 73dB while the phase margin is now infinity. The peak armature current has been reduced by 25% from 0.2A to 0 0.16 A. The peak torque required in the drive system increased from 1.5mN.m to 2.7mNm. It could therefore be concluded that the pole cancellation control strategy improved significantly the dynamic performance of the 3KW solar power platform; improving the damping factor from 0.0121 to 0.71; reducing the settling time from 105 sec. to 2 sec. Peak overshoot of the rotor angular position was reduced by 95%. Both the gain margin and the phase margin were also substantially improved. The peak torque requirement however was doubled. The controller structure is simple and may be easily implemented. However, the design neglected nonlinear sensor characteristics which may restrict the usefulness of the control strategy presented here. These effects on the performance of the pole-cancellation controller are discussed in details in subsection (5.4).

5.4 A Critical analysis of the shortcomings of the pole-cancellation control strategy for platform control

In practical systems, the efficacy of the pole-cancellation control strategy is limited by the effects of parameter variation/uncertainty and the nonlinear sensor behaviour. These two effects are discussed further in the following two subsections.

5.4.1 Effect of parameter uncertainty on performance of platform controlled by pole cancellation

Consider the effect of addititive parameter change on the open-loop platform model of equation (6). The resulting platform representation will admit the form in equation (36). Hence,

$$G(s)+\Delta G(s)=\frac{b+\Delta b}{s^3+(a_3+\Delta a_3)s^2+(a_2+\Delta a_2)s+(a_1+\Delta a_1)} \tag{36}$$

Or, more specifically:

$$G(s)+\Delta G(s)=\frac{1559.2+\Delta b}{s^3+(1666.7+\Delta a_3)s^2+(109.87+\Delta a_2)s+(10965+\Delta a_1)} \tag{37}$$

Numerical investigation of the platform systems with variable parameters in MATLAB yielded the roots variations shown in Table 7. It is evident from Table 7 that the open-loop poles of the platform change significantly with variation in system parameters. Under such circumatances, a basic pole-cancellation control strategy will be ineffective, except where on-line adaption is introduced. Such additional complexities in the structure of the controller would make the above control strategy more expensive and hence, less attractive.

Moreover, the increased complexity in the systems would lead to the more complex platform block diagram shown in Figure 16, with uncancelled dynamics, where:

Parameter	Variation	s_1	s_2	s_3
Δa_3	+5%	-1750	-0.035+2.5i	-0.035–2.5i
Δa_3	+10%	-1833.3	-0.036+2.4i	-0.036–2.4i
Δa_2	+5%	-1670	-0.037+2.6i	-0.037–2.6i
Δa_2	+10%	-1670	-0.040+2.6i	-0.040–2.6i
Δa_1	+5%	-1670	-0.039+ 2.6i	-0.039–2.6i
Δa_1	+10%	-1670	-0.041 +2.7i	-0.041-2.7i

Table 7. Dependence of platform system poles on parameter variation

$$\eta(s)=\frac{\left\{s^3+(1667.7+\Delta a_3)s^2+(109.87+\Delta a_2)s+(10965+\Delta a_1)\right\}\left\{s^3+1666.75s^2+109.87s+10965\right\}}{\Delta b\left(s^3+1666.75s^2+109.87s+10965\right)-1559.2\left(s^2+\Delta a_2 s+\Delta a_1\right)} \quad (38)$$

This, in turn, will modify the closed-loop response as shown in Figure 17-18. Note from Figure 17.b, that with a 10% increase in a_3, a non-positive defenite beahiour of the closed-loop platform is indicated. Figure 18 shows that, under some variations in a_2, the pole-cancellation controller loses function in steady state, and sustained oscillation of the plant is produced.

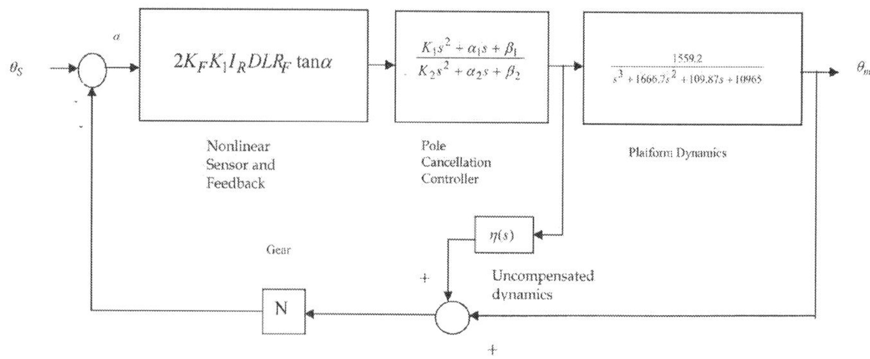

Fig. 16. Complex block diagram of platform due to failure of pole-cancellation

5.4.2 Effect of nonlinear measurement on linear controller performance

Recall that, while the approximate sensor characteristics was used for the design of the pole-cacellation controller, exact sensor characteristics are nonlinear, as in equation (23). In this subsection, a few comments are made as to how the true sensor characteristics affect the dynamics of the platform under pole-cancellation control.

Figure 19 compares the exact tangent characteristics with the linear approximation $\tan\alpha \approx \alpha$; together with the third-order polynomial approximation, $\tan\alpha \approx \alpha + \alpha/3!$. It is evident that none of the approximate representation is useful beyond $\alpha \geq 26^\circ$. Hence, the linear model is only valid within a very narrow window of the complete domain of operation of the solar power opearation. This will lead to significant model mismatches, for which the performance of the pole-cancellation controlle would be significantly inadequate.

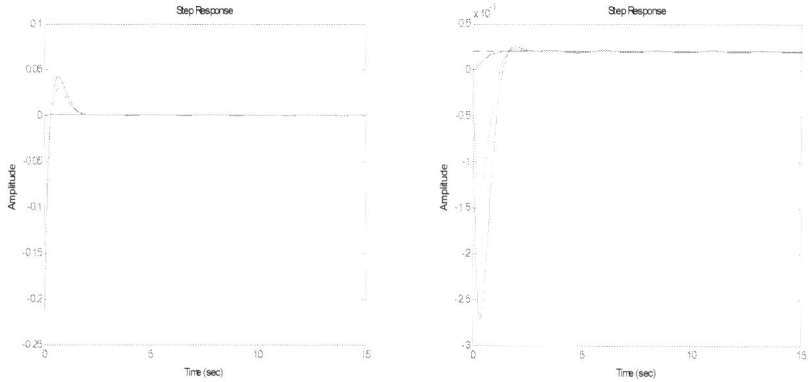

(a) (b)

Fig. 17. (a) Step response with a 5% variation of a_3 (b) Step response with 10% variation of a_3.

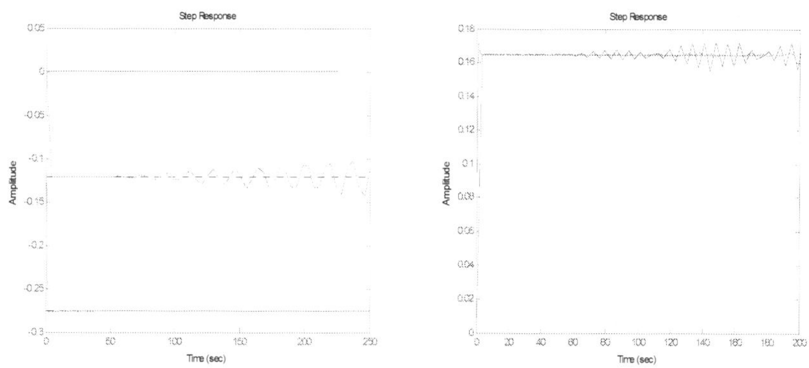

(a) (b)

Fig. 18. (a) Step response with 5% increase in a_2 (b) Step response with 10% increase in a_2

5.4.3 A summary perspective on the linear control of solar power platform

It is evident from the critical analysis of the effect of parameter variations and nonlinearity, that the suitability of linear control strategies for the polar-axis solar power platform with nonlinearities is limited. In particular, whereas the structure of the pole-cancellation controller makes controller implementation simple, the viablility of this strategy in the practical environments of parameter variation and nonlinearities is not guaranteed. Because of the very high forward gains of the platform, linear control was also not robust. On the other hand, the linear controlled system offers a suitable reference model for the comparison of the performance of possible nonlinear control strategies that may be employed on the platform system. In the rest of the chapter, the design and simulation of a nonlinear controller, the feedback-linearised control of the platform, is presented. Results are compared with those from the linear systems as presented above.

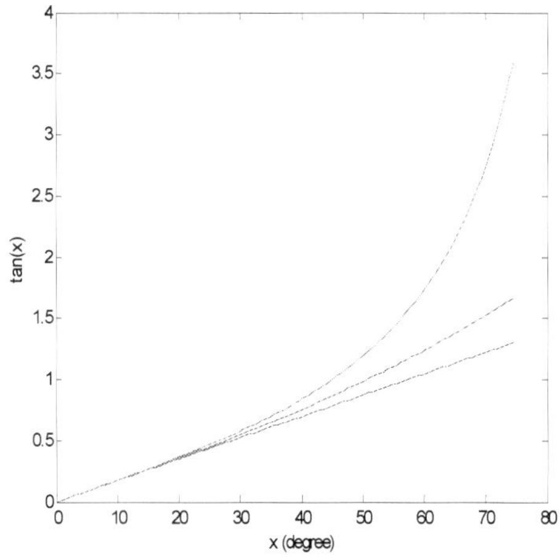

Fig. 19. Approximations of sensor tangent characteristics

5.5 Nonlinear control of platform

Feedback measurements used for the control of the tracking system lead to an overall nonlinear behaviour in the platform. This resulting nonlinear dynamics is much richer in complexity than the dynamics of the linear platform system. Consequently, nonlinear control strategies may be required to optimize the dynamic performance of the platform. The design of the nonlinear controller for the platform is presented in this section of the chapter. Controller discussions make comparisons of the nonlinear controller with the pole cancellation controller earlier designed for the tracking system.

5.5.1 The state space model of open-loop solar platform

Combining (1)-(3), it is straightforward to obtain the state-space representation for the open-loop systems, as shown in (39). Notice from the equation that, the basic open loop-loop platform, without feedback measurements is here again confirmed to be linear.

$$\begin{aligned} \dot{\theta}_1 &= \theta_2 \\ \dot{\theta}_2 &= \theta_3 \\ \dot{\theta}_3 &= -a_1\theta_1 - a_2\theta_2 - a_3\theta_3 + bV_a \\ y &= \theta_1 = \theta_m \end{aligned} \tag{39}$$

where,

$$a_1 = \frac{KR_a}{L_a J_t}, a_2 = \frac{KL_a + BR_a + K_b K_m}{L_a J_t}, a_3 = \frac{BL_a + J_t R_a}{L_a J_t}, b = \frac{K_m}{L_a J_t} \tag{40}$$

and $\theta = [\theta_1, \theta_2, \theta_3]^T \subset \Re^3 = [\theta_m, \dot{\theta}_m, \ddot{\theta}_m]^T$.

5.5.2 Nonlinear state-space model of platform with feedback measurements

The over all plant model, including the feedback measurement, will modify (39) to yield (41):

$$\begin{aligned}\dot{\theta}_1 &= \theta_2 \\ \dot{\theta}_2 &= \theta_3 \\ \dot{\theta}_3 &= -a_1\theta_1 - a_2\theta_2 - a_3\theta_3 + 2bK_F K_1 I_R DLR_F \tan(\theta_s - \theta_m))\end{aligned} \tag{41}$$

It is thus evident from (35) that where as the basic system has a linear model, the measurements used for control make the overall system nonlinear

5.5.3 The non-linear input-state feedback control of polar-axis solar power platform

For the linearisation of the platform systems using feedback, a function of state $x_1 = \psi(\theta_1, \theta_2, \theta_3)$ is required; where the $r \leq n$ derivatives of x_1 exist; and ψ is invertible such that all states of the platform $\theta_i; i = 1,2,..r$ and its input are functions of x_1 and its r derivatives. Recollect the fact that, for linear systems expressible in the controllable canonical form (as in equation (39), the input-state feedback linearising variable x_1 is the output of the canonical plant representation (Kuo & Golnaragh, 2003). It is possible to write that:

$$\begin{aligned}\theta_1 &= x_1 \\ \theta_2 &= \dot{x}_1 \\ \theta_3 &= \ddot{x}_1 \\ V_a &= (1/b)(\dddot{x}_1 + a_3\ddot{x}_1 + a_2\dot{x}_1 + a_1 x_1); b \neq 0\end{aligned} \tag{42}$$

Consequently, the model of the nonlinear platform could now be written in terms of the linearising variable as

$$\begin{aligned}\theta_1 &= x_1 \\ \theta_2 &= \dot{x}_1 \\ \theta_3 &= \ddot{x}_1 \\ \theta_s - \theta_m &= \tan^{-1}\left\{\frac{1}{2bK_F DLI_R R_F}(\dddot{x}_1 + a_3\ddot{x}_1 + a_2\dot{x}_1 + a_1 x_1)\right\}; b \neq 0\end{aligned} \tag{43}$$

Let,

$$\dot{x}_3 = v \tag{44}$$

Then

$$\begin{aligned}\dot{x}_1 &= x_2 \\ \dot{x}_2 &= x_3 \\ \dot{x}_3 &= -K_I(x_1 - x_1^*) - K_P(x_2 - x_2^*) - K_D(x_3 - x_3^*) = v(t)\end{aligned} \tag{45}$$

where x_1^*, x_2^*, x_3^* are the respective steady states of x_1, x_2, x_3 and

$$v = -a_1 x_1 - a_2 x_2 - a_3 x_3 + 2K_F K_1 I_R DLR_F bK_F \tan(\theta_s - \theta_m) \quad (46)$$

The nonlinear error measurements could now be written as:

$$\begin{aligned} \mu &= \theta_s - \theta_m = \tan^{-1}\left\{\frac{(\rho+\sigma)}{2bK_F K_1 LI_R R_F}\right\} \\ \rho &= -K_I(x_1 - x_1^*) - K_P(x_2 - x_2^*) - K_D(x_3 - x_3^*) \\ \sigma &= a_1 x_1 + a_2 x_2 + a_3 x_3 \end{aligned} \quad (47)$$

5.5.4 Tuning of the parameters of the nonlinear controller

For the evaluation of the v in equation (45), the controller parameters, K_I, K_P, K_D are chosen such that the stability of the linear system

$$\dot{x}_3 = -K_I(x_1 - x_1^*) - K_P(x_2 - x_2^*) - K_D(x_3 - x_3^*) \quad (48)$$

is guaranteed.

Now, define the error between the states and their references in the following manner:

$$\begin{aligned} \dot{x}_3(t) &= \dddot{e}(t) \\ x_3(t) - x_3^*(t) &= \ddot{e}(t) \\ x_2(t) - x_2^*(t) &= \dot{e}(t) \\ x_1(t) - x_1^*(t) &= e(t) \end{aligned} \quad (49)$$

And recast equation (48) in the form of equation (50):

$$(s^3 + K_D s^2 + K_P s + K_I)E(s) = 0 \quad (50)$$

Thus, select the tuning parameters K_I, K_P, K_D to ensure the asymptotic elimination of the error, such that.

$$\begin{aligned} \dot{x}_3(\infty) &\to 0 \\ x_3(\infty) &\to x_3^*(\infty) \\ x_2(\infty) &\to x_2^*(\infty) \\ x_1(\infty) &\to x_1^*(\infty) \end{aligned} \quad (51)$$

Here we proceed to chose K_I, K_P, K_D applying the Routh-Hurtwits criterion to the equivalent s polynomial:

$$s^3 + K_D s^2 + K_P s + K_I = 0 \quad (52)$$

Where

The nonlinear controller given by (43)-(47) shall be simulated and its impact on the dynamics of the platform compared with that of the linear controller.

5.6 Comparative simulation of the impact of nonlinear controller

Simulations comparing the performance of the platform under the impact of the feedback-linearised controller, with that under the impact of the pole-cancellation controller are shown in Figure 20 to Figure 22. The nonlinear system brings the system to rest within two seconds, as would the reference linear model. Overshoots are virtually eliminated in the dynamics of θ_m. Whereas the pole-cancellation strategy also brought the system to rest within two seconds, the accelerations observed in other system variables were very high. Overshoots remained significant. The nonlinear controller being reported in this paper achieves the same settling time and eliminates overshoots without causing excessive accelerations.

6. Conclusions and recommendations

6.1 Conclusions

The high overshoots associated with the dynamic response of the pole-cancellation controller could significantly add to the hardware costs of the tracking systems. Moreover, parameter uncertainty/variations could significantly compromise the performance of the platform under linear control. Feedback-linearised control of the platform yield exact linearization of platform dynamics, by feedback, without any approximations. It is demonstrated that this exact linearization leads to a better performance of the platform under the action of the feedback-linearised controller.

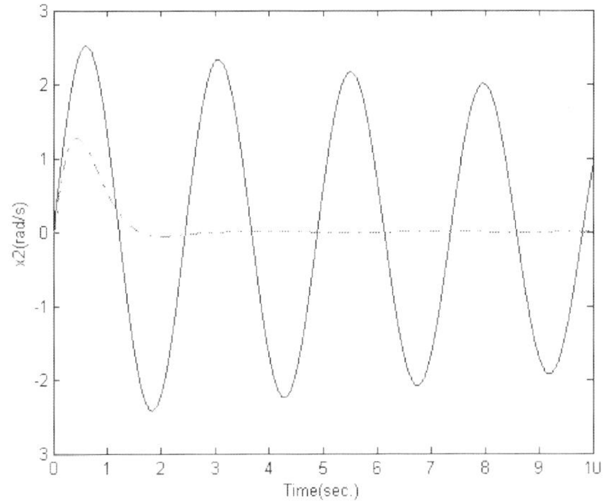

Fig. 20. Platform position under nonlinear control (_._) and linear control (---)

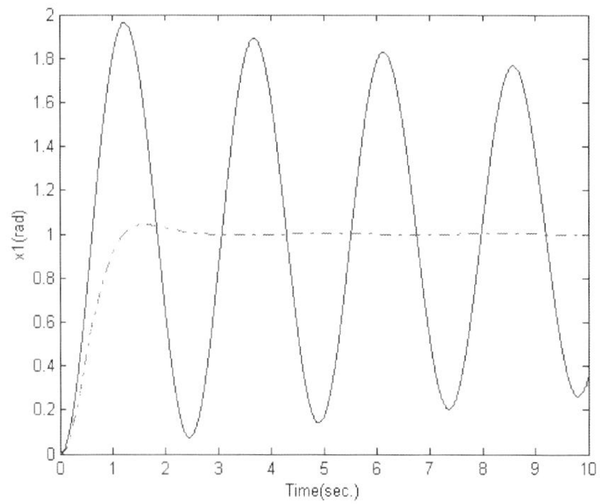

Fig. 21. Platform velocity under nonlinear control (_._) and linear control (----)

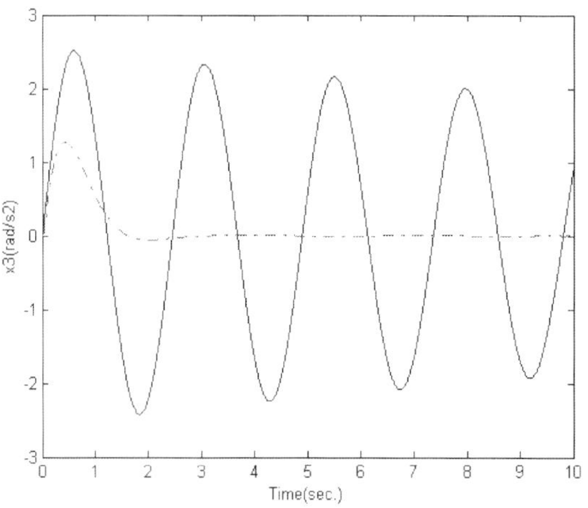

Fig. 22. Platform acceleration under nonlinear control (_._._) and linear control (----)

6.2 Recommendation for further studies

Further research directions here would consider the methods and costs of controller implementation. The feedback-linearised control strategy leads to a nonlinear controller whose structure is more complex than that of the linear pole-cancellation controller. It is recommended to explore neural networks for the implementation of the nonlinear controller. The basic structure of the solar tracker system is linear. The subsequent nonlinearity in the system is due to the measurement. It is recommended to further explore the use of a linear sensor or even sensor-less tracking strategies to simplify controller design.

7. References

Anderson, G. O. & Abkenari, M. H., (1990) Application of solar energy technology in Botswana. *Proceeding of IASTED Int. Conf. on Power and Energy Systems*. Las Vegas, Nevada, 1999, 141-150.

Agee, J. T., Masupe, S, Jeffrey, M. & Jimoh A. A.Enhancing the Output Characteristics of a Photovoltaic Position Snesor Using a feed-Forward Neural network, Advance Materials Research, Vol. 62-62, pp. 506-511. 2009

J. T. Agee, A. A. Jimoh (2007) Feedback linearised Control of a Solar Power Platform. IEEE Africon 2007. Namibia.

J. T. Agee, M. de Lazzer an M. K. Yanev, "A Pole cancellation strategy for stabilising a 3KW solar power platform. *Int. Conf. Power and Energy Systems(EuroPES 2006)*, Rhodes, Greece. June 26-28.

J. T. Agee, S. Obok-Opok and M. de lazzer, "Solar tracker technologies: market trends and field applications. *Int. Conf. on Eng. Resaerch and Devlopemnt: Impact on Industries*. 5-7th September, 2006.

Alternative Energy Store, 300W Shott Solar panel specifications. http://shop.altenergy store.com, 2005.

Archer, M. and Hill, R. (2001), Clean Electricity from Photovoltaics Botswana Power. http://www.wn.com /s/ Botswana power . 2004.

Southern Africa and the Southern African Development Community. http://www. eia.doe.gov /meu/cabs/sadc .hmtl. 2005.

Consultancy on Identifying and Overcoming Barriers to Widespread Adoption of Renewable Energy - Based Rural Electrification in Botswana: *Final report*. 2003.Customer Tips http://www.bpc.bw 10-2005

Daily energy consumption http://www.epsic.ch/pagesperso/ schneiderd/ Apelm /Sources/Solaire.htm 10-2005

De Lazzer, M. (2005). A positioning System for an Array of Solar Panels. Unpublished M.Sc Thesis University of Botswana.

Geche, J. & Irvine J. (1996). Photovoltaic Lighting in Rural Botswana: A Pilot Project. Renewable Energy for Development, Vol. 9, No. 2, pp. http://www.sei.se/red /red9609e.html 21st July, 2009

B. C. Kuo & F. Golnaraghi, *Automatic Control Systems* (eight edition, John Wiley and Sons, Inc., 2003).

Lasschuit, P, Westra, C. & van Roekel G., (xxxx). Financial Suatainability of PV Implementation in Swaziland. http://roo.undp.org/gef/solarpv/docs/bgmaterial /Misc%20PV%20Papers/ECN%20-%20Financial%20sustainability%20of%20PV %20implementation %20in%20Swaziland.pdf. 21st July, 2009

N. Matenge & V. Masilo, *Feasibility Study of Botswana Electricity Generation*(HND Project, *Power Point Presentation*, University of Botswana, 2004).

Mogotsi, B. (2002). Energy and Sustainable Development in Botswana. Sustainable Energy Watch Report. HELIO-Botswana. Helio International. http://www.helio-international.org /reports/2002/botswana.cfm. 21st July, 2009

Nelso, J (2003), The Physics of Solar Cells. Barnes and Noble.

N. S. Norman, *Control Systems Engineering* (John Willey and Sons, Inc. USA , 2004).

Shell SQ 80W solar panel specifications. http://www.shell.com, 2005.

TRACSTAR (Small Power systems) Solar Tracking for Architects Source: http://www.pacificsites.com/ sps/trackforarc.html, 2006.

4

Silicon Solar Cells: Recombination and Electrical Parameters

Saïdou Madougou[1], Mohamadou Kaka[1] and Gregoire Sissoko[2]
[1]*University Abdou Moumouni of Niamey, BP 10 963 - Niamey*
[2]*Université Cheikh Anta Diop de Dakar,*
[1]*Niger*
[2]*Senegal*

1. Introduction

Nowadays, the world's energy needs are growing steadily. However, the conventional sources of energy are limited.

Solar energy such as photovoltaic energy (PV) is the most available energy source which is capable to provide this world's energy needs. The conversion of sunlight into electricity using solar cells system is worthwhile way of producing this alternative energy. The history of photovoltaic energy started in 1839 when Alexandre-Edmond Becquerel discovered the photovoltaic effect (S.M. SZE 1981, W. Shockley 1949). Photovoltaic system uses various materials and technologies such as crystalline Silicon (c-Si), Cadmium telluride (CdTe), Gallium arsenide (GaAs), chalcopyrite films of Copper-Indium-Selenide ($CuInSe_2$), etc (W. Shockley 1949, W. Shockley et al. 1952). Now, silicon solar cells represent 40 % of the world solar cells production and yield efficiencies well higher than 25 % (A. Wang et al 1990). In solar technology, the main challenge of researchers is to improve solar cells efficiency. Due to this challenge, several investigations have been developed to characterize the solar cells by the determining their parameters. Indeed, it is important to know these parameters for estimating the degree of perfection and quality of silicon solar cells.

This chapter first describes the device physics of silicon solar cells using basic equations of minority carriers transport with its boundary conditions, the illumination mode and the recombination mechanisms. Then, a silicon solar cells recombination and electrical parameters are presented (S. Madougou et al 2005a, 2007b). Finally, some methods of determination of these parameters are described.

2. Overview of silicon material

In most cases, solar cells are manufactured on a silicon material. Its proportion represents 40% of world-wide semiconductor solar cells production. Pure silicon material is founded directly in solid silica by electrolysis. The production of silicon by processing silica (SiO_2) needs very high energy and more efficient methods of synthesis. Also, the most prevalent silicon solar cell material is crystalline silicon (c-Si) or amorphous silicon (a-Si).

Crystalline silicon can be separated into multiple categories according to its crystallinity and its crystal size. These include: monocrystalline silicon, poly or multicrystalline silicon, Ribbon silicon and new structures.
Compared to amorphous silicon, crystalline silicon absorbs the visible part of the solar spectrum more than infrared portion of spectrum. Crystalline silicon has a smaller band gap (Eg = 1.1 eV) than amorphous silicon (Eg = 1.75 eV) in live with Shockley-Hall-Read's recombination process experiment (W. Shockley 1949).
Crystalline silicon solar cells generate approximately 35 mA/cm^2 of current, and voltage 550 mV. Its efficiency is above 25 %. Amorphous silicon solar cells generate 15 mA/cm2 density of current and the voltage without connected load is above 800 mV. The efficiency is between 6 and 8% (S. W. Glunz et al. 2006).
But, all solar cells require a light absorbing material contained within the cell structure to absorb photons and generate electrons (G. Sissoko et al. 1996).

3. Device physics of silicon solar cells

3.1 Silicon solar cells

Commonly, most silicon solar cells are configured in N-P junctions or vice versa (S.M. SZE 1981) in one side and N^+-N-P^+ structure (or vice versa) for double sides named bifacial silicon solar cell (S. Madougou et al. 2004, 2005a, 2005b, 2007a et 2007b). Silicon solar cells have all contacts on the back of the cell. Figure 1 shows an example of silicon solar cell with its contacts.

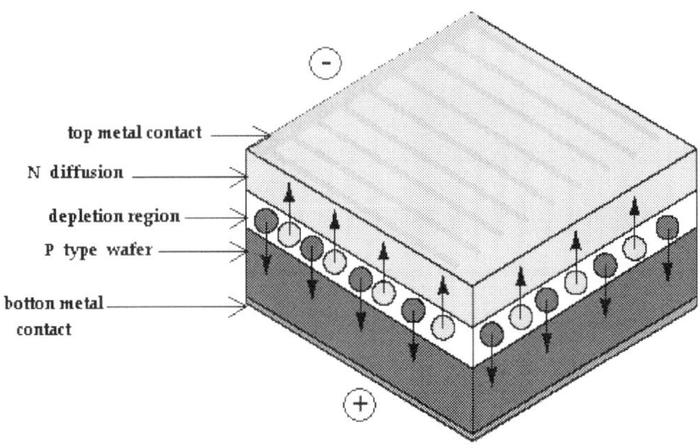

Fig. 1. Silicon Solar cell with its contacts

In this section, we will study the structure and the operation of N-P junction (monofacial and bifacial silicon solar cells).

3.1.1 Monofacial silicon solar cell

N-P junction or a P-N junction is a one side solar cell (W. Shockley 1949). When a P-type is placed in intimate contact with an N-type, a diffusion of electrons occurs from the region of high electron concentration (N-type) into the region of low electron concentration (P-type). Figure 2 shows the N-P junction and its forward biased with its corresponding diode schematic symbol and its I-V characteristic curve.

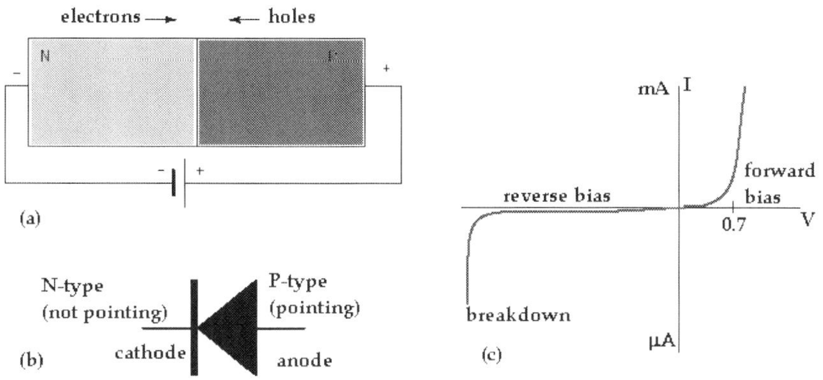

Fig. 2. N-P junction: (a) Forward biased N-P junction, (b) Corresponding diode schematic symbol (c) silicon diode I versus V characteristic curve.

3.1.2 Bifacial silicon solar cell

Bifacial silicon solar cell is a double sided silicon solar cell with N^+-N-P^+ or N^+-P-P^+ structure or vice versa (S. Madougou et al. 2004).

A bifacial silicon solar cell with N^+-P-P^+ structure has an N^+-P front side (surface) and P^+-P back side (surface). This back surface (P^+-P) is an important seat of a Back Surface Field (J. D. Alamo et al. 1981). In some solar cells, the front surface doping density ranges from 10^{17} to 10^{19} cm^{-3}. In the base, the doping ranges 10^{15} to 10^{17} cm^{-3}. The bifacial silicon solar cell can be illuminated from the front side, the back side or simultaneously from both sides as shown in figure 3.

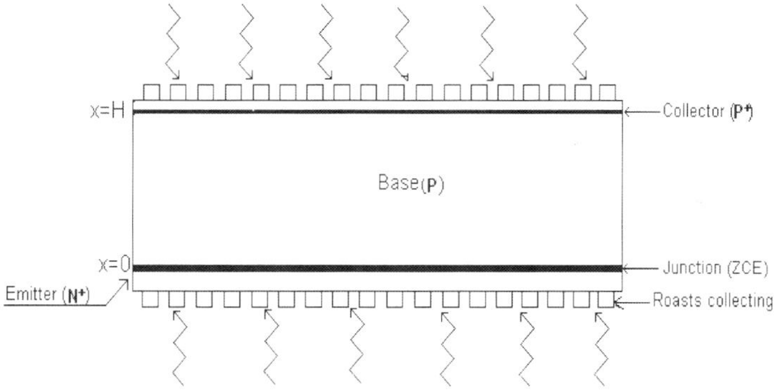

Fig. 3. Bifacial silicon solar cell with n+-p-p+ structure.

3.2 Recombination mechanisms

In a conventional silicon solar cell, recombination can occur in five regions (W. Shockley 1949):

- at the front surface;
- at the emitter region (N^+);
- at the junction (the depletion region of the junction);
- at the base region (P);

- at the back surface.

They are mainly two types of recombination: the recombination mechanisms in bulk (volume) and the surface recombination.

3.2.1 Recombination mechanisms in bulk of the silicon solar cell

In a bulk of the silicon solar cell, three fundamental recombination mechanisms are produced.

- *Auger recombination.* We have Auger recombination when the energy of the electron which falls in the valence band is transferred as kinetic energy to:
  - another free electron which will be transferred to a higher level in the conduction band;
  - a hole on a deep level of the valence band.
- *Shockley-Read-Hall recombination (through defect).* They are twofold:
  - those due to defects in the crystal lattice or chemical impurities;
  - those that occur when a deep level captures an electron.
- *Radiative or band-to-band recombination.* When a junction is forward biased, the recombination can be radiative. It is the opposite of the absorption phenomenon. The free carriers go directly from the conduction band to the valence band by emission of photon. This recombination is related by the lifetime τ of the excess minority carriers. They also intervene in the diffusion of charge carriers through the term $U(x)$ given by the expression:

$$U(x) = \frac{\delta n(x)}{\tau} \tag{1}$$

Where, $\delta n(x)$ is the excess minority carriers density, and τ is its lifetime.

3.2.2 Surface recombination

The many faults that characterize the surface of a semiconductor disrupt its crystalline structure. The Surface recombination corresponds to a phenomenon where excited electrons in the conduction band recombine with holes in the valence band via defect levels at the surface, called surface states. These surface states are the result of the abrupt discontinuity of a crystalline phase at the surface, which forms unsatisfied dangling silicon bonds.

3.3 Illumination mode

The solar cells can be under monochromatic light (single wavelength), constant multispectral light or variable and intense light concentration (more than 50 suns) or under other mode.
The optical generate rate $G(x)$ of monochromatic light is given by (S. Madougou 2007b):

$$G(x) = \alpha(1-R)\Phi_0 \exp(-\alpha x) \tag{2}$$

Where, R is the reflectivity coefficient, α is the optical absorption coefficient and Φ_0 is the incident flux of monochromatic light.
When the solar cell is illuminated by the multispectral light the generated rate $G(x)$ is given by (S. Madougou 2007a):

$$G(x) = \sum_{i=1}^{3} a_i \left[\exp(-b_i x)\right] \tag{3}$$

Where, $\mathbf{a_i}$ and $\mathbf{b_i}$ are the coefficients deduced from the modelling of the generation rate considered for over all solar radiation spectrums.
Figure 4 shows an experimental set-up of silicon solar cell illuminated with a direct light from a lamp.

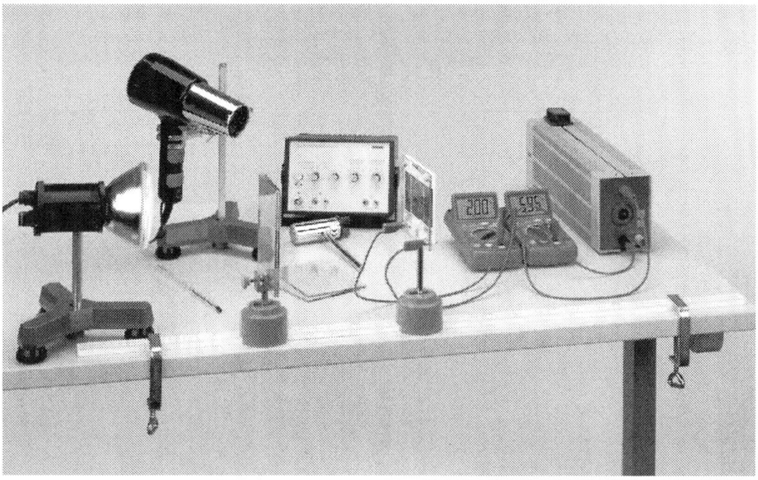

Fig. 4. Experimental set-up of silicon solar cell illuminated with a direct light from a lamp.

3.4 Basic equations of minority carriers transport

The basic equations describe the behaviour of the excess minority carriers in the base of the solar cell under the influence of an electric field and/or under illumination; both cause deviations from thermal equilibrium conditions. These equations can be expressed on one or two dimensions. In the following, we will work in one dimension.

3.4.1 Poisson equation

The Poisson equation relates the gradient of the electric field E to the space charge density ρ. According to W. Shockley(1949), it is given as:

$$-\frac{d^2\phi(x)}{dx^2}=\frac{dE(x)}{dx}=\frac{\rho}{\varepsilon_0\varepsilon} \tag{4}$$

Where, ϕ is the electrostatic potential, ε_0 is the permittivity of free space and ε is the static relative permittivity of medium.
In the same conditions, the electrons' current density I_n and the holes' current density I_p are obtained as follows:

$$I_n=+q\,\mu_n\,n(x)E(x)+q\,D_n\frac{dn(x)}{dx} \tag{5}$$

$$I_p=-q\,\mu_p\,p(x)E(x)+q\,D_p\frac{dp(x)}{dx} \tag{6}$$

Where, q is the elementary charge, μ_n and μ_p are the mobility's of electrons and holes, D_n and D_p are the diffusion constants related through the Einstein relationships: $D_n = \frac{kT}{q\mu_n}$; $D_p = \frac{kT}{q\mu_p}$.

k is the Boltzmann constant.

3.4.2 Continuity equation

When the solar cell is illuminated, the continuity equation related to photogenerated excess minority carriers density $\delta n(x)$ in the base region of the cell is given by (G. Sissoko et al 1996):

$$\frac{\partial^2 \delta n(x)}{\partial x^2} - \frac{\delta n(x)}{L^2} + \frac{G(x)}{D} = 0 \tag{7}$$

Where, D is the excess minority carriers diffusion constant and L is their diffusion length. $G(x)$ is the carriers generation rate in the base.
The solution $\delta n(x)$ of the continuity equation is well defined by the boundary conditions.

3.4.3 Boundary conditions

According to G. Sissoko et al. (1996), the boundary conditions defined by the minority carriers recombination velocities are:

- The emitter-base junction at x = 0:

$$\left.\frac{\partial \delta n(x)}{\partial x}\right|_{x=0} = \frac{Sf}{D}\delta n(0) \tag{8}$$

- The back-surface of the base at x = H:

$$\left.\frac{\partial \delta n(x)}{\partial x}\right|_{x=H} = -\frac{Sb}{D}\delta n(H) \tag{9}$$

Where, Sb is the minority carriers recombination velocity at the back-surface and Sf is the minority carriers recombination velocity at the junction.

3.5 Equivalent circuit of the solar cell

To understand the electronic behaviour of a solar cell, it is useful to create its model which is electrically equivalent at the solar cell. Because no solar cell is ideal, a shunt resistance and a series resistance component are therefore added to the model to have the equivalent circuit. This equivalent circuit of the solar cell is based on discrete electrical components. Figure 5 shows an example of an equivalent circuit of a solar cell with one diode.
For the practical analysis of the solar cell performance the dark current-voltage (I-V) characteristics curve is shifted down by a light generated current I_L resulting in the illuminated I-V characteristics.
The $I - V$ characteristic of a single-junction P-N under illumination can be written as follows:

$$I = I_0\left[\exp\left(\frac{qV}{kT}\right) - 1\right] - I_L \tag{10}$$

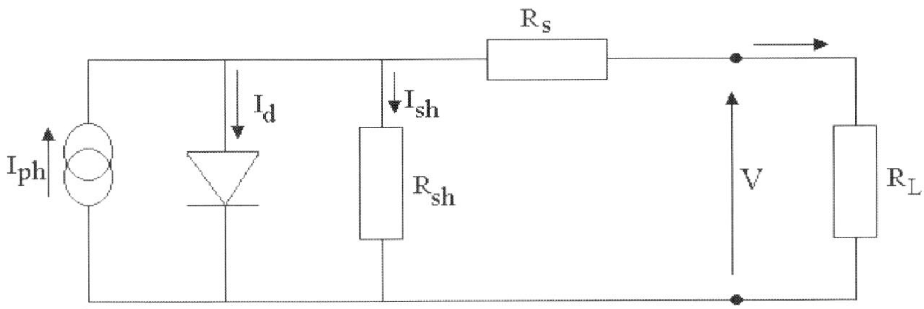

Fig. 5. Equivalent circuit of the solar cell with one diode.

And the dark current density of the P-N junction by:

$$I_{dark} = I_0 \left[\exp\left(\frac{qV}{kT} \right) - 1 \right] \tag{11}$$

Where, I_0 is the reverse saturation current density, V the voltage and T is the absolute temperature.
Figure 6 below shows the dark and illuminated current-voltage (I-V) characteristics.

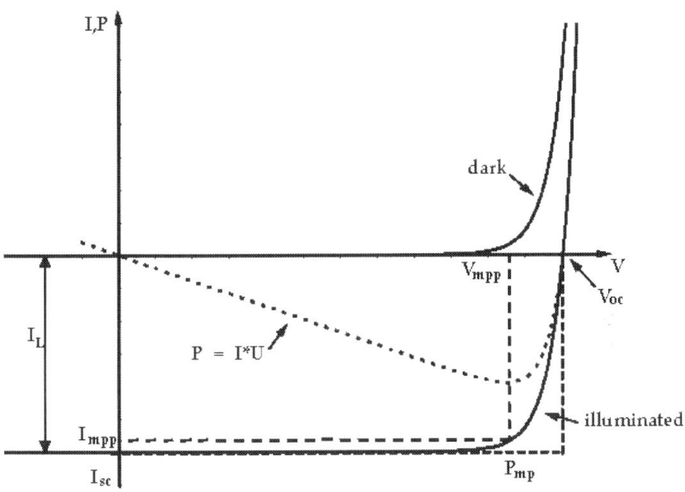

Fig. 6. Dark and illuminated current-voltage I-V curves of silicon solar cell.

Solar cells can be also usually connected in series in modules, creating an additive voltage. Connecting cells in parallel will yield higher amperage. Modules are then interconnected, in series or parallel, or both, to create the desired peak DC voltage and current.

4. Silicon solar cells electrical and recombination parameters

The solar cells have two categories of parameters: electrical parameters and recombination parameters.

4.1 Electrical parameters

4.1.1 Photocurrent

The photocurrent density I at the junction of solar cell is obtained from the excess minority carriers density of each illumination mode as follows (S. Madougou et al. 2004 et 2007a):

$$I = qD \left. \frac{\partial \delta n(x)}{\partial x} \right|_{x=0} \tag{12}$$

Where q and D are constants defined above.

4.1.2 Voltage

By means of Boltzmann's relation, the voltage V can be expressed as (J.F. Phyllips 1997):

$$V = V_T Log\left(N_B \frac{\delta n(0)}{n_i^2} + 1 \right) \tag{13}$$

Where, $V_T = \frac{kT}{q}$ is the thermal voltage, n_i the intrinsic carriers density and N_B the base doping density.

4.1.3 Power

The Power generated for the cell is given by:

$$P = VI \tag{14}$$

Where, I and V are the photocurrent and the voltage defined above.

4.1.4 Fill factor

The fill factor is given by (W. Shockley 1949):

$$FF = \frac{V_{\max} I_{\max}}{V_{co} I_{sc}} \tag{15}$$

Where, $V_{\max}$ and $I_{\max}$ are voltage and current at maximum power point respectively. V_{CO} is the open-circuit voltage and I_{SC} is the short-circuit current.

4.1.5 Series and shunt resistances

The series resistance is given by (M. Wolf et al. 1963):

$$R_S = \frac{V_{co} - V}{I} \tag{16}$$

The shunt resistance is given by (M. Wolf et al. 1963):

$$R_{Sh} = \frac{V}{I_{SC} - I} \tag{17}$$

I, V, V_{CO} and I_{SC} are defined in sections 4.1.3 and 4.1.4 above.

4.1.6 Capacitance

The capacitance is given by (Edoardo Barbisio 2000; S. Madougou et al 2004):

$$C = \frac{qn_0}{V_T}\exp\left(\frac{V}{V_T}\right) \tag{18}$$

Where, $n_0 = \frac{n_i^2}{N_B}$, V and V_T are the voltage and the thermal voltage defined above.

4.1.7 Internal quantum efficiency

The internal quantum efficiency η_{IQE} is expressed by (S. Madougou et al. 2007b):

$$\left(\eta_{IQE}\right) = \frac{I_{SC}}{q[1-R]\Phi_0} \tag{19}$$

I_{SC}, R and Φ_0 are defined above.

4.2 Recombination parameters

The different recombination parameters include: the diffusion length L, the minority carriers lifetime τ and the recombination velocities (Sf, Sb).

4.2.1 Diffusion length

The excess minority carriers diffusion length is given by:

$$L = \sqrt{\tau D} \tag{20}$$

4.2.2 Lifetime

The excess minority carriers lifetime is given by:

$$\tau = \frac{L^2}{D} \tag{21}$$

4.2.3 Recombination velocities (Sf, Sb)

When Sb is higher, the photocurrent tends towards the open-circuit current. Thus, we have the relationship:

$$(Sf) = \left[\frac{\partial J}{\partial Sf}\right]_{Sb \to +\infty} \tag{22}$$

When, Sf is higher, the photocurrent density tends towards its maximum value (short-circuit current). Thus, we have the following relationship:

$$(Sb) = \left[\frac{\partial J}{\partial Sf}\right]_{Sf \to +\infty} \tag{23}$$

5. Methods for determining the solar cells parameters

Research indicates many techniques for determining electrical and recombination parameters of solar cells (S. Madougou et al. 2007b, 2005a; G. Sissoko et al 1996; S. K. Sharma et al. 1985).
In this part, we will present some methods of electrical and recombination parameters determination.

5.1 Methods of electrical parameters determination

For determining solar cells electrical parameters, several methods exist.

5.1.1 The method based on the current-voltage characteristics

In these categories of methods, the authors use an algorithm for extracting solar cell parameters from I-V-curve using a single or double exponential model (S. Dib et al. 1999; C.L. Garrido Alzar 1997). They also use a non linear least squares optimization algorithm based on the Newton model using the measured current-voltage data and the subsequently calculated conductance of the device (M. Chegaar et al. 2001). To extract the solar cells parameters, the authors also utilize an analysing method of the current-voltage (I-V) characteristics of silicon solar cells under constant multispectral illumination and under magnetic field (S. Madougou et al 2007a).

5.1.2 The analytical method

In this method, researchers use the data of some solar cells parameters (measured data of current, short circuit current, current at maximum power point, voltage, open circuit voltage, voltage at maximum power point) to determine others parameters.

5.1.3 Others methods

In those methods, we have:

- Methods using numerical techniques approaches;
- The vertical optimisation method;
- Etc...

All of these techniques can be extended or modified adequately to cover many cases of solar cells.

5.2 Methods of recombination parameters determination

Here, we present two methods of recombination parameters determination. The first method is a linear fit of Internal Quantum efficiency (IQE) reverse curves versus light penetration depth. The second one is a programming method. For each method, after determining the minority carriers diffusion length, we calculate the excess minority carriers lifetime and recombination velocities.
In this approach, new analytical expressions of recombination parameters were established.

5.2.1 Fitting method

It is a linear fit of the IQE reverse curves versus light penetration depth. From this linear fit, we extract the minority carriers diffusion length (S. Madougou et al. 2007b). Figure 7 presents an example of a linear fit of measured IQE reverse curves versus light penetration depth.

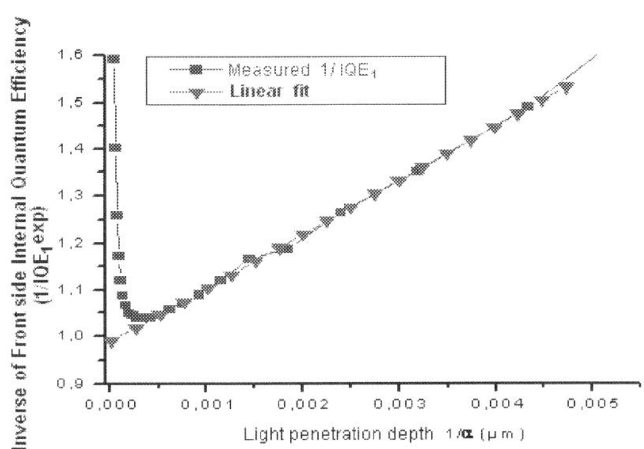

Fig. 7. Linear fit of the reverse of IQE with light penetration depth.

5.2.2 Programming method

Some researchers develop an algorithm to calculate the excess minority carriers diffusion length. This algorithm searches the weakness square type value between theoretical and measured internal quantum efficiency data. At this weakness square type value, we determine the corresponding diffusion length value of the solar cell (S. Madougou et al. 2007b).

6. Conclusion

This study shows that most of silicon solar cell is configured in N-P junctions. In these solar cells, recombination mechanisms occur in five regions and are mainly two types: recombination mechanisms in bulk and surface recombination. In practice, the solar cells can be illuminated using several illumination modes: under monochromatic light, under multispectral light, under intense light concentration or under other modes. Basic equations describe the behaviour of the excess minority carriers generated in the base of the solar cells. To understand the electronic behaviour in the study, the solar cell is modelled in an equivalent circuit containing a shunt resistance and a series resistance.

Silicon solar cells have two categories of parameters (electrical parameters and recombination parameters) which, the knowledge is very important to ameliorate the efficiency of the solar cells. Nowadays, many determination techniques of electrical and recombination parameters of solar cells exist. It is a great challenge for researchers to find a way to improve the solar cells efficiency. If this challenge is won, solar energy through photovoltaic energy can reveal itself to be a unique opportunity to solve energy and environmental problems simultaneously.

7. References

Alamo J. D., J. Eguren, and A. Luque (1981): "Operating limits of Al-alloyed high-low junctions for BSF solar cells". *Solid-State Electronics*. Vol. 24, pp. 415-420. ISSN: *0038-1101.*

Chegaar M., Z. Ouennoughi, and A. Hoffmann (2001): A new method for evaluating illuminated solar cell parameters, *Solid-state electronics*, Vol. 45, pp. 293. ISSN: 0038-1101.

Dib S., M. de la Bardonne, A. Khoury (1999), "A new method for extraction of diode parameters using a single exponential model". *Active and Passive Electronic Components*. Vol. 22, PP. 157. ISSN *0882-7516.*

Edoardo Barbisio (2000): "Diffusion capacitance identification of PV cells". *Proc. 16th European photovoltaic solar energy conference.* ISBN *1 902916 18 2.* Date 1 - 5 May 2000, Glasgow, UK, pp 919 - 922.

Garrido-Alzar C. L. (1997): "Algorithm for extraction of solar cell parameters from I-V-curve using double exponential model". *Renewable Energy.* Vol. 10, pp. 125. ISSN: 0960-1481.

Glunz S. W., S. Janz, M. Hofmann, T. Roth, and G. Willeke (2006): "Surface passivation of silicon solar cells using amorphous silicon carbide layers". *Proc. 4th WCPEC,* pp. 1016-1019. *ISBN*: 1-424-40017-1.

Madougou S., Nzonzolo, S. Mbodji, I. F. Barro, G. Sissoko (2004): Bifacial silicon solar cell space charge region width determination by a study in modelling: Effect of the magnetic field. *Journal des Sciences.* Vol.4, N°3, pp. 116-123. ISSN 0851 - 4631.

Madougou S., B. Dieng, A. Diao, I. F. Barro, G. Sissoko (2005a): Electrical parameters for bifacial silicon solar cell studied in modelling: space charge region width determination. *Journal des Sciences pour l'Ingénieur.* Volume 5, pp. 34 - 39. (Senegal). ISSN 0851 4453.

Madougou S., I.F. Barro, G. Sissoko (2005b): Effect of magnetic field on bifacial silicon solar cell studied in modeling: space charge region width determination. *Proc. 2005 Solar World Congress.* PaperN°1979. ISBN 089553 1771. Date 6-12 August 2005, Orlando (Florida, USA).

Madougou S. F. Made, M. S. Boukary, G. Sissoko (2007a): I -V characteristics for bifacial silicon solar cell studied under a magnetic field. *Advanced Materials Research: Trans Tech Publications Inc.* Volume 18 until 19 pp. 303 - 312; (Zurich - Switzerland). ISSN 1022 6680.

Madougou S., F. Made, M. S. Boukary, G. Sissoko (2007b): Recombination Parameters Determination by Using Internal Quantum Efficiency (IQE) Data Of Bifacial Silicon Solar Cells. *Advanced Materials Research. Trans Tech Publications Inc.* Volume 18 until 19 pp. 313 - 324; (Zurich - Switzerland). ISSN 1022 6680.

Phyllips J.E., T. Titus and D. Hofmann (1997): Determining the voltage dependence of the light generated current in CuInSe2-Based solar cells using I-V measurements made at different light intensities. *Proc 26th IEEE PVSC conference.* isbn 0080438652. Sept. 30 - Oct. 3, Anaheim, CA, USA.

Sharma S. K., S.N. Singh, B.C. Chakravarty, and B.K. Das (1986): "Determination of minority carrier diffusion length in a p-silicon wafer by photocurrent generation method" *Journal of Appl. Phys.* Vol 60, N°10. pp 3550 - 3552. ISSN*:* 0021-8979.

Shockley W., J. Bardeen and W. Brattain, "The Theory of P-N Junctions in Semiconductors and P-N Junction Transistors," *Bell System Technical Journal,* Vol 28: 1949, pp. 435 Semiconductive Materials Transistor. ISSN : 0005-8580.

Shockley W. and W. T. Read (1952): "Statistics of the recombination of holes and electrons," *Physical Review.* Vol. 87, pp. 835-842. ISSN*: 0031-9007.*

Sissoko G., C. Museruka, A. Correa, I. Gaye, A. L. Ndiaye (1996): "Light spectral effect on recombination parameters of silicon solar cell". *Proc. World Renewable Energy Congress.* Part III, pp 1487-1490. ISBN *0-7918-3763-7.* Date 15 - 21 June. Denver - USA (1996).

Sze S.M. (1981): *"Physics of semiconductors devices";* 2nd Edition, Wiley Interscience, Editor: John Wiley and Sons (WIE). ISBN-10 0471 0566 18. New York 11 p. 802.

Wang A., J. Zhao, and M. A. Green (1990): "24% efficient silicon solar cells," Applied Physics Letters, vol. 57, pp. 602-604. *ISSN* 0003-6951.

WOLF M. and Hans RAUSCHENBACH (1963): Series resistances effects on solar cell measurement. *Advanced energy conversion.* ISSN *0196-8904.* Volume 3, pp 445 - 479. Pergamon Press 1963. ISSN: 0196-8904.

5

Efficient Silicon Solar Cells Fabricated with a Low Cost Spray Technique

Oleksandr Malik and F. Javier De la Hidalga-W.
National Institute for Astrophysics, Optics and Electronics (INAOE)
Mexico

1. Introduction

Since the 1960s, studies of transparent and highly conducting semiconductor metal oxide films, such as tin-doped indium oxide (ITO) and fluorine-doped tin oxide (FTO), have attracted the interest of many researches due to their wide applicability in both industry and research, such as transparent heat reflecting films, gas sensors, protective coatings, and heterojunction solar cells. Excellent reviews of the subject can be found in the literature (Hamberg & Granquist, 1986, Granquist, 1993, Hartnagel et al, 1995, Dawar & Joshi, 2004). Several methods such as chemical vapour deposition, vacuum evaporation, sputtering techniques, magnetron sputtering, ion implantation, ion-beam sputtering, and spraying techniques have been tried out to fabricate coatings for the oxide materials (Dawar & Joshi, 2004). This chapter is intended to be a comprehensive review of the original results in the field of fabrication of ITO and FTO films by spray pyrolysis technique, as well as their applications for the fabrication of efficient silicon (Si) monocrystalline solar cells and modules with a low fabrication cost. Since the end of the 1970s, such solar cells have attracted the attention of the scientific community because of their fabrication simplicity (DuBow et al., 1976, Manifacier & Szepessy, 1977, Feng at al., 1979, Malik et all., 1979, 1980). It has been shown that those metal oxide films deposited on the silicon surface form heterojunctions, which at a first approximation, can be considered as Schottky "transparent" metal-semiconductor (MS) contacts due to the degeneracy of the electron gas occurring in the n-type highly conducting transparent metal oxide films, such as ITO and FTO. Depending on the method used for the deposition of the films and the type of the silicon substrate conductivity, such contact can work as a rectifier and the surface-barrier solar cell can be designed based on it. Thus, films deposited in vacuum on the p-silicon surface form rectifying contacts (DuBow et al., 1976), while electron-beam (Feng at al., 1979), whereas sprayed films deposited on n-type silicon form rectifying contacts (Manifacier & Szepessy, 1977, Malik et all., 1979). This phenomenon occurs because of the difference in electron affinity of the ITO and FTO films fabricated with the methods mentioned above. Further studies have shown that ITO/Si and FTO/Si solar cells operate almost always as majority-carriers MS or metal-insulator-semiconductor (MIS) diodes. The formation of a very thin SiO_x layer between the metal oxide film and the silicon leads to an increasing efficiency of the solar cells based on MS contacts (Ashok et al., 1980). However, if the potential barrier at the silicon surface is very high, an inversion p-n layer is formed at the silicon surface. In this case the properties of these solar cells will be similar to those of solar cells based on metallurgical p-n junctions obtained by diffusion of dopants in a silicon substrate. This was demonstrated with ITO/Si structures fabricated by sputtering of the ITO film on a p-type silicon substrate (DuBow et al., 1976), where the solar cells operated as minority-carriers (or p-n) diodes. On the other hand, numerous published works have shown that solar cells fabricated on n-type silicon operate only as majority (MS or MIS) diodes (Nagatomo et al., 1979, 1982). Nevertheless, in this work we show that combining a special treatment of the silicon surface with the electro-physical properties of the spray deposited films allows for the fabrication of inversion p-n solar cells based on n-type silicon substrates.

2. Fabrication of ITO and FTO films by spray pyrolysis

The spray pyrolysis technique was used for the deposition of thin ITO and FTO films on glass and sapphire substrates in order to investigate their structural, electric, and optical properties. A 10 Ω-cm n-type (100)-oriented silicon substrate, whose surface was chemically cleaned and specially treated, was used for the fabrication of the solar cells. The glass apparatus (atomizer) was designed in such a way that small-size droplets were obtained. The substrates were mounted on a heater covered with a carbon disk in order to assure a uniform temperature, and the spraying was conducted using compressed air. Periodical cycles of the deposition with durations of 1 sec and intervals of 5 sec were employed to prevent a rapid substrate cooling. Deposition rate was high, about 200 nm/min.

2.1 Deposition of ITO films

For the ITO films deposition, 13.5 mg of $InCl_3$ were dissolved in 170 ml of 1:1 water and ethylic alcohol mixture, with an addition of 5ml of HCl. Different ratios of Sn/In achieved in the ITO films were controlled by adding in the solution a calculated amount of tin chloride ($SnCl_4*5H_2O$). The substrate temperature was controlled with a thermocouple at a value of 480 ± 5^0C. The optimal distance from the atomizer to the substrate and the compressed air pressure were 25 cm and 1.4 kg/cm^2, respectively. We obtained a high deposition rate of about 200 nm/min.

2.2 Deposition of FTO films

The precursors for the deposition of the FTO film were prepared based on 0.2 M alcoholic solutions of $SnCl_4*5H_2O$ with different content of NH_4F for obtaining different F/Sn ratio in the films. The remaining deposition parameters were the same as those used for depositing the ITO films.

2.3 Characterization equipment and methods

The film thickness was measured with an Alpha Step 200 profilometer. The electrical resistivity, Hall mobility and carrier concentration were measured at room temperature using the van der Pauw method. Hall effect parameters were measured using a magnetic field of 0.25 Tesla. The optical transmission spectra were obtained with a spectrophotometer. The structural characterization was carried out with an X-ray diffractometer operating in the Bragg-Brentano Θ-Θ geometry with Cu K_α radiation. A JSPM 5200 atomic force microscope was used to study the film surfaces. The chemical composition of the films was determined using an UHV system of VG Microtech ESCA2000 Multilab with an Al- K_α X-ray source (1486.6 eV) and a CLAM4 MCD analyzer.

3. Brief description of the film properties

3.1 Tin-doped indium oxide (ITO) films

The X-ray diffraction (XRD) measurements shown in Figure 1 indicate that all deposited ITO films, with thickness 160-200 nm and fabricated from the chemical solutions with different Sn/In ratio, present a cubic bixebyte structure in a polycrystalline configuration with a (400) preferential grain orientation.

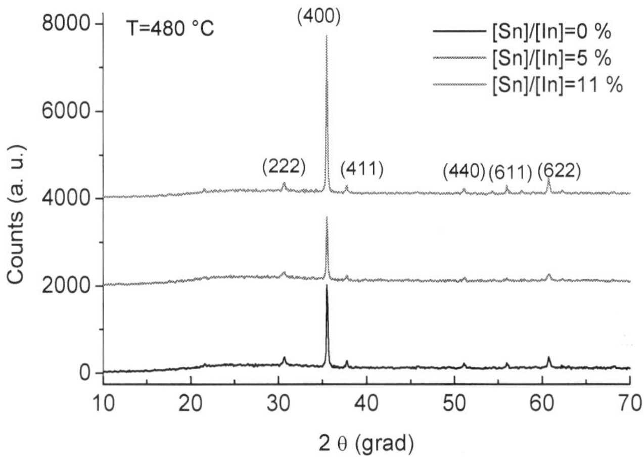

Fig. 1. XRD spectra of the ITO films fabricated from precursors with different Sn/In ratio

The average size of the grains, 30-50 nm, was determined using the classical Debye-Scherrer formula from the half-wave of the (400) reflections of the XRD patterns

A surface roughness about 30 nm was determined from images of the films surfaces obtained with the atomic force microscope (Figure 2).

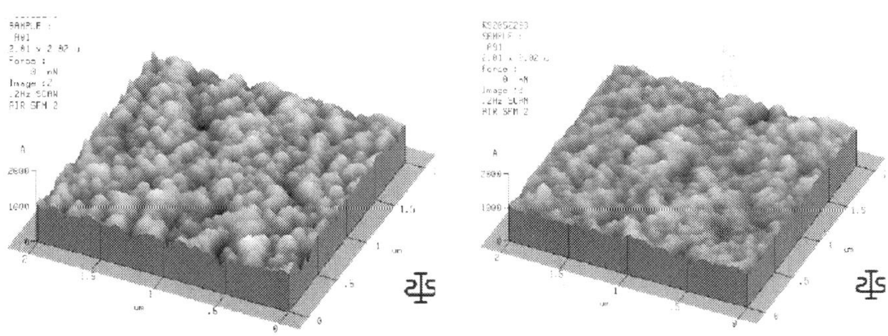

Fig. 2. AFM images of the In_2O_3 film (left) and the ITO film with 5% Sn/In (right)

Figures 3 and 4 show the dependence of electric parameters of the spray deposited ITO film on the ratio Sn/In. The sheet resistance R_s shown in Figure 3 presents a minimum of 12 Ω/□ the films prepared from the solution with a 5% Sn/In ratio.

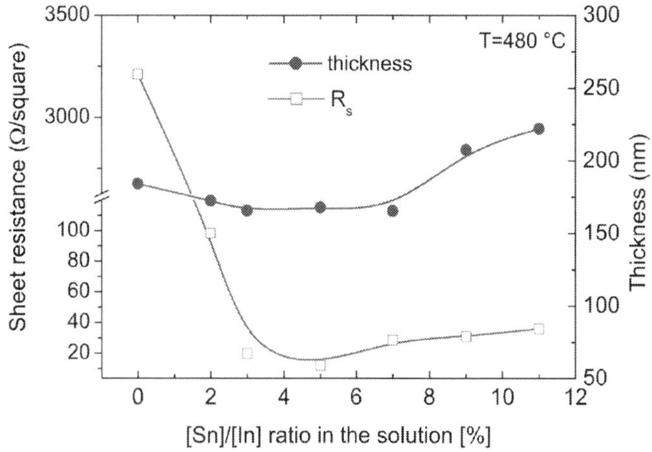

Fig. 3. The sheet resistance as a function of the Sn/In ratio in the precursor used for the film deposition. The thicknesses of the films are also shown

The minimal value of resistivity obtained for the films deposited for the solution with 5% Sn/In ratio is 2×10^{-4} Ω-cm. The variation of mobility and carrier concentration as a function of the Sn/In ratio are shown in Figure 4.

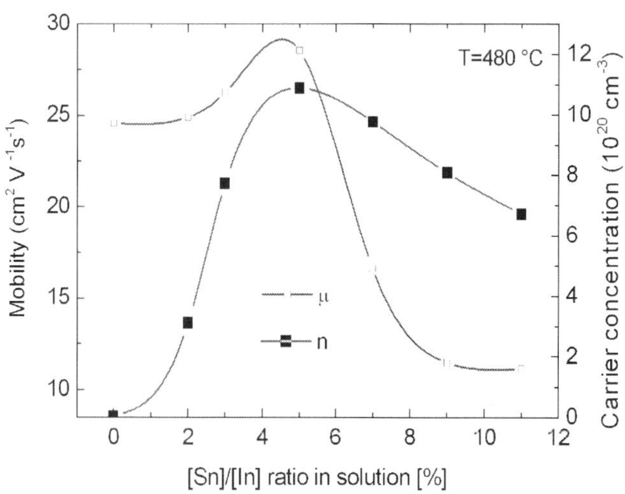

Fig. 4. Dependence of mobility (μ) and carrier concentration (n) on the Sn/In ratio

Figure 5 shows the optical transmission spectra for the ITO films spray-deposited on a sapphire substrate as a function of the wavelength for solutions with different Sn/In contents.

The use sapphire substrates allow for determining the optical energy gap of the ITO films by extrapolating the linear part of $\alpha^2(h\nu)$ curves to $\alpha^2=0$, where α is the absorption coefficient.

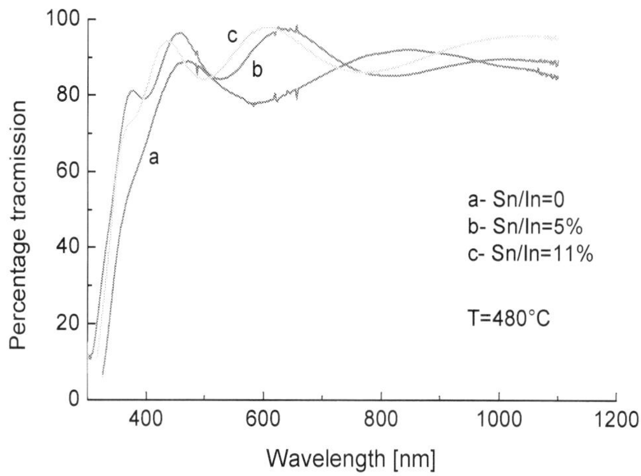

Fig. 5. Optical transmission spectra for the ITO films spray-deposited for different precursors as a function of the wavelength

The optical gap increases with the carrier concentration, corresponding to the well known Burstein-Moss shift. For the Ito films fabricated using the solution with a 5% Sn/In ratio this shift is 0.48 eV, and the optical gap is 4.2 ± 0.1 eV. Such high value for the optical gap offers transparency in the far ultraviolet range, which is important for the application of these films in solar cells.

Because of the opposite dependence of the conductivity (σ) and transmission (T) on the thickness (t) of the ITO, both parameters need to be optimized.

A comparison of the performance for different films is possible using the $\phi_{TC}=T^{10}/R_s=\sigma t \exp(-10\alpha t)$ figure of merit (Haacke, 1976). Table 1 compares the values of ϕ_{TC} for the spray deposited ITO films reported in this work with some results obtained by other authors using different deposition techniques.

Process	R_s, $\Omega/\square$	T (%)	ϕ_{TC}, $(\Omega^{-1})\times10^{-3}$	Author
spray	26.0	90	13.4	Gouskov, 1983.
spray	9.34	85	21.0	Vasu et al., 1990
spray	10.0	90	34.9	Manifacier, 1981
spray	4.4	85	44.7	Saxena, 1984
sputtering	12.5	95	47.9	Theuwissen, 1984
evaporation	25.0	98	32.6	Nath, 1980
spray	12.0	93.7	43.5	Present work

Table 1. Comparison of the values of ϕ_{TC} for ITO films

3.2 Fluorine-doped tin oxide (FTO) films

The X-ray diffraction (XRD) measurements indicate that all the spray-deposited FTO films present a tetragonal rutile structure in a polycrystalline configuration with a (200)

preferential grain orientation. The XRD spectra of the FTO films fabricated using precursors with different F/Sn ratios are shown in Figure 6.

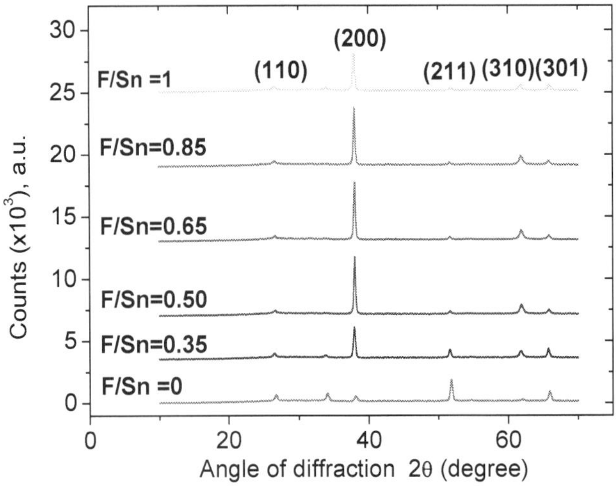

Fig. 6. The XRD spectra for the FTO films fabricated using precursors with different F/Sn ratio

The surface morphology of the films fabricated using precursors with different F/Sn ratio, and obtained with a scanning electron microscopy (SEM), is shown in Figure 7.

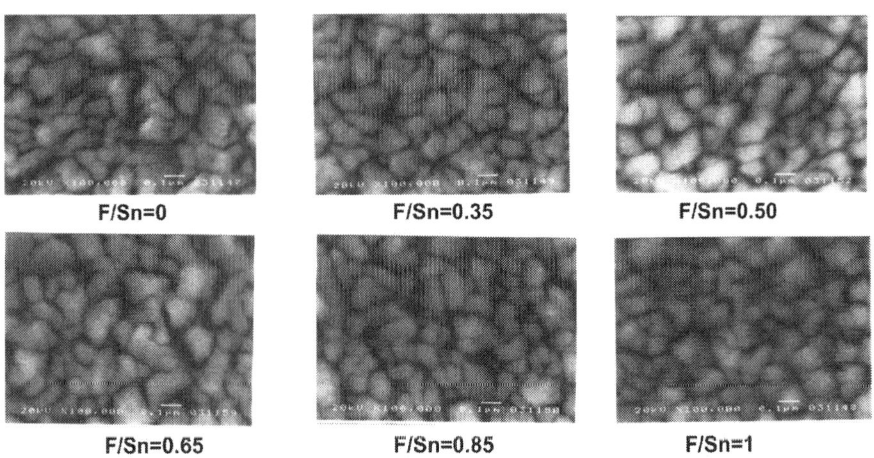

Fig. 7. The surface morphology obtained with a SEM for the films fabricated using precursors with different F/Sn ratios

The dependence of the average value of the grain size on the F/Sn ratio shows a maximum (~ 40 nm) for the films prepared using a precursor with F/Sn=0.5. The roughness variation

obtained with atomic force microscope for the FTO film fabricated using solutions with different F/Sn ratios presents a minimum of 8-9 nm at the F/Sn=0.5 ratio.
Figure 8 shows that the electrical characteristics also present some peculiarities for the films prepared using a precursor with this F/Sn ratio.

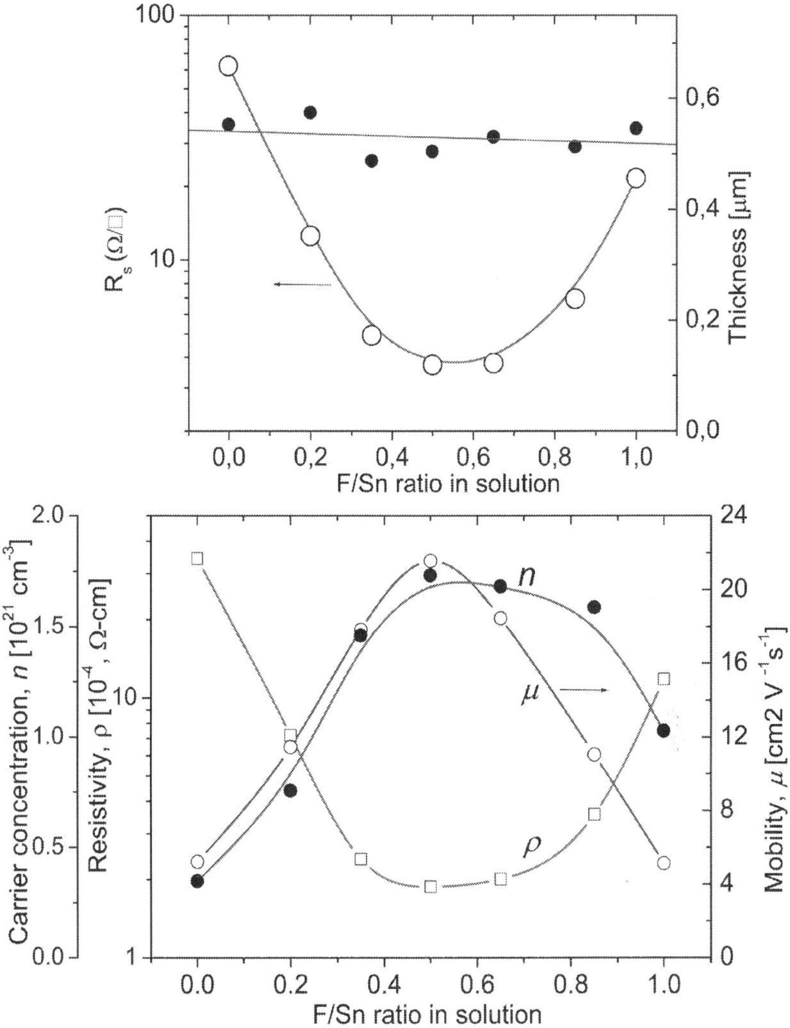

Fig. 8. Variation of the sheet resistance (above graph), resistivity (ρ), mobility (μ) and carrier concentration (n) (below graph) for the FTO films fabricated using precursors with different F/Sn ratios. The thicknesses of the films are also shown

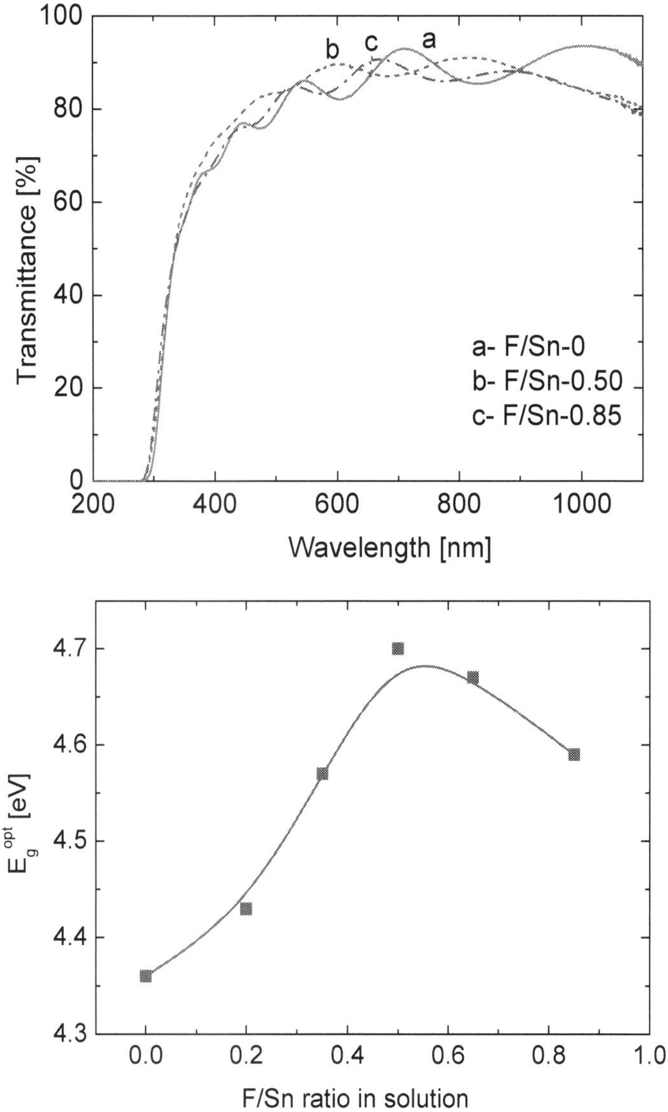

Fig. 9. Optical transmission (above graph) and dependence of the optical gap (below graph) for the FTO films fabricated using solutions with different F/Sn contents and spray-deposited on a glass substrate as a function of the wavelength

The optical energy gap (Fig. 9) was determined from the analysis of the absorption spectra for the films deposited on the sapphire substrate. The Burstein-Moss shift presents a maximum value of 0.6 eV for the films fabricated using the precursor with F/Sn =0.5, which also corresponds to the highest electron concentration (1.8×10^{21} cm^{-3}). Figure 10 shows the $\Phi=T^{10}/R_s$ figure of merit for the FTO films reported in this work.

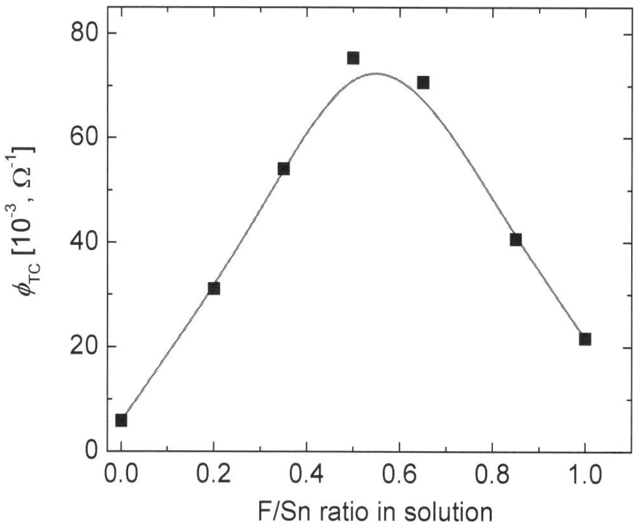

Fig. 10. Variation of the figure of merit $\Phi=T^{10}/R_s$ versus the F/Sn ratio used in the solution for the FTO films reported in this work

The value we obtained for this figure of merit was $\Phi=75\times10^{-3}\ \Omega^{-1}$ for the films prepared using a precursor with F/Sn =0.5; this is more than twice the value ($\Phi=35\times10^{-3}\ \Omega^{-1}$) reported in the literature (Moholkar et al., 2007) for spray deposited FTO films.

4. Solar cells based on ITO/n-Si heterojunctions

4.1 Physical model of the solar cells

When the ITO (or FTO) film is deposited on the silicon surface, a metal-semiconductor contact-like is formed due to the metallic electric properties of the degenerated metal oxide. Ideally, the barrier height (φ_b) formed between the metal and the n-type semiconductor is determined by the difference between the metal (or in our case the metal oxide) work function (φ_M) and the electron affinity (χ_s) in the semiconductor. Actually, the surface states present in the interface pin the Fermi level, which makes the barrier height less sensitive to the metal work function (Sze, 2007). The surface has to experiment a reconstruction due to the discontinuity of the lattice atoms on the surface. Each surface atom present a dangling bond and shares a dimer bond with its neighbor atom, thus giving place to surface states inside the Si band gap (Trmop, 1985).

Recently, it has been shown that the barrier height in a metal-silicon junction can take an almost ideal value if the n-Si surface is passivated with sulfur (Song, 2008). Also the open-circuit voltage of an Al/ultrathin SiO_2/n-Si solar cell (Fujiwara, 2003) was improved when the silicon surface was passivated by a cyanide treatment.

In this chapter we will discuss the properties of the ITO/n-Si solar cells presenting extremely high values of the potential barrier at the silicon interface obtained by passivating the surface with a hydrogen-peroxide solution.

If the ITO film is deposited on cleared n-type silicon, the barrier height not exceeds 0.76 eV. For this value of the barrier height, the ITO/nSi heterojunctions fabricated on silicon substrates with a resistivity of a few Ω-cm, operate as majority carrier devices, whose characteristics are well described by the Schottky theory. Usually, such type of devices present a high value for the dark current originated by the thermo-ionic mechanism, and the open circuit voltage for these structures designed as solar cells shows a sufficiently low value. The introduction of a very thin (~ 2 nm) intermediate SiO_x layer (Feng, 1979) decreases the dark current and increases the open-circuit voltage. However, the use of this approach to improve the characteristics of the surface-barrier solar cells requires a simultaneous and careful control of the intermediate oxide thickness. Furthermore, the thermal grown intermediate SiO_x layer always presents a positive fixed charge located at the SiO_x/Si interface, which decreases the barrier height in the case of n-type silicon.

Using known data for the work function of ITO films deposited by spray pyrolysis, whose average value is reported as 5.0 eV (Nakasa et al., 2005, Fukano, 2005), and the electron affinity of silicon as 4.05 eV, the ideal barrier height between ITO and n-type silicon is 0.95 eV according to the Mott-Schottky theory. After a treatment of the n-type silicon surface in the hydrogen-peroxide (H_2O_2) solution with a controlled temperature (60 ^{0}C) during 10 minutes, a barrier height of 0.9 eV was obtained with capacitance-voltage measurements. This value exceeds by 0.14 eV the barrier height obtained after the deposition of the ITO film on the silicon surface cleaned in HF without the treatment in an H_2O_2 solution.

It is worth discussing the possible reason for this increment of the barrier height after the treatment of the silicon surface, as well as the operation of the ITO/n-Si junctions with an extremely high barrier height. Obviously, a junction with such barrier height fabricated on the silicon substrates with moderate resistivity could behave as p-n junctions, in which a surface p-layer is induced by the high surface band bending.

Such situation was obtained (Shewchun, 1980) in solar cells ITO/ultrathin SiO_x/p-Si structures. However, in this case the inversion of the conductivity type of the p-Si at the surface was caused by other factors, such as the low work function of the sputtered ITO film and the presence of positive charge at the SiO_x/p-Si interface.

What is the physical reason for the increment of the barrier height in the ITO/n-Si heterojunctions after the treatment of the silicon substrate in heated 30% H_2O_2 solutions? It has been shown (Verhaverbeke, 1997) that the treatment of the silicon in H_2O_2 leads to the growth of oxide on the silicon surface. The analysis shows that the main oxidant responsible for this oxide growth is the peroxide anion, HO_2^-. It was also found that the oxide thickness is limited to a value around 0.8-1.0 nm due to the presence of localized negative charge (HO_2^-) at the silicon surface. From this point of view the HO_2^- at the silicon surface can play a double role. First, these ions can form a chemical composition with the silicon atoms having dangling bounds in the surface. This can be thought as a passivation of the silicon surface, which leads to an increment of the potential barrier during the formation of the ITO/Si heterostructure. On the other hand, the negative charge of these ions can produce a band-bending (φ_s) at the silicon surface due to an outflow of electrons under the influence of the electrostatic force. Under such conditions, the electron affinity (χ_s) of the silicon at the surface will be lower than that at the bulk by $\Delta\chi=\chi_s-\varphi_s$. The presence of a depletion layer at the silicon surface plays an important role for the formation of the potential barrier during the deposition of the ITO film. The barrier will prevent an electron flow from the silicon to the ITO film. The surface barrier between the ITO and the silicon will be formed by the flow of valence electrons from the silicon valence band into the ITO film, creating a hole excess at the silicon surface. Taking into account the initial band-bending at the silicon surface, the formation of an inversion layer is possible. As it was already mentioned, the experimentally determined barrier height at the ITO/Si interface is 0.9 eV. Schematically, the energy diagram of the ITO/n-Si heterojunction is shown in Figure 11.

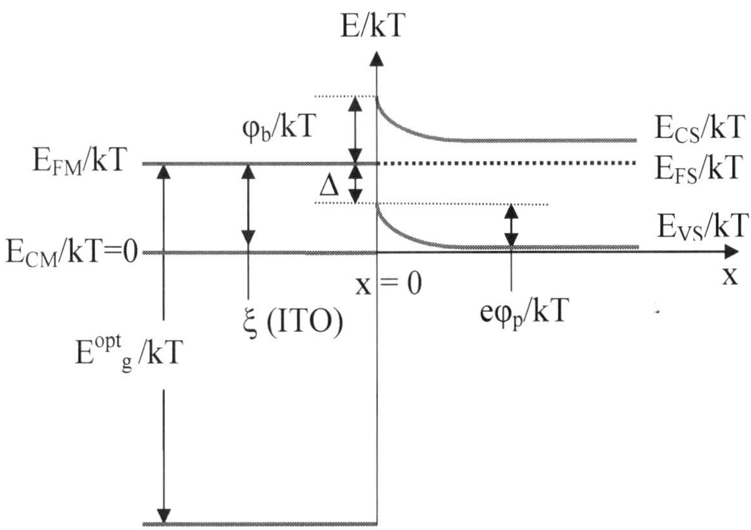

Fig. 11. Energy diagram (in kT units) of the heavy doped ITO/n-Si heterojunction

For sake of simplicity, we do not show the very thin (around 1 nm) intermediate SiO_x layer present between the ITO film and the silicon, because at this thickness it does not present any effect on the electro-physical characteristics of the heterojunction. Since the heavily doped ITO film is a degenerated semiconductor, in which the Fermi level lies above the minimum of the conduction band, we can consider this ITO film as a "transparent metal." The inversion layer at the silicon interface appears when the barrier height φ_b is higher than one-half of the Si energy gap. If such inversion p-n junction were connected in a circuit, which source of holes would be present in order to form an inversion p-layer that complicates the current flow across the forward-biased structure working as a solar cell? To answer this question we calculated the number of empty energy states in the conduction band of a heavy doped ITO, which are available to accept the electrons transferred from the top of the silicon valence band located at a distance Δ below the Fermi level (Malik et al., 2006). The probability that an energy state E below the Fermi level E_{FM} in the degenerated ITO is empty was calculated using the Fermi-Dirac distribution. Using a barrier height φ_n=0.9 eV, Δ=0.3 eV, and three different values for (E_{FM}-E_{CM}), which is the distance between the Fermi level an the conducting band of the ITO. This characterizes the degree of degeneration of the ITO film. The calculated number of empty states available to accept the electrons from the silicon valence band forming the additional amount of the holes is shown in Figure 12 as triangles. For comparison the number of empty states in the case of a gold/silicon contact with the same barrier height is also shown. For such calculations, the difference between the effective mass of electrons in the ITO and that in gold has been taken into account.

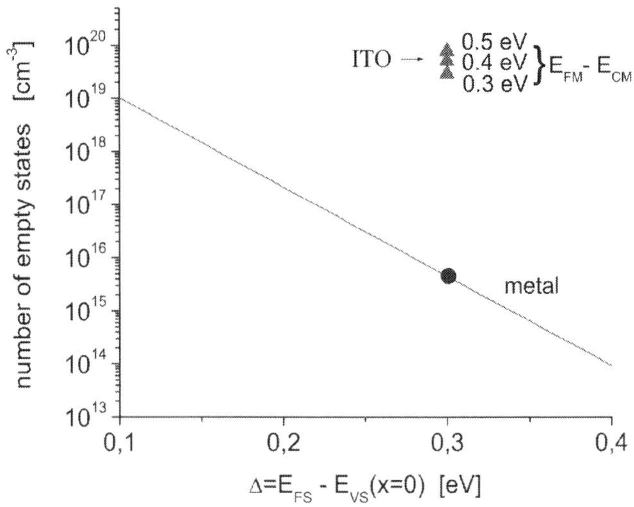

Fig. 12. Calculated number of empty states available to accept the electrons from the silicon valence band (Malik et al., 2006)

From the discussion presented above, and the amount of the calculated number of empty states in the ITO, leads to the important conclusion that a heavy doped ITO layer serves as an efficient source of holes necessary to form the inversion p-layer in the ITO/n-Si structures.

4.2 Evidence of the inversion in the type conductivity in the ITO/n-Si heterostructures

Based on the barrier height (0.9 eV) obtained from the measured C-V characteristics for the ITO/n-Si heterostructures on 10 Ω-cm monocrystalline silicon, one can discuss about the physical nature of such heterojunctions. Because the barrier height exceeds one half of the silicon band gap, the formation of an inversion p-layer at the silicon surface is obvious from the band diagram. To avoid any speculations on this issue and in order to present a clear evidence for the existence of a minority (hole) carrier transport in these heterojunctions, a bipolar transistor structure was fabricated on a 10 Ω-cm monocrystalline silicon substrate, in which the emitter and the collector areas, on opposite sides of the silicon substrate, were fabricated based on the ITO/n-Si junctions. The ITO film was deposited using the spray deposition technique described in section 2.1 followed by a photolithographic formation of the emitter and the collector areas. The treatment in the H_2O_2 solution described above was applied to the silicon substrate. An ohmic n^+-contact (the base) was formed using local diffusion of phosphorous in the silicon substrate. The dependence of the collector current versus the collector-base voltage, using the emitter current as a parameter, are shown in

Figure 13, together with the emitter injection efficiency of the ITO/n-Si/ITO transistor (Malik et al., 2004).

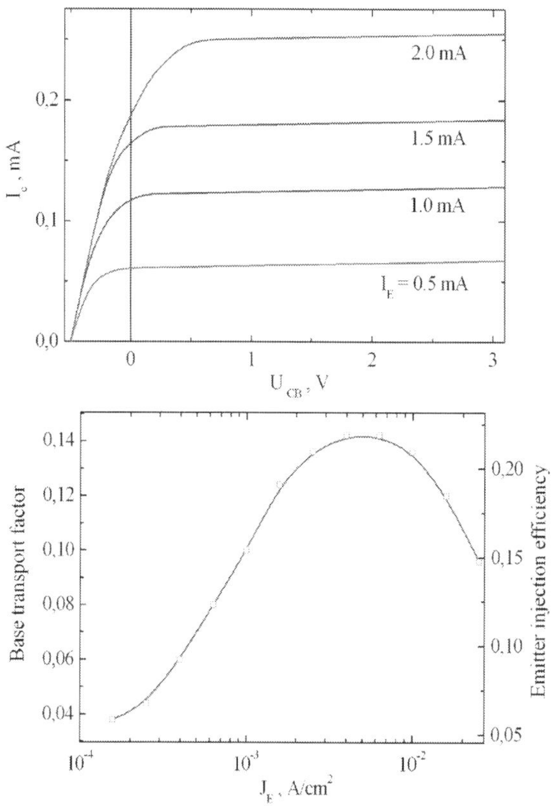

Fig. 13. Dependence of the collector current on the collector-base voltage (the emitter current is used as a parameter). The emitter injection efficiency of the ITO/n-Si/ITO transistor fabricated on a 10 Ω-cm silicon substrate is also shown. (Malik et al., 2004)

Hence, even in non-optimized transistors (wide base), an efficient hole injection of around 0.2 was observed. This is an obvious evidence of the existence of an inversion layer in the ITO/n-Si heterostructures with a barrier height of 0.9 eV. We can also present two indirect evidences of the p-n nature of the ITO/n-Si heterojunctions. The first one is based on the observation of an efficient radiation emission from the ITO/n-Si structures under a forward bias (Malik et al., 2004). In metal-semiconductor contacts operated as majority carriers' devices (described by the Schottky theory), the injection ratio does not exceed 10^{-4}. Thus, an efficient electroluminescence, in contrast to our devices, is not possible to observe. The next evidence is based on the observed modulation of the conductivity in the forward-biased ITO/n-Si diodes fabricated on high resistivity silicon, which operate as p-i-n diodes. So, the 0.9 eV barrier height belongs to an inversion ITO/n-Si heterojunction. This gives us the possibility to analyze theoretically such structures based on the well-known theory of p-n junctions.

4.3 Limit of applicability of the p-n model for the ITO/n-Si solar cells

Once we know the physical nature of the ITO/n-Si heterojunctions with extremely high potential barrier, it is possible to apply correctly the theory for their modelling, which is known as the theory of p-n based solar cells. The problem now is to find the range of resistivity of the silicon substrate on which the p-n theory can be applied to the ITO/n-Si heterojunction with extremely high potential barrier. Based on results published recently (Malik et al., 2008), the condition for strong inversion in the ITO/n-Si heterojunction requires that

$$\varphi_s \geq 2(E_F - E_i), \tag{1}$$

where

$$E_F - E_i = kT \ln(N_d / n_i), \tag{2}$$

φ_s is the surface potential at the Si/SiO$_x$ interface, k is the Boltzmann constant, T is the temperature, n_i is the intrinsic carrier concentration, and N_d is the donor concentration in the n-Si substrate. On the other hand,

$$\varphi_s = \varphi_b - (E_C - E_F), \tag{3}$$

$$E_C - E_F = kT \ln(N_C / N_d), \tag{4}$$

where φ_b is the potential barrier for carriers from the ITO side of the structure, and N_C is the effective density of states in the conduction band.
Moreover, the surface hole concentration is

$$p_s(x=0) = (n_i^2 / N_d)\exp(\varphi_s / kT) \tag{5}$$

Combining equations (2)-(5), it is possible to obtain the surface concentration of the minority carriers at the Si/SiO$_x$ interface under strong inversion of the conductivity type:

$$p_s(x=0) = (n_i^2 / N_C)\exp(\varphi_b / kT). \tag{6}$$

This concentration depends only on the barrier height and not on N_d. Figure 14 shows the two possible models in the space $p_s(x=0)/N_d$ vs. N_d in the substrate for different barrier heights.
The two shaded areas are related to the two possible models: a Schottky model for $p_s(x=0)/N_d < 0.01$ and an induced *p-n* junction, in which $p_s(x=0)/N_d > 10$. For instance, at a barrier height of 0.7 V, the green line takes two intercepts: one with the border of the area that is related to the Schottky barrier model, and the other one with the border of the area that is valid for the *p-n* inversion model. Thus, for $N_d > 3\text{x}10^{14}$ cm^{-3} the structures behave as Schottky-barrier structures, whereas the structures with $N_d < 4\text{x}10^{13}$ cm^{-3}, behave as p-n

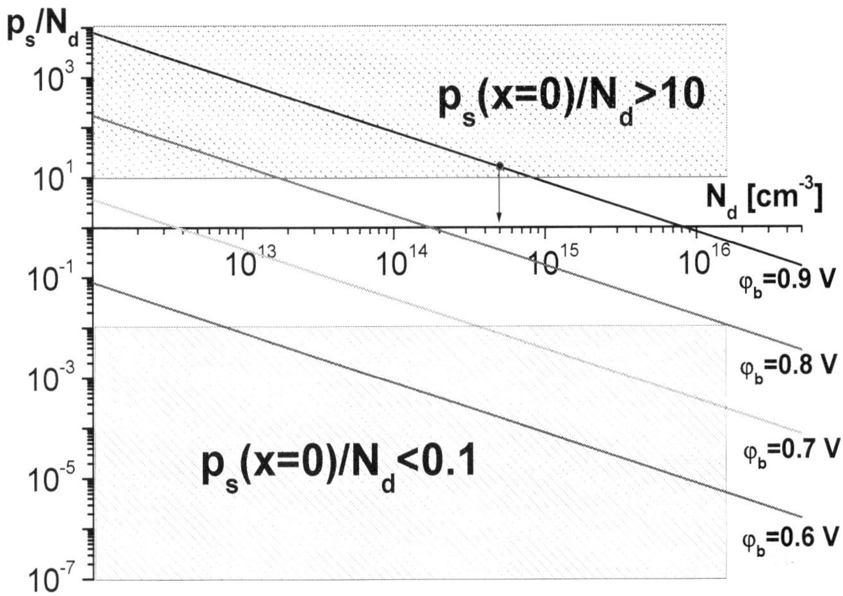

Fig. 14. Schematically representation of two possible models of the ITO/n-Si heterojunction in coordinates of p_s(x=0)/N_d vs. concentration N_d in the silicon substrate. The different barrier height serves as a parameter (Malik et al., 2008)

junctions. With the potential barrier height of 0.9 eV achieved in this work, the structures may be considered as a symmetrical *p-n* ($p_s = N_d$) for N_d=8x10^{15} cm^{-3} (0.3 Ω-cm resistivity of the substrate), or as an asymmetrical p^+-*n* junctions ($p_s \geq 10N_d$) for N_d=8x10^{14} cm^{-3} (5 Ω-cm resistivity of the substrate). Due to the substrate resistivity used in this work, 10 Ω-cm (N_d=5x10^{14} cm^{-3}), our solar cells with a barrier height of 0.9 eV present an asymmetrical p^+-*n* junctions, and the theoretical analysis of such structures will be conducted based on the theory of p+-n junctions.

We underline again that the intermediate SiO_x layer formed after the treatment of the silicon substrate in the H_2O_2 solution is sufficiently "transparent" for the carriers; then the tunneling current through this layer provides an ohmic contact between the ITO film and the surface-induced p^+-Si layer.

Thus, we can apply the diffusion theory of the p-n junction based solar cells for modelling the ITO/n-Si solar cells with a barrier height of 0.9 eV (the barrier height does not depend on the substrate carrier concentration) for a silicon substrate resistivity higher than 0.5 Ω-cm (or a carrier concentration lower than 8×10^{15} cm^{-3}).

4.4 ITO/n-Si solar cells: design, fabrication and characterization

The solar cells were fabricated using (100) n-type (phosphorous doped) single-crystalline silicon wafers with a 10 Ω-cm resistivity. Both sides of the wafer were polished. Standard wafer cleaning procedure was used. To form the barrier, an 80 nm-thick ITO film with a sheet resistance of 30 Ω/□ was deposited by spray pyrolysis on the silicon substrate treated

in the H_2O_2 solution. This ITO thickness was chosen in order to obtain an effective antireflection action of the film. Metal, as an ohmic contact in the back side of the wafer, was deposited on an n^+-layer previously created by diffusion. The device area for measurements was 1-4 cm^2. Approximately 1 μm-thick Cr/Cu/Cr film was evaporated through a metal mask to create a grid pattern (approximately 10 grid-lines/cm). After fabrication, the capacity-voltage characterization was conducted to control the value of the potential barrier. Then the following parameters were measured under AMO and AM1.5 illumination: open circuit voltage V_{oc}, short circuit current I_{sc}, fill factor FF, and efficiency. No attempt was made to optimize the efficiency of the cells by improving the collection grid. The series resistance (R_s) of the cell was measured using the $R_s=(V-V_{oc})/I_{sc}$ relationship (Rajkanan, Shewchun, 1979), where V is the voltage from the dark *I-U* characteristics evaluated at $I=I_{sc}$.
It was shown above that the ITO/n-Si heterostructures with a potential barrier height at the silicon surface of 0.9 eV behave as pseudo classical diffusion p-n junctions. Thus, it is expected that the diffusion of holes in the silicon bulk dominates the carrier transport instead of the dominance of the thermo-ionic emission in the Schottky and the metal/tunnel oxide/semiconductor structures. A straightforward measurement of the dependence of the dark current on temperature is, in principle, sufficient to identify a bipolar device in which the thermo-ionic current is negligible in comparison to the minority-carrier diffusion current J_d (in units of current density). A simple Shockley's analysis of the p-n diode including the temperature dependence of the silicon parameters (diffusion length, diffusion coefficient, minority carrier life-time, and the intrinsic concentration) (Tarr, Pulfrey, 1979) shows that

$$J_d = J_{0d}[\exp\ (qV/kT)-1] \tag{7}$$

and

$$J_{0d} \propto T^{\gamma} \exp\ (-E_{g0}/kT), \tag{8}$$

where $\gamma = 2.4$ and $E_{g0} = 1.20$ eV.
From Eq.(8) it can be seen that the plot $\log(J_{0d}/T^{\gamma})$ vs. $1/T$ should produce a straight line, and that the slope of this line should be the energy E_{g0}. In the case of MS and MIS devices this slop must be equal to the value of the barrier φ_b.
Usually, the series resistance of the device affects the *I-V* characteristics at high forward current densities. To prevent this effect, we must measure the J_{sc} vs V_{oc} dependences (Rajkanan, Shewchun, 1979). The photogenerated current is equal to the saturation photocurrent. For minority-carrier MIS diode with a thin insulating layer (Tarr, Pulfrey, 1979)

$$J_{sc} = J_{rg}(V_{oc}) + J_d(V_{oc}). \tag{9}$$

For an increasing bias, J_d increases faster than the recombination current density J_{rg} ; in the high illumination limit we should have

$$J_{sc} - J_{0d}\exp(qV_{oc}/nkT), \tag{10}$$

which gives an n factor approximately equal to 1.
Figure 15 shows the measured dependence of J_{sc} on V_{oc} at room temperature. The value of J_{0d} in (10) was determined by measuring J_{sc} and V_{oc} at different temperatures, and under

illumination with a tungsten lamp. An optical filter was used to prevent the heating of the cell by the infra-red radiation. For each J_{sc} - V_{oc} pair lying in the range where $n \approx 1$, $J_{0d}=J_{02}$ was calculated from (10). After making the correction for the T^{γ} factor appearing in Eq.(8), the J_{0d} values were plotted as a function of the reciprocal temperature, as shown in the insert of Figure 15.

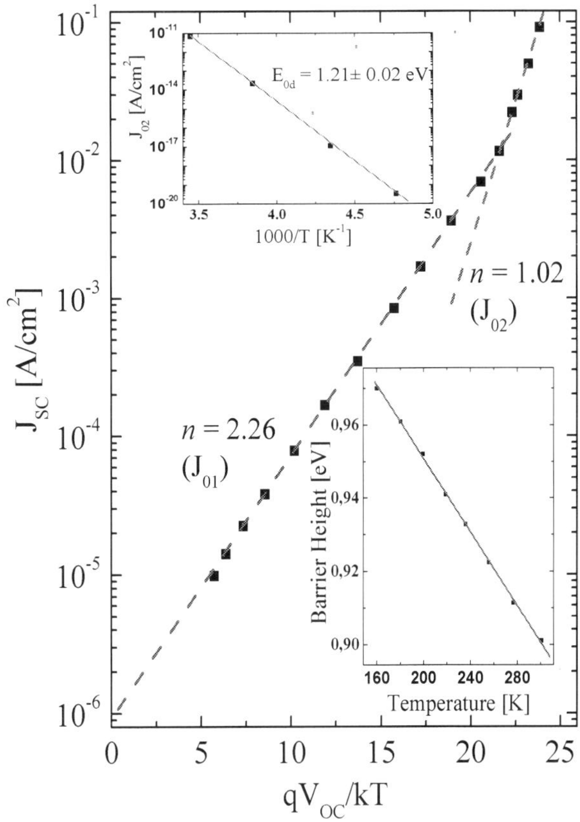

Fig. 15. Measured dependence of J_{sc} on V_{oc} at room temperature, and calculated dependence of the current density $J_{02}=J_{0d}$ at high illumination level corrected for the T^{γ} factor, as function of reciprocal temperature for ITO/n-Si solar cells with the barrier height of 0.9 eV. The dependence of the barrier height on temperature is also shown in the insert

The slope of the J_{02} vs. $1/T$ line was found to correspond to an energy E_{g0} from Eq.(8). It can be concluded that for high current densities the current in the cell is carried almost exclusively by holes injected from the ITO contact that later diffuse into the base of the cell. The output characteristics of the ITO/n-Si solar cell measured under AM0 and AM1.5 illumination conditions, as well as the calculated dependence of output power of the cell versus the photocurrent, are shown in Figure 16.

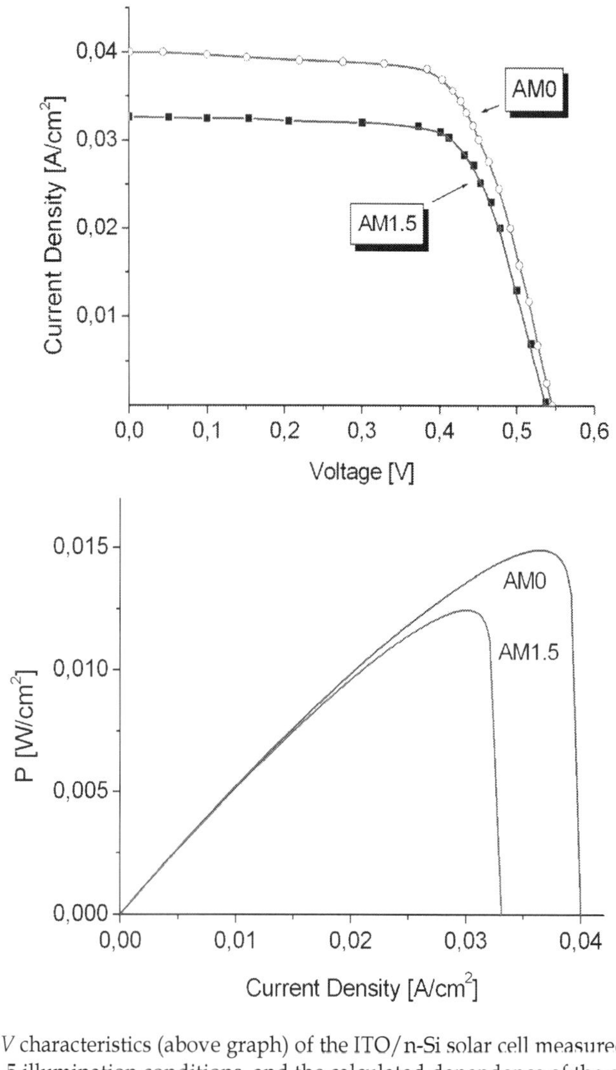

Fig. 16. *I-V* characteristics (above graph) of the ITO/n-Si solar cell measured under AM0 and AM1.5 illumination conditions, and the calculated dependence of the output power of the cell (below graph) versus the photocurrent

The fill factor (FF) and the efficiency calculated from these characteristics are 0.68 and 10.8% for AM0 illumination conditions; and 0.68 and 12.1% for AM1.5 illumination conditions. The fill factor and efficiency obtained are not optimized because of the cell design and the used silicon substrate with relatively high resistivity. Below, a theoretical analysis followed for increasing the output parameters of the cells using silicon substrates with lower resistivity is presented.

4.5 Optimization of the output characteristics of the cells: theoretical analysis

Recently, a detailed theoretical modelling of the ITO/n-Si solar cells has been reported (Malik et al, 2008). Based on these published results, here we show the most important conclusions; further details can be consulted in that work.

For all the calculations, the thickness of the silicon substrate, sheet resistance and thickness of the ITO film were taken as d=500 μm, 30 Ω/□, and t=80 nm, respectively. We considered the case when the diffusion length of minority carriers is shorter than the thickness of the silicon substrate, and assumed that the carrier recombination rate at the back contact of the silicon substrate is infinite. The total series resistance of the cell with an area of 1 cm^2 used for these calculations was taken as 1.8 Ω.

In order to calculate the theoretical parameters of the solar cells we assumed also that the equation for the *I-V* characteristic for an illuminated cell (Sze & Ng, 2007) is

$$\ln\left(\frac{J+J_{sc}}{J_0}-\frac{V-JR_s}{J_0R_{sh}}+1\right)=\frac{q}{\gamma kT}\left(V-JR_s\right), \tag{11}$$

where J is the current density, J_0 the saturation dark current density, J_{sc} is the short-circuit current density, V is the output voltage, R_s and R_{sh} are the series and shunt resistances, and γ is the "ideality" factor of the solar cell. According to our experimental results, γ was taken as 1 for the calculations.

In order to calculate the photocurrent density we integrated the next equation based on the spectral distribution of the incident solar radiation, and the parameters of silicon (absorption coefficient $\alpha(\lambda)$, diffusion length for minority carriers L_p, and thickness of the silicon substrate d):

$$J_{sc}=q\int_{\lambda 1}^{\lambda 2}\left\{(1-R)_\lambda F_\lambda\left(\frac{\alpha L_p}{\alpha^2L_p^2-1}e^{-\alpha W}\right)\times\left(\alpha L_p-\frac{\cosh\left(\frac{d}{L_p}\right)-e^{-\alpha d}}{\sinh\left(\frac{d}{L_p}\right)}\right)+(1-R)_\lambda F_\lambda(1-e^{-\alpha W})\right\}d\lambda \tag{12}$$

where q is the electron charge, W is the depletion width in the silicon substrate, and $R(\lambda)$ is the spectral reflectance from the ITO/Si interface calculated from the optical constants of silicon and ITO (Malik et al, 2008).

The spectral distribution F_λ of the solar radiation, which are related to the AM0 (136 mW/cm^2) and AM1.5 (100 mW/cm^2) conditions, have been used in the calculations according to the 2000 ASTME-490-00 and ASTM G-173-03 standards, respectively.

The values of the open-circuit voltage under AM0 and AM1 conditions were calculated according to the equation

$$V_{oc}=\frac{\gamma kT}{q}\ln\left(\frac{J_{sc}}{J_0}+1\right), \tag{13}$$

where the saturation dark current density J_0 is calculated from the equation

$$J_0 = qn_i^2 \frac{D_p}{N_d L_p} coth \frac{d}{L_p} \tag{14}$$

Here, n_i and N_d are the intrinsic and donor concentrations in the silicon substrate, respectively, and D_p is the diffusion coefficient for holes.

Figure 17 shows the experimental (dots) and calculated (lines) *I-V* characteristics of the solar cell (using equation (11)) with an area of 1 cm² fabricated on 10 Ω-cm silicon under both AM0 and AM1.5 illumination conditions. Initially, these characteristics were calculated using J_{sc} = 40 mA/cm², R_s = 1.8 Ω, and $R_{sh} = \infty$. Then, in order to improve the fitting with the experimental results, the calculated characteristics were corrected using R_{sh}= 300 Ω. One can see an excellent coincidence between the experimental and calculated characteristics, as well as for the parameters of the cell (fill factor *F.F.* and conversion efficiency η).

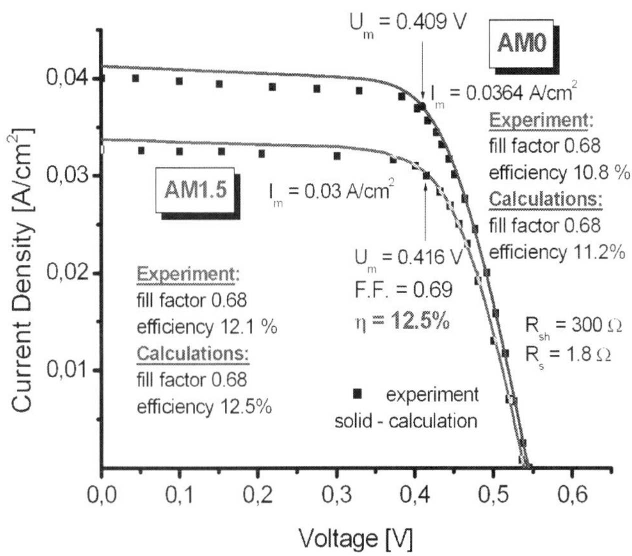

Fig. 17. Experimental (dots) and calculated (solid lines, using equation (11)) *I-V* characteristics of the solar cell with an area of 1 cm², and fabricated on a 10 Ω-cm silicon substrate, under both AM0 and AM1.5 illumination conditions

Figure 18 (above graph) shows the dependence of the short-circuit current density J_{sc} on the diffusion length in the silicon substrate, under AM0 and AM1.5 illumination conditions.

The values of the open-circuit voltage under the same conditions were calculated according to equations. (13) and (14). From equation (14), the value of J_0 decreases with the resistivity ρ of the silicon substrate. The calculated dependence for the open-circuit voltage (V_{OC}) on the resistivity of the Si substrate is also shown in Figure 18 (below graph).

The calculations show that the conversion efficiency of the ITO-SiO_x-*n*Si solar cells can be improved by using silicon with a lower resistivity. Under the AM1.5 conditions, the calculated dependences of the open circuit voltage, fill factor, and efficiency on the resistivity of the silicon substrate, are shown in Figure 19. The surface recombination velocity (S_p) was taken into account for these calculations. The value of S_p for the ITO/n-Si heterostructures under investigation, and determined from the analysis of the transistor structures, was 500 cm/s approximately.

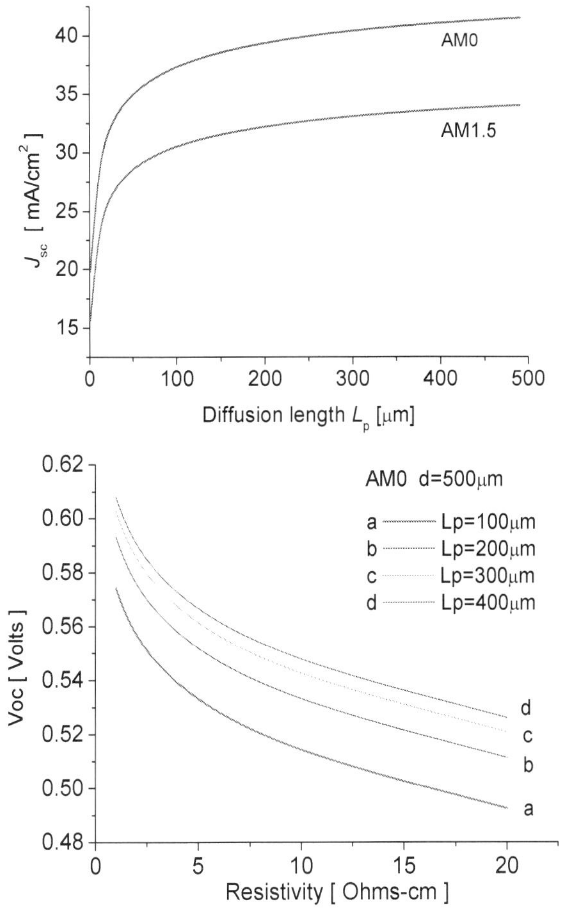

Fig. 18. Calculated dependence of J_{sc} and V_{OC} for the ITO/n-Si solar cells

Solar cells fabricated on silicon substrates with a resistivity of 1Ω-cm and a hole diffusion length of L_p= 200 μm may present an efficiency of 14 %. For an experimentally found potential barrier of 0.9 eV it is not possible to achieve a further reduction of the silicon resistivity for structures with a *p-n* inversion layer or minority carrier devices. Such structures are majority carrier devices, and their properties are described by the theory of Schottky barriers. In such cases, a lower efficiency is expected due to a higher saturation current. Solar cells using FTO films present similar characteristics.

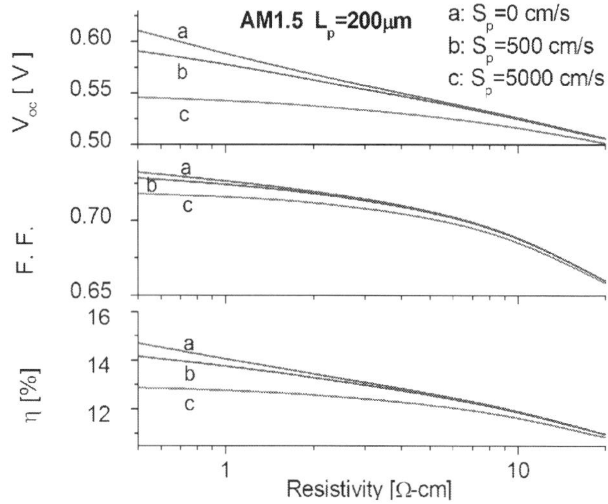

Fig. 19. Calculated dependences of the series resistance, fill factor, and efficiency of the cells, on the resistivity of the silicon substrate.

5. Conclusions

ITO-nSi solar cells have been produced using a spraying technique. Transparent and conductive tin-doped indium oxide films, as well as fluorine-doped tin oxide films, presenting excellent structural, optical and electrical parameters, were fabricated using a very simple, low cost, and no-time consuming method. The cells obtained in such a way can be considered as structures presenting an inversion *p-n* junction. Under the AM0 and AM1.5 solar illumination conditions, the efficiency is 10.8% and 12.2%, respectively. The theoretical modelling based on *p-n* solar cells show an excellent coincidence between the theoretical and the experimental results. It is also shown that using 1 Ω-cm silicon substrates is a promising alternative for obtaining solar cells with 14% efficiency under AM1.5 illumination conditions. The use of substrates with a lower resistivity leads to a reduction of the conversion efficiency due to the formation of Schottky barriers, which gives place to a higher saturation dark current than that presented by *p-n* structures. The fabrication of reported solar cells is more controllable than that needed for obtaining metal-insulator-silicon solar cells because of the necessity of controlling a very thin (nearly 2 nm) intermediate oxide layer on the silicon substrate. Moreover, a detailed theoretical analysis (Shewchun et al., 1980) shows a higher efficiency for p-n inversion solar cells in comparison with those based on majority-carrier MIS structures.

6. References

Ashok, S.; Sharma, P. & Fonash, S. (1980). Spray-deposited ITO-silicon SIS heterojunction solar cells. *IEEE Trans. Electron. Dev.*, Vol.ED-27, N.4, 725-730, ISSN 0018-9383

Dawar, A. & Joshi J. (1984). Semiconducting transparent thin films: their properties and applications. *J. Mater. Sci.*, Vol.19, 1-23, ISSN 0022-2461

DuBow, J.; Burk, D. & Sites, J. (1976). Efficient photovoltaic heterojunctions of indium tin oxides on silicon. *Appl. Phys. Lett.*, Vol.29, N.8, 494-496, ISSN 0003-6951

Feng, T.; Ghosh, A. & Fishman, G. (1979). Efficient electron-beam-deposited ITO/n-Si solar cells, *J. Appl. Phys.*, Vol.50, N.7, 4972-4974, ISSN 0022-3727

Fukano, T.; Motohiro, T. & Ida, T. (2005). Ionization potential of transparent conductive indium oxide films covered with a single layer of fluorine-doped tin oxide nanoparticles grown by spray pyrolysis deposition, *J. Appl. Phys.*, Vol.97, N.8, Nanoscale science and design, ISSN 0022-3727

Fujiwara, N.; Fujinaga, T.; Niinobe, D.; Maida, O.; Takahashi, M. & Kobayashi, H. (2003). Passivation of defect states in Si and Si/SiO_2 interface states by cyanide treatment: improvement of characteristics of pin-junction amorphous Si and crystalline Si-based metal-oxide-semiconductor junction solar cells. *Acta Phys. Slovaca*, Vol.53, N.3, 195-205, ISSN 0323-0465

Granqvist, C. (1993). Transparent Conductive Electrodes for Electrochromic Devices: A Review. *Appl. Phys.*, Vol.A57, 19-24, ISSN 0947-8396

Gouskov, L.; Saurel, J.; Gril, C.; Boustani, M. & Oemry, A. (1983). Sprayed indium tin oxide layers: Optical parameters in the near-IR and evaluation of performance as a transparent antireflecting and conducting coating on GaSb or $Ga_{1-x}Al_xSb$ for IR photodetection. *Thin Solid Films*, Vol.99, N.4, 365-369, ISSN 0040-6090

Haacke, J. (1976). New figure of merit for transparent conductors. *J. Appl. Phys.*, Vol.47, 4086-4089, ISSN 0022-3727

Hamberg, J. & Granqvist, C. (1986). Evaporated Sn-doped In2O3 films: basic optical properties and applications to energy-efficient windows. *J. Appl. Phys.*, V.60, n.11, R13, ISSN 0022-3727

Hartnagel, H.; Dawar, A.; Jain, A. & Jagadish, C. (1995). *Semiconducting Transparent Thin Films*, IOP Publishing Ltd., ISBN 0 7503 03220, Bristol UK

Malik, A.; Baranyuk, V. & Manasson, V. (1979). Solar cells based on the SnO_2–SiO_2-Si heterojunction. *Appl. Sol. Energy*, N. 2, 83-84, ISSN 0003-701X

Malik, A.; Baranyuk, V. & Manasson, V. (1980). Improved model of solar cells based on the In_2O_3/ SnO_2- SiO_x-nSi structure. *Appl. Sol. Energy*, N.1, 1-2, ISSN 0003-701X

Malik, O.; Grimalsky, V.; Torres-J., A. & De la Hidalga-W, J. Room Temperature Electroluminescence from Metal Oxide-Silicon. *Proceedings of the 16th International Conference on Microelectronics (ICM 2004)*, pp. 471-474, ISBN 0-7803-8656-6, Tunis, December 06-08, 2004, IEEE, Tunisia

Malik, O.; Grimalsky, V. & De la Hidalga-W, J. (2006). Spray deposited heavy doped indium oxide films as an efficient hole supplier in silicon light-emitting diodes. *J. Non-Cryst. Sol.*, Vol.352, 1461-1465, ISSN 0022-3093

Malik, O.; De la Hidalga-W, J.; Zúñiga-I, C. & Ruiz-T, G. (2008). Efficient ITO-Si solar cells and power modules fabricated with a low temperature technology: results and perspectives. *J. Non-Cryst. Sol.*, Vol.354, 2472-2477, ISSN 0022-3093

Manifacier, J. & Szepessy, L. (1977). Efficient sprayed In_2O_3:Sn n-type silicon heterojunction solar cell. *Appl. Phys. Lett.*, Vol.31, N.7, 459-462, ISSN 0003-6951

Manifacier, J.; Fillard, J. & Bind J. (1981). Deposition of In_2O_3-SnO_2 layers on glass substrates using a spraying method. *Thin Solid Films*, Vol. 77, N.1-3, 67-80, ISSN 0040-6090

Moholkar, A.; Pawar, S.; Rajpure, K.; & Bhosale, C. (2007). Effect of solvent ratio on the properties of highly oriented sprayed fluorine-doped tin oxide thin films. *Mater. Lett.*, Vol.61, N.14-15, 3030-3036, ISSN 0167-577X

Nagatomo, T.; Endo, M. & Omoto, O. (1979). Fabrication and characterization of SnO2/n-Si solar cells, *Jpn. J. Appl. Phys.*, Vol.18, 1103-1109, ISSN 0021-4922

Nagatomo, T.; Inagaki, Y.; Amano, Y. & Omoto, O. (1982). A comparison of spray deposited ITO/n-Si and SnO2/n-Si solar cells, *Jpn. J. Appl. Phys.*, Vol.21, N. 21-2, 121-124, ISSN 0021-4922

Nakasa, A.; Adachi, M.; Suzuki, E.; Usami, H. & Fujimatsu, H. (2005). Increase in conductivity and work function of pyrosol indium tin oxide by infrared irradiation, *Thin Solid Films*, Vol.84, N.1-2, 272-277, ISSN 0040-6090

Nath, P. & Bunshah, R. (1980). Preparation of In_2O_3 and tin-doped In_2O_3 films by novel activated reactive evaporation technique. *Thin Solid Films*, Vol.69, N.1, 63-68, ISSN 0040-6090

Rajkanan, K. & Shewchun, J.(1979) A better approach to the evaluation of the series resistance of solar cells, *Sol. St. Electron.*, Vol.22, N.2-E, 193-197, ISSN 0038-1101

Saxena, A.; Singh, S.; Thangaraj, R. & Agnihotri O. (1984). Thickness dependence of the electrical and structural properties of In_2O_3:Sn films, *Thin Solid Films*, Vol.117, N.2, 95-100, ISSN 0040-6090

Shewchun, J.; Burc, D. & Spitzer, M. (1980). MIS and SIS solar cells, *IEEE Trans. Electron. Dev.*, Vol.ED-27, N.4, 705-716, ISSN 0018-9383

Song, G.; Ali, M. & Tao, M. (2008). A high Schottky barrier between Ni and S-passivated n-type Si (100) surface. *Sol. St. Electron.*, Vol.52, 1778-1781, ISSN 0038-1101

Sze, S. & Ng, K. (2007). *Physics of semiconductor devices*, 3rd ed., John Wiley and Sons, ISBN 9780471143239, N.Y.

Tarr, N. & Pulfrey, D. (1979). New experimental evidence for minority-carrier MIS diodes. *Appl. Phys. Lett.*, V.34, N.4, 15 February 1979, 295-297, ISSN 0003-6951

Theuwissen, A. & Declerck, G. (1984). Optical and electrical properties of reactively d. c. magnetron-sputtered In_2O_3:Sn films. *Thin Solid Films*, Vol.121, N.2, 109-119, ISSN 0040-6090

Trmop, R.; Hamers, R. & Demuth, J. (2005). Si (100) dimmer structure observed with scanning tunneling microscopy, *Phys. Rev. Lett.*, Vol.55, N.12, 1303- 1308, ISSN 0031-9007

Vasu, V. & Snbrahmanyam, A. (1990). Reaction kinetics of the formation of indium tin oxide films grown by spray pyrolysis. *Thin Solid Films*, Vol.193-194, n.2, 696-703, ISSN 0040-6090

Verhaverbeke, S.; Parker, J. & McConnell, C. (1997). The role of HO_2^-: in SC-1 cleaning solutions. In: Mat. Res. Soc. Symp. Proc., Vol. 477: Science and Technology of Semiconductor Surface Preparation, 47-56, MRS, ISBN 1-55899-381-9, N.Y.

6

Efficiency of Thin-Film CdS/CdTe Solar Cells

Leonid Kosyachenko
Chernivtsi National University
Ukraine

1. Introduction

Over the last two decades, polycrystalline thin-film CdS/CdTe solar cells fabricated on glass substrates have been considered as one of the most promising candidates for large-scale applications in the field of photovoltaic energy conversion (Surek, 2005; Goetzberger et al., 2003; Romeo et al., 2004). CdTe-based modules have already made the transition from pilot scale development to large manufacturing facilities. This success is attributable to the unique physical properties of CdTe which make it ideal for converting solar energy into useful electricity at an efficiency level comparable to traditional Si technologies, but with the use of only about 1% of the semiconductor material required by Si solar cells.

To date, the record efficiencies of laboratory samples of CdS/CdTe solar cells and large-area modules are ~ 16.5 % and less than 10 %, respectively (Britt & Ferekides, 1993; Hanafusa et al., 1997; Meyers & Albright, 2000; Wu et al., 2001; Hanafusa et al., 2001; Bonnet, 2003). Thus, even the record efficiency of such type solar cells is considerable lower than the theoretical limit of 28-30% (Sze, 1981). Next challenge is to improve the performance of the modules through new advances in fundamental material science and engineering, and device processing. Further studies are required to reveal the physical processes determining the photoelectric characteristics and the factors limiting the efficiency of the devices.

In this chapter, we present the results of studying the losses accompanying the photoelectric conversion in the thin-film CdS/CdTe heterostructures and hence reducing the efficiency of modules on glass substrate coated with a semitransparent ITO or SnO_2 conducting layer. We discuss the main parameters of the material used and the barrier structure determining the photoelectric conversion efficiency in CdS/CdTe solar cell: (i) the width of the space-charge region, (ii) the lifetime of minority carriers, (iii) their diffusion length and drift length, (iv) the surface recombination velocity, and (v) the thickness of the CdTe absorber layer.

Among other factors, one of the important characteristics determining the efficiency of a solar cell is the spectral distribution of the quantum efficiency which accounts for the formation of the drift and diffusion components of the photocurrent and ultimately the short-circuit current density. In the paper particular attention is given to this aspect of solar cell. We demonstrate the possibility to describe quantitatively the quantum efficiency spectra of the thin-film CdS/CdTe solar cells taking into account the recombination losses at the CdS-CdTe interface and the back surface of the CdTe absorber layer.

Charge collection efficiency in thin-film CdS/CdTe solar cells are also discussed taking into consideration losses caused by a finite thickness of the p-CdTe layer, recombination losses at the front and back surfaces as well as in the space-charge region. The dependences of the

drift and diffusion components of short-circuit current on the uncompensated acceptor concentration, charge carrier lifetime, recombination velocities at the interfaces are evaluated and discussed.
The mechanism of the charge transport in the CdS/CdTe heterostructure determining the other photoelectric parameters of the solar cell, namely, the open-circuit voltage and fill factor is also considered. It is shown that the above-barrier (diffusion) current of minority carriers is important only at high bias voltage, and the dominant charge transport mechanism is the generation-recombination occurring in the depletion layer. The observed I–V characteristics in the dark and the light are described mathematically in the context of the Sah-Noyce-Shockley theory.

2. Spectral distribution of quantum efficiency of CdS/CdTe heterostructure

In this section we will describe mathematically the spectral distribution of quantum efficiency of the thin-film CdS/CdTe solar cells taking into account the main parameters of the material used and the barrier structure, recombination in the space-charge region, at the CdS-CdTe interface and the back surface of the CdTe absorber layer.
Quantum efficiency η_{ext} is the ratio of the number of charge carriers collected by the solar cell to the number of photons of a given energy (wavelength λ) shining on the solar cell. Quantum efficiency relates to the response (A/W) of a solar cell to the various wavelengths in the spectrum. In the case of monochromatic radiation (narrow spectral range) $\eta_{ext}(\lambda)$ relates to the radiation power P_{opt} and the photocurrent I_{ph} by formula

$$\eta_{\rm ext}(\lambda) = \frac{I_{\rm ph} / q}{P_{\rm opt} / h\nu}, \qquad (1)$$

where q is the electronic charge, $h\nu$ is the photon energy.

2.1 Experimental

Fig. 1(a) shows the quantum efficiency spectra of the CdS/CdTe solar cell taken at different temperatures. The substrates used for the development of thin film layers were glass plates coated with a semitransparent ITO (SnO_2 + In_2O_3) layer. The window layer CdS (~ 0.1 μm) was developed by chemical bath deposition (CBD); the absorber layer CdTe (1-3 μm) was deposited on top of CdS by close-space sublimation (CSS) (Mathew et al., 2007). Non-rectifying ohmic contact to the CdTe layer was fabricated by sputtering Ni in vacuum after bombarding the CdTe surface by Ar ions with energy ~ 500 eV. The electrical characteristics of two neighboring Ni contacts on the CdTe surface were linear over the entire range of measured currents.
The spectral characteristics of the samples in the 300-900 nm range were recorded with a photoresponse spectral system equipped with a quartz halogen lamp. The spectral distribution of the photon flux at the outlet slit of the system was determined using a calibrated Si photodiode.
As can be seen from Fig. 1(a), compared with the literature data, the measured curves seem to reflect the most common features of the corresponding spectral curves for these devices (Sites et al., 2001; McCandless et al., 2003; Ferekides et al., 2004).
In the long-wavelength region, the spectra are restricted to the value λ_g corresponding to the band gap of CdTe which is equal to 1.46 eV at 300 K ($\lambda_g = hc/E_g$ = 845 nm). In the short-

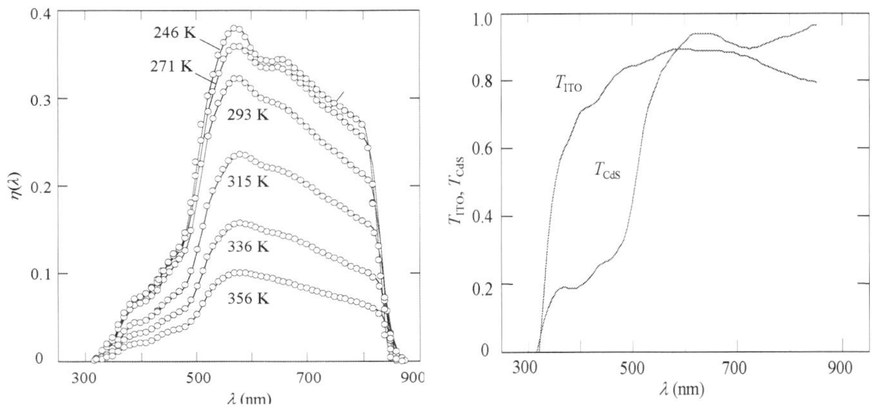

Fig. 1. (a) Spectral distribution of the quantum efficiency of CdS/CdTe device measured at different temperatures. (b) The transmission curves of the ITO coated glass (T_{ITO}) and the CdS layer (T_{CdS}) as a functions of the wavelength λ.

wavelength side, the quantum efficiency decays due to the lower transmission through the thin film layers: CdS in the range $\lambda < 500$-520 nm and ITO at $\lambda < 350$ nm (Fig. 1(b)).
The external quantum efficiency η_{ext} is related with the quantum efficiency of photoelectric conversion in the CdTe absorber layer, the transmission of the glass plate coated by ITO, T_{ITO}, and the transmission of the CdS layer, T_{CdS}, by the expression:

$$\eta_{\mathrm{ext}} = T_{\mathrm{ITO}} T_{\mathrm{CdS}} \eta_{\mathrm{int}} \tag{2}$$

where η_{int} is the ratio of photogenerated carriers collected to the photon flux that arrives at the CdTe absorber layer.
In order to describe the external quantum efficiency spectrum η_{ext} we used the measured spectral dependences $T_{\mathrm{ITO}}(\lambda)$ and $T_{\mathrm{CdS}}(\lambda)$ shown in Fig. 1(b). The quantum efficiency η_{int} will be determined in the following by considering the photoelectric processes in the diode structure.

2.2 Width of the space-charge region and energy diagram of thin-film CdS/CdTe heterostructure

One of the parameters of a solar cell that determines the electrical and photoelectric characteristics is the width of the space-charge region W. It is known that in CdS/CdTe solar cells only the CdTe is contributing to the light-to-electric energy conversion and the window layer CdS absorbs light in the range $\lambda < 500$-520 nm thereby reducing the photocurrent. Therefore in numerous papers where the energy band diagram of a CdS/CdTe junction is discussed a band bending in the CdS layer (and hence a depletion layer) is not depicted (see, for example, Goetzberger et al, 2003; Birkmire & Eser, 1997; Fritsche et al., 2001). Analyzing the efficiency of CdS/CdTe solar cells, however, one is forced to assume the concentration of uncompensated acceptors in the CdTe layer to be 10^{16}-10^{17} cm^{-3} and even higher (a narrow depletion layer is assumed). It may appear that the latter comes into conflict with the commonly accepted model of CdS/CdTe as a sharply asymmetrical p-n heterojunction. In fact, this is not the case because the width of space-charge region of a diode structure is determined by the concentration of uncompensated impurity (N_d - N_a for CdS) rather than by the free carrier concentration (n for n-CdS). These values coincide only in an uncompensated semiconductor with a shallow donor level whose ionization energy is less than the average thermal energy kT. However, the CdS layer contains a large number of background impurities (defects) of donor and acceptor types, which introduce in the band gap both shallow and deep levels. The effect of self-compensation is inherent in this material: a donor impurity introduces p-type compensating defects in just as high concentration as needed to annihilate virtually the electrical activity of donor (Desnica et al, 1999). As a result, CdS is a compensated semiconductor to a greater or lesser extent. In this case, the Fermi level is known to be pinned by the level (so-called pinning effect), which is partially compensated with the degree of compensation N_a / N_d close to 0.5 (if N_a / N_d = 0.5 the Fermi level exactly coincides with the deep impurity level) (Mathew et al., 2007). Assuming N_d - N_a equal to n, one can make serious mistake concerning the determination of the space-charge region width in CdS. Thus, due to a large number of background impurities (defects), the uncompensated donor concentration in the CdS layer can be much higher than the electron concentration in the conduction band of CdS. When such is the case the depletion layer of the CdS/CdTe diode structure is virtually located in the p-CdTe layer even in the case of a CdS layer with comparable high-resistivity (Fig. 2). This is identical to the case of an asymmetric abrupt p-n junction or a Schottky diode; the width of the space-charge region in the CdS/CdTe heterojunction can be expressed as (Sze, 1981):

$$\varphi(x,V) = (\varphi_o - qV)\left(1 - \frac{x}{W}\right)^2, \tag{3}$$

$$W = \sqrt{\frac{2\varepsilon\varepsilon_o(\varphi_o - qV)}{q^2(N_a - N_d)}}, \tag{4}$$

where ε_o is the electric constant, ε is the dielectric constant of the semiconductor, $\varphi_o = qV_{bi}$ is the barrier height at the semiconductor side (V_{bi} is the built-in potential), V is the applied voltage, and $N_a - N_d$ is the uncompensated acceptor concentration in the CdTe layer.

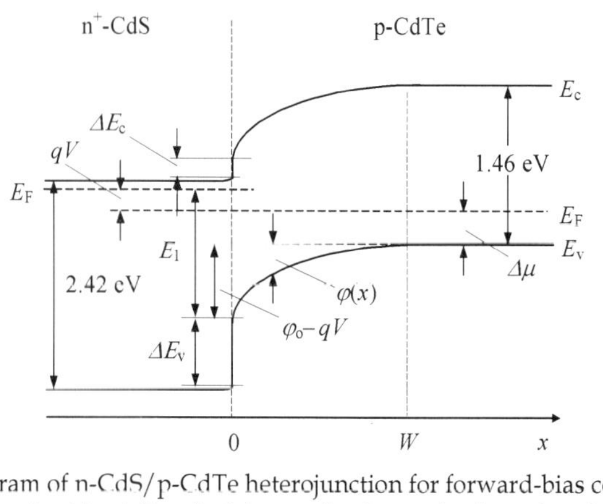

Fig. 2. Energy diagram of n-CdS/p-CdTe heterojunction for forward-bias condition. ΔE_c and ΔE_v show the discontinuities (offsets) of the conduction and valence bands, respectively.

2.3 Theoretical description of quantum efficiency

The quantum efficiency η_{int} can be found from the continuity equation which is solved using the boundary conditions. The exact solution of this equation with account made for the drift and diffusion components as well as surface recombination at the CdS/CdTe interface leads to rather cumbersome and non-visual expression for η_{int} (Lavagna et al, 1977):

$$\eta_{int} = \frac{1+\dfrac{S}{D_n}\exp\left(-\dfrac{W^2}{W_o^2}\right)\left[A(\alpha)-D_1(\alpha)\right]}{1+\dfrac{S}{D_n}\exp\left(-\dfrac{W^2}{W_o^2}\right)B} - \frac{\exp(-\alpha W)}{1+\alpha L_n} - D_2(\alpha)\,, \qquad (5)$$

where S is the velocity of recombination at the front surface, D_n is the diffusion coefficient of electrons, and $L_n = (\tau_n D_n)^{1/2})$ is the electron diffusion length. The values A, B, D_1 and D_2 in Eq. (5) are the integral functions of the absorption coefficient α, the width of the depletion layer W and the effective Debye length $W_o = (\varepsilon\varepsilon_o kT/q^2(N_a - N_d))^{1/2}$.

Eq. (5) may be essentially simplified. At the boundary between the depletion and neutral regions ($x = W$), the photogenerated electrons are entrained by strong electric field and, hence, one may put $\Delta n(W) = 0$. This means that the terms D_1 and D_2 in Eq. (5) can be neglected. When calculating the values $A(\alpha)$ and B, one can replace the integration by multiplication of the maximum value of the integrands by their "half-widths". The half-widths are determined by the value of x at the point where the value of the integrand is smaller than the peak value by a factor of $e = 2.71$.

After such simplification, instead of Eq. (5) one can write (Kosyachenko et al., 1999):

$$\eta_{int} = \frac{1+\dfrac{S}{D_n}\left(\alpha+\dfrac{2}{W}\dfrac{\kappa_o - qV}{kT}\right)^{-1}}{1+\dfrac{S}{D_n}\left(\dfrac{2}{W}\dfrac{\varphi_o - qV}{kT}\right)^{-1}} - \frac{\exp(-\alpha W)}{1+\alpha L_n}\,. \qquad (6)$$

Comparison of the dependences $\eta_{int}(\lambda)$ calculated in a wide range of the parameters and the absorption coefficient α shows that equation (6) approximates the exact equation (5) very well (Kosyachenko, 2006).

It should be emphasized that Eqs. (5) and (6) do not take into consideration the recombination at the back surface of the CdTe layer (when deriving Eq. (5) the condition $\Delta n = 0$ at $x \to \infty$ was used) which can result in significant losses in the case of a thin CdTe layer with large diffusion length of the minority carriers. However, we can use the Eq. (5) to find the expression for the drift component of the photoelectric quantum yield. This can be done as follows.

In the absence of recombination at the front surface, equation (6) transforms into the known Gartner formula (Gartner, 1959)

$$\eta_{int} = 1 - \frac{\exp(-\alpha W)}{1+\alpha L_n} \qquad (7)$$

which ignores recombination at the back surface of the CdTe layer. In this case, the photoelectric quantum yield caused by processes in the space-charge region is equal to the absorptivity of this layer, that is, $1 - \exp(-\alpha W)$. Thus, subtracting the term $1 - \exp(-\alpha W)$ from the right side of Eq. (7), we obtain the expression for the diffusion component of the photoelectric quantum yield

$$\eta^{o}_{\text{diff}} = \exp(-\alpha W)\frac{\alpha L_n}{1+\alpha L_n}, \tag{8}$$

which, of course, ignores recombination at the back surface of the CdTe layer. Subtracting $\exp(-\alpha W)\alpha L_n/(1 + \alpha L_n)$ from the right side of Eq. (6) we come to the expression for the *drift* component of the photoelectric quantum yield taking into account surface recombination at the CdS-CdTe interface:

$$\eta_{\text{drift}} = \frac{1+\dfrac{S}{D_n}\left(\alpha+\dfrac{2}{W}\dfrac{\varphi_o - qV}{kT}\right)^{-1}}{1+\dfrac{S}{D_n}\left(\dfrac{2}{W}\dfrac{\varphi_o - qV}{kT}\right)^{-1}} - \exp(-\alpha W). \tag{9}$$

For the *diffusion* component of the photoelectric quantum yield that takes into account surface recombination at the back surface of the CdTe layer, we can use the exact expression obtained for the p-layer in a p-n junction solar cell (Sze, 1981)

$$\eta_{\text{dif}} = \frac{\alpha L_n}{\alpha^2 L_n^2 - 1}\exp(-\alpha W)\times$$
$$\times\left\{\alpha L_n - \frac{\dfrac{S_b L_n}{D_n}\left[\cosh\left(\dfrac{d-W}{L_n}\right) - \exp(-\alpha(d-W))\right] + \sinh\left(\dfrac{d-W}{L_n}\right) + \alpha L_n \exp(-\alpha(d-W))}{\dfrac{S_b L_n}{D_n}\sinh\left(\dfrac{d-W}{L_n}\right) + \cosh\left(\dfrac{d-W}{L_n}\right)}\right\} \tag{10}$$

where d is the thickness of the CdTe absorber layer, S_b is the recombination velocity at the back surface of the CdTe layer.

The *total* quantum yield of photoelectric conversion in the CdTe absorber layer is the sum of the two components:

$$\eta_{\text{int}} = \eta_{\text{drift}} + \eta_{\text{dif}}. \tag{11}$$

2.4 Comparison of calculation results with experiment

Fig. 3(a) shows the computed spectra of the external quantum efficiency $\eta_{ext}(\lambda)$ illustrating the effect of the uncompensated donor impurities in a CdS/CdTe heterojunction. In this calculation the absorption curve $\alpha(\lambda)$ was used from publication Toshifumi et al., 1993, the values of S and τ_n were taken as 10^7 cm/s and 10^{-10} s, respectively.

It can be seen from Fig. 3(a) that, as the uncompensated donor concentration varies, the shape of the $\eta_{ext}(\lambda)$ curves undergo significant changes. If the $N_a - N_d$ decreases from 10^{17} cm^{-3} to 10^{13} cm^{-3}, the external quantum efficiency increases first and then decreases. The increase in $\eta_{ext}(\lambda)$ is due to the expansion of the depletion layer and, hence, to more efficient collection of photogenerated carriers from the bulk of the film. However, if the depletion

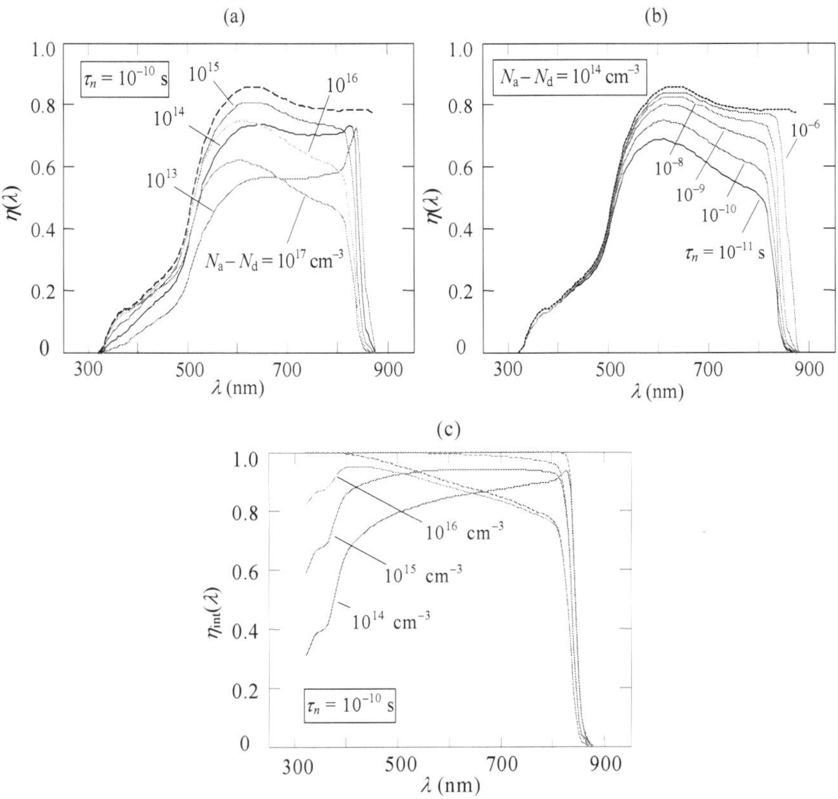

Fig. 3. (a) The external quantum efficiency spectra η_{ext} calculated using equations. (4),(9)-(11) at $\tau_n = 10^{-10}$ s and different uncompensated acceptor concentrations $N_a - N_d$, (b) at $N_a - N_d = 10^{16}$ cm^{-3} and different electron lifetimes τ_n and (c) the internal quantum efficiency spectra η_{int} calculated at different uncompensated acceptor concentrations and surface recombination velocity $S = 7\times10^7$ cm/s (solid lines) and $S = 0$ (dashed lines).

layer widens, the electric field becomes weaker which is favorable for the surface recombination. This effect is clearly demonstrated by the graphs for which N_a - N_d is 10^{14} cm^{-3} and 10^{13} cm^{-3}: the surface recombination causes a significant decay with decreasing the wavelength. Evidently, the decay in photosensitivity in the $\lambda < 400$ nm region is also caused by absorption by the CdS layer and the conducting glass substrate. Absorption in the CdS layer masks the influence of surface recombination, however, that can be revealed in the spectra of "internal" quantum efficiency η_{int} shown in Fig. 3(c). As seen, the recombination losses are significant for rather low concentration of uncompensated acceptors 10^{14}-10^{15} cm^{-3}. As can be seen from Fig. 3(b), the variation in the carrier lifetime τ_n has practically no influence on the spectral curves of the device (N_a - N_d = 10^{16} cm^{-3}) in the wavelength range $\lambda < 500$ nm. This is because in this spectral range, the depth of penetration of photons α^{-1} ($\alpha > 10^5$ cm^{-1}) is equal to or even smaller than the width of the space-charge region W.

Thus, only a small portion of the incident radiation is absorbed outside the space charge region and therefore the dependence of diffusion component of photocurrent on the electron lifetime is negligible.

On the other hand, in the wavelength range $\lambda > 500$ nm a considerable portion of the radiation is absorbed outside the space charge region and consequently as the electron lifetime increases the photoresponse also increases. When $N_a - N_d$ increases the effect of the electron lifetime increases and when $N_a - N_d$ decreases the effect of the electron lifetime becomes weaker so that at $N_a - N_d \leq 10^{14}$ cm^{-3} the photosensitivity is practically independent of the electron lifetime except the long-wavelength edge of the spectrum. It follows from Fig. 3(a) and (b) that, by varying the values of $N_a - N_d$ and τ, one can obtain the photosensitivity spectra of various shapes including those similar to the experimental curves shown in Fig. 1(a).

Fig. 4 illustrates the comparison of the calculated curves $\eta_{ext}(\lambda)$ using Eqs. (4),(9)-(11) with the measured spectrum taken at 300 K (we can not do it for other temperatures due to the lack of α values). The figure shows a quite good fit of the experimental data with the calculated values. Note that only the two adjustable parameters, the uncompensated acceptor concentration $N_a - N_d$ and electron lifetime τ_n, have been used to fit the calculation results with the experimental data and which were found to be 7×10^{16} cm^{-3} and 8×10^{-11} s, respectively. As can be seen from Fig. 3(c), the surface recombination velocity is not relevant at such high uncompensated acceptor concentration owing to the effect of a high electric field.

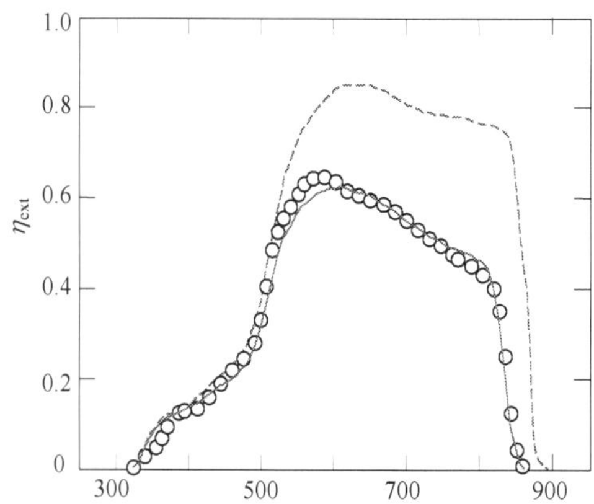

Fig. 4. Comparison of the measured (circles) and calculated (solid line) quantum efficiency spectrum η_{ext} using Eqs. (4), (9)-(11) at $N_a - N_d = 7\times10^{16}$ cm^{-3}, $\tau_n = 8\times10^{-11}$ s. The dashed line shows the spectrum of 100 % internal efficiency ($N_a - N_d = 10^{16}$ cm^{-3}, $\tau_n = 10^{-6}$ s, $d = 100$ μm).

3. Short-circuit current in CdS/CdTe heterostructure

The obtained expressions for quantum efficiency spectra can be used to calculate the short-circuit current density J_{sc} which is a quantitative solar cell characteristic reflecting the charge collection efficiency under radiation. The calculations will be done for AM1.5 solar radiation using Tables ISO 9845-1:1992 (Standard ISO, 1992). If Φ_i is the spectral radiation power density (in mW cm^{-2} nm^{-1}) and $h\nu$ is the photon energy (in eV), the spectral density of the incident photon flux is $\Phi_i/h\nu_i$ (in s^{-1}cm^{-2}), and then

$$J_{sc} = q\sum_{i}\eta_{int}(\lambda)\frac{\Phi_i(\lambda)}{h\nu_i}\Delta\lambda, \tag{12}$$

where $\Delta\lambda_i$ is the wavelength range between the neighboring values of λ_i (the photon energy $h\nu_i$) in the table and the summation is over the spectral range $\lambda < \lambda_g = hc/E_g$.

3.1 The drift component of the short-circuit current

Let us first consider the drift component of the short-circuit current density J_{drift} using Eq. (12). Fig. 5 shows the calculation results for J_{drift} depending on the space-charge region width W. In the calculations, it was accepted $\varphi_o - qV = 1$ eV, $S = 10^7$ cm/s (the maximum possible velocity of surface recombination) and $S = 0$. The Eq. (9) was used for $\eta_{int}(\lambda)$.
Important practical conclusions can be made from the results presented in the figure.
If $S = 0$, the short-circuit current gradually increases with widening of W and approaches a maximum value of J_{drift} = 28.7 mA/cm^2 at $W > 10$ μm (the value J_{drift} = 28.7 mA/cm^2 is obtained from equation (12) at $\eta_{drift} = 1$).

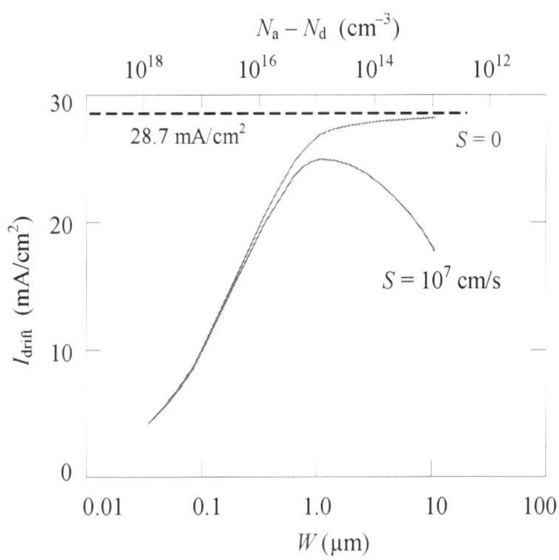

Fig. 5. Drift component of the short-circuit current density J_{drift} of a CdTe-based solar cell as a function of the space-charge region width W (the uncompensated acceptor concentration $N_a - N_d$) calculated for the surface recombination velocities $S = 10^7$ cm/s and $S = 0$.

Such result should be expected because the absorption coefficient α in CdTe steeply increases in a narrow range $h\nu \approx E_g$ and becomes higher than 10^4 cm^{-1} at $h\nu > E_g$. As a result, the penetration depth of photons α^{-1} is less than ~ 1 μm throughout the entire spectral range and in the absence of surface recombination, all photogenerated electron-hole pairs are separated by the electric field acting in the space-charge region.
Surface recombination decreases the short-circuit current only in the case if the electric field in the space-charge region is not strong enough. The electric field decreases as the space-charge region widens, i.e. when the uncompensated acceptor concentration $N_a - N_d$ decreases. One can see from Fig. 5 that the influence of surface recombination at $N_a - N_d = 10^{14}$-10^{15} cm^{-3} is quite significant. However, as $N_a - N_d$ increases and consequently the electric field strength becomes stronger, the influence of surface recombination becomes

weaker, and at $N_a - N_d \geq 10^{16}$ cm^{-3} the effect is virtually eliminated. However in this case, the short-circuit current density decreases with increasing $N_a - N_d$ because a significant portion of radiation is absorbed outside the space-charge region.

It should be noted that the fabrication of the CdTe/CdS heterostructure is typically completed by a post-deposition heat treatment. The annealing enables grain growth, reduces defect density in the films, and promotes the interdiffusion between the CdTe and CdS layers. As a result, the CdS-CdTe interface becomes alloyed into the $CdTe_xS_{1-x}$-CdS_yTe_{1-y} interface, and the surface recombination velocity is probably reduced to some extent (Compaan et al, 1999).

3.2 The diffusion component of the short-circuit current

In order to provide the losses caused by recombination at the CdS-CdTe interface and in the space-charge region at a minimum we will accept in this section $N_a - N_d \geq 10^{17}$ cm^{-3}. On the other hand, to make the diffusion component of the short-circuit current J_{dif} as large as possible, we will set $\tau_n = 3\times10^{-6}$ s, i.e. the maximum possible value of the electron lifetime in CdTe. Fig. 6(a) shows the calculation results of J_{dif} (using Eqs. (10) and (12)) versus the CdTe layer thickness d for the recombination velocity at the back surface $S = 10^7$ cm/s and $S = 0$ (the thickness of the neutral part of the film is $d - W$).

One can see from Fig. 6(a) that for a thin CdTe layer (few microns) the diffusion component of the short-circuit current is rather small. In the case $S_b = 0$, the total charge collection in the neutral part (it corresponds to $J_{dif} = 17.8$ mA/cm^2 at $\eta_{dif} = 1$) is observed at $d = 15$-20 μm. To reach the total charge collection in the case $S_b = 10^7$ cm/s, the CdTe thickness should be 50 μm or larger. Bearing in mind that the thickness of a CdTe layer is typically between 2 and 10 μm, for $d = 10$, 5 and 2 μm the losses of the diffusion component of the short-circuit current are 5, 9 and 19%, respectively. The CdTe layer thickness can be reduced by shortening the electron lifetime τ_n and hence the electron diffusion length $L_n = (\tau_n D_n)^{1/2}$. However one does not forget that it leads to a significant decrease in the value of the diffusion current itself. This is illustrated in Fig. 6(b), where the curve $J_{dif}(\tau_n)$ is plotted for a thick CdTe layer (50 μm) taking into account the surface recombination velocity $S_b = 10^7$ cm/s. As it can be seen, shortening of the electron lifetime below 10^{-7}-10^{-6} s results in a significant lowering of the diffusion component of the short-circuit current density. Thus, when the space-charge region width is narrow, so that recombination losses at the CdS-CdTe interface can be neglected (as seen from Fig. 5, at $N_a - N_d > 10^{16}$-10^{17} cm^{-3}), the conditions for generation of the high diffusion component of the short-circuit current are $d > 25$-30 μm and $\tau_n > 10^{-7}$-10^{-6} s.

In connection with the foregoing the question arises why for total charge collection the thickness of the CdTe absorber layer d should amount to several tens of micrometers. The value d is commonly considered to be in excess of the effective penetration depth of the radiation into the CdTe absorber layer in the intrinsic absorption region of the semiconductor. As mentioned above, as soon as the photon energy exceeds the band gap of CdTe, the absorption coefficient α becomes higher than 10^4 cm^{-1}, i.e. the effective penetration depth of radiation α^{-1} becomes less than 10^{-4} cm = 1 μm. With this reasoning, the absorber layer thickness is usually chosen at a few microns. However, all that one does not take into the account, is that the carriers arisen outside the space-charge region, diffuse into the neutral part of the CdTe layer penetrating deeper into the material. Carriers reached the back surface of the layer, recombine and do not contribute to the photocurrent. Losses

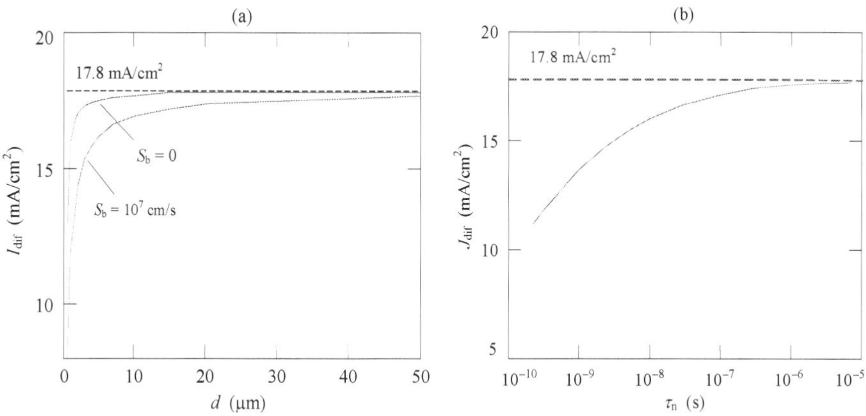

Fig. 6. Diffusion component of the short-circuit current density J_{dif} as a function of the CdTe layer thickness d calculated at the uncompensated acceptor concentration $N_a - N_d = 10^{17}$ cm^{-3}, the electron lifetime $\tau_n = 3\times10^{-6}$ s and surface recombination velocity $S_b = 10^7$ cm/s and $S_b = 0$ (a) and the dependence of the diffusion current density J_{dif} on the electron lifetime for the CdTe layer thickness $d = 50$ μm and recombination velocity at the back surface $S_b = 10^7$ cm/s (b).

caused by the insufficient thickness of the CdTe layer should be considered taking into account this process.

Consider first the spatial distribution of excess electrons in the neutral region governed by the continuity equation with two boundary conditions. At the depletion layer edge, the excess electron density Δn can be assumed equal zero (due to electric field in the depletion region), i.e.

$$\Delta n = 0 \text{ at } x = W. \qquad (13)$$

At the back surface of the CdTe layer we have surface recombination with a velocity S_b:

$$S_b \Delta n = -D_n \frac{d\Delta n}{dx} \text{ at } x = d, \qquad (14)$$

where d is the thickness of the CdTe layer.

Using these boundary conditions, the exact solution of the continuity equation is (Sze, 1981):

$$\Delta n = T(\lambda)N_o(\lambda)\frac{\alpha\tau_n}{\alpha^2L_n^2-1}\exp[-\alpha W]\left\{\cosh\left(\frac{x-W}{L_n}\right)-\exp[-\alpha(x-W)]-\right.$$
$$\left.-\frac{\frac{S_bL_n}{D_n}\left[\cosh\left(\frac{d-W}{L_n}\right)-\exp[-\alpha(d-W)]\right]+\sinh\left(\frac{d-W}{L_n}\right)+\alpha L_n\exp[-\alpha(d-W)]}{\frac{S_bL_n}{D_n}\sinh\left(\frac{x-W}{L_n}\right)+\cosh\left(\frac{d-W}{L_n}\right)}\times\sinh\left(\frac{x-W}{L_n}\right)\right\} \qquad (15)$$

where $T(\lambda)$ is the optical transmittance of the glass/TCO/CdS, which takes into account reflection from the front surface and absorption in the TCO and CdS layers, N_o is the number of incident photons per unit time, area, and bandwidth ($cm^{-2}s^{-1}nm^{-1}$), $L_n = (\tau_n D_n)^{1/2}$ is the electron diffusion length, τ_n is the electron lifetime, and D_n is the electron diffusion coefficient related to the electron mobility μ_n through the Einstein relation: $qD_n/kT = \mu_n$.

Fig. 7 shows the electron distribution calculated by Eq. (15) for different CdTe layer thicknesses. The calculations have been carried out at $\alpha = 10^4$ cm^{-1}, $S_b = 7\times10^7$ cm/s, $\mu_n = 500$ $cm^2/(V\cdot s)$ and typical values $\tau_n = 10^{-9}$ s and $N_a - N_d = 10^{16}$ cm^{-3} (Sites & Xiaoxiang, 1996). As it is seen from Fig. 7, even for the CdTe layer thickness of 10 μm, recombination at back surface leads to a remarkable decrease in the electron concentration. If the layer thickness is reduced, the effect significantly enhances, so that at d = 1-2 μm, surface recombination "kills" most of the photo-generated electrons. Thus, the photo-generated electrons at 10^{-9} s are involved in recombination far away from the effective penetration depth of radiation (~ 1 μm). Evidently, the influence of this process enhances as the electron lifetime increases, because the non-equilibrium electrons penetrate deeper into the CdTe layer due to increase of the diffusion length. Calculation using Eq. (15) shows that if the layer thickness is large (~ 50 μm), the non-equilibrium electron concentration reduces 2 times from its maximum value at a distance about 8 μm at $\tau_n = 10^{-8}$ s, 20 μm at $\tau_n = 10^{-7}$ s, 32 μm at $\tau_n = 10^{-6}$ s.

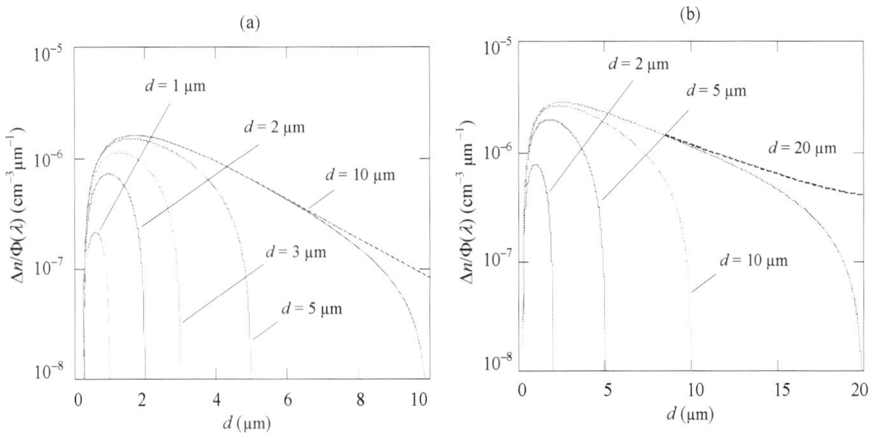

Fig. 7. Electron distribution in the CdTe layer at different its thickness d calculated at the electron lifetime $\tau_n = 10^{-9}$ s (a) and $\tau_n = 10^{-8}$ s (b). The dashed lines show the electron distribution for d = 10 and 20 μm if recombination at the back surface is not taken into account.

3.3 The density of total short-circuit current

It follows from the above that the processes of the photocurrent formation within the space-charge region and in the neutral part of the CdTe film are interrelated. Fig. 8 shows the *total* short-circuit current J_{sc} (the sum of the drift and diffusion components) calculated for different parameters of the CdTe layer, i.e. the uncompensated acceptor concentration, minority carrier lifetime and layer thickness. As the space-charge region is narrow (i.e., $N_a - N_d$ is high), a considerable portion of radiation is absorbed *outside* the space-charge region. One can see that when the film thickness and electron diffusion length are large enough (the top curve in Fig. 8(a) for d = 100 μm, $\tau_n > 10^{-6}$ s), practically the total charge collection takes place and the density of short-circuit current J_{sc} reaches its maximum value of 28.7 mA/cm^2 (note, the record experimental value of J_{sc} is 26.7 mA/cm^2 (Holliday et al, 1998)). However if the space-charge region is too wide ($N_a - N_d < 10^{16}$-10^{17} cm^{-3}) the electric field becomes weak and the short-circuit current is reduced due to recombination at the front surface.

For d = 10 μm, the shape of the curve J_{sc} versus $N_a - N_d$ is similar to that for d = 100 μm but the saturation of the photocurrent density is observed at a smaller value of J_{sc}. A significant lowering of J_{sc} occurs after further thinning of the CdTe film and, moreover, for d = 5 and 3 μm, the short-circuit current even decreases with increasing $N_a - N_d$ due to incomplete charge collection in the neutral part of the CdTe film.

It is interesting to examine quantitatively how the total short-circuit current varies when the electron lifetime is shorter than 10^{-6} s. This is an actual condition because the carrier lifetimes in thin-film CdTe diodes can be as short as 10^{-9}-10^{-10} s and even smaller (Sites & Pan, 2007).

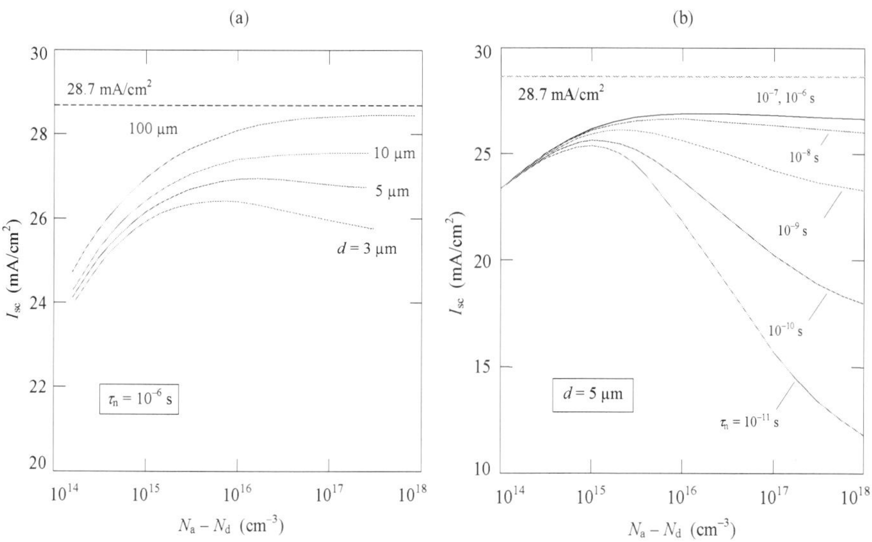

Fig. 8. Total short-circuit current density J_{sc} of a CdTe-based solar cell as a function of the uncompensated acceptor concentration $N_a - N_d$ calculated at the electron lifetime $\tau_n = 10^{-6}$ s for different CdTe layer thicknesses d (a) and at the thickness d = 5 μm for different τ_n (b).

Fig. 5(b) shows the calculation results of the total short-circuit current density J_{sc} versus the concentration of uncompensated acceptors $N_a - N_d$ for different electron lifetimes τ_n. Calculations have been carried out for the CdTe film thickness d = 5 μm which is often used in the fabrication of CdTe-based solar cells (Phillips et al., 1996; Bonnet, 2001; Demtsu & Sites, 2005; Sites & Pan, 2007). As it can be seen, at $\tau_n \geq 10^{-8}$ s the short-circuit current density is 26-27 mA/cm^2 when $N_a - N_d > 10^{16}$ cm^{-3}. For shorter electron lifetime, J_{sc} peaks in the $N_a - N_d$ range (1-3)×10^{15} cm^{-3}. As $N_a - N_d$ is in excess of this concentration, the short-circuit current decreases since the drift component of the photocurrent reduces. In the range of the uncompensated acceptor concentration $N_a - N_d <$ (1-3)×10^{15} cm^{-3}, the short-circuit current density also decreases, but because of recombination at the front surface of the CdTe layer. Anticipating things, it should be noted, that at N_a - N_d < 10^{15} cm^{-3}, recombination in the space-charge region becomes also significant (see Fig. 9). Thus, in order to reach the short-circuit current density 25-26 mA/cm^2 when the electron lifetime τ_n is shorter than 10^{-8} s, the uncompensated acceptor concentration N_a - N_d should be equal to (1-3)×10^{15} cm^{-3} (rather than N_a - N_d > 10^{16} cm^{-3} as in the case of $\tau_n \geq 10^{-8}$ s).

4. Recombination losses in the space-charge region

In analyzing the photoelectric processes in the CdS/CdTe solar cell we ignored the recombination losses (capture of carriers) in the space-charge region. This assumption is based on the following considerations.

The mean distances that electron and hole travels during their lifetimes along the electric field without recombination or capture by the centers within the semiconductor band gap, i.e. the electron drift length λ_n and hole drift length λ_p, are determined by expressions

$$\lambda_n = \mu_n E \tau_{no}, \tag{16}$$

$$\lambda_p = \mu_p E \tau_{po}, \tag{17}$$

where E is the electric-field strength, μ_n and μ_p are the electron and hole mobilities, respectively.

In the case of uniform field (E = const), the charge collection efficiency is expressed by the well-known Hecht equation (Eizen, 1992; Baldazzi et al., 1993):

$$\eta_c = \frac{\lambda_n}{W}\left[1-\exp\left(-\frac{W-x}{\lambda_n}\right)\right]+\frac{\lambda_p}{W}\left[1-\exp\left(-\frac{x}{\lambda_p}\right)\right]. \tag{18}$$

In a diode structure, the problem is complicated due to nonuniformity of the electric field in the space-charge region. However, due to the fact that the electric field strength decreases linearly from the surface to the bulk of the semiconductor, the field nonuniformity can be reduced to the substitution of E in Eqs. (16) and (17) by its average values $E_{(0,x)}$ and $E_{(x,W)}$ in the portion (0, x) for electrons and in the portion (x, W) for holes, respectively:

$$E_{(x,W)} = \frac{(\varphi_o - eV)}{eW}\left(1-\frac{x}{W}\right), \tag{19}$$

$$E_{(0,x)} = \frac{(\varphi_o - eV)}{eW}\left(2-\frac{x}{W}\right). \tag{20}$$

Thus, with account made for this, the Hecht equation for the space-charge region of CdS/CdTe heterostructure takes the form

$$\eta_c = \frac{\mu_p E_{(x,W)}\tau_{po}}{W}\left[1-\exp\left(-\frac{W-x}{\mu_p E_{(x,W)}\tau_{po}}\right)\right]+\frac{\mu_n E_{(0,x)}\tau_{no}}{W}\left[1-\exp\left(-\frac{x}{\mu_n E_{(0,x)}\tau_{no}}\right)\right]. \tag{21}$$

Fig. 9(a) shows the curves of charge-collection efficiency $\eta_c(x)$ computed by Eq. (21) for the concentration of uncompensated acceptors 3×10^{16} cm^{-3} and different carrier lifetimes $\tau = \tau_{no} = \tau_{po}$. It is seen that for the lifetime 10^{-11} s the effect of losses in the space-charge region is remarkable but for $\tau \geq 10^{-10}$ s it is insignificant (μ_n and μ_n were taken equal to 500 and 60 cm^2/(V·s), respectively). For larger carrier lifetimes the recombination losses can be neglected at lower values $N_a - N_d$.

Thus, the recombination losses in the space charge-region depend on the concentration of uncompensated acceptors $N_a - N_d$ and carrier lifetime τ in a complicated manner. It is also seen from Fig. 9(a) that the charge collection efficiency η_c is lowest at the interface CdS-CdTe ($x = 0$). An explanation of this lies in the fact that the product $\tau_{no}\mu_n$ for electrons in CdTe is order of magnitude greater than that for holes. With account made for this, Fig. 9(b) shows the dependences of charge-collection efficiency on $N_a - N_d$ calculated at different carrier lifetimes for the "weakest" place of the space-charge region concerning charge collection of photogenerated carriers, i.e. at the cross section $x = 0$. From the results presented in Fig. 9(b), it follows that at the carrier lifetime $\tau \geq 10^{-8}$ s the recombination losses can be neglected at the uncompensated acceptor concentration $N_a - N_d \geq 10^{14}$ cm^{-3} while at $\tau = 10^{-10}$-10^{-11} s it is possible if $N_a - N_d$ is in excess of 10^{16} cm^{-3}.

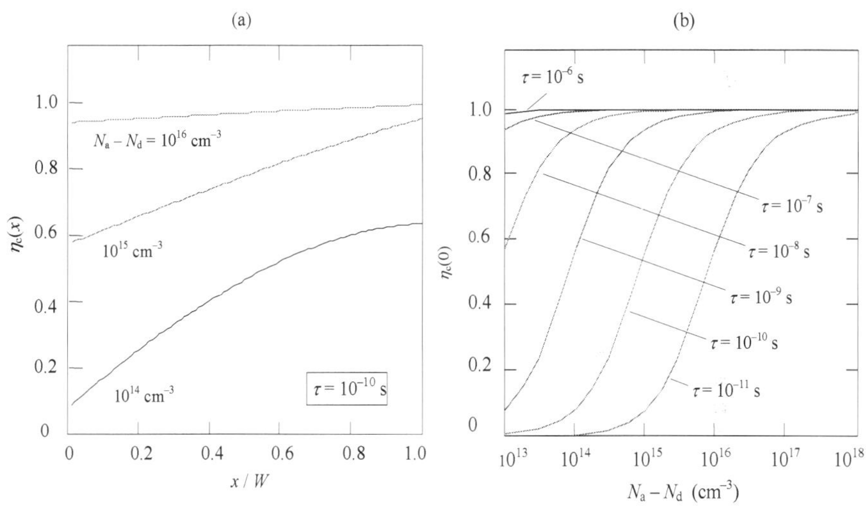

Fig. 9. (a) The coordinate dependences of the charge-collection efficiency $\eta_c(x)$ calculated for the uncompensated acceptor concentrations $N_a - N_d = 3\times10^{16}$ cm^{-3} and different carrier lifetimes τ. (b) The charge-collection efficiency η_c at the interface CdS-CdTe ($x = 0$) as a function of the uncompensated acceptor concentration $N_a - N_d$ calculated for different carrier lifetimes τ.

5. Open-circuit voltage, fill factor and efficiency of thin-film CdS/CdTe solar cell

In this section, we investigate the dependences of the open-circuit voltage, fill factor and efficiency of a CdS/CdTe solar cell on the resistivity of the CdTe absorber layer and carrier

lifetime with the aim to optimize these parameters and hence to improve the solar cell efficiency. The open-circuit voltage and fill factor are controlled by the magnitude of the forward current. Therefore the *I-V* characteristic of the device is analyzed which is known to originate primarily by recombination in the space charge region of the CdTe absorber layer.
The *I-V* characteristic of CdS/CdTe solar cells is most commonly described by the semi-empirical formulae which consists the so-called "ideality" factor and is valid for some cases. Contrary to usual practice, in our calculations of the current in a device, we use the recombination-generation Sah-Noyce-Shockley theory developed for p-n junction (Sah et al., 1957) and adopted to CdS/CdTe heterostructure (Kosyachenko et al., 2005) and supplemented with over-barrier diffusion flow of electrons at higher voltages. This theory takes into account the evolution of the *I-V* characteristic of CdS/CdTe solar cell when the parameters of the CdTe absorber layer vary and, therefore, reflects adequately the real processes in the device.

5.1 *I-V* characteristic of CdS/CdTe heterostructure

The open-circuit voltage, fill factor and efficiency of a solar cell is determined from the *I-V* characteristic under illumination which can be presented as

$$J(V) = J_{\mathrm{d}}(V) - J_{\mathrm{ph}}, \tag{22}$$

where $J_d(V)$ is the dark current density and J_{ph} is the photocurrent density.
The dark current density in the so-called "ideal" solar cell is described by the Shockley equation

$$J_{\mathrm{d}}(V) = J_{\mathrm{s}}\left[\exp\left(\frac{qV}{kT}\right) - 1\right], \tag{23}$$

where J_s is the saturation current density which is the voltage independent reverse current as qV is higher than few kT.
An actual *I-V* characteristic of CdS/CdTe solar cells differs from Eq. (23). In many cases, a forward current can be described by formula similar to Eq. (23) by introducing an exponent index qV/AkT, where A is the "ideality" factor lied in the range 1 to 2. Sometimes, a close correlation between theory and experiment can be attained by adding the recombination component $I_0[\exp(qV/2kT) - 1]$ to the dark current in Eq. (23) (I_0 is a new coefficient). Our measurements show, however, that such generalizations of Eq. (23) does not cover the observed variety of *I-V* characteristics of the CdS/CdTe solar cells. The measured voltage dependences of the forward current are not always exponential and the saturation of the reverse current is never observed. On the other hand, our measurements of *I-V* characteristics of CdS/CdTe heterostructures and their evolution with the temperature variation are governed by the generation-recombination Sah-Noyce-Shockley theory (Sah al., 1957). According to this theory, the dependence $I \sim \exp(qV/AkT)$ at $n \approx 2$ takes place only in the case where the generation-recombination level is placed near the middle of the band gap. If the level moves away from the midgap the coefficient A becomes close to 1 but only at low forward voltage. If the voltage elevates the *I-V* characteristic modified in the dependence where $n \approx 2$ and at higher voltages the dependence I on V becomes even weaker (Sah et al., 1957; Kosyachenko et al., 2003). At higher forward currents, it is also necessary to take into account the voltage drop on the series resistance R_s of the bulk part of the CdTe layer by replacing the voltage V in the discussed expressions with $V - I{\cdot}R_s$.

The Sah-Noyce-Shockley theory supposes that the generation-recombination rate in the section x of the space-charge region is determined by expression (Sah et al., 1957)

$$U(x,V) = \frac{n(x,V)p(x,V) - n_i^2}{\tau_{po}\left[n(x,V)+n_1\right] + \tau_{no}\left[p(x,V)+p_1\right]}, \tag{24}$$

where $n(x,V)$ and $p(x,V)$ are the carrier concentrations in the conduction and valence bands, n_i is the intrinsic carrier concentration. The values n_1 and p_1 are determined by the energy spacing between the top of the valence band and the generation-recombination level E_t, i.e. $p_1 = N_\upsilon\exp(-E_t/kT)$ and $n_1 = N_c\exp[-(E_g - E_t)/kT]$, where $N_c = 2(m_n kT/2\pi\hbar^2)^{3/2}$ and $N_v = 2(m_p kT/2\pi\hbar^2)^{3/2}$ are the effective density of states in the conduction and valence bands, m_n and m_p are the effective masses of electrons and holes, τ_{no} and τ_{po} are the effective lifetime of electrons and holes in the depletion region, respectively.
The recombination current under forward bias and the generation current under reverse bias are found by integration of $U(x, V)$ throughout the entire depletion layer:

$$J_{gr} = q\int_0^W U(x,V)dx\,, \tag{25}$$

where the expressions for the electron and hole concentrations have the forms (Kosyachenko et al., 2003):

$$p(x,V) = N_c \exp\left[-\frac{\Delta\mu + \varphi(x,V)}{kT}\right], \tag{26}$$

$$n(x,V) = N_\upsilon \exp\left[-\frac{E_g - \Delta\mu - \varphi(x,V) - qV}{kT}\right]. \tag{27}$$

Here $\Delta\mu$ is the energy spacing between the Fermi level and the top of the valence band in the bulk of the CdTe layer, $\varphi(x,V)$ is the potential energy of hole in the space-charge region.
Over-barrier (diffusion) carrier flow in the CdS/CdTe heterostructure is restricted by high barriers for both majority carriers (holes) and minority carriers (electrons) (Fig. 2). For transferring holes from CdTe to CdS, the barrier height in equilibrium ($V = 0$) is somewhat lower than $E_{g\,CdS} - (\Delta\mu + \Delta\mu_{CdS})$, where $E_{g\,CdS} = 2.42$ eV is the band gap of CdS and $\Delta\mu_{CdS}$ is the energy spacing between the Fermi level and the bottom of the conduction band of CdS, $\Delta\mu$ is the Fermi level energy in the bulk of CdTe equal to $kT\ln(N_v/p)$, p is the hole concentration which depends on the resistivity of the material. An energy barrier impeding electron transfer from CdS to CdTe is also high but is equal to $E_{g\,CdTe} - (\Delta\mu + \Delta\mu_{CdS})$ at $V = 0$. Owing to high barriers for electrons and holes, under low and moderate forward voltages the dominant charge transport mechanism is recombination in the space-charge region. However, as qV nears φ_o, the over-barrier currents become comparable and even higher than the recombination current due to much stronger dependence on V. Since in CdS/CdTe junction the barrier for holes is considerably higher than that for electrons, the electron component dominates the over-barrier current. Obviously, the electron flow current is analogous to that occurring in a p-n junction and one can write for the over-barrier current density (Sze, 1981):

$$J_{\mathrm{n}} = q\frac{n_{\mathrm{p}}L_{\mathrm{n}}}{\tau_{\mathrm{n}}}\left[\exp\left(\frac{qV}{kT}\right) - 1\right], \tag{28}$$

where $n_p = N_c \exp[-(E_g - \Delta\mu)/kT]$ is the concentration of electrons in the p-CdTe layer, τ_n and $L_n = (\tau_n D_n)^{1/2}$ are the electron lifetime and diffusion length, respectively (D_n is the diffusion coefficient of electrons).

Thus, according to the above discussion, the dark current density in CdS/CdTe heterostructure $J_d(V)$ is the sum of the generation-recombination and diffusion components:

$$J_{\mathrm{d}}(V) = J_{\mathrm{gr}}(V) + J_{\mathrm{n}}(V)\,. \tag{29}$$

5.2 Comparison with the experimental data

The current-voltage characteristics of CdS/CdTe solar cells depend first of all on the resistivity of the CdTe absorber layer due to the voltage drop across the series resistance of the bulk part of the CdTe film R_s (Fig. 10(a)). The value of R_s can be found from the voltage dependence of the differential resistance R_{dif} of a diode structure under forward bias. Fig. 10 shows the results of measurements taken for two "extreme" cases: the samples No 1 and 2 are examples of the CdS/CdTe solar cells with low resistivity (20 Ω·cm) and high resistivity of the CdTe film (4×10^7 Ω·cm), respectively. One can see that, in the region of low voltage, the R_{dif} values decrease with V by a few orders of magnitude. However, at $V > 0.5$-0.6 V for sample No 1 and $V > 0.8$-0.9 V for sample No 2, R_{dif} reaches saturation values which are obviously the series resistances of the bulk region of the film R_s.

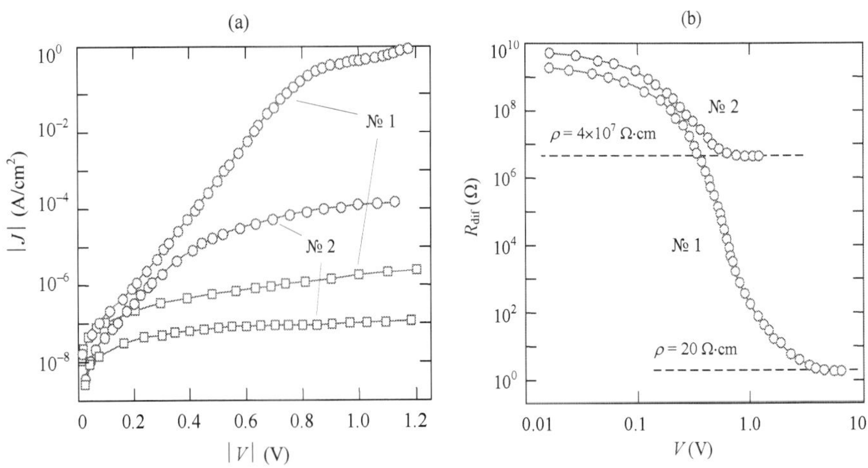

Fig. 10. *I-V* characteristics (a) and dependences of differential resistances R_{dif} on forward voltage (b) for two solar cells with different resistivities of CdTe layers: 20 and 4×10^7 Ω·cm (300 K).

Because the value of R_s for a sample No 1 is low, the presence of R_s does not affect the shape of the diode *I-V* characteristic. In contrast, the resistivity of the CdTe film for a sample No 2 is ~ 6 orders higher, therefore at moderate forward currents ($J > 10^{-6}$ A/cm²), the

experimental points deviate from the exponential dependence which is strictly obeyed for sample No 1 over 6 orders of magnitude.
The experimental results presented in Fig. 11 reflect the common feature of the *I-V* characteristic of a thin-film CdS/CdTe heterostructure (sample No 1). The results obtained for this sample allow interpreting them without complications caused by the presence of the series resistance R_s. Nevertheless, in this case too, the forward *I-V* characteristic reveals some features which are especially pronounced. As one can see, under forward bias, there is an extended portion of the curve ($0.1 < V < 0.8$ V) where the dependence $I \sim \exp(qV/AkT)$ holds for $A = 1.92$. At higher voltages, the deviation from the exponential dependence toward lower currents is observed. It should be emphasized that this deviation is not caused by the voltage drop across the series resistance of the neutral part of the CdTe absorber layer R_s (which is too low in this case). If the voltage elevates still further (> 1 V), a much steeper increase of forward current is observed.
Analysis shows that all of varieties of the thin-film *I-V* characteristics are explained in the frame of mechanism involving the generation-recombination in the space-charge region in a wide range of moderate voltages completed by the over-barrier diffusion current at higher voltage.
The results of comparison between the measured *I-V* characteristic of the thin-film CdS/CdTe heterostructure (circles) and that calculated using Eqs. (25), (28) and (29) (lines) are shown in Fig. 11.

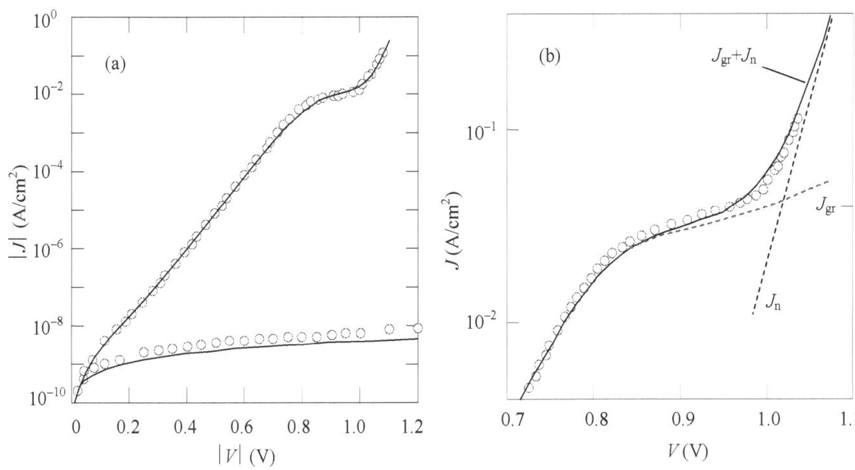

Fig. 11. (a) *I-V* characteristic of thin-film CdS/CdTe heterostructure. The circles and solid lines show the experimental and calculated results, respectively. (b) Comparison of the calculated and measured dependences in the range of high forward currents (J_{gr} and J_n are the recombination and diffusion components, respectively).

To agree the calculated results with experiment, the effective lifetimes of electrons and holes in the space-charge region were taken $\tau_{no} = \tau_{po} = \tau = 1.2\times10^{-10}$ s (τ determines the value of current but does not affect the shape of curve). The ionization energy E_t was accepted to be 0.73 eV as the most effective recombination center (the value E_t determines the rectifying coefficient of the diode structure), the barrier height φ_o and the uncompensated acceptor concentration $N_a - N_d$ were taken 1.13 eV and 10^{17} cm^{-3}, respectively. One can see that the *I-V* characteristic calculated in accordance with the above theory (lines) are in good agreement with experiment both for the forward and reverse connection (circles).

Attention is drawn to the fact that the effective carrier lifetime in the space charge region $\tau = (\tau_{n0}\tau_{p0})^{1/2}$ was taken equal to 1.5×10^{-8} s whereas the electron lifetime τ_n in the crystals is in the range of 10^{-7} s or longer (Acrorad Co, Ltd., 2009). Such a significant difference between τ and τ_n appears reasonable since τ_n is proportional to $1/N_t f$, where N_t is the concentration of recombination centers and f is the probability that a center is empty. Both of the values τ_{n0} and τ_{p0} in the Sah-Noyce-Shockley theory are proportional to $1/N_t$. At the same time, since the probability f in the bulk part of the diode structure can be much less than unity, the electron lifetime τ_n can be far in excess of the effective carrier lifetime τ in the space-charge region.

5.3 Dependences of open-circuit voltage, fill factor and efficiency on the parameters of thin-film CdS/CdTe solar cell

The open-circuit voltage V_{oc}, fill factor *FF* and efficiency η of a solar cell is determined from the *I-V* characteristic under illumination which can be presented as

$$J(V) = J_d(V) - J_{ph}, \tag{30}$$

where $J_d(V)$ and J_{ph} are the dark current and photocurrent densities, respectively.

Calculations carried out for the case of a film thickness $d = 5$ μm which is often used in the fabrication of CdTe-based solar cells and a typical carrier lifetime of 10^{-9}-10^{-10} s (Sites et al., 2007) in thin-film CdTe/CdS solar cells show that the maximum value of $J_{sc} \approx 25$-26 mA/cm^2 (Fig. 8(b)) is obtained when the concentration of noncompensated acceptors is $N_a - N_d = 10^{15}$-10^{16} cm^{-3}. Therefore, in the following calculations a photocurrent density $J_{sc} \approx 26$ mA/cm^2 will be used.

In Fig. 12(a) the calculated *I-V* characteristics of the CdS/CdTe heterojunction under illumination are shown. The curves have been calculated by Eq. (30) using Eqs. (25), (28), (29) for $\tau = \tau_{no} = \tau_{po} = 10^{-9}$ s, $N_a - N_d = 10^{16}$ cm^{-3} and various resistivities of the p-CdTe layer. As is seen, an increase in the resistivity ρ of the CdTe layer leads to decreasing the open-circuit voltage V_{oc}. As ρ varies, $\Delta\mu$ also varies affecting the value of the recombination current, and especially the over-barrier current. The shape of the curves also changes affecting the fill factor *FF* which can be found as the ratio of the maximum electrical power to the product $J_{sc}V_{oc}$ (Fig. 12(a)). Evidently, the carrier lifetime τ_n also influences the *I-V* characteristic of the heterojunction under illumination. In what follows the dependences of these characteristics on ρ and τ are analyzed.

The dependences of open-circuit voltage, fill factor and efficiency on the carrier lifetime calculated at different resistivities of the CdTe absorber layer are shown in Fig. 13. As is seen, V_{oc} considerably increases with lowering ρ and increasing τ. In the most commonly encountered case, as $\tau = 10^{-10}$-10^{-9} s, the values of $V_{oc} = 0.8$-0.85 V are far from the maximum possible values of 1.15-1.2 V, which are reached on the curve for $\rho = 0.1$ Ω·cm and $\tau > 10^{-8}$. A remarkable increase of V_{oc} is observed when ρ decreases from 10^3 to 0.1 Ω·cm.

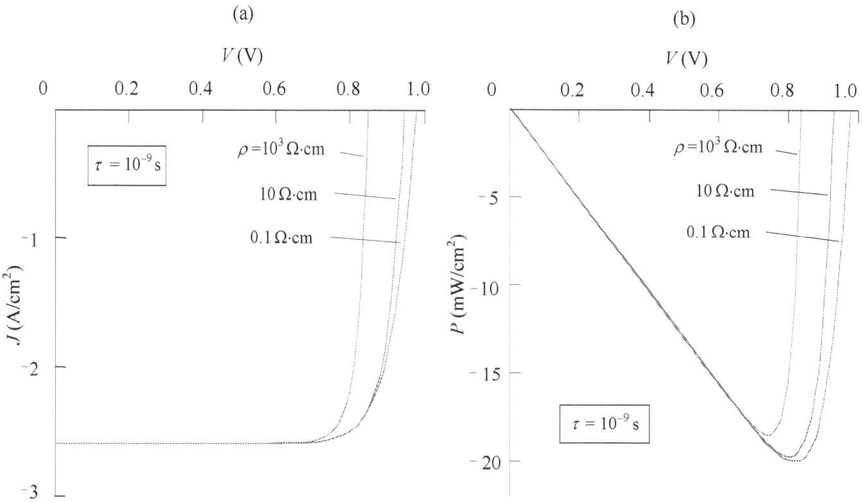

Fig. 12. *I*-*V* characteristics (a) and voltage dependence of the output power (b) of CdS/CdTe heterojunction under AM1.5 solar irradiation calculated for J_{sc} = 26 mA/cm^2, $\tau = 10^{-9}$ s and different resistivities ρ of the CdTe absorber layer.

Fig. 13(b) illustrates the dependence of the fill factor $FF = P_{max}/(J_{sc}\cdot V_{oc})$ on the parameters of the CdS/CdTe heterostructure within the same range of ρ and τ (P_{max} is the maximal output power found from the illuminated *I*-*V* characteristic). As it is seen, the fill factor increases from 0.81-0.82 to 0.88-0.90 with the increase of the carrier lifetime from 10^{-11} to 10^{-7} s. The non-monotonic dependence of *FF* on τ for ρ = 0.1 Ω·cm is caused by the features of the *I*-*V* characteristics of the CdS/CdTe heterostructures, namely, the deviation of the *I*-*V* dependence from exponential law when the resistivity of CdTe layer is low (see Fig. 11, $V > 0.8$ V).

Finally, the dependences of the efficiency $\eta = P_{out} / P_{irr}$ on the carrier lifetime τ_n calculated for various resistivities of the CdTe absorber layer are shown in Fig. 13(c), where P_{irr} is the AM 1.5 solar radiation power over the entire spectral range which is equal to 100 mW/cm^2 (Standard IOS, 1992). As it is seen, the value of η remarkably increases from 15-16% to 21-27.5% when τ and ρ changes within the indicated limits. For $\tau = 10^{-10}$-10^{-9} s, the efficiency lies near 17-19% and the enhancement of η by lowering the resistivity of CdTe layer is 0.5-1.5% (the shaded area in Fig. 13(c)).

Thus, assuming $\tau = 10^{-10}$-10^{-9} s, the calculated results turn out to be quite close to the experimental efficiencies of the best samples of thin-film CdS/CdTe solar cells (16-17%).

The conclusion followed from the results presented in Fig. 13(c) is that in the case of a CdS/CdTe solar cell with CdTe thickness 5 μm, enhancement of the efficiency from 16-17% to 27-28% is possible if the carrier lifetime increases to $\tau \geq 10^{-6}$ s and the resistivity of CdTe reduces to $\rho \approx 0.1$ Ω·cm. Approaching the theoretical limit η = 27-28% requires also an increase in the short-circuit current density. As it is follows from section 3.3, the latter is possible for the thickness of the CdTe absorber layer of 20-30 μm and more.

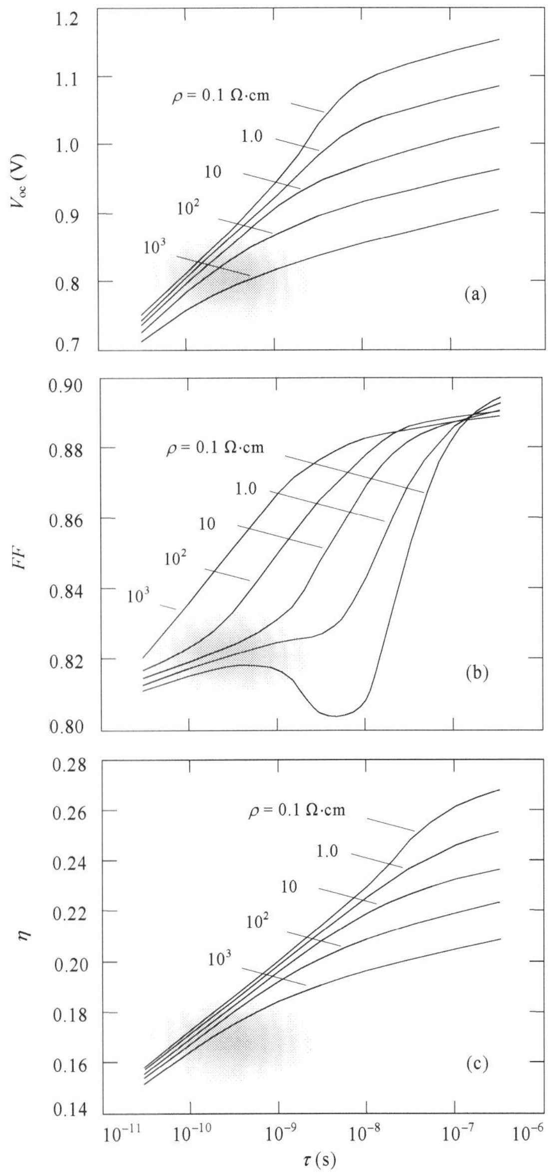

Fig. 13. Dependences of the open-circuit voltage V_{oc} (a), fill factor *FF* (b) and efficiency η (c) of CdS/CdTe heterojunction on the carrier lifetime τ calculated by Eq. (30) using Eqs. (24)-(29) for various resistivities ρ of the CdTe layer. The experimental results achieved for the best samples of thin-film CdS/CdTe solar cells are shown by shading.

6. Conclusion

The findings of this paper give further insight into the problems and ascertain some requirements imposed on the CdTe absorber layer parameters in a CdTe/CdS solar cell, which in our opinion could be taken into account in the technology of fabrication of solar cells.

The model taking into account the drift and diffusion photocurrent components with regard to recombination losses in the space-charge region, at the CdS-CdTe interface and the back surface of the CdTe layer allows us to obtain a good agreement with the measured quantum efficiency spectra by varying the uncompensated impurity concentration, carrier lifetime and surface recombination velocity. Calculations of short-circuit current using the obtained efficiency spectra show that the losses caused by recombination at the CdTe-CdS interface are insignificant if the uncompensated acceptor concentration $N_a - N_d$ in CdTe is in excess of 10^{16} cm^{-3}. At $N_a - N_d \approx 10^{16}$ cm^{-3} and the thickness of the absorbing CdTe layer equal to around 5 μm, the short-circuit current density of 25-26 mA/cm^2 can be attained. As soon as $N_a - N_d$ deviates downward or upward from this value, the short-circuit current density decreases significantly due to recombination losses or reduction of the photocurrent diffusion component, respectively. Under this condition, recombination losses in the space-charge region can be also neglected, but only when the carrier lifetime is equal or greater than 10^{-10} s.

At $N_a - N_d \geq 10^{16}$ cm^{-3}, when only a part of charge carriers is generated in the neutral part of the p-CdTe layer, *total* charge collection can be achieved if the electron lifetime is equal to several microseconds. In this case the CdTe layer thickness d should be greater than that usually used in the fabrication of CdTe/CdS solar cells (2-10 μm). However, in a common case where the minority-carrier (electron) lifetime in the absorbing CdTe layer amounts to 10^{-10}–10^{-9} s, the optimum layers thickness d is equal to 3–4 μm, i.e., the calculations support the choice of d made by the manufacturers mainly on an empirical basis. Attempts to reduce the thickness of the CdTe layer to 1-1.5 μm with the aim of material saving appear to be unwarranted, since this leads to a considerable reduction of the short-circuit's current density J_{sc} and, ultimately, to a decrease in the solar-cell efficiency. If it will be possible to improve the quality of the absorbing layer and, thus, to raise the electron lifetime at least to 10^{-8} s, the value of J_{sc} can be increased by 1-1.5 mA/cm^2.

The Sah-Noyce-Shockley theory of generation-recombination in the space-charge region supplemented with over-barrier diffusion flow of electrons provides a quantitative explanation for all variety of the observed *I-V* characteristics of thin-film CdS/CdTe heterostructure. The open circuit voltage V_{oc} significantly increases with decreasing the resistivity ρ of the CdTe layer and increasing the effective carrier lifetime τ in the space charge region. At $\tau = 10^{-10}$-10^{-9} s, the value of V_{oc} is considerably lower than its maximum possible value for $\rho \approx 0.1$ Ω·cm and $\tau > 10^{-8}$ s and the calculated efficiency of a CdS/CdTe solar cell with a CdTe layer thickness of 5 μm lies in the range 17-19%. An increase in the efficiency and an approaching its theoretical limit (28-30%) is possible in the case when the electron lifetime $\tau_n \geq 10^{-6}$ s and the thickness of CdTe absorber layer is 20-30 μm or more. The question of whether an increase in the CdTe layer's thickness is reasonable under the conditions of mass production of solar modules can be answered after an analysis of economic factors.

7. Acknowledgements

I thank X. Mathew, Centro de Investigacion en Energia-UNAM, Mexico, for the CdS/CdTe thin-film heterostructures for measurements, V.M. Sklyarchuk for sample preparation to study, V.V. Motushchuk and E.V. Grushko for measurements carried out, and all participants of the investigation for helpful discussion. The study was supported by the State Foundation for Fundamental Investigations (Ministry of Education and Science, Ukraine) within the Agreement Ф14/259-2007.

8. References

Acrorad Co, Ltd., 13-23 Suzaki, Gushikawa, Okinawa 904-2234, Japan. Available: http://www.acrorad.jp/us/cdte.html.

Aramoto T., Kumazawa S., Higuchi H., Arita T., Shibutani S., Nishio T., Nakajima J., Tsuji M., Hanafusa A., Hibino T., Omura K., Ohyama H. &, Murozono M., (1997). 16.0% Efficient Thin-Film CdS/CdTe Solar Cells. *J. Appl. Phys.* 36, 6304-6305.

Baldazzi G., Bollini D., Casali F., Chirco P., Donati A., Dusi W., Landini G., Rossi M. & Stephen J. B., (1993). Timing response of CdTe detectors, *Nucl. Instr. and Meth.* A326 319-324.

Birkmire, R.W. & Eser, E. (1997). Polycrystalline thin film solar cells: Present status and future potential, *Annu. Rev. Mater. Sc.* 27, 625.

Britt, J. & Ferekides, C., (1993), Thin-film CdS/CdTe solar cell with 15.8% efficiency. Appl. Phys. Lett. 62, 2851-2853.

Bonnet, D. (2001). Cadmium telluride solar cells. In: Clean Electricity from *Photovoltaic.* Ed. by M.D. Archer, R. Hill. Imperial College Press, pp. 245-276.

Bonnet, D. (2003). CdTe thin-film PV modules, In: *Practical Handbook of Photovoltaic: Fundamentals and Applications.* Ed. by T. Makkvart and L. Castaner. Elseivier, Oxford.

Compaan A.D., Sites J.R., Birkmire R.W., Ferekides C.S. and Fahrenbruch A.L. (1999). Critical Issues and Research Needs for CdTe-Based Solar Cells, *Proc. 195th Meeting of the Electrochemical Society),* PV99-11, Seattle, WA, pp. 241-249.

Demtsu S.H. & Sites J.R., (2005). Quantification of losses in thin-film CdS/CdTe solar cells, *Proc. 31rd IEEE Photovoltaic Specialists Conf.* pp. 3-7, Florida, Jan. 347-350.

Desnica, U.V., Desnica-Frankovic I.D., Magerle R., Burchard A. & Deicher M.. (1999). Experimental evidence of the self-compensation mechanism in CdS, *J. Crystal Growth,* 197, 612-615.

Eizen Y. (1992). Current state-of-the-art applications utilizing CdTe derectors, *Nucl. Instr. and Meth.* A322, 596-603.

Ferekides, C.S., Balasubramanian, U., Mamazza, R., Viswanathan, V., Zhao, H. & Morel, D.L. (2004) CdTe thin-film solar cells: device and technology issues, *Solar Energy* 77, 823-830.

Fritsche, J., Kraft, D., Thissen, A., Mayer, Th., Klein & A., Jaegermann W. (2001). Interface engineering of chalcogenide semiconductors in thin film solar cells: CdTe as an example, *Mat. Res. Soc. Symp.* Proc. , 668, 601-611.

Gartner W.W., (1959). Depletion-layer photoeffects in semiconductors, *Phys. Rev.* 116, 84-87.

Goetzberger, A. , Hebling, C. & Schock, H.-W. (2003). Photovoltaic materials, history, status and outlook, Materials Science and Engineering R40, 1-46.

Grasso, C., Ernst, K., R. Könenkamp, Lux-Steiner, M.C. & Burgelman, M. (2001). Photoelectrical Characterisation and Modelling of the Eta-Solar Cell. *Proc. 17th European Photovoltaic Solar Energy Conference*, vol. 1. pp. 211-214, Munich, Germany, 22-26 October.

Hanafusa, A., Aramoto, T., Tsuji, M., Yamamoto, T., Nishio, T., Veluchamy, P., Higuchi, H., Kumasawa, S., Shibutani, S., Nakajima, J., Arita T., Ohyama, H., Hibino T., Omura & K. (2001). Highly efficient large area (10.5%, 1376 cm2) thin-film CdS/CdTe solar cell, *Solar Energy Materials & Solar Cells*. 67, 21-29.

Holliday D. P., Eggleston J. M. and Durose K., (1998). A photoluminescence study of polycrystalline thin-film CdTe/CdS solar cells. *J. Cryst. Growth*. 186, 54-549.

Kosyachenko, L.A., Sklyarchuk, V.M. , Sklyarchuk, Ye.F. & Ulyanitsky, K.S. (1999). Surface-barrier p-CdTe-based photodiodes, *Semicond. Sci. Technol.*, 14, 373-377.

Kosyachenko L.A., Rarenko I.M., Zakharuk Z.I., Sklyarchuk V.M., Sklyarchuk Ye.F., Solonchuk I.V., Kabanova I.S. & Maslyanchuk E.L. (2003). Electrical properties of CdZnTe surface-barrier diodes. *Semiconductors*. 37, 238-242.

Kosyachenko L.A., Mathew X., Motushchuk V.V. & Sklyarchuk V.M. (2005). The generation-recombination mechanism of charge transport in a thin-film CdS/CdTe heterojunction, *Semiconductors*, 39, 539–542.

Kosyachenko, L.A. (2006). Problems of Efficiency of Photoelectric Conversion in Thin-Film CdS/CdTe Solar Cells, Semiconductors. 40, 710-727.

Lavagna, M., Pique, J.P. & Marfaing, Y. (1977). Theoretical analysis of the quantum photoelectric yield in Schottky diodes, *Solid State Electronics*, 20, 235-240.

Mathew, X., Kosyachenko, L.A., Motushchuk, V.V. & Sklyarchuk. O.F. (2007). Requirements imposed on the electrical properties of the absorbed layer in CdTe-based solar cells. *J. Materials Science: Materials in Electronics*. 18, 1021-1028.

McCandless, B.E., Hegedus, S.S., Birkmire, R.W. & Cunningham, D. (2003). Correlation of surface phases with electrical behavior in thin-film CdTe. devices. *Thin Solid Films* 431–432, 249-256.

Meyers, P.V. & Albright, S.P. (2000). Photovoltaic materials, history, status and outlook. Prog. Photovolt.: Res. Appl. 8, 161- 168.

Phillips J.I., Brikmire R.W., McCandless B.E., Mayers P.V. & Shaparman W.N., (1996). Polycrystalline heterojunction solar cells: a device perspective. *Phys. Stat. Sol.* (b) 31, 31-39.

Reference solar spectral irradiance at the ground at different receiving conditions. Standard of International Organization for Standardization ISO 9845-1:1992.

Romeo, N., Bosio, Canevari, A. V. & Podesta A., (2004). Recent progress on CdTe/CdS thin film solar cells, *Solar Energy* , 77, 795-801.

Sah C., Noyce R. & Shockley W. (1957). Carrier generalization recombination in p-n junctions and p-n junction characteristics, *Proc. IRE*. 45, 1228–1242.

Sites, J.R. & Xiaoxiang Liu, (1996). Recent efficiency gains for CdTe and $CuIn_{1-x}Ga_xSe2$ solar cells: What has changed? *Solar Energy Materials & Solar cells*. 41/42 373-379.

Sites J.R. & Pan J., (2007). Strategies to increase CdTe solar-cell voltage. *Thin Solid Films*, 515, 6099-6102.

Surek, T. (2005). Crystal growth and materials research in photovoltaics: progress and challenges, Journal of Crystal Growth. 275, 292-304.

Sze, S. (1981). *Physics of Semiconductor Devices,* 2nd ed. Wiley, New York.

Toshifumi, T., Adachi, S., Nakanishi, H. & Ohtsuka M. (993). K. Optical constants of Zn1-xCdxTe Ternary alloys: Experiment and Modeling. *Jpn. Appl. Phys.* 32, 3496-3501.

Wu, X., Keane, J.C., Dhere, R.G., Dehart, C., Albin, D.S., Duda, A., Gessert, T.A., Asher, S., Levi, D.H. & Sheldon, P. (2001). 16.5%-efficient CdS/CdTe polycrystalline thin-film solar cell, In: *Proceedings of the 17th European Photovoltaic Solar Energy Conference,* Munich, Germany, October 2001, p. 995-1000.

7

Energy Control System of Solar Powered Wheelchair

Yoshihiko Takahashi, Syogo Matsuo, and Kei Kawakami
Department of Mechanical System Engineering
Department of Vehicle System Engineering
Kanagawa Institute of Technology
Japan

1. Introduction

Independence is a major concern for individuals with severe handicaps. Welfare assistance robotic technology is a popular solution to this concern (e.g. Hashino, 1996; Takahashi, Ogawa, and Machida, 2002 and 2008). Assistance robotic technologies offer potential alternatives to the need for human helpers. People bound to wheelchairs have limited mobility reliant on battery life, which only allows for short distance travel between charges. In addition, recharging batteries is time consuming.

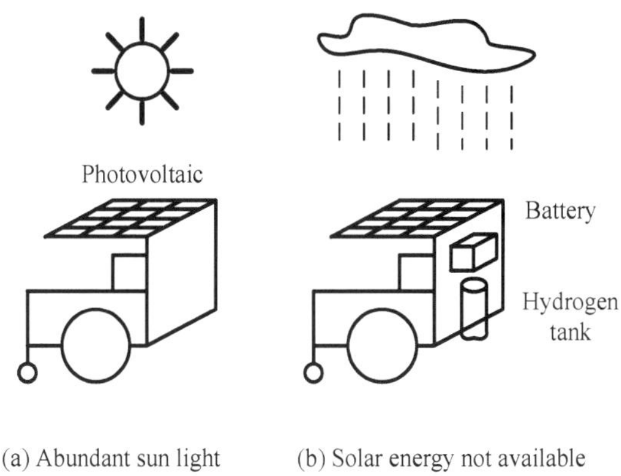

Fig. 1. Running conditions of proposed robotic wheelchair

The aim of this paper is to propose a system which will increase the moving distance of an electrical wheelchair by adding two solar powered energy sources; a small photovoltaic cell and a fuel cell. Fig.1 displays the running conditions of the proposed robotic wheelchair. The control system will ideally give priority to the photovoltaic cell, next to the fuel cell and finally to the battery. When sufficient sun light is available, the photovoltaic cell on the wheelchair roof is used, when it is limited, the fuel cell or the battery is used. The energy control system is designed using a micro computer, and the energy source is quickly

changeable. Our objective is that the proposed robotic solar wheelchair will enable users to enjoy increased independence when they are outdoors.
The advantage of using a solar powered energy source is that it produces power without requiring use of fossil fuels. A photovoltaic cell is installed on the roof of the wheelchair, which produces enough power to operate the apparatus when enough sun light is available. The battery is charged using a large photovoltaic cell on the roof of the setup. Hydrogen is produced using a water electrolysis hydrogen generator, and the fuel cell utilizes the produced hydrogen. The large photovoltaic cell also sends electricity to the hydrogen generator.
Photovoltaic cells and fuel cells are representative sustainable technologies (Bialasiewicz, 2008; Okabe et al., 2009; KE Jin et al., 2009). We are able to use two methods to produce hydrogen using these sustainable technologies for our wheelchair. The first method is to generate hydrogen from the electrolysis of water. The next is to use waste biomass which produces biomass ethanol. Hydrogen is produced by steam reforming the ethanol (Takahashi, & Mori, 2006; Essaki et al., 2008; Saxena et al., 2009; Rubin, 2008; Sugano, & Tamiya, 2009). Standard sized fuel cell models are developed with the aim to develop a commercially viable vehicle (Tabo et al., 2004; Kotz, et al., 2001; Rodatz, et al., 2001). Hybrid vehicles using photovoltaic cells and fuel cells are developed in two universities (Konishi, et al., 2008; Obara, 2004). Small fuel cell vehicles were developed (Nishimura, 2008; Takahashi, 2009a and 2009b). A wheelchair with a fuel cell has been developed (Yamamuro, 2003).
This paper will present a robotic wheelchair using solar powered energy sources of the photovoltaic and fuel cell, detail the energy flow concept for charging electricity to the battery and for storing hydrogen to the tank, the mechanical construction, the energy control system, and the experimental results of the running test.

2. Energy flow of proposed robotic wheelchair

A schematic explanation and block diagram of the energy flow used in the proposed robotic wheelchair are shown in Figs. 2 and 3. The energy system used in the robotic wheelchair does not exhaust carbon dioxide as it does not utilize fossil fuels.
The first energy flow line in the schematic diagrams is the line from the photovoltaic cells on the roof of the wheelchair. A cascade connection of two photovoltaic cells (Kyosera, KC-40TJ) of 17.4 V and 43 W in nominal value is utilized as the energy source. The output voltage is reduced to 24 V using a DC-DC converter.
The second energy line is the line from the water electrolysis hydrogen generator. The photovoltaic cell on the setup roof (approximately 10 kW) sends electricity to the water electrolysis hydrogen generator. The generated hydrogen is stored in a metal hydride hydrogen tank of 60 NL. The output pressure of the hydrogen generator is approximately 0.3 MPa. The hydrogen tanks are installed on the wheelchair body after storing hydrogen. A metal hydride tank is used for safety concerns. A fuel cell (Daido Metal, HFC-24100) producing 24 V and 100 W in nominal values is used to generate electricity to the motor.
The third energy flow line is the battery line. The battery is charged with electricity from the photovoltaic cell producing approximately 10 kW on the setup roof, and then installed on the wheelchair body.
The fourth energy flow line is the biomass line. Ethanol is produced from waste biomass. Hydrogen is then generated from the ethanol using a steam reforming hydrogen generator. The generated hydrogen is stored in a 60 NL metal hydride hydrogen tank. The hydrogen

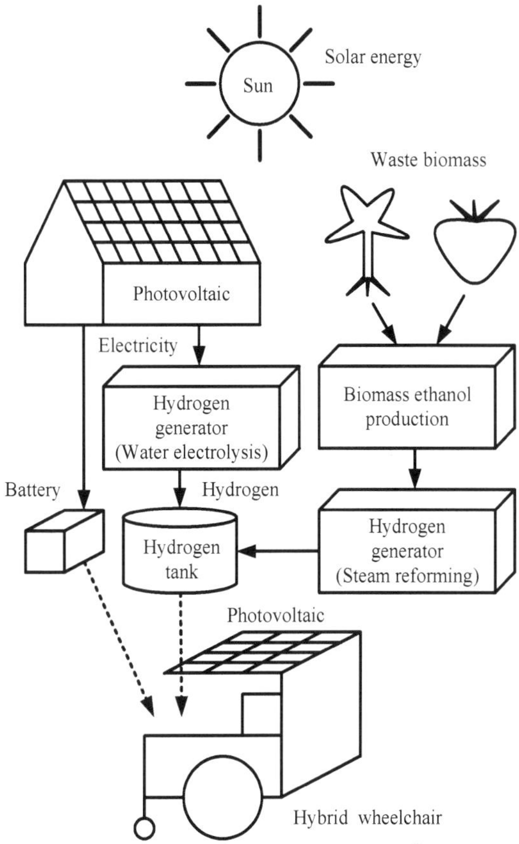

Fig. 2. Schematic explanation of energy flow

tanks are installed on the wheelchair body in the same manner as the second line. Ethanol is safe to handle, and is easy to carry, however, the fourth energy flow line is still a matter under consideration.

The High-Tech Research Center Project for Solar Energy System at the Kanagawa Institute of Technology is conducting research on applications of solar energy. The development of the robotic wheelchair is conducted as a part of the High-Tech Research Center Project. The battery charging and hydrogen storing to the metal hydride are conducted using the facility at the High-Tech Research Center Project.

3. Mechanical construction

Fig. 4 displays the fabricated robotic wheelchair with the photovoltaic and fuel cell. A reinforced YAMAHA JW-1 wheelchair was used as the main body of the experimental set up. In this configuration, the photovoltaic cell, the fuel cell, and the battery are installed on the top, on the back, and under the wheelchair, respectively. The energy control system and

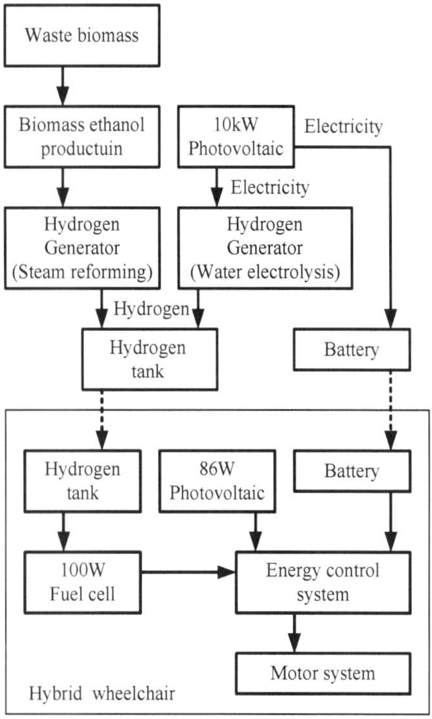

Fig. 3. Block diagram of energy flow

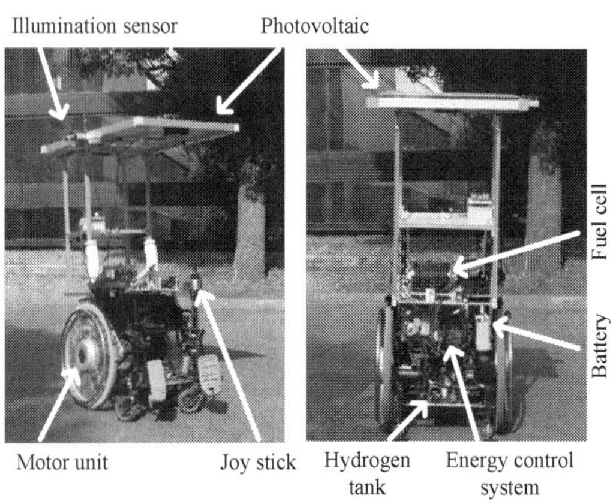

Fig. 4. Fabricated robotic wheelchair with photovoltaic and fuel cell

hydrogen tanks are installed on the back of the wheelchair. Figs.5 (a) and (b) show the photovoltaic and illumination sensor. Fig.6 (a) exhibits the fuel cell (Daido Metal, HFC-24100) and the vibration isolator. Fig.6 (b) shows the metal hydride tanks of 60 NL and 0.3

MPa. The hydrogen pressure is adjusted to 0.08 MPa using a regulator. The main specifications are as follows.

Wheelchair mechanism (Yamaha, JW-1)
    Weight : 13 kg
    Running operation : Joy stick
    Motor : DC24V, 90Wx2
Photovoltaic (Kyosera, KC-40TJ)
    Type : Multi crystal
    Nominal power : 43 W
    Maximum voltage : 17.4 V
    Dimensions : 526x652x54 mm
    Weight : 4.5 kg
Fuel cell (Daido Metal, HFC-24100)
    Nominal power : 100 W
    Nominal voltage : 24 V
    Dimensuions : 160x110x240 mm
    Weight : 3 kg
    Air fans : DC24, 0.94Wx 24

(a) Photovoltaic (b) Illumination sensor

Fig. 5. Photovoltaic and illumination sensor

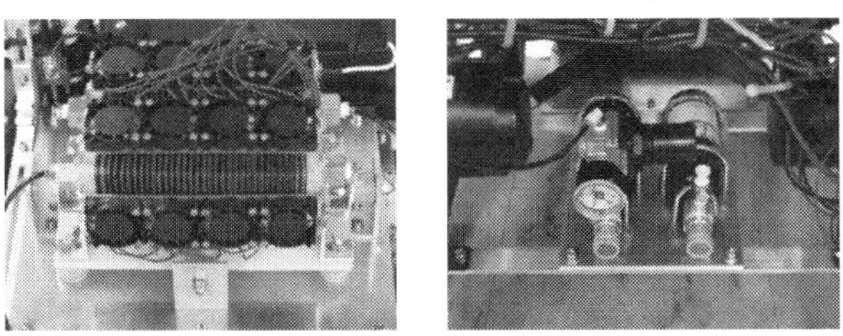

(a) Fuel cell and vibration isolator (b) Hydrogen tank and regulator

Fig. 6. Fuel cell and hydride tank

4. Energy control system

Fig.7 shows the concept of the energy control system where a micro computer determines the wheelchair condition, and selects the optimum energy source from the three energy sources: the photovoltaic on the wheelchair roof; the fuel cell; or the battery. Solid lines indicate energy flow lines, and dotted lines indicate the control signal flow lines. Fig.8 displays energy control architecture in detail. The switching control system inputs the voltages of the photovoltaic cell, the fuel cell, and the motor drive current, and selects the energy source determined by the wheelchair condition.

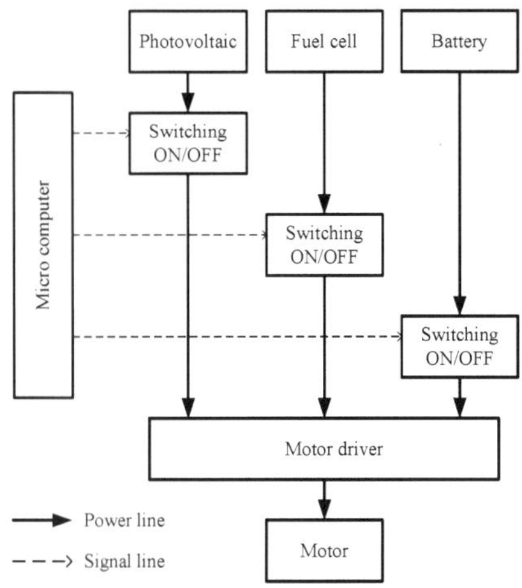

Fig. 7. Concept of energy control system

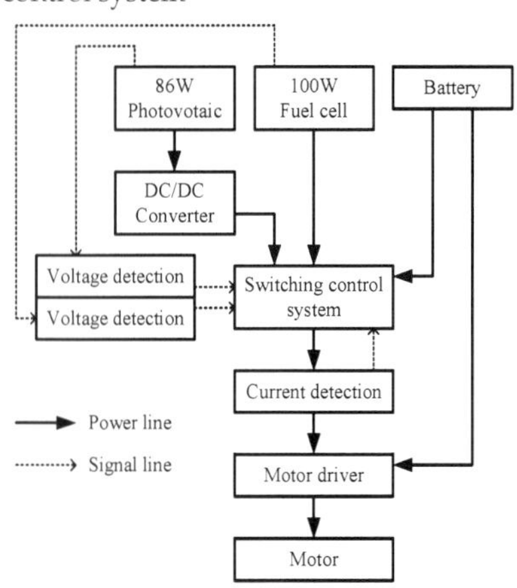

Fig. 8. Detailed energy control architecture

Fig.9 is the software control algorithm of the energy control system. Fig.10 shows the fabricated switching control system of the energy control system where a micro computer controls the entire energy control system, and FETs are used to switch the energy flow. Performance of energy source switching is also tested as this is the first attempt to develop a solar powered wheelchair. The electricity acquired from the photovoltaic cell on the wheelchair roof will be utilized to charge with the battery. Instant power increase using a capacitor will also be required. Improvement of the energy control system must be addressed in future research.

The control system will ideally give priority to the photovoltaic cell then to the fuel cell and then to the battery. Essentially, the switching control is conducted on the motor driving current considering the condition of the photovoltaic and fuel cells. If the motor driving current is below 2.5 A and the photovoltaic voltage is above 30 V, then the photovoltaic is selected. If the motor driving current is below 4.0 A and the fuel cell voltage is above 24 V, then the fuel cell is selected. When the motor driving current is below 20.0 A, then the battery is selected. The following details the software control algorithm of the energy control system.

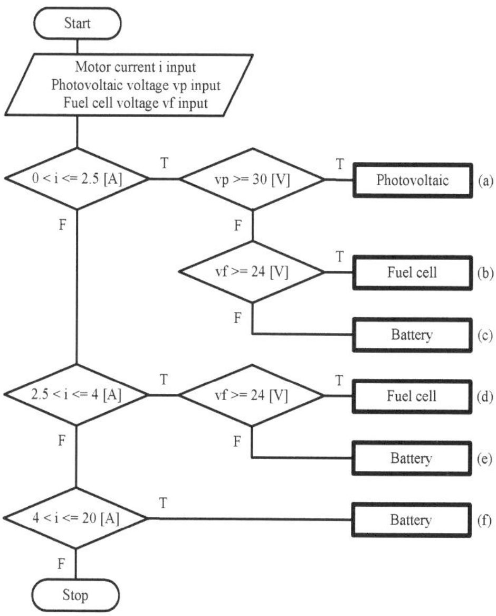

Fig. 9. Software control algorithm of energy control system

Condition (a) :

When the motor current is over 0.0 A and less than 2.5 A,
and the photovoltaic voltage is over 30 V,
then the photovoltaic is selected.

Condition (b) :

When the motor current is over 0.0 A and less than 2.5 A,
and the photovoltaic voltage is less than 30 V,
and the fuel cell voltage is over 24 V,
then the fuel cell is selected.

Condition (c) :

When the motor current is over 0.0 A and less than 2.5 A,
and the photovoltaic voltage is less than 30 V,
and the fuel cell voltage is less than 24 V,
then the battery is selected.

Condition (d) :

When the motor current is over 2.5 A and less than 4.0 A,
and the fuel cell voltage is over 24 V,
then the fuel cell is selected.

Condition (e) :

When the motor current is over 2.5 A and less than 4.0 A,
and the fuel cell voltage is less than 24 V,
then the battery is selected.

Condition (f) :

When the motor current is over 4.0 A and less than 20.0 A,
then the battery is selected.

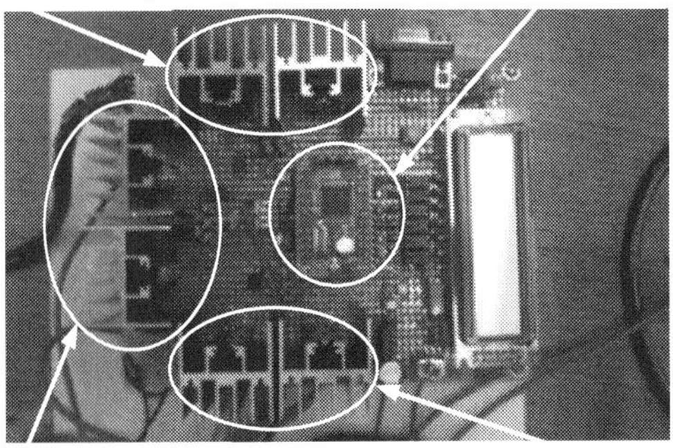

Fig. 10. Fabricated switching control system of energy control system

5. Experimental results

We conducted experiments on energy source selection between the solar panel, the fuel cell, and the battery.

5.1 Experimental conditions

The wheelchair was tested on a flat, straight course. It maintained the same speed maneuvering between two turning points during which joystick operation was required.

Figs.11 to 13 present the experimental results of the running tests using the fabricated wheelchair. Each figure will be detailed later. In each figure, the top graph shows the results of the motor driving current, the photovoltaic (PV) current, and the fuel cell (FC) current. Second, the results of the photovoltaic (PV) voltage, and the fuel cell (FC) voltage. The third figure from the top shows the results of the photovoltaic (PV) power and fuel cell (FC) power. Finally, the last chart shows illumination.

The following four test patterns are conducted;

1. Low Speed mode, Low (approximately 1.4 km/h) (Fig.11)
2. Low Speed mode, High (approximately 2.4 km/h) (Fig.12)
3. High Speed mode, Low (approximately 3.0 km/h) (Fig.13)

4. High Speed mode, High (approximately 3.4 km/h) (Fig.14)
Low Speed and High Speed modes are selected using a switch on the wheelchair frame. "Low" refers to the joy stick declined approximately 10 degrees, "High" refers to the joy stick declined completely. For example, the condition of the "Low Speed mode, Low" denotes that the switch was selected to low speed mode and the joy stick was inclined by approximately 10 degrees. The charts from Figs. 11 through 14 show the results of the conditions: Low Speed mode, Low; Low Speed mode, High; High Speed mode, Low; and High Speed mode, High, respectively.

5.2 Results of "Low Speed mode, Low" condition

Fig.11 (a) shows the experimental results of the current at the "Low Speed mode, Low" condition. The speed was approximately 1.4 km/h. The experimental results show that the average value of the motor driving current was approximately 1.5 A while moving straight on the course. The photovoltaic current was approximately the same value of the motor driving current. The fuel cell current was small. At the turning point, the motor driving current was over 2.5 A, therefore the current of the photovoltaic and fuel cell were lower than the required motor driving current. Energy from the battery was necessary.
Fig.11 (b) shows the experimental results of the voltage. While moving on the straight course, the photovoltaic voltage was reduced, and the fuel cell did not change greatly.

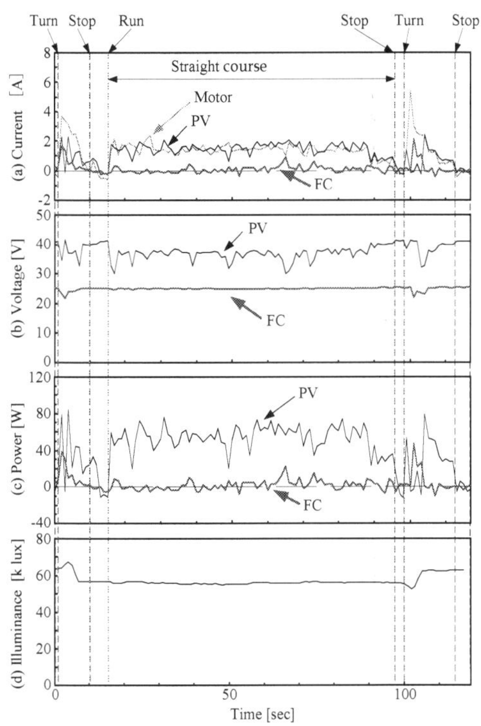

Fig. 11. Experimental results (Low Speed mode, Low)

Fig. 11 (c) shows the experimental results of the power. While moving on the straight course, the photovoltaic power was larger than the fuel cell power. Fig.11 (d) shows the experimental results of the illumination. Illumination was approximately 60 k lux while moving on the straight course. The illumination value changed while turning at the course ends.

5.3 Results of "Low Speed mode, High" condition

Fig.12 (a) shows the experimental results of the current at the "Low Speed mode, High" condition. The speed was approximately 2.4 km/h. The experimental results show that the average value of the motor driving current was approximately 1.8 A while moving on the straight portion of the course. The photovoltaic current was approximately the same value as the motor driving current. The photovoltaic current was occasionally reduced to 0 A. Fuel cell current was negligible. During turns, the motor driving current was over 4.0 A. The photovoltaic and fuel cell current were less than the motor driving current, therefore battery energy was required.

Fig.12 (b) shows the experimental results of the voltage. While moving on the straight part of the course, the photovoltaic voltage was lowered, the change in the fuel cell was negligible. Fig.12 (c) shows the experimental results of the power. The photovoltaic power

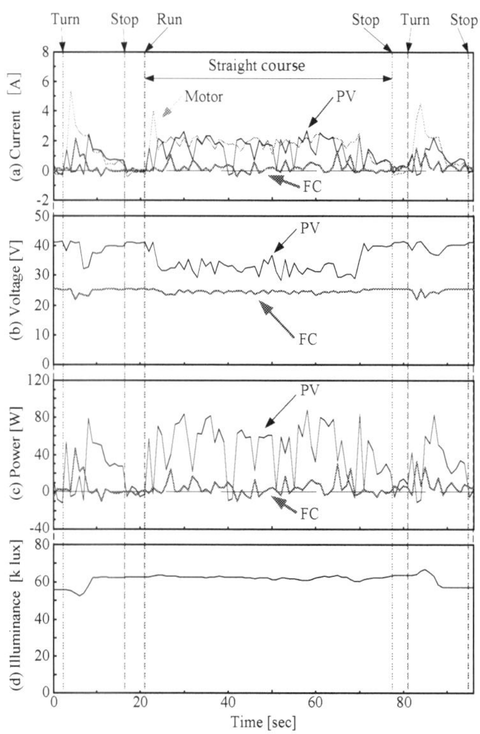

Fig. 12. Experimental results (Low Speed mode, High)

was larger than the fuel cell power. Fig.12 (d) shows the experimental results of the illumination. The illumination was approximately 60 k lux while moving on the straight part of the course. The illumination value changed while turning at the course ends.

5.4 Results of "High Speed mode, Low" condition

Fig.13 (a) shows the experimental results of the current at the "High Speed mode, Low" condition. The speed was approximately 3.0 km/h. The experimental results show that the average value of the motor driving current was approximately 2.0 A while moving straight. The photovoltaic current was approximately the same value as the motor driving current. The photovoltaic current occasionally fell to 0 A. Fuel cell current was minimal. At the turning point, the motor driving current was over 4.0 A, and the current of the photovoltaic and fuel cells were smaller than the motor driving current. Battery energy was required.
Fig.13 (b) shows the experimental results of the voltage. While moving straight, photovoltaic voltage decreased, and the fuel cell did not change prominently. Fig.13 (c) shows the experimental results of the power. While moving straight, photovoltaic power was larger than fuel cell power. Fig.13 (d) shows the experimental results of the illumination. Illumination was approximately 60 k lux while moving straight. It changed while turning at the course ends.

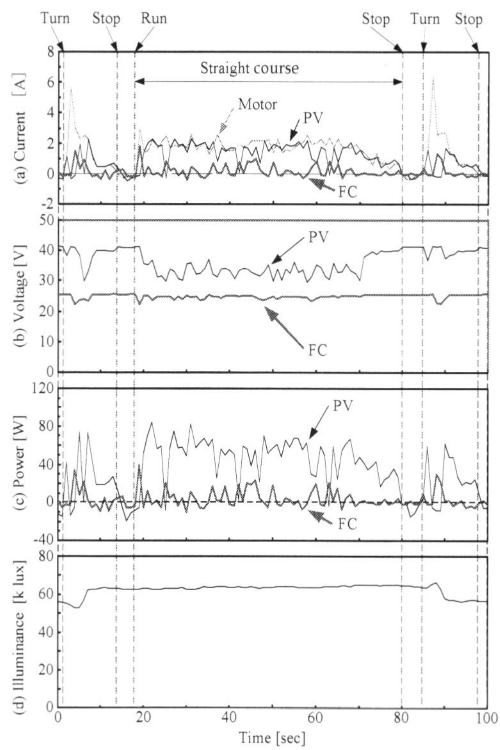

Fig. 13. Experimental results (High Speed mode, Low)

5.5 Results of "High Speed mode, High" condition

Fig.14 (a) shows the experimental results of the current at the "High Speed mode, High" condition. The speed was approximately 3.4 km/h. The experimental results show that the average value of the motor driving current was approximately 2.8 A while moving straight. Photovoltaic current was approximately 0 A. Fuel cell current increased, however the value was smaller than the motor driving current. Combined energy from the fuel cell and the battery was used. At the turning point, the motor driving current was over 4.0 A, and the current of the photovoltaic and fuel cell were smaller than the motor driving current. Energy from the battery was required.

Fig.14 (b) shows the experimental results of the voltage. While moving straight, the photovoltaic voltage was not reduced, however the fuel cell was. Fig.14 (c) shows the experimental results of the power. While moving the straight course, the photovoltaic power was not significant, and the fuel cell power increased. Fig.14 (d) shows the experimental results of the illumination. Illumination was approximately 60 k lux while moving straight. The illumination value changed while turning at the course ends.

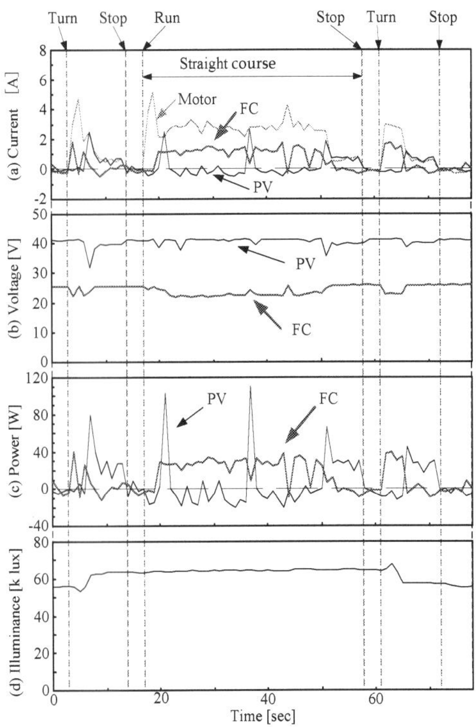

Fig. 14. Experimental results (High Speed mode, High)

5.6 Summary of the experimental results

The proposed robotic solar powered wheelchair uses a small powered photovoltaic of 86 W (at nominal value), and a small powered fuel cell of 100 W (at nominal value). When ample sun light is available, a flat and straight course is used, and the wheelchair travels at a low speed, the robotic wheelchair is able to move primarily powered by the photovoltaic cell. As a result, the solar powered wheelchair is able to travel further distances. When the wheelchair travels at higher speeds, and turns, it requires greater power, therefore it uses the energy from the fuel cell and the battery.

6. Conclusions

A new robotic solar powered wheelchair using three energy sources, a small photovoltaic cell, a small fuel cell, and a battery is proposed in this paper. All three energy sources use solar energy. The photovoltaic cell uses sun light directly. The battery is charged with electricity provided by the large photovoltaic cell installed on the setup roof. Hydrogen for the fuel cell is generated by a water electrolysis hydrogen generator, which is also powered by the same large photovoltaic cell on the building roof. The energy control system selects the optimal energy source to use based on various driving conditions.

It was confirmed from the experimental results that the robotic wheelchair is able to maneuver mainly using the photovoltaic cell when good moving conditions are available (i.e. abundant sun light, a flat and straight course, and low speed). The experimental results demonstrate that the robotic wheelchair is able to increase its moving distance. When moving conditions are not optimal, the robotic solar wheelchair uses energy from the fuel cell and the battery.

Improvements to the energy control system such as charging to the battery from the photovoltaic cell on the wheelchair roof, power increase using a capacitor, and hydrogen generation from waste biomass, must be addressed in future research.

7. Acknowledgments

The authors would like to express their deepest gratitude to the research staff of the High-Tech Research Center Project for Solar Energy Systems at the Kanagawa Institute of Technology for their kind cooperation with the experiments and for their kind advice.

8. References

Hashino, H. (1996); Daily Life Support Robot, *Journal of Robotics Society of Japan,* Vol.14, No.5, pp.614-618

Takahashi, Y., Ogawa, S., and Machida, S., (2002); Mechanical design and control system of robotic wheelchair with inverse pendulum control, Trans. Inst. Meas. Control, vol.24, no.5, pp.355-368.

Takahashi, Y., Ogawa, S., and Machida, S., (2008); Experiments on step climbing and simulations on inverse pendulum control using robotic wheelchair with inverse pendulum control, Trans. Inst. Meas. Control, vol.30, no.1, pp.47-61.

Takahashi, J., And Mori, T., (2006); Hydrogen Production from Reaction of Apple Pomace with Water over Commercial Stream Reforming Ni Catalysis, Journal of Japan Petroleum Institute, vol.49, no.5, pp.262-267.

Essaki, K., Muramatsu, T., and Kato, M., (2008); Hydrogen Production from Ethanol by Equilibrium Shifting Using Lithium Silicate Pellet as CO2 Absorbent, Journal of Japan Institute of Energy, vol.87, no.1, pp.72-75.

Saxena, R.C., Adhikari, D.K. and Goyal, H.B., (2009); Biomass-Based Energy Fuel Cell through Biochemical Routes, Renew. Sust. Energ. Rev. Vol.13, pp.167-178.

Rubin, E.M., (2008); Genomics of Cellulosic Biofuels, Nature, vol.454, pp.841-845.

Sugano, Y., and Tamiya, E., (2009); A direct Cellulose-Based Fuel Cell System, Journal of Fuel Cell Technology, vol.9, no.1, pp.114-119.

Bialasiewicz, J.T., (2008); Renewable Energy Systems with Photovoltaic Power generations: Operation and Modeling, IEEE Trans. on Industrial Electronics, vol.55, no.7, pp.2752-2758.

Okabe, M., Nakazawa, K., Taruya, K., and Handa, K., (2008); Verification test of solar-powered hydrogen station (SHS) with photovoltaic modules, Honda R&D Technical Review, Vol.20, No.1, pp.67-73.

Ramos-Paja, C.A., Bordons, C., Romero, A., Giral, R., and Martinez-Salamero, L., (2009); Minimum Fuel Cell Consumption Strategy for PEM Fuel Cell, Trans. on Industrial Electronics, vol.56, no.3, pp.685-696.

KE Jin, Xinbo Ruan, MengxiongYang, and Min Xu, (2009); A Hybrid Fuel Cell Power System, Trans on Industrial Electronics, vol.56, no.4, pp.1212-1222.

Tabo, E., Kuzuoka, K., Takada, M., and Yoshida, H., (2004); Fuel cell vehicle technology trends and MMC initiatives, Mitsubishi Motors Technical Review, No.16, pp.51-55.

Kotz, R., Muller, S., Bartschi, M., Schnyder, B., Dietrich, P., Buchi, F.N., Tsukada, A., Scherer, G., Rodatz, P., Garcia, O., Barrade, P., Hermann, V., and Gallay, R., (2001); Supercapacitors for peak-power demand in fuell-cell-driven cars, Electrochemical Society Proceedings, Vol.2001-21, pp.564-575.

Rodatz, P., Garcia, O., Guzzella, L., Buchi, F., Bartschi, M., Tsukada, A., Dietrich, P., Kotz, R., Scherer, and G., Wokaun, A., (2001); Performance and operation characteristics of a hybrid vehicle powered by fuel cells and supercapacitors, Soc. of Automotive Eng. 2003 Congress, SAE Paper 2003-01-0418, pp.1-12.

Konishi, H., Akizuki, M., Ogawa, T., Kojima, H., Yamada, Y., Fujii, H., Matsunaga, N,. Yoshida, Y., Ishida, T., nad Warashina, T., (2008); Development of a Solar and Fuel Cell Powered Hybrid Electrical Vehicle Cocoon 2007, Proc. of 2008 JSME Conf. on Robotics and Mechatronics, 2P1-A18, pp.1-4.

Obara, H., (2004); Progress of Development on the Hybrid Solar Car in Tamagawa University, Journal of Fuel Cell Technology, vol.4, no.2, pp.103-107.

Nishimura, I., (2008); Design and Fabrication of Fuel Cell Vehicle Regarding Manufacturing Education, Proc. of 2008 JSME Conf. on Robotics and Mechatronics, 2P1-A13, pp.1-4.

Takahashi, Y., (2009a); Ultra Light Weight Fuel Cell Electrical Vehicle (UL-FCV), Proc. of IEEE Int. Symp. on Industrial Electronics, pp.189-194.

Takahashi, Y., (2009b); Environmental System Education using Small Fuel Cell Electrical Vehicle, Journal of Fuel Cell Technology, vol.9, no.1, pp.128-131.

Yamamuro, S., (2003); Development of Fuel Cell Powered Wheelchair, Kuromoto Kihou, no.52, pp.40-44.

8

Uses of Concentrated Solar Energy in Materials Science

Gemma Herranz and Gloria P. Rodríguez
University of Castilla La Mancha. ETSII.
Metallic Materials Group
Avda. Camilo José Cela s/n. 13071. Ciudad Real.
Spain

1. Introduction

In recent decades tremendous advances have been made in the development of new materials capable of working under increasingly extreme conditions. This advance is linked to the development of Materials Surface Engineering. The utilisation of techniques based on high density energy beams (laser, plasma, electron beam or arc lamps) in surface modification and metallic material treatment allow for the creation of non-equilibrium microstructures which can be used to manufacture materials with higher resistance to corrosion, high temperature oxidation and wear, among other properties.

These techniques, despite their multiple possibilities, have one inconvenient property in common: their low overall energy efficiency. While it is true that the energy density obtained through a laser is three to four magnitudes greater than that which is obtained by solar energy concentration facilities, Flamant (Flamant et al. 1999) have carried out a comparison of the overall energy and the capital costs of laser, plasma and solar systems and came to the conclusion that solar concentrating systems appear to offer some unique opportunities for high temperature transformation and synthesis of materials from both the technical and economic points of view.

It is important to bear in mind that the use of this energy could lower the cost of high temperature experiments. Combined with the wide array of superficial modifications that can be carried out at solar facilities, there are numerous other advantages to using this energy source. The growing (and increasingly necessary) trend towards the use of renewable clean energy sources, which do not contribute to the progressive deterioration of the environment, is one compelling argument. Solar furnaces are also excellent research tools for increasing scientific knowledge about the mechanisms involved in the processes generated at high temperatures under non-equilibrium conditions. If, in addition, the solar concentration is carried out using a Fresnel lens, several other positive factors come into play: facility costs are lowered, adjustments and modifications are easy to carry out, overall costs are kept low, and the structure is easy to build, which makes the use of this kind of lens highly attractive for research, given its possible industrial applications.

These are the reasons that justify the scientific community's growing interest in researching the possible uses of highly concentrated solar energy in the field of materials. But this interest is not new. At the end of the 18th century, Lavoisier (Garg, 1987) constructed a concentrator based on a lens system designed to achieve the melting point temperature for platinum (1773ºC). But it was not until the twentieth century that the full range of possibilities of this energy source and its applications to the processing and modification of materials started to be explored in depth. The first great inventor was Felix Trombe who transformed German parabolic searchlights used for anti-aerial defence during WW II into a solar concentrator. Using this device he was able to obtain the high temperatures needed to carry out various chemical and metallurgic experiments involving the fusion and purification of ceramics (Chaudron 1973). In 1949 he was able to melt brass resting in the focal area of a double reflection solar furnace which he constructed using a heliostat or flat mirror and a parabolic concentrator (50kW Solar Furnace of Mont-Louis, France). But his greatest achievement was the construction of the largest solar furnace that currently exists in the world, which can generate 100kW of power. The "Felix Trombe Solar Furnace Centre" is part of the Institute of Processes, Materials and Solar Energy (PROMES-CNRS) and is a leader in research on materials and processes.

Another of the main figures in the use of solar energy in the materials field and specifically in the treatment and surface modification of metallic materials is Prof. A.J. Vázquez of CENIM-CSIC. His research in this field started at the beginning of the 1990's, using the facilities at the Almería Solar Plant (Vazquez & Damborenea, 1990). His role in encouraging different research groups carrying out work in material science to experiment with this new solar technology has also been very important.

Our group's main focus at the ETSII-UCLM involved using concentrated solar energy (CSE) from a Fresnel lens to propose new sintering processes and surface modifications of metallic components. The aim was to increase the resistance of metallic materials (mainly ferrous and titanium alloys) to wear, corrosion and oxidation at high temperatures.

The initial studies with CSE at the ETSII-UCLM involved characterising a Fresnel lens with a diameter of 900 mm, for its use as a solar concentrator (Ferriere et al. 2004). The characterisation indicated that the lens concentrated direct solar radiation by 2644 times, which meant that on a clear day with an irradiance of 1kW/m^2 the density of the focal area would be 264.4 W/cm^2 (Figure 1). This value is much lower than this obtained with other techniques based on high density beams, but is sufficiently high to carry out a large number of processes on the materials, and even a fusion of their surfaces.

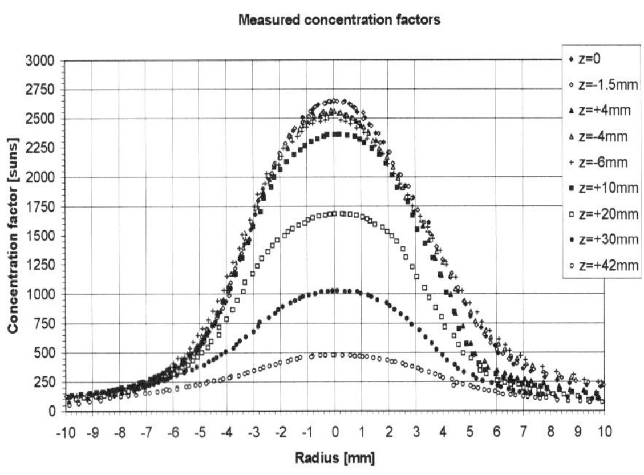

Fig. 1. Concentration factor of the Fresnel lens

The investigations carried out to date include processes involving the sintering of metallic alloys, surface treatment of steel and cast irons, cladding of stainless steel and intermetallic compound, high temperature nitriding of titanium alloys and NiAl intermetallic coating processing through a SHS reaction (Self-propagating high temperature synthesis). This research has been carried out in European and national programmes for Access to Large-Scale Facilities which allowed us to collaborate with the groups of A. J. Vázquez (CENIM-CSIC, Spain), A. Ferriere (PROMES-CNRS, France) and I. Cañadas (PSA-CIEMAT, Spain) and to use higher powered solar facilities such as the solar furnaces of PSA and the PROMES laboratory.
The aim of our research was not just to make inroads on the use of new non-contaminating technologies, which resolve environmental issues arising from high temperature metallurgy, but also to increase scientific knowledge about the mechanisms involved in these processes carried out at high temperatures under non-equilibrium conditions. In the studies we have conducted to date we have seen a clear activating effect in CSE which results in treatment times that are shorter, and which add to the efficiency of the process as well as increase in the quality of the modified surface. This is due to, among other factors, the properties of solar radiation. The visible solar spectrum extends from the wavelengths between 400 and 700 nm where most metals present greater absorbance, making the processes more energy efficient. In figure 2 (Pitts et al., 1990) the solar spectrum is compared to the absorbance values of the different wavelengths of iron and copper. The figure also includes the wavelength at which certain lasers (those which are habitually used in treating materials) operate. Here we see the high absorbance of iron for the more energetic wavelengths of the solar spectrum, and that its absorbance is low at the wavelengths, which the most common lasers use.

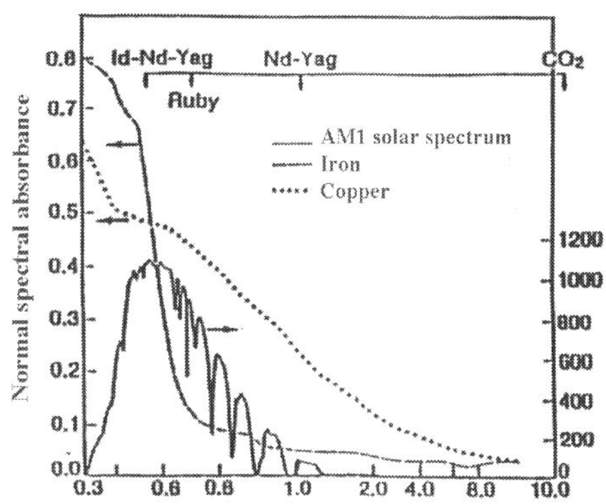

Fig. 2. Solar spectrum (Pitts el al. 1990).

Although the use of solar energy for industrial applications suffers a disadvantage due to its intermittent nature, it should be noted that according to Gineste (Gineste et al. 1999) in Odeillo where the Felix Trombe Solar Furnace Centre is located, the peak value of the direct normal irradiation is 1100 W.m^{-2} and it exceeds 700 W.m^{-2} during 1600 hours per year and 1000 W.m^{-2} during only 200 hours per year. In Ciudad Real, Spain, at latitude 38°, the availability of the solar energy reported by the Spanish "Instituto Nacional de

Meteorologia" (Font Tullot, 1984), is 11% higher than in Odeillo. Direct solar radiation measured with a pyrheliometer between 19 June and 31 August, 2009, at the ETSII-UCLM (Ciudad Real, Spain) registered values higher than 950W.m^{-2} for 20% of the days and higher than 800 W.m^{-2} for 97% of the days. The peak value has been attained in this period was 976 W.cm^{-2} .

2. Experimental Installations

There are various types of installations for concentrating solar energy. One way of classifying these installations uses the concentration process as a reference for differentiating between the different types. In this manner we can distinguish between installations which use reflection and those which use refraction.

Reflection installations

Reflection installations use mirrors to concentrate solar energy producing one diversion (direct concentrators) or several diversions (indirect concentrators) of the radiation. The light is reflected along the entire spectrum of wavelengths, since the mirror does not absorb anything. Direct concentrators are cylindrical parabolic mirrors and dish parabolic reflectors. First one uses the heat energy generated mainly to heat the fluids which circulate through the conduit located in the reflector focal line (Figure 3). Dish parabolic reflector may be full-surface parabolic concentrators when the entire surface forms an approximately parabolic shape or multifaceted concentrators composed of various facets arranged in a parabolic structure that reflects the solar radiation concentrating it in its focal point. The concentration factor depends on the size, aperture and quality of the surface. The solar radiation hitting the focal point has a Gaussian distribution and its energy efficiency is very high due to the high concentration.

Fig. 3. Cylindrical parabolic concentrators at the PSA (Almería Solar Plant).

The indirect concentrators are mainly the solar furnaces. They are systems that take advantage of the thermal energy generated by the sun for use in applications requiring medium to high temperatures. They are indirect concentrators that produce several diversions of the radiation through optical systems specially designed to deflect the incident light. To deflect the radiation, they use mirrored heliostats, completely flat surfaces that deflect the direct solar radiation. They are composed of flat reflective facets and have a sun-tracking system on two axes. Given that a single heliostat is usually totally flat, it does not concentrate. Therefore, a field of heliostats pointed towards a parabolic concentrator is used for this purpose (Fig. 4). The power concentrated may be regulated through an attenuator which adjusts the amount of incident solar light entering.

Fig. 4. Parabolic reflector at the PSA (Almería Solar Plant, Spain)

When the heliostat field is pointed towards a tower (Figure 5) is a direct concentrator because this system produces only one diversion of the solar radiation.

Fig. 5. Heliostat field with a central tower Solar Two, in Barstow, California

Refraction installations

In these installations solar light travels through a concentrator device that redirects the light towards its axis. These types of installations absorb part of the wavelength of the solar light. The most common way of concentrating solar radiation is through the use of converging lenses, which concentrate radiation in its focal point. Conventional lenses would need to be too large and too expensive to make them worthwhile for concentrating solar radiation at the required levels. An alternative to these types of lenses are Fresnel lenses, which serve the same function, but are much lighter and cheaper.

In Fresnel lenses, the curve of the surface is composed of a series of prisms or facets, in such a way that each of them refracts the radiation in the same manner as the surface of which they are a part. This is why a Fresnel lens functions like a conventional lens. The different polymers used in the manufacture of the lens determine the part of the spectrum in which it will be effective, and therefore, its applications. The lenses used for concentrating solar radiation are made of acrylic, rigid vinyl, and polycarbonate. Figure 6 shows how the facets of a Fresnel lens can be created from a conventional lens.

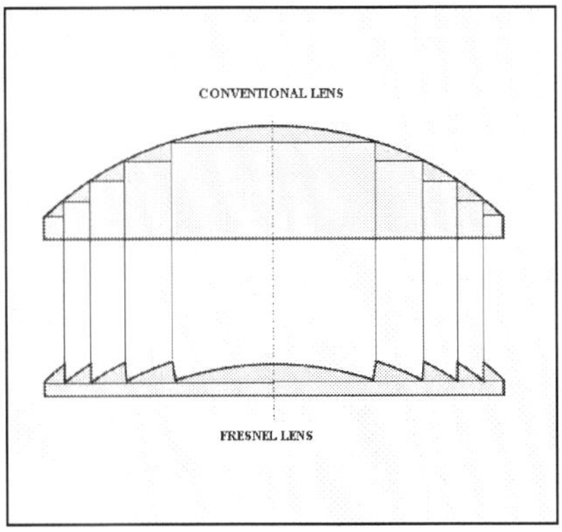

Fig. 6. Diagram of Fresnel lens

There are several research laboratories that use solar installations to experiment and study materials at high temperatures (higher than 1000°C). Table 1 lists the solar installations in operation across the globe, among which is the installation at ETSII in Ciudad Real.

Country	Location	Technology	Maximum power density (kW/m²)	Power (kW)
China	Guangzhou	Parabolic concentrator*	30000***	1.7
France	Odeillo, CNRS	Solar Furnace*	16000	1.5
		Solar Furnace	10000	1000
		Solar Furnace	4700	6
	Odeillo, DGA	Solar Furnace *	6000	45
Germany	Cologne, DLR	Solar Furnace	5200	22
Spain	Almería, PSA-CIEMAT	Solar Tower*	1000-2000	3360-7000
		Solar Furnace *	2500	60
	Madrid, CENIM-CSIC	Fresnel lens*	2640	0.6
	Ciudad Real, UCLM	Fresnel lens*	2640	0.6
Switzerland	Villigen, PSI	Solar Furnace	5000	45
		Solar Furnace	4000	15
		Parabolic concentrator	4000	70
Ukraine	Ac. of Science	Parabolic concentrator *	2500	-
USA	Alburquerque, Sandia	Solar Furnace *	3000	25
	Denver, NREL	Solar Furnace *	2500-20000**	10
	Minneapolis, Univ. Minn.	Solar Furnace	7000	6
Uzbekistan	Tashkent	Solar Furnace	17000	1000

*Used in the surface modification of materials (papers published), **Used as secondary concentrator, ***Calculated values

Table 1. Solar Installations in the World (Rodríguez, 2000).

2.1 Fresnel lens

The installation is on the roof of the Escuela Técnica Superior de Ingenieros Industriales building in the UCLM in Ciudad Real (Figure 7). The lens is affixed in a metal structure, and has a single-axis sun tracking system, connected to a software system in which the different data generated by the experiment can be collected, such as the values of different thermocouples. It also has a pyrheliometer which measures the direct incident solar radiation over the course of the day. The geometry of the lens is circular, with a 900 mm diameter and centre that is 3,17 mm thick. It is made out of acrylic material, which gives it a long useful life with low maintenance. The specification of the lens was determined in previous studies (Ferriere et al., 2004) which allowed the measurement of the concentration factor along the focal axis. The focal point of the lens is 757 mm from its centre. This is the point where the greatest density of energy is reached. The lens concentrates direct solar energy by up to 2644 times (maximum value at the focal point), which means that for exposure of 1000 W/m^2 the maximum power density at the focal point is 264 W/cm^2.

Fig. 7. Fresnel lens at the ETSII (Ciudad Real, Spain).

The density of the solar radiation has a Gaussian distribution in function of the distance from the focal point within the focal plane. This variation is what allows us to choose the temperature to be used for the experiment. We can control the energy density of the solar radiation, adjusting the distance of the sample in the Z axis. (Figure 1).

The Fresnel lens has a reaction chamber where experiments can be carried out in a controlled atmosphere. The reaction chamber is features a quartz window and a refrigeration system. In order to measure the temperature a thermocouple is welded to the bottom of the samples.

2.2 Solar Furnace

The second installation used on a regular basis for generating concentrated solar energy is the Solar Furnace of Almería Solar Plant (PSA), which belongs to the Centro de Investigaciones Energéticas, Medioambientales y Tecnológicas (CIEMAT, in English, Centre of Energy, Environmental and Technological Research). The solar furnace consists of a heliostat which tracks the sun and reflects the solar rays onto a parabolic mirror. The furnace of PSA has a heliostat of 160m^2 composed of 28 flat facets which reflect solar rays perpendicular and parallel to the optic axis of the concentrator and continuously tracks the sun through a tracking system with two axes (Fig. 8). The mirrors have reflectivity of 90%.

Fig. 8. Heliostat of the PSA solar furnace (Almería Solar Plant, Spain).

The concentrator disk is the main component of the solar furnace (Fig. 9). It concentrates the incident light of the heliostat, multiplying the radiant energy in the focal zone. Its optic properties especially affect the distribution of the distribution of the flow on the focal zone. It is composed of 89 spherical facets covering a total surface area of 98,5 m^2 and with a reflectivity of 92%. Its focal distance is 7,45 m. The parabolic surface is achieved with spherically curved facets, distributed along five radii with different curvatures, depending on their distance from the focal point.

Fig. 9. Concentrator disc of PSA

The attenuator (Fig.10) consists of a set of horizontal louvers that rotate on their axes regulating the entry of incident solar light hitting the concentrator. The total energy on the focal zone is proportional to the radiation that passes through the attenuator. The concentration and distribution of the power density hitting the focal point is key factor in a solar furnace. The characteristics of the focus with the aperture 100% opened and solar radiation of 1000 W/m^2 are: peak flux: 3000 kW/m^2, total power: 58 kW, and a focal diameter of 25 mm. In this case, the reaction chamber also allows work to take place in a controlled atmosphere. The chamber also has a quartz window which allows concentrated solar energy to enter and also allows researchers to monitor the experiment using different kinds of cameras (digital and IR).

Fig. 10. Attenuator of the solar furnace of PSA (Almería Solar Plant)

2.3 Solar Furnace at PROMES-CNRS

Another solar facility used in our research is the 2kW parabolic solar furnace at the PROMES-CNRS laboratory (France). The furnace is composed of one heliostat and a parabolic reflector with a 2 m diameter. The parabolic concentrator has a vertical axis which allows the samples used in the experiments to rest in a horizontal position without the need to add more optical diversion systems to the device. Furthermore, given that the parabolic reflector is the only mirror which is not faceted, it has a higher quality optical properties and allows a greater concentration factor that that which is obtained with the faceted reflectors. The focal zone behind the second reflection has a diameter of 15 mm and Gaussian distribution with the maximum energy at the centre, of 16.000 times the impinging solar radiation.

3. Surface hardening of steels by martensitic transformation.

Surface quenching is a widely used treatment by the industry to harden and improve wear resistance of the steel pieces. These types of treatments may be carried out using conventional heating methods (flame and electromagnetic induction) and high-density energy beams (laser, electron beams, plasmas, etc.). In all cases the source should be sufficiently powerful to guarantee that only the surface layer of the piece heats up to a higher temperature than the austenizing temperature. After cooling off, only the zones which were previously austenized will have undergone the martensitic transformation that results in the hardening. In the internal zones, where no microstructural transformations would have taken place, the mechanical properties would remain unchanged. Therefore, the end result is pieces that combine a high degree of hardness and toughness and greater resistance to wear.

Of all the different types of modifications and treatment of materials carried out in the solar furnaces, surface hardening of ferrous alloys has been the most widely studied. Since the first study was published by Yu and others in 1982, several research groups have been exploring the possibilities of this process (Maiboroda et al., 1986; Stanley et al., 1990;

Ferriere, 1999). This first study showed how the high concentrations obtained in the solar focal area of a parabolic concentrator with a 1.5 m diameter produced self-quenching in a surface zone 0.5 mm deep and 5 mm in diameter in a steel piece after a second of exposure to solar radiation (Yu et al. 1982). In addition, the initial investigations show how localised treatments can be carried out on industrial pieces with complicated geometries by moving the sample with respect to the focal area of the furnace (Yu et. al, 1983) (Zong et al., 1986).

In Europe, the first experiments were carried out in the 1990's by a group led by Prof. Vázquez of CENIM-CSIC (Spain). The research carried out was highly important because it demonstrated the viability of using the different types of solar facilities available at the Almeria Solar Plant to surface harden steel pieces. The experiments were carried out using the SSPS-CRS facility, which comprises a heliostat field and a central tower (Rodríguez et al., 1995), and the Parabolic Solar Furnace which comprises a group of heliostats and a faceted parabolic concentrator (Rodríguez et al., 1997). The results indicated that under the best conditions of direct solar radiation it was possible to obtain homogeneous quenched layers between 1 and 3 mm thick with heating times of between 30 and 60 seconds. The study was completed using a Fresnel lens with a 900 mm diameter which was available at the Instituto de Energías Renovables of CIEMAT in Madrid (Rodríguez et al., 1994). The study added to knowledge regarding the advantages and limitations of each one of the facilities. With the facility comprising the central tower and the heliostat field, a surface of 10 cm^2 can be quenched, much more than what is possible using other techniques and types of solar facilities. But with the PSA Solar Furnace and the Fresnel lens it is possible to obtain greater energy densities in the focal area (250-300$W.cm^{-2}$) depending on the incident solar radiation, which allows for self-quenching in steel alloys.

Using the ETSII-UCLM Fresnel lens described above, our group carried out research on surface hardening steels and cast iron, with the ultimate aim being the discovery of industrial applications for this process. The first experiments consisted of surface hardening through martensitic transformation of three types of steel and a nodular cast iron piece.

In all cases the influence on the treatment of the different variables was assessed: heating rate, maximum temperature reached, cooling medium, size of the treated pieces (diameters: 10 and 16 mm, height: 10 and 15 mm). The study entailed determining the microstructural transformations, the profile of the hardness and the depth of the quenching (total and conventional).

Figure 11 shows the results obtained using a sample with a 10 mm diameter and 10 mm high of tool steel AISI 02, where the homogeneous quenching can be seen along the entire diameter of the test sample, as well as the surface hardness values obtained. Figure 12 shows the results obtained after carrying out a surface quenching treatment of a nodular cast iron piece.

In addition, studies were carried out to assess the possibility of carrying out localised treatments on the surfaces of pieces that required greater hardness and resistance to wear. The microhardness curves in Figure 13 show the effect of heating time on the diameter of the quenching zone of a 1 mm plate of martensitic stainless steel AISI 420.

Due to the fact that the heating conditions depend on the direct solar radiation it is necessary to have a predictive tool that can set the treatment conditions in function of the direct solar radiation present. To this end, a finite element model (FEM) (Serna & Rodríguez, 2004) has been developed which gives the distribution of the temperatures of the pieces during treatment. The model takes into account both the Gaussian distribution of the energy density in the focal area and the variation in the temperature of the phase transformation of the steel in function of heating speed.

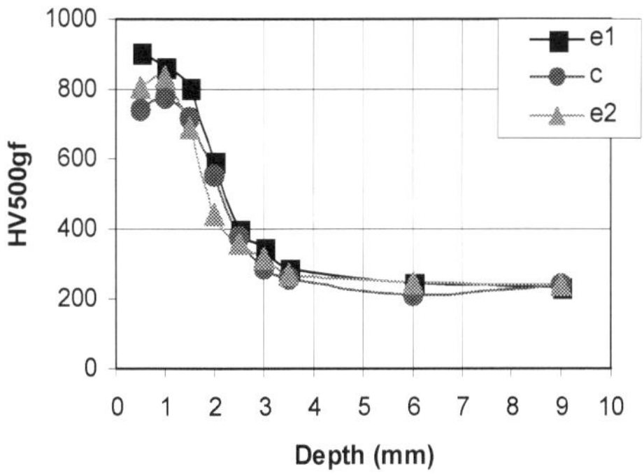

Fig. 11. Surface quenching of 10 mm high test sample after 45 seconds of heating.

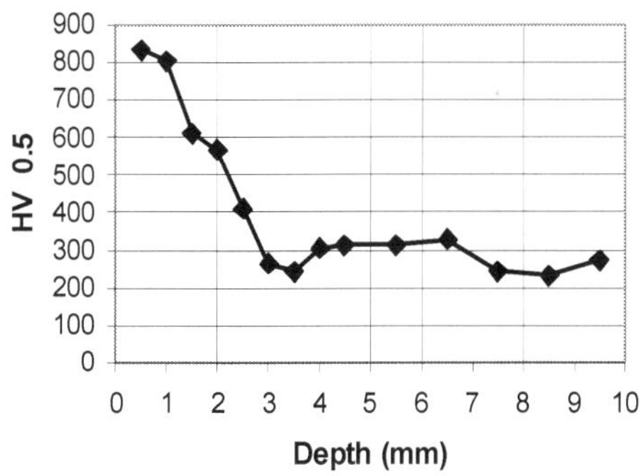

Fig. 12. Microhardness profile of a nodular cast iron piece heated for 40 seconds in the focal area of a Fresnel lens.

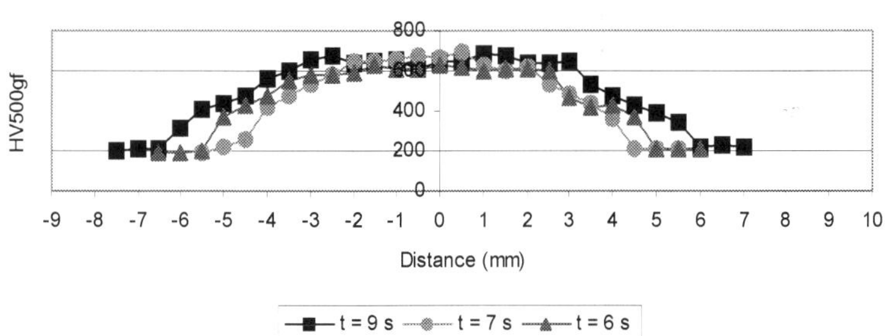

Fig. 13. Influence of heating time on the diameter of the quenching zone of a 1 mm thick plate of martensitic stainless steel.

The size of the focal area of the lens limits the possible application to small pieces (surface areas of between 50 mm^2 and 100 mm^2, depending on the maximum required temperature), or to localised treatments on larger sized pieces. Obviously treatments and modifications of metallic materials with small surface areas are carried out on an industrial scale for a large number of applications, but there are few bibliographic references concerning specific applications for localised treatments using solar facilities.

4. Hardening through surface melting of cast iron

The surface melting treatment of grey cast iron leads to the formation of superficial layers with excellent resistance to wear. The rapid cooling from the melted state gives rise to the formation of extremely hard white cast iron with great resistance to wear. Given this profile, this type of melting process is an excellent candidate for machine pieces that are subject to movement and vibrations and which are in contact with other components. At present, industrial processes are using various heating techniques such as TIG, electromagnetic induction, and electron or laser beams, among others, in order to carry out this type of surface melting treatment.

Recent experiments at the ETSII show how it is possible to use concentrated solar energy to carry out at the melting treatments on industrial cast iron pieces. The study to date has centred on hardening through surface melting and quenching of camshafts used in the automobile industry and manufactured with grey cast iron. In order to carry out the treatment a specimen support device was created which allowed the cam to spin in the focal plane of the Fresnel lens. The results obtained are compared with those of a camshaft manufactured using conventional industrial processes (TIG). Figure 14 shows the hardness profiles of the hardened area of the two pieces, one that was hardened with concentrated solar energy, and the other through TIG (both made of the same cast iron). The graph shows us that the camshaft treated with CSE attained a far greater hardness and had a smaller heat affected zone. In addition, the surface finishing attained from the solar treatment is better than that obtained after treatment with TIG, which would translate into lower costs (Figure 15). Current research is focused on automating the system in such a way that the entire camshaft may be treated continuously.

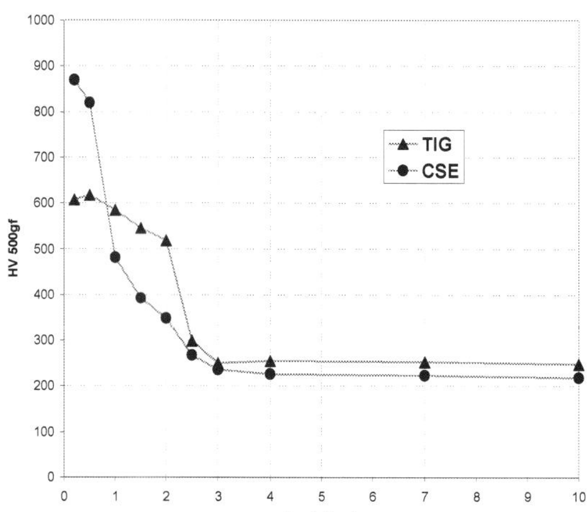

Fig. 14. Hardness profile of samples treated by TIG vs. CSE

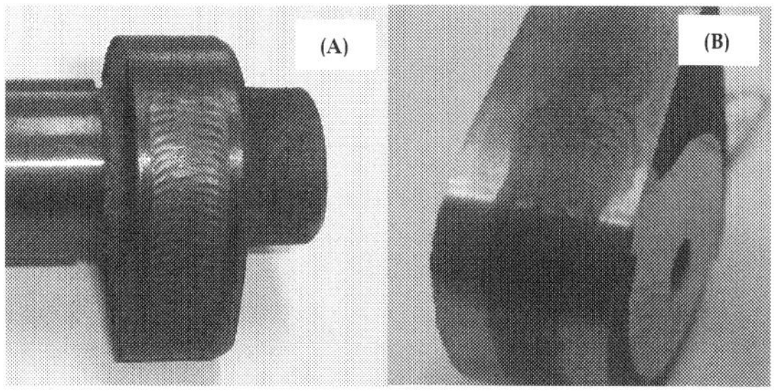

Fig. 15. Surface finishing after TIG (A) or CSE (B) treatment.

5. Cladding of stainless steel and intermetallic compounds onto steel substrates.

There are many types of surface modification techniques that aim to improve the corrosion and oxidation resistance. One widely studied surface modification technique is cladding. A material with the desired properties is melted on the base metal by means of an energy beam. The mixture between the coating material and the base metal must be as small as possible in order to guarantee the original properties of the coating. In this way it is possible to use a cheap structural material and to coat it with another that confers its surface the desired properties.

The literature contains several references that describe the use of different solar installations to obtain cladding coatings. In the USA Pitts et al., (Pitts et al., 1990) obtained cladding coatings on stainless steel. Subsequently, in Spain, Fernandez et al. performed Ni cladding on steel at the Almeria Solar Plant (Fernandez et al., 1998).

We have studied the possibility of obtaining cladding coatings using the parabolic solar furnace of the PROMES-CNRS laboratory previously described (Ferriere et al., 2006). The high energy densities obtained with the solar beam allowed stainless steel and NiAl to be cladding coated through rapid melting-solidification of powders pre-deposited on carbon steel samples. Coatings have been processed in tracks by scanning the concentrated solar beam across the specimen surface with the aim of modifying larger areas than are possible with a stationary treatment. The scanning process is performed by moving the specimen at a controlled speed that depends on the direct solar irradiation. The coatings processed are homogeneous, adherent and have low porosity. In addition, the formation of dendritic microstructures results in increased electrochemical corrosion resistance.

The fundamental disadvantage that has been encountered in this research is the difficulty of achieving a coating with a composition close to that of the initial powder while at the same time guaranteeing good adhesion to the substrate. A possible solution to this problem consists of using a powder injector (nozzle), in order to carry out the process in one single step as is habitual in the case of laser cladding.

6. Salt-bath nitriding of steels.

It is possible to harden the surface of different kind of steels using a novel technology that combines the use of non-contaminant salts with the activator effect of the concentrated solar energy. Groundbreaking research (Shen et al. 2006a), (Shen et al. 2006b) has studied the

possibility of the substituting highly contaminating cyanide salts used in liquid nitriding for common salt KNO_3, which avoids this high toxicity. In line with this innovative research vector, our research has tried to explain the different mechanisms through which nitrogen from the salt is introduced into the steel matrix, thereby achieving the desired surface hardening. In addition, we are researching the use of concentrated solar power as an energy source (Herranz & Rodríguez 2008). The experiments were carried out using the Fresnel lens at the ETSII and the solar furnace at the Almeria Solar Plant. The results were compared to those that were obtained using an electric muffle furnace. Steels with a wide range of characteristics were selected for the research. One was a relatively cheap, low alloy steel with low carbon content, the AISI 1042. The other steel used was the high-speed tool steel M2 which is commonly used in industry. These two types of steel were used to highlight the different characteristics made evident during the nitriding process. The treatment resulted in the surface hardening of the two steels through the interstitial diffusion of nitrogen in the steel network and in some samples nitrides were formed. The exhaustive study of these results interprets the nitriding mechanism that occurred in each steel. In addition, the traditional treatment times were reduced to a large degree. This evidences the viability of this nitriding process using concentrated solar energy (CSE).
The maximum surface hardening of AISI 1042, which is not a conventional steel to use for nitriding, increased 61% with respect to its nominal value. Nitrided M2 steel attained a surface hardness of 900HK (Herranz & Rodríguez, 2007) (Fig. 16).

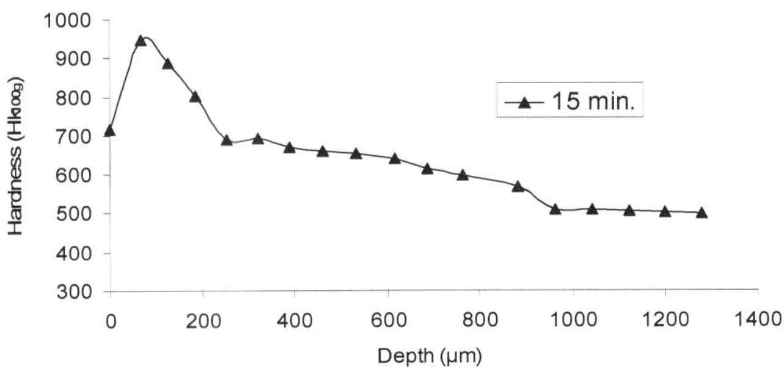

Fig. 16. Hardness profile obtained in a nitrided M2 piece using a treatment of nitrate salts during 15 minutes using concentrated solar energy.

The diffusion layer obtained in both steels submitted to the nitriding process in KNO_3 salts are greater than those obtained in earlier experiments with other nitriding methods. Therefore, the maximum diffusion layer obtained in AISI 1042 steel has an approximate thickness of 1300 µm and 550 µm in the M2 steel.
The other fundamental conclusion of our research is that concentrated solar energy activates the process, notably reducing the treatment times. Treatment time was greatly reduced in the treatment of the AISI 1042 using the Fresnel lens. Only 35 minutes was needed, versus the 90 minutes needed in a muffle furnace for the same treatment. This reduction is especially notable if we compare it with other conventional nitriding methods, such as with plasma, which requires 3 hours. In the case of the nitriding of the M2 sample, the reduction in treatment time using concentrated solar energy is more significant. With a Fresnel lens, only 60 minutes was needed (a muffle furnace requires 3 hours). The reduction was even

more remarkable in the case of PSA (Almería Solar Plant), which only needed 15 minutes of treatment.
The nitriding of low alloy steel characterised by its free interstices, such as AISI 1042 steel, occurred primarily through the interstitial diffusion of nitrogen coming from the salts towards the interior part of the steel piece. In the nitriding mechanism for the M2 high-speed steel, high alloy tool steel, part of the nitrogen originating from the salts was introduced through interstitial diffusion towards the interior of the steel, while another part formed iron nitrides and/or iron alloy nitrides. These compounds contribute to the surface hardness.

7. Gas nitriding of titanium alloys.

In recent years, Ti alloys have been widely studied due to their properties: low density, high melting point, good mechanical properties, high corrosion and oxidation resistance and biocompatibility. These properties explain their appeal to the aerospace industry and as biomaterials. However their use in these fields remains limited because of their poor tribological properties. These problems can be overcome by changing the nature of the surface using thermochemical treatments. In the case of the Ti alloys the most widely surface treatment used is the nitriding. There are different ways to nitride the surface. One of the most frequently method is the gas nitriding.
Gas nitriding is a diffusion process that involves heating the surface at high temperatures for a long time. In most cases, the equipment is expensive both start up and to maintenance. For this reason there is significant interest in developing more economic and energy efficient systems. Nitrogen is soluble in titanium and forms a interstitial solid solution, resulting in a hardening by solid solution caused by deformation in the crystalline network. It is important to assure that no oxygen is present in the process in order to prevent the formation of TiO_2 oxides on the piece. The types of dissolution of nitrogen in the titanium may be seen in the equilibrium diagram of the Ti-N system (Fig. 17) (Wriedt &

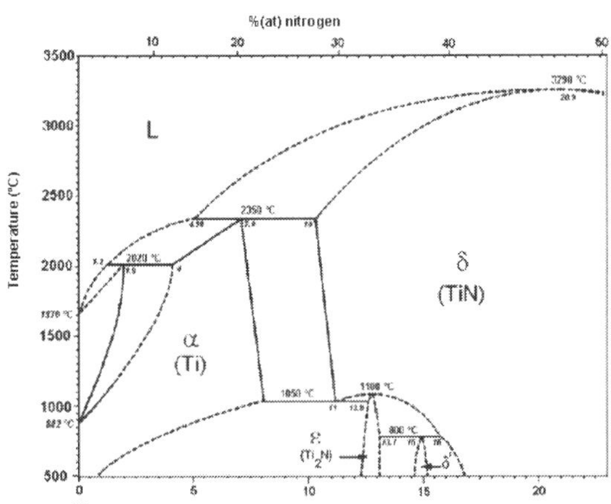

Fig. 17. Phase diagram Ti-N. (Wriedt & Murray, 1987).

Murray, 1987). We observe a solid phase α solution with a high degree of solubility of nitrogen (from 0 to 8% in weight) which remains stabilised at high temperatures. At levels of over 8% of nitrogen (in weight) the alpha phase saturates and intermetallic compounds are obtained, such as Ti_2N and TiN. These nitrides have very high hardness values and give the maximum hardness to the surface layer.

Existing studies indicate that there are some problems related to the use of this treatment in the traditional manner, which uses electric furnaces, such as the need for long treatment times of over 16 hours and at high temperatures of over 1000° C. In contrast, the use of concentrated solar energy obtains high temperatures at high heating and cooling rates (that allow the creation of non-equilibrium microstructures) with short treatment times. We have found that due to the photoactivation capability of the concentrated solar energy, the duration of the process can be greatly reduced. To our best knowledge, the study by Vazquez in 1999 is the only significant research that has been conducted in this area (Sánchez Olías et al, 1999).

Our research focuses on the gas nitriding of the Ti6Al4V alloy. The base material has a biphasic microstructure (α+β) and a hardness of 400 HK. Tests involved applying heat of between 1000° C and 1200°C for between 5 to 30 min. In order to measure the temperature, a thermocouple was welded to the bottom of the samples and the samples were situated in the reaction chamber where the nitrogen atmosphere was controlled (Fig. 18). The chamber has a quartz window which permits the entrance of the concentrated solar energy and also it allows to observe and record the experiment by different kind of cameras (digital and IR).

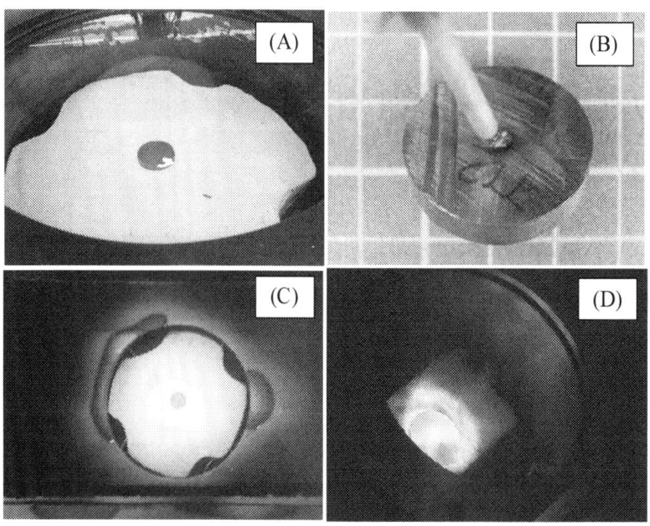

Fig. 18. (A) The samples in the reaction chamber. (B) Sample with the thermocouple. (C) Digital record of the experiment in the solar furnace. (D) Direct observation of the experiment in the Fresnel lens.

The results obtained indicate that with the solar facilities significant hardening can be obtained from processes carried out at 1050°C for only 5 minutes. In the microstructure we observed a layer of a different kind of nitrides identified by X-Ray diffraction that increase the hardness of the samples. In addition, nitride layer growth occurred where the treatment time was increased. Besides the hardening due to the formation of nitrides, we also observed a deeper hardening due to the interstitial solid solution of the nitrogen in titanium matrix. As shown in Figure 19 we have been able to distinguish between the different hardened layers. The first is a compound layer, then a diffusion layer that can achieve 400 µm for the longest treatment times and then, the base material.

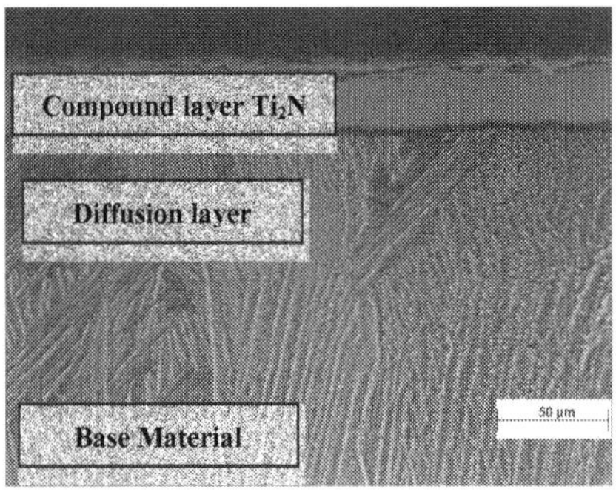

Fig. 19. Microstructure of sample nitrided during 10 min at 1050°C using concentrated solar energy.

The compound layer was identified through X-ray diffraction as Ti_2N, which was the maximum hardness detected (1200HK). The presence of TiN was also detected, but its density was so scant that it was impossible to measure its hardness. In addition, it was evident that the process did not result in a completely continuous layer at this temperature.

Compound layers grew considerably as the treatment temperature rose or time was increased. These layers covered the pieces more homogeneously and the diffusion layer was thicker. As shown in figure 20, at 1200°C, after 15 minutes of treatment, two totally continuous layers were formed. According to the X-ray diffraction, the exterior layer was TiN, and the second layer was Ti_2N. Under these parameters, the TiN layer had a maximum hardness of 2600 HK.

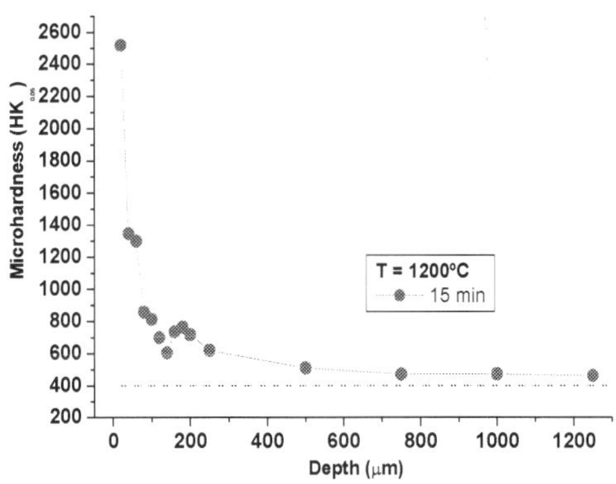

Fig. 20. Microhardness evolution in the samples nitrided during 15 min at 1200°C using CSE

After these experiments, wear resistance was evaluated using a pin on a disc test machine. The tests were made in dry conditions against a ball of aluminium oxide. In the case of the samples nitrided in the Fresnel lens, the samples showed a lower coefficient than the as-received material, with a friction coefficient of 0.5, showing values of 0.2 for treatment of only 15 minutes at 1200°C. In regard to the wear rate, compared with the base material it decreased by two orders of magnitude after only 15 minutes of treatment. The track width was decreased from 1.890 µm in the Ti6Al4V alloy to 180 µm in the nitrided sample. These experiments show the significant potential of this new modification process consisting of gas nitriding with concentrated solar energy. The great reduction of nitriding time can be explained by the photo-activation effect of the concentrated solar energy.

8. Solar sintering of metallic powders.

The feasibility of using a solar furnace for the sintering-consolidation of green parts previously obtained by compaction has been tested. The use of solar furnaces allows materials to be processed at much higher heating and/or cooling rates, which results in better mechanical properties and a quite significant reduction in the total cycle time. The demand for faster and less expensive techniques for processing pieces has resulted in the development of high energy techniques for manufacturing prototypes and carrying out short series tooling, such as laser sintering (Asgharzadeh & Simchi 2005) and capacitor discharge sintering (CDS) (Fais & Maizza 2008). The use of these faster technologies has usually resulted in finer microstructures, while improving mechanical properties and the sintering window. In the search for new sintering systems that overcome this issue, the use of concentrated solar energy (CSE) appears to be an interesting candidate. The CSE is clean, renewable and pollutant free. In spite of its apparent limitations for industrial applications, due to the unpredictable availability of solar radiation, CSE shows a clear activator effect in different processes (Herranz & Rodriguez 2007); (Herranz & Rodríguez 2008).
Earlier research focused on studying the technique's viability for manufacturing WC-10%Co ceramic components and complex ceramics (cordierite) (Guerra Rosa et al., 2002); (Almeida Costa Oliveira et al., 2005); (Cruz Fernandes et al., 2000) using concentrated solar energy. These studies have proven that it is possible to use CSE to obtain pieces with very similar characteristics to those obtained using conventional electric furnaces. In the same line, a preliminary study of a copper system has also been published (Cañadas et al., 2005). The results of these studies all share two characteristics: the good mechanical properties obtained, and the major reduction in total treatment time required.
We have assessed the feasibility of concentrated solar energy for sintering-consolidation of green parts previously obtained using compacted metallic powder. The experiments were carried out using the two facilities described in this chapter; a parabolic solar furnace of the Almeria Solar Plant (PSA) and a Fresnel lens at the Castilla-La Mancha University (ETSII-UCLM). The research focused on the microstructural evolution of the samples treated in both facilities. In this work we compared the results of the sintering process in a N_2-H_2 atmosphere using concentrated solar energy with the conventional furnace results. The main materials we worked with were M2 tool steel, copper-based alloys (such as bronze), and carbon-based steels with alloys such as Astalloy. The energy density obtained in the focal area of the solar furnace or in the focal area of the Fresnel lens is high enough to raise the sample's temperature to sintering levels. To process the samples at different sintering temperatures, experiments were conducted changing the focal length. The samples were

then heated up to the maximum temperature for 30 minutes. The total duration of the sintering process (including the heating and the cooling) was less than 70 minutes in all cases. In order to compare results, the sintering experiments were carried out in a conventional tubular furnace in the same atmosphere, applying conventional sintering temperatures for 30 min (dwell time). (Figure 21).

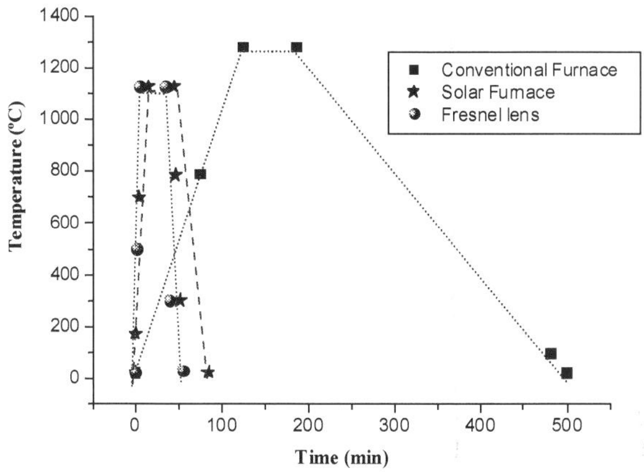

Fig. 21. Cycle in the conventional furnace and in the solar installations comparing the duration of the sintering processes.

We made several interesting findings. For example, we found that the high-speed M2 steel showed major differences in the microstructures obtained, depending on the sintering process used. The microstructure of M2 sintered in a conventional furnace in N_2-H_2 atmosphere at the optimum sintering temperature, around 1290°C, presented a ferrite matrix with some retained austenite, Figure 22. We also observed a significant amount of homogeneously distributed bright rounded M6C carbides (identified as rich in W and Mo). The sintering atmosphere has an important influence on the development of the microstructure because the nitrogen content of the steel slightly increases the optimum sintering temperature. The increase in nitrogen content has no effect on the M_6C carbides but, as pointed out in a previous study by Jauregi (Jauregi et al 1992), the MC carbides (rich in V) transform into MX carbonitrides that appear with grey contrast inside the grains and at the grain boundaries. In the conventional furnace an increase of the temperature up to 1300°C produces an over-sintered microstructure in which a continuous M_6C carbides film around the grain boundaries is observed. The EDX analysis revealed that at high temperatures the MX carbonitrides were able to transform themselves into black-contrast square VN nitrides detected at the grain boundaries. The maximum hardness achieved was 550 HV.

In the case of the samples treated in the PSA solar furnace and with the Fresnel lens, we observed a completed densification. The most remarkable discovery was that for the processes carried out using concentrated solar energy, the process activated at lower temperatures. Figure 23 shows the densification process and the microhardness evaluation. The feasibility of the sintering process using CSE was demonstrated as the parts displayed well-defined necks and greatly reduced porosity.

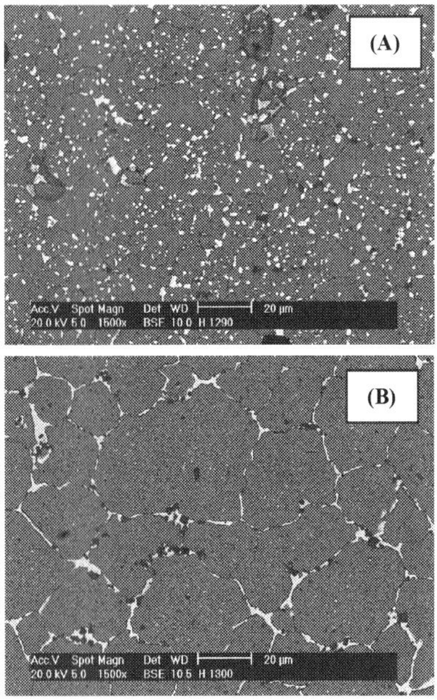

Fig. 22. SEM micrograph of M2 sample sintered at (A) 1290°C and (B)1300°C in nitrogen-hydrogen atmosphere in the conventional furnace.

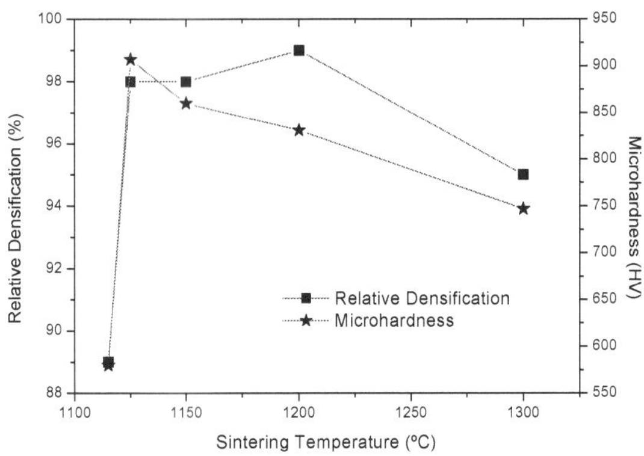

Fig. 23. Densification process of the samples treated in the solar furnace of PSA.

As can be seen in the Fig. 24 (A), at only 1115 °C, some rounded and isolated porosity was observed and a homogeneous distribution of carbides indicates that the sintering process really took place. Microhardness measurements indicate that, at this temperature, the distribution of the carbides was homogeneous and the porosity had disappeared in most of the sample. The microhardness achieved values similar to those of the samples sintered in the conventional furnace (580 HV). The microstructure obtained at this temperature was the same as that obtained at 1260°C in the electric furnace. When the sintering temperature was increased to 1125 °C, Fig. 24 (B), full density was achieved and there was a sharp increase in grain growth. In addition, massive bright carbide segregation was observed, as well as formations corresponding to the eutectic phase, but only in some areas. Most of the grain boundaries did not show carbides, indicating that they underwent a partial dissolution during the sintering process. Based on EDX analysis, we ascertained that the increase of nitrogen in the matrix due to the sintering gas allowed the formation of VN nitrides both inside the particles and along the grain boundaries. In spite of the grain growth, this sample's hardness was over 900 HV.

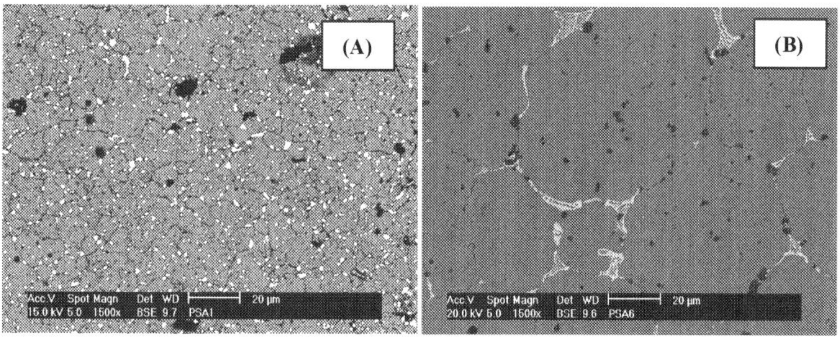

Fig. 24. SEM micrograph of M2 sample sintered at (A) 1115°C and (B)1125°C in nitrogen-hydrogen atmosphere in the solar furnace.

As part of our efforts to explain these results, we carried out several preliminary studies using transmission electron microscopy (TEM). Initial analysis corroborated that the high velocities obtained had caused the formation of vanadium-rich nanometric particles on the order of 400 nanometres, which in certain cases may be complex and contain small 30 nanometre particles rich in other steel alloy elements (W, Mo and V) in their interior. The presence of these particles would explain the high hardness values obtained. However, it would be necessary to carry out a more in depth analysis to fully understand the mechanism responsible for the formation of these small particles.

The results obtained using the Fresnel lens are similar. In figure 25, the large darker particles (identified as VN) stand out in particular and are precipitated at the grain boundaries or close to them, with an average size of 2 µm. In addition, there are a large number of small nanometric particles dispersed throughout the matrix, which once again explain the high hardness values obtained, despite the rapid growth observed in the grain. Lighter contrast carbides (rich in W and Mo) are concentrated at the grain boundaries while larger particles are distributed homogenously throughout the sample. In addition, eutectic M_6C "fishbone" structures can also be observed at some of the grain boundaries.

We would highlight that certain pieces were totally densified at lower temperatures, 150°C lower than with a conventional furnace. Almost full density was attained after 75 minutes in the PSA and after 50 minutes in the Fresnel lens installation, with higher hardness than the conventional microstructures (760-900 HV). The higher heating and cooling rates could explain these results, since the equilibrium phase diagram changes with the heating rate and in these new heating conditions new sintering mechanisms have occurred. The microstructure and the hardness measurements were similar to those found in the M2 system treated by selective laser sintering (Niu & Chang, 2000).

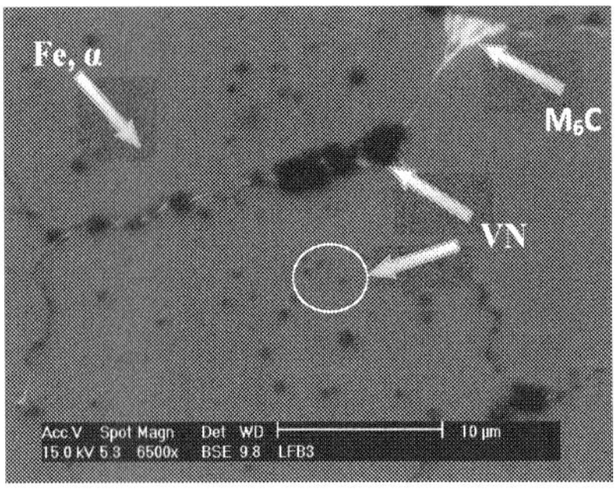

Fig. 25. SEM micrograph of M2 sample sintered in nitrogen-hydrogen atmosphere in the Fresnel lens.

9. Processing of intermetallic coatings through a self-propagating high temperature synthesis process initiated with solar energy (SHS-CSE).

The intermetallic compound NiAl shows attractive properties such as low density, high melting temperatures, high thermal conductivity and good mechanical behaviour at high temperature. Although its low ductility at room temperature could limit its applications as structural material its good tribological properties and excellent oxidation resistance justify its use as a protective coating for metallic components. Cladding, and self-propagating high temperature synthesis (SHS) processes are among the other techniques used currently for coating this intermetallic compound (Matsuura et al., 2000).

SHS is an energy efficient method to process advanced ceramic materials and intermetallic compounds. Discovered by Merzhanov in 1967 (Merzhanov, 1967) it uses the highly exothermic properties of a chemical reaction that are sustained and propagated through a mix of reactants (usually in powder form) in the form of a combustion wave. The energy savings and the economic benefits obtained from SHS are widely acknowledged (see, for example, Deevi & Sikka, 1997).

Concentrated solar energy was first used for the coating process using an SHS reaction in a study by G.P. Rodríguez et al. (Rodríguez et al., 1999) in which they were able to process coatings of NiAl on circular samples with a diameter of 30 mm, using a Fresnel lens. The solar energy triggered a exothermic reaction between the Ni and the Al at ignition temperature (the melting point of Al). The heat released by this reaction was transferred through conduction to the adjacent zones, initiating the reaction in these zones again, for which reason the reaction is considered self-propagated. The heat released (combined with that of the solar beam) melts the compound obtained, resulting in a dense non-sinterized material. This process was optimised by Sierra and others (Sierra & Vázquez, 2005), where the adherence of the coating to the substrate was improved through the electrodeposition of a nickel layer prior to the treatment. The study carried out with the Fresnel lens of the ETSII-UCLM used elemental Ni and Al powders and resulted in adherent NiAl coatings over cast iron. In order to increase the adherence and decrease the porosity that frequently occurs in coating processes that use SHS, a preheating system was designed for the substrate, which also uses solar power.

In addition, the 2 kW solar furnace at the PROMES-CNRS laboratory was used (under the collaboration framework between CENIM-CSIC, PROMES-CNRS and ETSII-UCLM) to obtain NiAl coatings in track forms, processing larger surface areas than those that were processed using the Fresnel lens (Sánchez Bautista et al., 2006). The scanning method is similar to that used when coating surfaces using solar cladding, although the beam-sample interaction times are lower (due to faster movement of the sample). The main difficulty arising when coating surfaces using a solar assisted SHS process lies in controlling the process variables so that the resultant coating is of low porosity, high adherence and with the required chemical composition. A possible solution may involve preheating the substrate. Another challenge is to optimise the processing of coatings over large areas.

10. Conclusions

Concentrated solar energy (CSE) represents an alternative to other types of energy beams for treating and modifying the surfaces of metallic materials. The research conducted by the Metallic Materials group of ETSII-UCLM (Spain) is interesting for two reasons. First, it is breaking new ground in the use of new non-contaminating and environmentally acceptable technologies for processes involving the surface modification of metallic materials at high temperatures. Second, it is increasing scientific knowledge about the mechanisms involved when processing materials at high temperatures under non-equilibrium conditions. In the studies carried out to date, it has been observed that CSE has a clear activator effect, which results both in shorter treatment times and therefore, in increased processing efficiency, and in improved quality of the modified surface.

For high-temperature processes, concentrated solar energy has shown to be highly energy efficient and also competitive in terms of cost. It is especially suitable for countries such as Spain, which has a high number of sunny days per year. Going forward, one of the main challenges for the scientific community will be to develop industrial applications for solar technology, especially in the current context, where energy-efficiency and environmental preservation have become top social priorities.

11. References

Almeida Costa Oliveira, F.; Shohoji, N. ; Cruz Fernandes, J. & Guerra Rosa, L. (2005). Solar sintering of cordierite-based ceramics at low temperaturas, *Solar Energy*, 78, 3, 351-361 ISSN 0038-092X

Asgharzadeh, H. & Simchi A. (2005). Effect of sintering atmosphere and carbon content on the densification and microstructure of laser-sintered M2 high-speed steel powder, *Materials Science and Engineering A*, 403, 290-298, ISSN 0921-5093

Cañadas, I.; Martínez, D.; Rodríguez, J. & Gallardo, J.M. (2005). Viabilidad del uso de la radiación solar concentrada al proceso de sinterización de cobre, *Rev. Metal. Madrid*, Vol. Extr. 165-169, ISSN 0034-8570

Chaudron, G.& Trombe, F. (1973). *Les hautes températures et leurs utilisations en physique et en chimie*, Masson et Cie, ISBN 2225360698; Paris, France.

Cruz Fernandes, J.; Amaral, P.M.; Guerra Rosa, L. & Shohoji, N. (2000). Weibull statistical analysis of flexure breaking performance for alumina ceramic disks sintered by solar radiation heating, *Ceram. Int.* 26, 203–206 ISSN 0272-8842

Deevi, S.C. & Sikka, V.K. (1997) Exo-Melt_ process for melting and casting intermetallics, *Intermetallics*; 5, 17-27, ISSN 0966-9795

Fais, A. & Maizza, G. (2008). Densification of AISI M2 high speed steel by means of capacitor discharge sintering (CDS), *Journal of Materials Processing Technology*, 202, 70-75, ISSN: 0924-0136

Fernández, B.J. ; López, V.; Vázquez, A.J. & Martínez, D. (1998). Cladding of Ni superalloy powders on AISI 4140 steel with concentrated solar energy, *Solar Energy Materials & Solar Cells*, 53, 1-2, 153-161, ISSN 0927-0248.

Ferriere, A.; Faillat, C.; Galasso, S.; Barallier, L.& Masse, J.E. (1999). Surface hardening of steel using highly concentrated solar energy process, *Journal of Solar Energy Engineering*, 21, 36-39, ISSN 0199-6231

Ferriere, A.; Rodríguez, G.P. & Sobrino, J.A. (2004). Flux distribution delivered by a fresnel lens used for concentrating solar energy, *Journal of Solar Energy Engineering*, 126, 1, 654-660, ISSN 0199-6231

Ferriere, A. ; Sanchez Bautista, C.; Rodriguez, G.P. & Vazquez, A.J. (2006). Corrosion resistance of stainless steel coatings elaborated by solar cladding process, *Solar Energy* 80, 10, 1338-1343, ISSN 0038-092X

Flamant, G.; Ferriere, A.; Laplaze, D. & Monty, C. (1999). Solar processing materials: opportunities and new frontiers. *Solar energy*, 66, 2, 117-132, ISSN 0038-092X

Font Tullot, I. (1984). *Atlas de la radiación solar en España*, Instituto Nacional de Meteorología, ISBN 8450505011, Madrid, Spain.

Guerra Rosa, L.; Amaral, P.M.; Anjinho, C.; Cruz Fernandes J. & Shohoji, N. (2002) Fracture toughness of solar-sintered WC with Co additive, *Ceram. Int.* 28, 345–348. ISSN 0272-8842

Gineste, J.M.; Flamant, G.& Olalde, G. (1999). Incident solar radiation data at Odeillo solar furnaces., *Journal de Physique IV*, 9, 623-628, ISSN 1155-4339.

Herranz, G. & Rodriguez, G. (2007). Surface Nitriding of M2 HSS using Nitrate Salt Bath, *Proceedings of Euro PM2007*, Vol. 3, 425-428, ISBN 978-1-899072-30-9, EURO PM 2007, International Powder Metallurgy Conference & Exhibition, Toulouse, Oct. 2007, Ed. The European Powder Metallurgy Association (EPMA)

Herranz, G.& Rodríguez, G. (2008). Solar power drives improved wear resistance in HSS, *Metal Powder Report* 63, 4, (April) 28-29, 31, ISSN 0026-0657

Hunt, L.B. (1982). The First Real Melting of Platinum: Lavoisier's ultimate success with oxygen, *Platinum Metals Rev.*, 26, 2, 79-86, ISSN 0032-1400

Jauregi, S.; Fernández, F.; Palma, R.H.; Martínez, V. & Urcola, J.J. (1992). Influence of atmosphere on sintering of T15 and M2 steel powders, *Metallurgical and Materials Transactions A*, 23, 2, 389-400, ISSN 1073-5623

Maiboroda, V.P.; Pasichniy, V.V.; Palaguta, N.G.; Stegnii, A.I.& Krivenko, V.G. (1986). Special features of local heat treatment of steel 34KhN3MFA in the focal spot of a solar furnace, *Metalloved. i Term. Obrab. Met.*, 1, 59-60, ISSN 0026-0819.

Matsuura, K. ; Jinmon, H. & Kudoh, M. (2000). Fabrication of NiAl/Steel Cladding by Reactive Casting, *ISIJ International*, 40, 2, 167–171, ISSN 0915-1559

Merzhanov, AG. (1967). Thermal explosion and ignition as a method for formal kinetic studies of exothermic reactions in the condensed phase, *Combustion and Flame*, 11, 3, 201-211, ISSN 0010-2180

Niu, H. J. & Chang, I.T.H. (2000). Selective laser sintering of gas atomized M2 high speed steel powder, *Journal of Materials Science*, 35, 1, 31-38, ISSN: 0022-2461

Pitts, J.R.; Stanley, J.T.& Fields, C.L. (1990). Solar Induced Surface Transformation of Materials, In: *Solar Thermal Technology-Research-Development and Applications*, B.P. Gupta & W.H. Trangott, (Eds.), 459-470, Hemisphere Publishing Corporation, ISBN 1560320958, New-York, USA

Rodríguez, G.P.; Vázquez, A.J.& Damborenea, J.J. (1994). Steel heat treatment with Fresnel lenses, *Materials Science Forum*, 163, 133-138, ISSN: 0255-5476

Rodríguez, G.P.; López, V.; Damborenea, J.J. & Vázquez, A.J. (1995). Surface transformation hardening on steels treated with solar energy in central tower and heliostat field, *Solar Energy Materials and Solar Cells*, 37, 1-12, ISSN 0927-0248

Rodríguez, G.P. ; Damborenea, J.J. & Vázquez, A.J. (1997). Surface Hardening of Steel in a Solar Furnace, *Surface and Coatings Technology* 92, 165-170, ISSN 0257-8972

Rodríguez, G.P.; García, I. & Vázquez, A.J. (1999). Materials and coating processing by self-propagating high-temperature synthesis (SHS) using a Fresnel lens, *Journal de Physique IV*, 9; 411-416, ISSN 1155-4339

Rodríguez, G.P. (2000) Modificación de superficies con energía solar. *Ciencia e ingeniería de superficie de los materiales metálicos.* Ediciones CSIC, Colección Textos Universitarios, No 31, 211-226, Madrid. ISBN: 84-00-07920-5.

Sánchez Bautista, C.; Ferriere, A.; Rodríguez, G.P; López-Almodovar, M.; Barba, A.; Sierra, C. & Vázquez, A.J. (2006). NiAl intermetallic coatings elaborated by a solar assisted SHS process, *Intermetallics*, 14 ,10-11, 1270-1275, ISSN 0966-9795

Sánchez Olías, J.; García, I.& Vázquez, A.J. (1999). Synthesis of TiN with solar energy concentrated by a Fresnel lens, *Materials Letters*, 38, 379-385, ISSN 0167-577X

Serna Moreno, M.C. & Rodríguez Donoso, G.P. (2004). Predicciones numéricas del endurecimiento superficial de aceros mediante energía solar concentrada, In: *Métodos Computacionais em Engenharia*, Soares, Batista, Bugida, Casteleiro, Goicolea, Martins, Pina y Rodríguez, (Eds.) 33, APMTAC y SEMNI, ISBN 972-49-2008-9, Lisboa, Portugal.

Shen, Y. Z.; Oh, K. H. & Lee, D. N. (2006a). Nitriding of Interstitial Free Steel in Potassium-Nitrate Salt Bath, *ISIJ International*, 46, 1, 111–120, ISSN 0915-1559

Shen, Y.Z., Oh, K.H. & Lee, D.N. (2006b). Nitrogen strengthening of interstitial-free steel by nitriding in potassium nitrate salt bath, *Materials Science and Engineering A* 434, 314-318, ISSN 0921-5093

Sierra, C. & Vázquez, A.J. (2005). NiAl coatings on carbon steel by self-propagating high temperature synthesis assisted with concentrated solar energy: mass influence on adherence and porosity, *Solar Energy Materials & Solar Cells, 86,* 33-42, ISSN 0927-0248

Stanley, J.T.; Fields, C.I.& Pitts, J.R. (1990). Surface treating with sunbeams, *Adv. Mat. Proc.*, 12, 16-21; ISSN 0882-7958

Vázquez, A.J. & Damborenea, J.J. (1990). Aplicaciones de la energía solar al tratamiento de materiales metálicos. Resultados preliminares. *Rev. Metal. Madrid,* 26, 3, 157-163, ISSN 0034-8570

Wriedt, H.A. & Murray, J.L. (1987). The N-Ti system, *Bulletin of Alloy Phase Diagrams,* 8, 4, 378, ISSN 0197-0216

Yu, Z.K.; Zong, Q.Y.& Tam, Z.T. (1982). A preliminary investigation of surface hardening of steel and iron by solar energy, *Journal of Heat Treating*, 2, 4, 344-350, ISSN 0190-9177

Yu, Z.K.; Zong, Q.Y.& Tam, Z.T. (1983). A further investigation of surface hardening of iron and steel by solar energy, *Journal of Heat Treating*, 3, 2, 120-125, ISSN 0190-9177

Zong, Q.Y.; Tam, Z.T.; Cao, M.D.& Yu, Z.K. (1986). Localized hardening of a machine gun bolt by means of high-intensity solar beam, *Heat Treat. Met.*, 6, 15-18.

9

Solar Chimney Power Plants – Developments and Advancements

Marco Aurélio dos Santos Bernardes
Centro Federal de Educação Tecnológica de Minas Gerais - CEFET-MG
Brazil

1. Introduction

A wide range of existing power technologies can make use of the solar energy reaching Earth. Basically, all those ways can be divided into two basic categories: transformed for use elsewhere or utilized directly - direct - and involving more than one transformation to reach a usable form - indirect. The Solar Chimney Power Plant (SCPP) is part of the solar thermal group of indirect solar conversion technologies.

More specifically, a natural phenomenon concerning the utilization of the thermal solar energy involves the earth surface heating and consequently the adjacent air heating by the sun light. This warm air expands causing an upward buoyancy force promoting the flow of air that composes the earth atmosphere. The amount of energy available due to the upward buoyancy force associated with the planet revolution is so vast that can generate catastrophic tropical cyclones with disastrous consequences.

From another standpoint, such phenomenon can be enhanced and used in benefit of the human well-being. In this way, the SCPP is a device developed with the purpose to take advantage of such buoyancy streams converting them into electricity. For that, a greenhouse - the collector - is used to improve the air heating process, a tall tube - the chimney - promotes the connection between the warm air nearby the surface and the fresh air present in higher atmosphere layers and a system to convert the kinetic energy into electricity - the generator-turbine system (Fig. 1).

2. First steps and recent developments

One of the earliest descriptions of a solar chimney power station was written in 1903 by Isidoro Cabanyes, a Spanish artillery colonel. He made public the proposition "Proyecto de motor solar" (solar engine project) introducing an apparatus consisting of an air heater attached to a house with a chimney. In the house interior, a kind of wind propeller was placed with the purpose of electricity production, as shown in Fig. 2, (Cabanyes, 1903).

In 1926 Prof Engineer Bernard Dubos proposed to the French Academy of Sciences the construction of a Solar Aero-Electric Power Plant in North Africa with its solar chimney on the slope of the high height mountain, (Fig. 3., (Günther, 1931)). The author claims that an ascending air speed of 50 m/s can be reached in the chimney, whose enormous amount of energy can be extracted by wind turbines. Fig. 4 shows an solar chimney futurist representation presented by (Günther, 1931). Fig. 5 shows a simple experiment proposed by

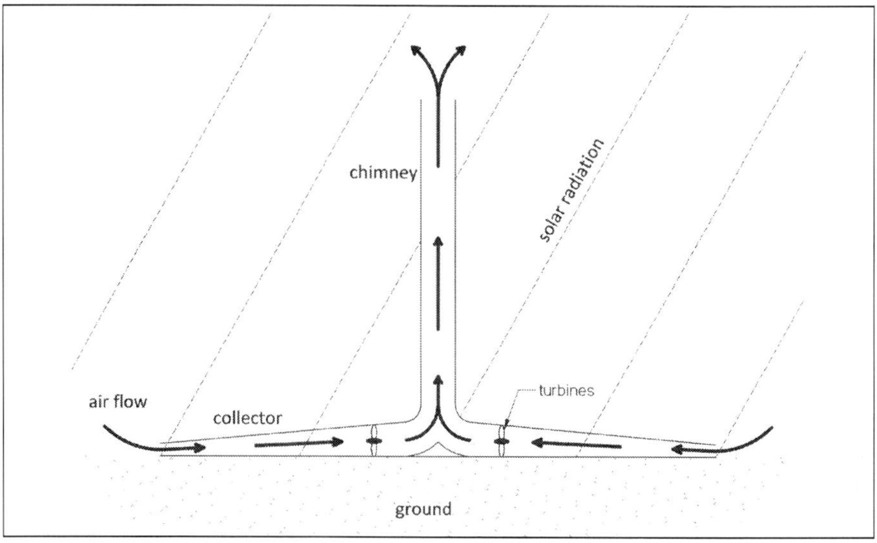

Fig. 1. SCPP components.

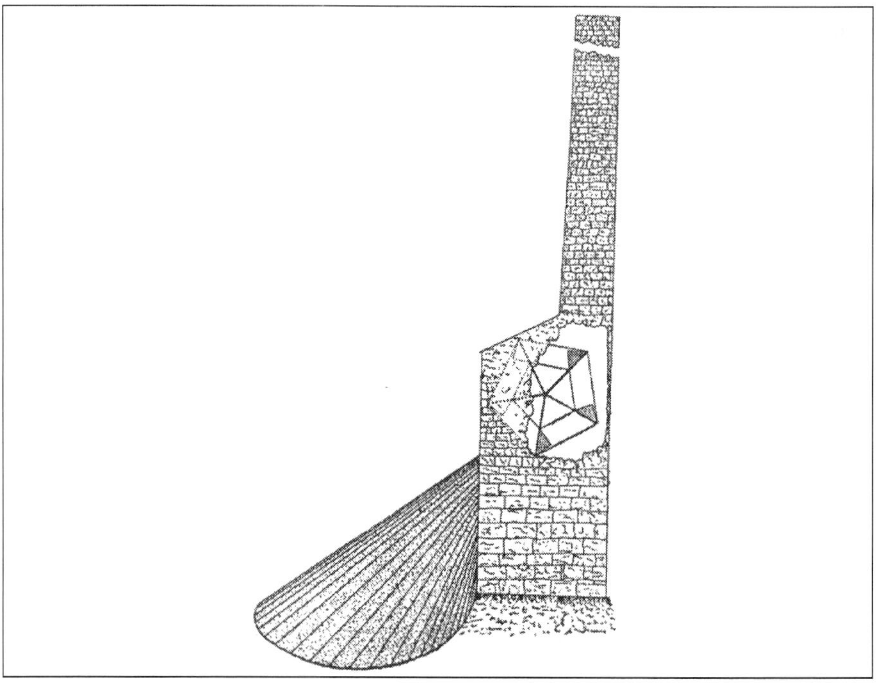

Fig. 2. Solar engine project proposed by Isidoro Cabanyes..

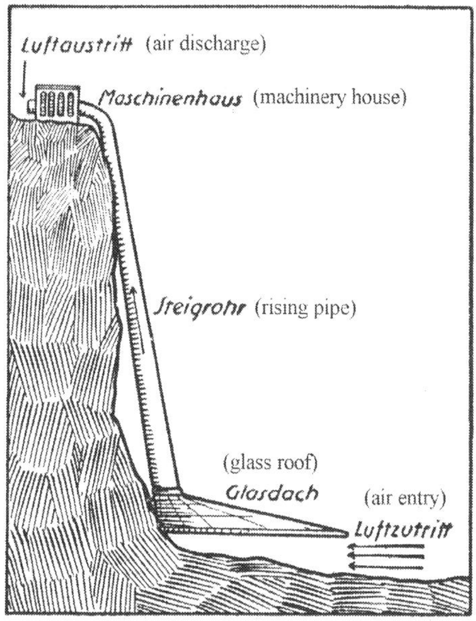

Fig. 3. Solar chimney proposal presented by (Günther, 1931).

Fig. 4. Solar chimney futurist representation presented by (Günther, 1931).

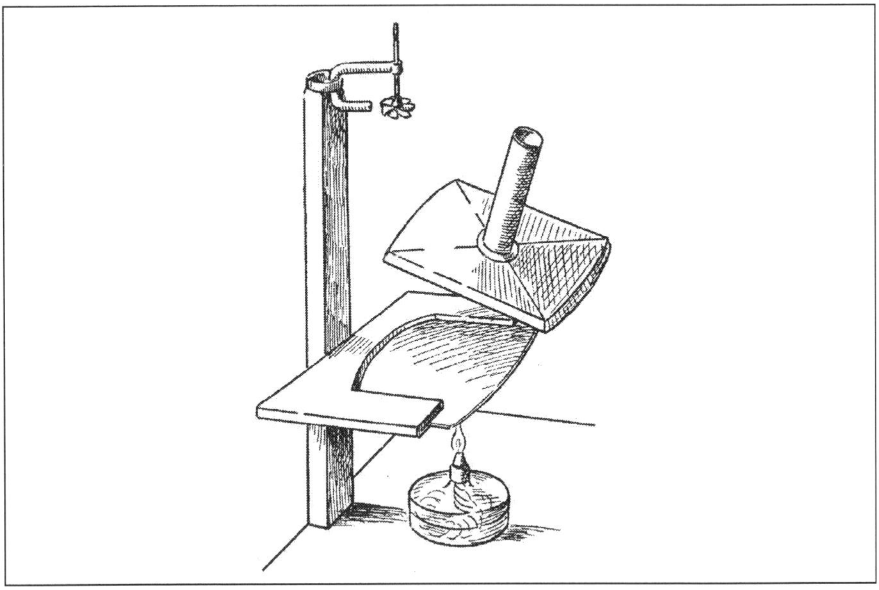

Fig. 5. Solar chimney proposal presented by (Günther, 1931).

Dubos confirming its concept. According to Günther (1931), the plate and the spirit lamp represent the Sahara desert and the solar heat, respectively. The small wind wheel placed on the top represents the wind turbines. If the spirit lamp is positioned under the plate, warm air flows concentrically through the plate reaching the tube. Consequently, the ascendant flow impels the wind wheel.

In the face of the original concepts, the first outstanding action for the SCPP development was the prototype erection in 1982 in Manzanares, Ciudad Real, 150 km south of Madrid, Spain. The chimney height was 195 m and its diameter 10 m. The collector area was 46,000 m^2 (about 11 acres, or 244 m diameter). Regardless of its dimensions, this prototype was considered as a small-scale experimental model. As the model was not intended for power generation, the peak power output was about 50 kW. Different glazing materials were tested, as well as, collector sections were used as an actual greenhouse, growing plants under the glass. The construction of the pilot plant were commissioned by the Minister of Research and Technology of the Federal Republic of Germany, (Haaf, *et al.*, 1983). The work was supervised by the energy research project management department of *Kernforschungsanalage (KFA) Jülich GmbH* (the Jülich Nuclear Energy Research Establishment) and fundamental principles of SCPP were found in the form of simple estimates. (Haaf, 1984) divulged preliminary test results including energy balances, collector efficiency values, pressure losses due to friction and losses in the turbine section. (Castillo, 1984) suggested a new "soft" structure approach to the chimney building instead of the conventional "rigid" one.

The SCPP has notable advantages in comparison with other power production technologies, namely - (Schlaich, 1995):

- the collector uses both direct und diffuse radiation;

- the ground provides a natural heat storage;
- the low number of rotating parts ensure its reliability;
- no cooling water is necessary for its operation;
- simple materials and known technologies are used in its construction;
- non OECD countries are able to implement such technology without costly technological efforts

After the prototype operation kick-off, several studies can be found about SCPP. They include transient and steady state fluid dynamic and thermal models, as well as, structural analysis for chimney, collector (including the ground natural heat storage capability) and turbine setups.

2.1 SCPP theoretical models

(Mullett, 1987) started the SCPP operational theoretical models development by deriving overall efficiency and relevant performance data. In his calculation, the overall efficiency is proportional the chimney height, returning about 1% for a height of 1000 m. He concluded that the solar chimney is essentially a power generator of large scale. The chimney efficiency is given by the equation (1).

$$\eta_t = \frac{gH}{c_p T_0} \tag{1}$$

Here, g is the gravity [m/s], H is the chimney height [m], c_p is the air heat capacity [J/kg·K] and T_0 is the ambient temperature [K]. For instance, with a chimney height of 1000 m and standard conditions for temperature and pressure, the chimney efficiency achieves the maximal value of 3 %. Considering a collector efficiency (η_c) of 60 % and a turbine efficiency (η_{tur}) of 80 %, the total system efficiency (η_{tot}) reaches 1.4%, as shown by equation (2).

$$\eta_{tot} = \eta_t \cdot \eta_c \cdot \eta_{tur} == 0.03 \cdot 0.6 \cdot 0.8 = 0.014 \tag{2}$$

Based on the data from the prototype of Manzanares, (Padki & Sherif, 1989) elaborated extrapolated SCPP models for medium-to-large scale power generation. (Yan, *et al.*, 1991) described a more comprehensive analytical model for SCPP by using practical engineering correlations obtaining equations for air velocity, airflow rate, power output, and the thermo-fluid efficiency and (Padki & Sherif, 1992) also presented a mathematical model for SCPP.

In the end of the 90's, (Pasumarthi & Sherif, 1998a) built a SCPP small-scale demonstration prototype to study the effect of various geometric parameters on the air temperature, air velocity, and power output of the solar chimney. Further studies conducted by (Pasumarthi & Sherif, 1998b) exploited the collector performance by extending the collector base and by introducing an intermediate absorber. According to them, both enhancements helped to increase the overall chimney power output. In addition, a brief economic assessment of the system costs is presented.

The first known attempt to solve by CFD (Computational Fluid Dynamics) the convective flow in a SCPP is showed by (Bernardes, *et al.*, 1999). He presented a solution for Navier-Stokes and Energy Equations for the natural laminar convection in steady state, predicting its thermo-hydrodynamic behavior. The approach Finite Volumes Method in Generalized

Coordinates was employed allowing a detailed visualization of the effects of geometric of optimal geometric and operational characteristics.
(Kröger & Blaine, 1999) evaluated the influence of prevailing ambient conditions. Their work shows that the air moisture can enhance the driving potential and that under certain conditions condensation may occur. In the meanwhile, (Kröger & Buys, 1999) developed analytical relations for determining the pressure differential due to frictional effects and heat transfer correlations for developing radial flow between the roof and the collector.
A set of differential equations for SCPP is deducted and integrated by (Padki & Sherif, 1999). Expressions for the power generated and the efficiency were obtained by making simplifying assumptions.
(Gannon & von Backström, 2000) introduced a study including chimney friction, system, turbine and exit kinetic energy losses in the analysis. For that, a simple model of the solar collector is used to include the coupling of the mass flow and temperature rise in the solar collector. This work, verified by comparing the simulation of a small-scale plant with experimental data, is useful to predict the performance and operating range of a large-scale plant. A one-dimensional compressible flow approach for the calculation of all the thermodynamic variables as dependent on chimney height, wall friction, additional losses, internal drag and area change was developed by von Backström and Gannon (2000). They concluded that the pressure drop associated with the vertical acceleration of the air is about three times the pressure drop associated with wall friction. For a flared chimney (14%, in order to keep the through-flow Mach number constant virtually), the vertical acceleration pressure drop can be eliminated. (Kröger & Buys, 2001) developed relevant equations for a SCPP.
(Gannon & Von Backström, 2002a) proposed a turbine design based on the design requirements for a full-scale solar chimney power plant integrating the turbine with the chimney. In this way, the chimney base legs are offset radially and act as inlet guide vanes introducing pre-whirl before the rotor reducing the exit kinetic energy. By employing a three-step turbine design method and a free vortex analysis method, the major turbine dimensions are determined. The flow path through the inlet guide vanes and rotor is predicted by a matrix throughflow method. The blade profiles are optimized by using the scheme coupled to a surface vortex method to achieve blades of minimum chord and low drag. The authors stated that the proposed turbine design can extract over 80% of the power available in the flow. Additionally, an experimental investigation of a SCPP design was undertaken by (Gannon & Von Backström, 2002b) and (Gannon & Von Backström, 2003). Results of the experimental model turbine revealed a total-to-total efficiency of 85-90% and total-to-static efficiency of 77-80% over the design range.
(Bernardes, *et al.*, 2003) developed a comprehensive SCPP analysis including analytical and numerical models to describe its performance, i.e., to estimate power output of solar chimneys as well as to examine the effect of various ambient conditions and structural dimensions on the power output. The mathematical model was validated with experimental results and the model was used to predict the performance characteristics of large-scale commercial solar chimneys. It turns out that the height of chimney, the factor of pressure drop at the turbine, the diameter and the optical properties of the collector are important parameters for the SCPP design.
(Schlaich, *et al.*, 2003) describe the functional principle of the Solar Tower and give some results from designing, building and operating a small scale prototype in Mazanares. Future

commercial Solar Tower systems are presented, as well as technical issues and basic economic data.

In his technical brief, (von Backström, 2003) develops calculation methods for the pressure drop in very tall chimneys. Equations for the vertical pressure and density distributions in terms of Mach number allowing density and flow area change with height, wall friction and internal bracing drag are presented. To solve the equations, two simplifications are presented, namely, the adiabatic pressure lapse ratio equation to include flow at small Mach numbers and extension of the hydrostatic relationship between pressure, density, and height to small Mach numbers. Accurate value of the average density in the chimney can be obtained by integration.

(Schlaich, *et al.*, 2004) reintroduces the SCPP concept reinforcing its role as a sustainable option to produce solar electricity at low costs. (Koonsrisuk & Chitsomboon, 2004) analyzed the frictional effect on the flow in a SCPP.

The influence of the atmospheric winds on performance of SCPP was studied by (Serag-Eldin, 2004). By means of a computational model - governing partial differential equations expressing conservation of mass, energy and balance of momentum in addition to a two equation model of turbulence - the flow pattern in the neighborhood of a small-scale SCPP model was computed. The analysis results shown a total degradation of performance with strong winds and significant degradation for low speed winds, except for collector with low inlet height.

(Kirstein, *et al.*, 2005) and (Kirstein & Von Backström, 2006) presented studies concerning the loss coefficient in the transition section between a SCPP turbine and the chimney as dependent on inlet guide vane stagger angle and collector roof height including scaled model experiments and commercial CFD simulations. The very good agreement between the experiments and the simulations permits predictions for a proposed full-scale geometry.

Using the solar chimney prototype in Manzanares as a practical example, (Ming, *et al.*, 2006) carried out numerical studies to explore the geometric modifications on the system performance, showing reasonable agreement with the analytical model. The SCPP performance was evaluated, in which the effects of various parameters on the relative static pressure, driving force, power output and efficiency were investigated.

(Pretorius & Kröger, 2006c) evaluated the performance of a large-scale SCPP. A particular reference plant under specified meteorological conditions at a reference location in South Africa was choose and developed convective heat transfer and momentum equations were employed. The authors claim that 24 hr plant power production is possible and that plant power production is a function of the collector roof shape and inlet height. Furthermore, more accurate turbine inlet loss coefficient, quality collector roof glass and various types of soil on the performance of a large scale SCPP are introduced ((Pretorius & Kröger, 2006b)). Results pointed out that the heat transfer correlations employed reduced the plant power output significantly. A more accurate turbine inlet loss coefficient has no noteworthy effect, at the same time as utilizing better quality glass enhances plant power production. Simulations employing Limestone and Sandstone soil gave virtually similar results to a Granite-based model.

(Von Backström & Fluri, 2006) investigated analytically the validity and applicability of the assumption that, for maximum fluid power, the optimum ratio of turbine pressure drop to pressure potential (available system pressure difference) is 2/3. They concluded that the constant pressure potential assumption may lead to overestimating the size of the flow passages in the plant, and designing a turbine with inadequate stall margin and excessive runaway speed margin.

The effects of the solar radiation on the flow of the SCPP were analyzed by (Huang, *et al.*, 2007). Boussinesq approximation and Discrete Ordinate radiation model (DO) were introduced in the model and simulations were carried out.

(Koonsrisuk & Chitsomboon, 2007) proposed the use of dimensionless variables to conduct the experimental study of flow in a small-scale SCPP for generating electricity. The similarity of the proposed dimensionless variables was confirmed by computational fluid dynamics.

(Ming, *et al.*, 2007) set up different mathematical models for heat transfer and flow for the collector, chimney and turbine. After the validation with the Manzanares prototype, they suggested that the output power of MW-graded can exceed 10 MW.

A new mathematical model based on the concept of relative static pressure was developed by (Peng, *et al.*, 2007). According to them, optimized local geometric dimensions between the collector outlet and the chimney inlet, the local velocity at this place ca increase 14% the SCPP performance, temperature profiles are more uniform, and the relative static pressure decreases about 50% improving the system energy conversion and reducing the energy losses.

(Pretorius & Kröger, 2007) presented a sensitivity analysis on the influence of the quality, thickness, reflectance, emissivity, shape, and insulation of the collector roof glass, the cross section of the collector roof supports, various ground types, ground surface roughness, absorptivity and emissivity, turbine and bracing wheel loss coefficients, and the ambient pressure and lapse rate on the performance of a large-scale (reference) solar chimney power plant. Computer simulation results point out that collector roof insulation, emissivity and reflectance, the ambient lapse rate, and ground absorptivity and emissivity all have a key effect on the power production of such a plant.

(Ming, *et al.*, 2008a) continued their work carrying out numerical simulations analyzing characteristics of heat transfer and air flow in the solar chimney power plant system with an energy storage layer including the solar radiation and the heat storage on the ground. They concluded that the ground heat storage depends on the solar radiation incidence. Higher temperature gradients also increase the energy loss from the ground ((Ming, *et al.*, 2008b)) and (Ming, *et al.*, 2008c)). Subsequently, (Tingzhen, *et al.*, 2008) included a 3-blade turbine in the model simulation and validated the model, turning out power out and turbine efficiency of 10 MW and 50%, respectively.

Due to the use of different heat transfer coefficients found in the literature, namely those of ((Bernardes, 2004b),(Pretorius & Kröger, 2006a)), (Bernardes, *et al.*, 2009) made a comparison of the methods used to calculate the heat fluxes in the collector and their effects on solar chimney performance. Notwithstanding the difference between the heat transfer coefficients, both approaches returned similar air temperature rises in the collector and therefore, comparable produced power.

(Koonsrisuk & Chitsomboon, 2009) went on with their previous study, namely (Koonsrisuk & Chitsomboon, 2007), trying to maintain dynamic similarity between the prototype and its theoretical model keeping the same solar heat flux. They showed that, for the same heat flux condition, all dimensional parameter, except the roof radius, must remain similar. They also revealed some engineering interpretations for the similarity variables.

SCPP energy and exergy balances were carried out by (Petela, 2009), turning out the energy distribution in the components for a input of 36.81 MW energy of solar radiation (or an equivalent input of 32.41 MW of radiation exergy), including also a sensibility analysis.

More recently (Bernardes and von Backström, 2010) performed a study regarding the performance of two schemes of power output control applicable to solar chimney power plants. It was revealed that the optimum ratio is not constant during the whole day and it is dependent of the heat transfer coefficients applied to the collector.

3. SCPP analysis for specific sites

(Dai, et al., 2003) analyzed a solar chimney power plant to provide electric power for remote villages in northwestern China. Three counties in Ning Xia Hui Autonomous region, namely, Yinchuan, Pingluo, and Helan with good solar radiation availability were selected as sites to simulate a SCPP model. In according to the authors, a SCPP consisting of a 200 m height and 10 m diameter chimney, and the 500 m diameter solar collector can produce 110 ~ 190 kW electric power on a monthly average all year.

(Bilgen and Rheault, 2005) developed a mathematical model based on monthly average meteorological data and thermodynamic cycle to simulate the SCPP power production at high latitudes. Three locations in Canada, namely Ottawa, Winnipeg and Edmonton were choose to evaluate a 5 MW nominal power production plant. The authors also suggested the construction of a sloped collector field at suitable mountain hills in order to work as a collector. Next, for this proposition, a short vertical chimney is added to install the vertical axis air turbine. They also shown that such plant can produce as much as 85% of the same plants in southern locations with horizontal collector field and the overall thermal performance of these plants is a little less than 0.5%.

SCPP for rural villages was studied by (Onyango and Ochieng, 2006) emphasizing some features for power generating. They disclosed that for a temperature ratio = 2.9 (i.e., the difference between the collector surface temperature and the temperature at the turbine to the difference between the air mass temperature under the roof and the collector surface temperature) an 1000 W of electric power can be generated. The minimum dimension of a practical by a reliable SCPP to assist approximately fifty households in a typical rural setting has been determined to be chimney length = 150 m, height above the collector = chimney radius = 1.5 m.

(Zhou, et al., 2007b) developed mathematical model to investigate power generating performance of a SCPP prototype. The steady state simulation returned power outputs for different global solar radiation intensity, collector area and chimney height. The simulation results were validated with the measurements.

4. SCPP turbine developments

(Gannon & Von Backström, 2003), (Gannon & Von Backström, 2002b) and (Gannon & Von Backström, 2002b) were the first to develop an experimental investigation of the performance of a solar chimney turbine. The design consisted of a single rotor by using inlet guide vanes to introduce pre-whirl. Such strategy reduces the turbine exit kinetic energy at the diffuser inlet and assists the flow turning in the IGV-to-rotor duct. Total-to-total efficiencies of 85-90% and total-to-static of 77-80% over the design range were measured.

(Von Backström & Gannon, 2004) presented analytical equations in terms of turbine flow and load coefficient and degree of reaction, to express the influence of each coefficient on turbine efficiency. Analytical solutions for optimum degree of reaction, maximum turbine efficiency for required power and maximum efficiency for constrained turbine size were

found. According to the authors, a peak turbine total-to-total efficiency of around 90% is achievable, but not necessarily over the full range of plant operating points.
Some structural aspects of classical wind energy turbines, like their high-cycle dynamic loading and reaction as well as their fatigue behavior were exposed by (Harte & Van Zijl, 2007). Structural challenges concerning wind action, eigenfrequencies, stiffening and shape optimization with special focus on the inlet guide vanes were discussed for SCPP's.
(Fluri & von Backström, 2008) analyzed many different layouts for the SCPP turbogenerator. Turbine layouts with single rotor and counter rotating turbines, both with or without inlet guide vanes were considered. They concluded that the single rotor layout without guide vanes performs very poorly; the efficiency of the other three layouts is much better and lies in a narrow band.
(Tingzhen, et al., 2008) performed numerical simulations on SCPP's coupled with turbine. The model was validated with the measurements from the Spanish prototype, obtaining a maximum power output higher than 50 kW. Subsequently, the authors presented the design and the simulation of a MW-graded solar chimney power plant system with a 5-blade turbine. The numerical simulation results show that the power output and turbine efficiency are 10 MW and 50%, respectively, which presents a reference to the design of large-scale SCPP's.

5. SCPP experimental analysis

(Pasumarthi & Sherif, 1998a; b): The solar chimney is a natural draft device which uses solar radiation to provide upward momentum to the in-flowing air, thereby converting the thermal energy into kinetic energy. A study was undertaken to evaluate the performance characteristics of solar chimneys both theoretically and experimentally. In this paper, a mathematical model which was developed to study the effect of various parameters on the air temperature, air velocity, and power output of the solar chimney, is presented. Tests were conducted on a demonstration model which was designed and built for that purpose. The mathematical model presented here, was verified against experimental test results and the overall results were encouraging. his paper describes details of the experimental program conducted to assess the viability of the solar chimney concept. A demonstration model was designed and built and its theoretical and experimental performance was examined. Two experimental modifications were tried on the collector: (1) extending the collector base and (2) introducing an intermediate absorber. The former modification helped in enhancing the air temperature, while the latter contributed to increasing the air temperature as well as the mass flow rate inside the chimney. Both enhancements helped to increase the overall chimney power output. Theoretical and experimental performance results of this demonstration model are presented in this paper, while the mathematical model developed in Part I was used to predict the performance of much larger systems. Mathematical model results were validated by comparing them to published data on the solar chimney system built in Manzanares, Spain. Also, an economic assessment of the system costs are presented.
(Gannon & Von Backström, 2002b): An experimental investigation of a solar chimney turbine design is undertaken. The aim of the program is to demonstrate and evaluate a proposed solar chimney turbine design. The measured results of an experimental model turbine are presented and the turbine efficiency calculated. The current turbine design has a total-to-total efficiency of 85-90% and total-to-static efficiency of 77-80% over the design

range. Secondary objectives are to compare the measured and predicted results and through investigation of the experimental results suggest improvements to the turbine design.
(Zhou, *et al.*, 2007a): A pilot experimental solar chimney power setup consisted of an air collector 10 m in diameter and an 8 m tall chimney has been built. The temperature distribution in the solar chimney power setup was measured. Temperature difference between the collector outlet and the ambient usually can reach 24.1 Â°C, which generates the driving force of airflow in the setup. This is the greenhouse effect produced in the solar collector. It is found that air temperature inversion appears in the latter chimney after sunrise both on a cool day and on a warm day. Air temperature inversion is formed by the increase of solar radiation from the minimum and clears up some time later when the absorber bed is heated to an enough high temperature to make airflow break through the temperature inversion layer and flow through the chimney outlet.
(Ferreira, *et al.*, 2008): Solar dryers use free and renewable energy sources, reduce drying losses (as compared to sun drying) and show lower operational costs than the artificial drying, thus presenting an interesting alternative to conventional dryers. This work proposes to study the feasibility of a solar chimney to dry agricultural products. To assess the technical feasibility of this drying device, a prototype solar chimney, in which the air velocity, temperature and humidity parameters were monitored as a function of the solar incident radiation, was built. Drying tests of food, based on theoretical and experimental studies, assure the technical feasibility of solar chimneys used as solar dryers for agricultural products. The constructed chimney generates a hot airflow with a yearly average rise in temperature (compared to the ambient air temperature) of 13 ± 1 °C. In the prototype, the yearly average mass flow was found to be 1.40 ± 0.08 kg/s, which allowed a drying capacity of approximately 440 kg.
(Ming, *et al.*, 2008c): A small scale solar chimney system model has been set up, and the temperature distribution of the system with time and space, together with the velocity variation inside the chimney with time, has been measured. The experimental results show that the temperature distributions inside the collector and the effects of seasons on the heat transfer and flow characteristic of system show great agreement with the analysis, while the temperature decrease significantly inside the chimney as the chimney is very thin which causes very high heat loss.

6. SCPP structural analysis

(Harte & Van Zijl, 2007) presented some structural aspects of classical wind energy turbines, like their high-cycle dynamic loading and reaction as well as their fatigue behaviour. Actual research results concerning pre-stressed concrete tower constructions for wind turbines will be focused on. For the solar chimney concept the structural challenges concerning wind action, eigenfrequencies, stiffening and shape optimization with special focus on the inlet guide vanes will be discussed.

7. SCPP ecological analysis

(Bernardes, 2004a) performed is a comprehensive evaluation of impacts caused by mass and energy flows of solar chimneys systems from its design through to production and then final disposal using the method of Life Cycle Assessment. The conventional Life Cycle Assessment method was improved by an additional sectoral analysis (input-output analysis), namely Hybrid Approach. The study was an important contribution for the integration of Life Cycle Assessment in the decision making process in the renewable energy sector and for an integrated evaluation of processes.

8. SCPP economical analysis

In their first approach (Haaf, *et al.*, 1983) concluded, from the relationships between the physical principles on the one hand and the scale and construction costs on the other, that economical power generation will be possible with large-scale plants designed for up to 400 MW/pk

(Pretorius & Kröger, 2008) undertook a study to stablish a thermoeconomically optimal plant configurations for a large-scale SCPP. For that, an approximated cost model was developed, giving the capacity for finding optimum plant dimensions for different cost structures. Thermoeconomically optimal plant configuration were obtained through multiple computer simulations and results comparison to the approximated cost of each specific plant.

A study developed by (Fluri, *et al.*, 2009) revealed that previous economical models may have underestimated the initial cost and levelised electricity cost of a large-scale solar chimney power plant. It also showed that carbon credits significantly can reduce the levelised electricity cost for such a plant.

9. Alternative concepts and applications

Probably (Ferreira, *et al.*, 2008) were the first to propose a solar chimney as a device to dry agricultural products. A small scale prototype solar chimney was built, in which the air velocity, temperature and humidity parameters were monitored as a function of the solar incident radiation. Based on theoretical and experimental studies Drying tests reveled the technical feasibility of solar chimneys used as solar dryers for agricultural products. A hot airflow with a yearly average rise in temperature (compared to the ambient air temperature) of 13 ± 1 °C could be achieved allowing a drying capacity of approximately 440 kg.

(Zhu, *et al.*, 2008) proposed different heat storage styles for SCPP. The experimental studies showed that the temperature difference in the sealed water system is the largest, while the open water system has the lowest one because of the latent heat consumed by water evaporation. The study also showed that there is the temperature distribution optimization of the system if the heat loss part of the collector to be avoided.

The concept for producing energy by integrating a solar collector with a mountain hollow is presented and described by (Zhou, *et al.*, 2009). As in a conventional SCPP, the hot air is forced by the pressure difference between it and the ambient air to move along the tilted segment and up the vertical segment of the 'chimney', driving the turbine generators to generate electricity. The author claimed that such concept provides safety and reduces a great amount of construction materials in the conventional chimney structure and the energy cost to a level less than that of a clean coal power plant.

The hypothesis of combining a salinity gradient solar pond with a chimney to produce power in salt affected areas is examined by (Akbarzadeh, *et al.*, 2009). The salinity in northern Victoria, Australia was analyzed and salinity mitigation schemes were presented. It was shown that a solar pond can be combined with a chimney integrating an air turbine for the production of power. A prototype of a solar pond of area 6 hectares and depth 3 m with a 200 m tall chimney of 10 m diameter was investigated and

10. Conclusion

The previous literature review about SCPP presents an outstanding technological development enlightening considerable advances in its construction, operation, including its technical economical and ecological relevant facets.

In contrast with other solar facilities, SCPPs can be used above and beyond power production. Very relevant byproducts are distilled water extracted from ocean water or ground water. Under certain conditions, agribusiness may be appropriate under the solar collector. It can involve fruits and vegetables, medicinal and aromatic essential oils from herbs and flowers, seaweeds and planktons, blue-green algae, ethanol and methane, biodiesel and all manner of vegetable and plant derivatives, etc. Besides, remaining biomass is useful creating additional heat during composting.

The insertion of SCPP in the power generation market requires scalability and base, shoulder and peak load electricity generation. Further developments should meet such localized requirements.

11. References

Akbarzadeh, A., et al. (2009), Examining potential benefits of combining a chimney with a salinity gradient solar pond for production of power in salt affected areas, *Solar Energy, 83,* 1345-1359.

Bernardes, M. A. d. S. (2004a), Life Cycle Assessment of solar Chimneys, paper presented at VIII World Renewable Energy Congress and Expo, Denver, USA, August 29-Sep 3.

Bernardes, M. A. d. S. (2004b), Technische, ökonomische und ökologische Analyse von Aufwindkraftwerken, PhD Thesis thesis, 230 pp, Universität Stuttgart, Stuttgart.

Bernardes, M. A. d. S., et al. (1999), Numerical Analysis of Natural Laminar Convection in a Radial Solar Heater, *International Journal of Thermal Sciences, 38,* 42-50.

Bernardes, M. A. d. S., et al. (2009), Analysis of some available heat transfer coefficients applicable to solar chimney power plant collectors, *Solar Energy, 83,* 264-275.

Bernardes, M. A. d. S., et al. (2003), Thermal and technical analyses of solar chimneys, *Solar Energy, 75,* 511-524.

Bilgen, E., and J. Rheault (2005), Solar chimney power plants for high latitudes, *Solar Energy, 79,* 449-458.

Cabanyes, I. (1903), Proyecto de Motor Solar, *La Energia Eléctrica - Revista General de Electricidad y sus Aplicaciones, 8,* 61-65.

Castillo, M. A. (1984), A New Solar Chimney Design to Harness Energy from the Atmosphere, in *Spirit of Enterprise - The Rolex 1984 Rolex Awards,* edited by M. Nagai and A. e. J. Heiniger.

Dai, Y. J., et al. (2003), Case study of solar chimney power plants in Northwestern regions of China, *Renewable Energy, 28,* 1295-1304.

Ferreira, A. G., et al. (2008), Technical feasibility assessment of a solar chimney for food drying, *Solar Energy, 82,* 198-205.

Fluri, T. P., et al. (2009), Cost analysis of solar chimney power plants, *Solar Energy, 83*, 246-256.

Fluri, T. P., and T. W. von Backström (2008), Comparison of modelling approaches and layouts for solar chimney turbines, *Solar Energy, 82*, 239-246.

Gannon, A. J., and T. W. von Backström (2000), Solar chimney cycle analysis with system loss and solar collector performance, *Journal of Solar Energy Engineering, Transactions of the ASME, 122*, 133-137.

Gannon, A. J., and T. W. Von Backström (2002a), Solar chimney turbine part 1 of 2: Design, paper presented at International Solar Energy Conference, Reno, NV.

Gannon, A. J., and T. W. Von Backström (2002b), Solar chimney turbine part 2 of 2: Experimental results, paper presented at International Solar Energy Conference, Reno, NV.

Gannon, A. J., and T. W. Von Backström (2003), Solar chimney turbine performance, *Journal of Solar Energy Engineering, Transactions of the ASME, 125*, 101-106.

Günther, H. (1931), *In hundert Jahren*, 78 pp., Kosmos - Gesellschaft der Naturfreunde, Stuttgart.

Haaf, W. (1984), Solar Chimneys - Part II: Preliminary Test Results from the Manzanares Pilot Plant, edited, pp. 141 - 161, Taylor & Francis.

Haaf, W., et al. (1983), Solar Chimneys - Part I: Principle and Construction of the Pilot Plant in Manzanares, edited, pp. 3 - 20, Taylor & Francis.

Harte, R., and G. P. A. G. Van Zijl (2007), Structural stability of concrete wind turbines and solar chimney towers exposed to dynamic wind action, *Journal of Wind Engineering and Industrial Aerodynamics, 95*, 1079-1096.

Huang, H., et al. (2007), Simulation Calculation on Solar Chimney Power Plant System, in *Challenges of Power Engineering and Environment*, edited, pp. 1158-1161.

Kirstein, C. F., and T. W. Von Backström (2006), Flow through a solar chimney power plant collector-to-chimney transition section, *Journal of Solar Energy Engineering, Transactions of the ASME, 128*, 312-317.

Kirstein, C. F., et al. (2005), Flow through a solar chimney power plant collector-to-chimney transition section, paper presented at International Solar Energy Conference, Orlando, FL.

Koonsrisuk, A., and T. Chitsomboon (2004), Frictional effect on the flow in a solar chimney paper presented at Proceedings of the 4th National Symposium on Graduate Research, Chiang Mai, Thailand.

Koonsrisuk, A., and T. Chitsomboon (2007), Dynamic similarity in solar chimney modeling, *Solar Energy, 81*, 1439-1446.

Koonsrisuk, A., and T. Chitsomboon (2009), Partial geometric similarity for solar chimney power plant modeling, *Solar Energy, 83*, 1611-1618.

Kröger, D. G., and D. Blaine (1999), Analysis of the Driving Potential of a Solar Chimney Power Plant, *South African Inst. of Mechanical Eng. R & D J., 15*, 85-94.

Kröger, D. G., and J. D. Buys (1999), Radial Flow Boundary Layer Development Analysis, *South African Inst. of Mechanical Eng. R & D J., 15*, 95-102.

Kröger, D. G., and J. D. Buys (2001), Performance Evaluation of a solar Chimney Power Plant, in *ISES 2001 Solar World Congress*, edited, Adelaide, south Australia.

Ming, T., et al. (2008a), Numerical analysis of flow and heat transfer characteristics in solar chimney power plants with energy storage layer, *Energy Conversion and Management, 49*, 2872-2879.

Ming, T., et al. (2008b), Numerical analysis of heat transfer and flow in the solar chimney power generation system, *Taiyangneng Xuebao/Acta Energiae Solaris Sinica, 29*, 433-439.

Ming, T., et al. (2006), Analytical and numerical investigation of the solar chimney power plant systems, *International Journal of Energy Research, 30*, 861-873.

Ming, T. Z., et al. (2008c), Experimental simulation of heat transfer and flow in the solar chimney system, *Kung Cheng Je Wu Li Hsueh Pao/Journal of Engineering Thermophysics, 29*, 681-684.

Ming, T. Z., et al. (2007), Numerical simulation of the solar chimney power plant systems with turbine, *Zhongguo Dianji Gongcheng Xuebao/Proceedings of the Chinese Society of Electrical Engineering, 27*, 84-89.

Mullett, L. B. (1987), Solar Chimney - Overall Efficiency, Design and Performance, *International Journal of Ambient Energy, 8*, 35-40.

Onyango, F. N., and R. M. Ochieng (2006), The potential of solar chimney for application in rural areas of developing countries, *Fuel, 85*, 2561-2566.

Padki, M. M., and S. A. Sherif (1989), Solar chimney for medium-to-large scale power generation, paper presented at Proceedings of the Manila International Symposium on the Development and Management of Energy Rexources, Manila, Philippines.

Padki, M. M., and S. A. Sherif (1992), A Mathematical Model for Solar Chimneys, paper presented at Proceedings of the 1992 International Renewable Energy Conference, in Renewable Energy: Research and Applications, University of Jordan, Faculty of Engineering and Technology, Amman, Jordan, June 22-26, 1992.

Padki, M. M., and S. A. Sherif (1999), On a simple analytical model for solar chimneys, *International Journal of Energy Research, 23*, 345-349.

Pasumarthi, N., and S. A. Sherif (1998a), Experimental and theoretical performance of a demonstration solar chimney model - Part I: Mathematical model development, *International Journal of Energy Research, 22*, 277-288.

Pasumarthi, N., and S. A. Sherif (1998b), Experimental and theoretical performance of a demonstration solar chimney model - Part II: Experimental and theoretical results and economic analysis, *International Journal of Energy Research, 22*, 443-461.

Peng, W., et al. (2007), Research of the optimization on the geometric dimensions of the solar chimney power plant systems, *Huazhong Keji Daxue Xuebao (Ziran Kexue Ban)/Journal of Huazhong University of Science and Technology (Natural Science Edition), 35*, 80-82.

Petela, R. (2009), Thermodynamic study of a simplified model of the solar chimney power plant, *Solar Energy, 83*, 94-107.

Pretorius, J. P., and D. G. Kröger (2006a), Critical evaluation of solar chimney power plant performance, *Solar Energy, 80*, 535-544.

Pretorius, J. P., and D. G. Kröger (2006b), Solar chimney power plant performance, *Journal of Solar Energy Engineering, Transactions of the ASME, 128*, 302-311.

Pretorius, J. P., and D. G. Kröger (2006c), Thermo-economic optimization of a solar chimney power plant, paper presented at CHISA 2006 - 17th International Congress of Chemical and Process Engineering, Prague.

Pretorius, J. P., and D. G. Kröger (2007), Sensitivity analysis of the operating and technical specifications of a solar chimney power plant, *Journal of Solar Energy Engineering, Transactions of the ASME, 129*, 171-178.

Pretorius, J. P., and D. G. Kröger (2008), Thermoeconomic optimization of a solar chimney power plant, *Journal of Solar Energy Engineering, Transactions of the ASME, 130*, 0210151-0210159.

Schlaich, J. (1995), *The Solar Chimney: Electricity from the Sun*, Edition Axel Menges, Stuttgart.

Schlaich, J., et al. (2003), Design of commercial solar tower systems - Utilization of solar induced convective flows for power generation, paper presented at International Solar Energy Conference, Kohala Coast, HI.

Schlaich, J., et al. (2004), Sustainable electricity generation with solar updraft towers, *Structural Engineering International: Journal of the International Association for Bridge and Structural Engineering (IABSE), 14*, 225-229.

Serag-Eldin, M. A. (2004), Computing flow in a solar chimney plant subject to atmospheric winds, paper presented at Proceedings of the ASME Heat Transfer/Fluids Engineering Summer Conference 2004, HT/FED 2004, Charlotte, NC.

Tingzhen, M., et al. (2008), Numerical simulation of the solar chimney power plant systems coupled with turbine, *Renewable Energy, 33*, 897-905.

von Backström, T. W. (2003), Calculation of pressure and density in solar power plant chimneys, *Journal of Solar Energy Engineering, Transactions of the ASME, 125*, 127-129.

von Backström, T. W., and T. P. Fluri (2006), Maximum fluid power condition in solar chimney power plants - An analytical approach, *Solar Energy, 80*, 1417-1423.

von Backström, T. W., and A. J. Gannon (2004), Solar chimney turbine characteristics, *Solar Energy, 76*, 235-241.

Yan, M. Q., et al. (1991), Thermo-fluid analysis of solar chimneys, paper presented at American Society of Mechanical Engineers, Fluids Engineering Division (Publication) FED, Publ by ASME, Atlanta, GA, USA.

Zhou, X., et al. (2009), Novel concept for producing energy integrating a solar collector with a man made mountain hollow, *Energy Conversion and Management, 50*, 847-854.

Zhou, X., et al. (2007a), Experimental study of temperature field in a solar chimney power setup, *Applied Thermal Engineering, 27*, 2044-2050.

Zhou, X., et al. (2007b), Simulation of a pilot solar chimney thermal power generating equipment, *Renewable Energy, 32*, 1637-1644.

Zhu, L., et al. (2008), Temperature rise performance in solar chimneyes with different heat storages, *Taiyangneng Xuebao/Acta Energiae Solaris Sinica, 29*, 290-294.

10

Floating Solar Chimney Technology

Christos D. Papageorgiou
National Technical University of Athens
Greece

1. Introduction

1.1 Floating Solar Chimney technology description

The purpose of this chapter is to present the Floating solar chimney (FSC) technology, look for the site www.floatingsolarchimney.gr, in order to explain its principles of operation and to point out its various significant benefits. This technology is the advisable one for candidacy for large scale solar electricity generation especially in desert or semi desert areas of our planet and a major technology for the global warming elimination.

The solar chimney power plants are usually referred to as solar updraft towers (http://en.wikipedia.org/wiki/Solar_updraft_tower) and the related solar chimneys are huge reinforced concrete structures. However due to the high construction cost of the concrete solar chimneys the solar up-draft tower technology is expensive demanding a high initial investment in comparison to its competitive solar technologies. Their solar up-draft towers are huge structures of high initial investment cost that can not be split into small units. That is possible for the relatively also expensive PV solar technology. Also the solar updraft technology is far more expensive compared to the conventional fossil fueled power plants of similar electricity generation. That is why the solar chimney technology has not yet been applied although it is a solar technology of many advantages.

The **Floating Solar Chimney (FSC)** is a fabric low cost alternative of the concrete solar chimney up-draft towers that can make the Floating Solar Chimney technology cost competitive in comparison not only with the renewable electricity generation technologies but also with the conventional fossil fueled electricity generation technologies. Also the FSC technology is cost effective to be split into small units of several MW each.

The Floating Solar Chimney Power Plant, named by the author as **Solar Aero-Electric Power Plant (SAEP)** due to its similarity to the Hydro-Electric power plant, is a set of three major components:

- **The Solar Collector**. It is a large greenhouse open around its periphery with a transparent roof supported a few meters above the ground.
- **The Floating Solar Chimney (FSC)**. It is a tall fabric cylinder placed at the centre of the solar collector through which the warm air of the greenhouse, due to its relative buoyancy to the ambient air, is up-drafting.
- **The Turbo-Generators**. It is a set of air turbines geared to appropriate electric generators in the path of up-drafting warm air flow that are forced to rotate generating electricity. The gear boxes are adjusting the rotation speed of the air turbines to the generator rotation speed defined by the grid frequency and their pole pairs.

An indicative figure of a solar chimney Power Plant with a circular solar collector and a Floating Solar Chimney inclined due to external winds is shown in next figure(1).

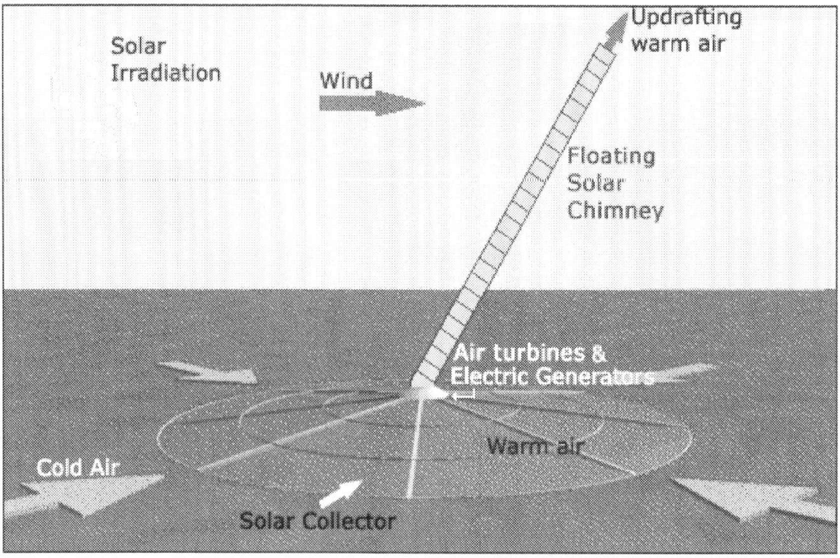

Fig. 1. Floating Solar Chimney Power Plant in operation

Because of its patented construction the FSC is a free standing lighter than air structure that can tilt when external winds appear. Low cost Floating Solar Chimneys up to 1000 m with internal diameters 25 m ÷ 40 m, can be constructed with an existing polyester fabric, giving to their respective Solar Aero-Electric Power Plants, low investment costs.
By this innovating Floating Solar Chimney Technology of heights of the FSCs up to1000m, up to 1.2 % of the arriving horizontal solar radiation on the solar collector surface, can be converted to electricity

1.2 Similarity to hydro-electric power plants

The Floating Solar Chimney power plants, due to their similarity to hydro-electric power plants, are named by the author Solar Aero Electric Power Plants (SAEPs).
Their similarity is due to the following facts:

- The hydro-electric PPs operate due to falling water gravity, while the solar aero-electric PPs operate due to the up-drafting warm air buoyancy.
- The electricity generation units of hydro-electric PPs are water turbines engaged to electric generators while the generation units of solar aero-electric PPs are air turbines engaged to electric generators.
- The energy produced by the hydro-electric PPs is proportional to the falling water height, while the energy produced by the solar aero-electric PPs is proportional to up-drafting height of warm air, which is equal to the height of the solar chimneys.
- That is why Prof J. Sclaigh in his book named the solar chimney technology power plants as the hydro-electric power plants of deserts.

1.3 Continuous operation

As it will be shown later the SAEPs operate continuously due to the ground thermal storage. The minimum electric power is generated when the sun is just starting rising, while the maximum electric power is achieved about 2 hours after the sun's maximum irradiation on ground. The power generation profile can become smoother if we increase the solar collector thermal capacity. This can be done by putting on its ground area closed tubes filled with water (as happens already in conventional greenhouses).

2. History

The Solar Chimney technology for electricity generation was inspired by several engineering pioneers early in the first decade of the 20th century.
In 1926 Prof Engineer Bernard Dubos proposed to the French Academy of Sciences the construction of a Solar Aero-Electric Power Plant in North Africa with its solar chimney on the slope of a sufficient height mountain. His proposal is shown in the following figure(2), found in a book of 1954 ("Engineer's Dream" Willy Ley, Viking Press 1954)

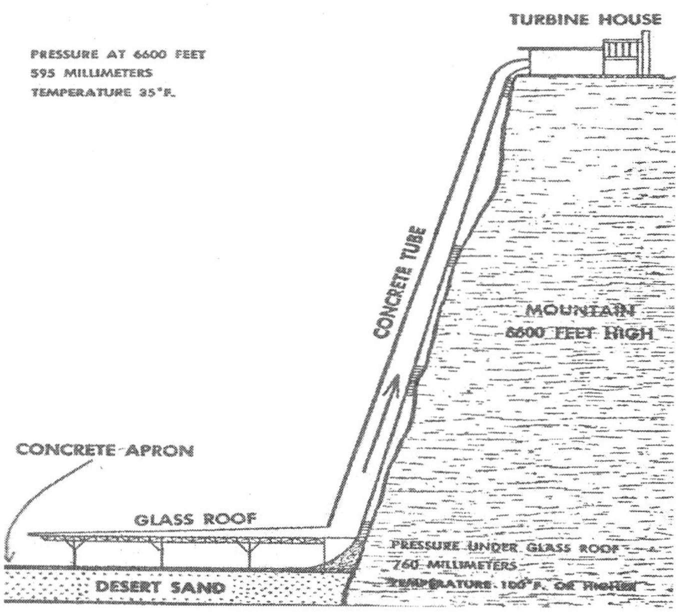

Fig. 2. (from the book: "Engineer's Dream"By: Willy Ley, Viking Press 1954)

Lately Schaich, Bergerman and Partners, under the direction of Prof. Dr. Ing. Jorg Schlaigh, built an operating model of a SAEPP in 1982 in Manzaranes (Spain), which was funded by the German Government.
This solar chimney power plant, shown in next figure (3) was of rating power 50 KW. Its greenhouse had a surface area of 46000 m^2 and its solar chimney was made out of steel tubes of 10 m diameter and had a height of 195 m.

This demo SAEP was operating successfully for approximately 6 years. During its operation, optimization data were taken.

The collected operational data were in accordance with the theoretical results of the scientific team of Prof Jorg Schlaigh.

Fig. 3. A view of the Manzanares Solar Chimney Power Plant

Prof. Jorg Schlaigh in 1996 published a book (Schlaigh 1995) presenting the solar chimney technology. He proposed in his book the huge reinforced concrete solar chimneys of heights of 500m-1000m.

The proposed concrete solar chimneys are huge and very expensive. Therefore the investment cost per produced KWh on the solar chimney technology with concrete chimneys is in the same cost range with the competitive solar thermal technologies. The generated KWh, by the CSP Parabolic Through for example, it has almost the same direct production cost, but the CSP power plants can be split into small units and developed using reasonable recourses.

However the proposed solar chimney technology had an important benefit in comparison with the major renewable technologies (Wind, SCP, PV).

That is its ability, equipping its solar collectors, with thermal storage facilities of negligible cost, to generate uninterrupted electricity of a controlled smooth profile for 24h/day, 365days/year.

The last decade several business plans and a series of scientific research papers have focused on the solar chimney technology, whereby the author with a series of patents and papers has introduced and scientifically supported the floating solar technology (Papageorgiou 2004, 2009).

3. Principles of operation of the solar chimney technology and its annual efficiency Information

3.1 Short description and principles of operation

A floating solar chimney power plant (SAEP) is made of three major components:

- A large solar collector, usually circular, which is made of a transparent roof supported a few meters above the ground (the greenhouse). The transparent roof can be made of glass or crystal clear plastic. A second cover made of thin crystal clear plastic is suggested to be hanged just underneath the roof in order to increase its thermal efficiency. The periphery of the solar collector is open in order that the ambient air can move freely into it.
- A tall fabric free standing lighter than air cylinder (the floating solar chimney) placed in the center of the greenhouse which is up drafting the warm air of the greenhouse, due to its buoyancy, to the upper atmospheric layers.
- A set of air turbines geared to appropriate electric generators (the turbo generators), placed with a horizontal axis in a circular path around the base of the FSC or with a vertical axis inside the entrance of the solar chimney. The air turbines are caged and can be just a rotor with several blades or a two stage machine (i.e. with a set of inlet guiding vanes and a rotor of several blades). The gear boxes are adjusting the rotation frequency of the air turbines to the electric generator rotation frequency defined by the grid frequency and the electric generator pole pairs.

The horizontal solar irradiation passing through the transparent roof of the solar collector is heating the ground beneath it. The air beneath the solar collector is becoming warm through a heat transfer process from the ground area to the air. This heat transfer is increased due to the greenhouse effect of the transparent roof.

This warm air becomes lighter than the ambient air. The buoyancy of the warm air is forcing the warm air to escape through the solar chimney. As the warm air is up drafting through the chimney, fresh ambient air is entering from the open periphery of the greenhouse. This fresh air becomes gradually warm, while moving towards the bottom of the solar chimney, and it is also up-drafting.

Thus a large quantity of air mass is continuously circulating from ground to the upper layers of the atmosphere. This circulating air mass flow is offering a part of its thermodynamic energy to the air turbines which rotate and force the geared electric generators also to rotate. Thus the rotational mechanical power of the air turbines is transformed to electrical power. An indicative diagram of the SAEP operation is shown in the next figure(4).

Thus the first two parts of the SAEPs form a huge thermodynamic device up drafting the ground ambient air to the upper atmosphere layers and the third part of the SAEP is the electricity generating unit.

The solar energy arriving on the horizontal surface area A_c of the greenhouse of the SAEP is given by $E_{IR}=A_c \cdot W_y$, where Wy is the annual horizontal solar irradiation in KWh/m^2, at the place of installation of the SAEP and is given by the meteorological data nearly everywhere.

The average annual horizontal solar irradiance is given by $G_{av}=W_y/A_c$.

The horizontal solar irradiation is offering thermal power $P_{Th}= \dot{m} \cdot c_p \cdot (T_{03}-T_{02})$ to the up drafting air mass flow $\dot{m}$ of the ambient air, $c_p \approx 1005$ and T_{02} is equal to the average ambient temperature T_0 plus ~ 0.5 ^{0}K, in order that it is taken into account the outer air stream increased inlet temperature due to its proximity to the ground on its entrance inside the solar collector.

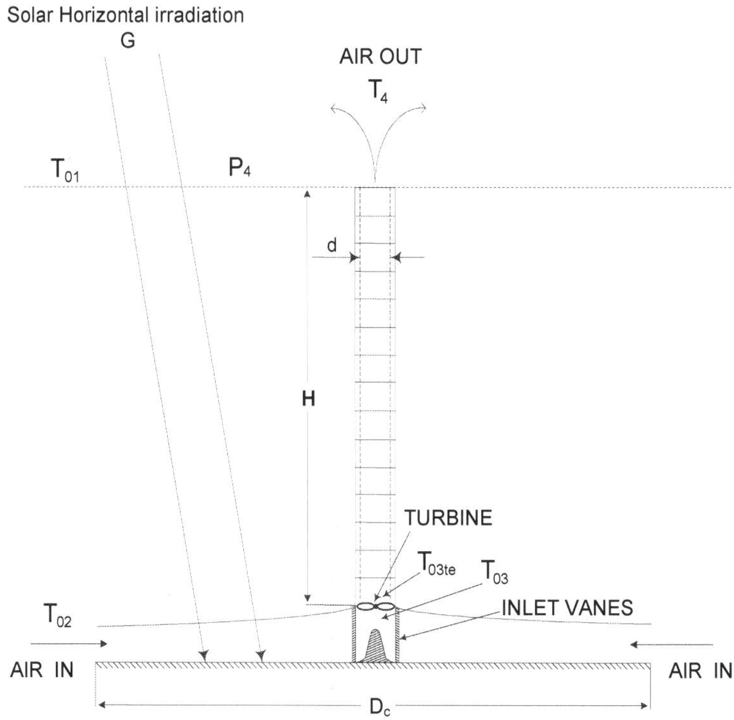

Fig. 4. Schematic diagram of the SAEP in operation

3.2 Annual average efficiency of SAEPs

The annual efficiency of the solar collector η_{sc} is defined as the average ratio of the thermal power P_{Th} absorbed by the air mass flow to the horizontal solar irradiation arriving on the greenhouse roof $G_{av} \cdot A_c$, where G_{av} is the average horizontal irradiance and A_c the greenhouse surface area.

The annual average double glazing solar collector efficiency η_{sc} is theoretically estimated to ~50%, while the annual efficiency for the single glazing solar collector is estimated to 2/3 of the previous figure i.e. ~33%.

Thus the average exit temperature T_{03} from the solar collector can be calculated by the equation $\dot{m} \cdot c_p \cdot (T_{03}-T_{02}) = \eta_{sc} \cdot G_{av} \cdot A_c$ where T_{02} is the average inlet air temperature.

The exit thermal power P_{Th} from the solar collector is transformed to electric power P, plus power thermal losses P_L (to the air turbines, gear boxes and electric generators), plus warm air kinetic power at the top exit of the solar chimney P_{KIN} and friction thermal losses inside the solar chimney P_{FR}.

The maximum efficiency of the solar chimney is the Carnot efficiency defined as the ratio of the temperature difference between the incoming and outcoming air temperatures of the up-drafting air divided by the ambient air temperature.

This maximum efficiency has been proven (Gannon & Backstrom 2000) to be equal to:

$$\eta_{FSC,max}=g \cdot H/(c_p \cdot T_0) \quad (1)$$

Due to friction and kinetic losses in the solar chimney the actual solar chimney efficiency η_{FSC} is for a properly designed SAEP approximately 90% of its maximum Carnot efficiency (close to the optimum point of operation of the SAEP).

The combined efficiency η_T of the air turbines, gear boxes and electric generators is within the range of 80%.

The average annual efficiency of the SAEP is the product of the average efficiencies of its three major components i.e. the solar collector, the floating solar chimney and the turbo-generators i.e. $\eta_{av}= \eta_{sc} \cdot \eta_{FSC} \cdot \eta_T$.

Thus the annual average efficiency of a SAEP of proper design, with a double glazing solar collector should be approximately:

$$\eta_{av} =(1.2 \cdot H/1000)\ \% \quad (2)$$

While for the SAEP with a single cover collector it is approximately:

$$\eta_{av}= (0.79 \cdot H/1000)\ \%. \quad (3)$$

The formulae have been calculated for $g=9.81, c_p=1005$ and $T_0 \approx 293.2^0K(20^0C)$.

This means that if the annual horizontal irradiation arriving on the place of installation of the SAEP is 2000 KWh/m^2, the solar collector surface area is 10^6 m^2 (one square Km) and the solar chimney height is 750 m the SAEP can generate approximately 18 million KWh. The same SAEP with a single glazing roof will generate approximately only 12 million KWh.

Following approximate analysis, for a SAEP with a double cover roof of given dimensions (Ac=Greenhouse area in m^2 and d=internal diameter of the Floating Solar Chimney in m) to be installed in a place of annual horizontal solar irradiation Wy in KWh/m^2 the diagram showing the relation between the annual efficiency of the SAEP and its FSC height H can be calculated.

The following figure (5) shows the annual efficiency as a function of FSC's height for a SAEP of $Ac=10^6m^2$, d=40m and Wy=1700 KWh/m^2 (Cyprus, South Spain).

The calculated efficiency curve is practically independent of the annual horizontal solar irradiation W_y. However it depends on the FSC internal diameter d. The reason is that a smaller diameter will increase the warm air speed at the top exit of the FSC and consequently will increase the kinetic power losses and decrease the average annual efficiency. If we vary the solar collector diameter of the SAEP its FSC internal diameter should vary proportionally in order to keep almost constant the air speed at the top exit of the FSC and consequently the annual efficiency of the SAEP.

Hence we should notice that in order to receive the efficiency diagram as shown in the following figure (5) figure the kinetic and friction losses of the Floating Solar Chimney should be approximately 10% of the total chimney power. This can be achieved if the internal diameter of the FSC is appropriate in order to keep the average air speed in the range of 7÷8 m/sec, and the FSC internal surface has a low friction loss coefficient.

The following figure (6) shows the variation of the annual efficiency of a SAEP of a FSC 500m high, installed in a place of annual horizontal solar irradiation 1700KWh/m^2 as function of the internal diameter of its FSC.

The annual electricity generated by the SAEP, E_y can be calculated as a product of the annual efficiency and the arriving horizontal solar irradiation on its greenhouse surface $A_c \cdot W_y$. Thus taking into consideration that the annual efficiency is proportional to the FSC

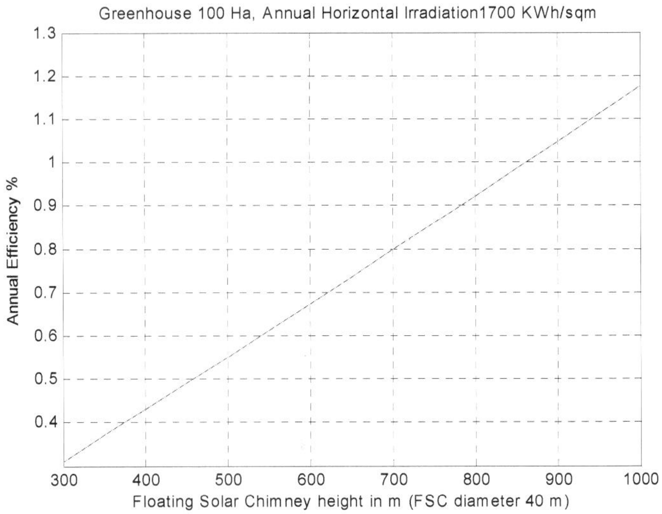

Fig. 5. Annual efficiency of a SAEP as function of its FSC height

height H, the annual generated electricity by the SAEPs is also proportional to the Floating Solar Chimney height H, is as follows:

$$E_y = c \cdot H \cdot A_c \cdot W_y \quad (4)$$

The constant c is mainly depending on the FSC's internal diameter d.

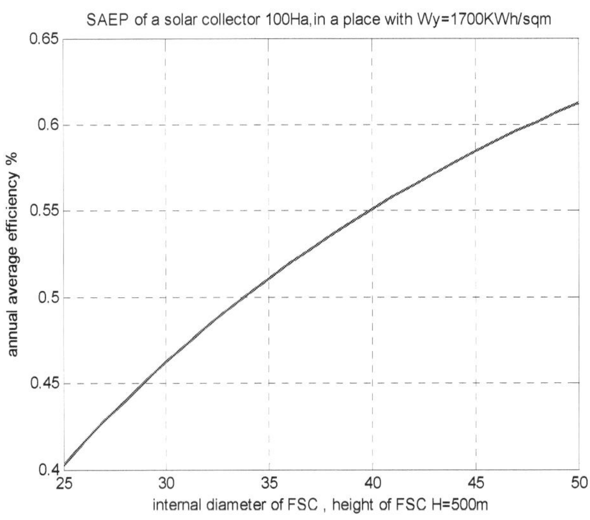

Fig. 6. variation of the annual efficiency of a SAEP with internal FSC diameter

4. Theoretical analysis of the Floating Solar Chimney technology

4.1 Annual average efficiency of SAEPs

The ground thermal storage effect and the daily electricity generation profile, have been studied by several authors (Bernades et.al 2003, Pretorius & Kroger 2006, Pretorious 2007).
The author has used an equivalent approach on the daily power profile study of the floating solar chimney SAEPs using the thermodynamic model see (Backstrom & Gannon 2000) and Fourier series analysis on the time varying temperatures and varying solar irradiance during the 24 hours daily cycle.
Following the code of the author analysis an evaluation of the sensitivity of the various parameters has been made leading to useful results for the initial engineering dimensioning and design of the SAEPs.
The important results of these studies are that the solar chimney power plant annual power production can be increased by using a second glazing below the outer glazing and its output power production can be affected by the ground roughness and ground solar irradiation absorption coefficients.
The thermodynamic cycle analysis proposed in ref. (Gannon Backstrom 2000) is an excellent way of engineering analysis and thermodynamic presentation of the solar chimney power plant operation.
The thermodynamic cycle of the solar chimney operation power plant using the same symbols of the study of ref (Backstrom & Gannon 2000) is shown in the following figure.

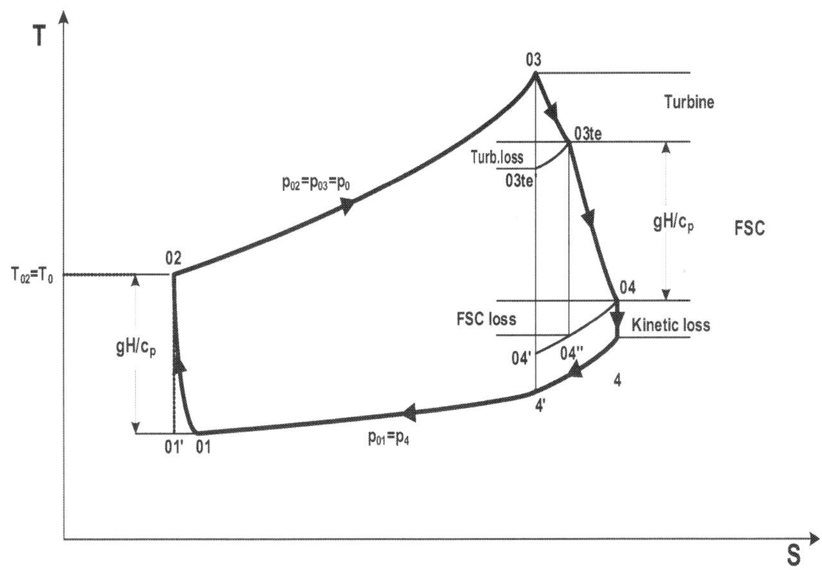

Fig. 7. The thermodynamic diagram of the SAEP

Temperatures, stagnation temperatures (marked with $_0$) and equivalent isentropic temperatures (marked with ') are shown in the indicative diagram on the previous figure.
The main thermodynamic cycle temperatures are defined in the following table:

T_{01}	Isendropic temperature of ambient air in height H (exit of solar chimney)
T_{02}	Ambient temperature in the ground around the solar collector
T_{03}	Inlet temperature in the air turbines
T_{03te}	Exit air temperature from the turbo generators
T_{04}	Stagnation temperature at the top of the solar collector
T_4	Exit temperature of the air mixed with the ambient air at the top of the exit layers

Table 1.Tthermodynamic cycle temperatures

The process {T_{02} to T_{o3}} is assumed as approximately isobaric. This assumption is very reasonable taking into consideration that the heat and expansion of moving air is taking place inside the solar collector.

The processes {T_4 to T_{01}}, {T_{03te} to T'_{03te}} and {T_{04} to T'_{04}} are definitely isobaric by nature.

By the analysis on the relations between the temperatures the following relationships can be derived:

$$\begin{cases} T_{04}=T_4+\dfrac{\alpha\cdot\upsilon_{ex}^2}{2\cdot c_p}=T_4+C_2\cdot T_4^2, T_{03te}=T_{04}+\dfrac{g\cdot H}{c_p} \\ T'_{03te}=T_{03}-\dfrac{T_{03}-T_{03te}}{n_T}, T''_{04}=T_{04}-k\cdot\dfrac{\alpha\cdot\upsilon_{ex}^2}{2\cdot c_p} \\ T'_{04}=\dfrac{T'_4\cdot T_{04}}{T_4} \text{ and } T''_{04}=T_{03}\cdot\dfrac{T'_{04}}{T'_{03te}} \end{cases} \tag{5}$$

Whereby the parameters participating in the relations are defined as follows:

H =solar chimney height

d=internal solar chimney diameter

$A_{ch}=\pi\cdot d^2/4$, is the solar chimney internal cut area

$\dot{m}$ =moving mass flow

α=kinetic energy correction coefficient, of a usual value of 1.058 calculated in (White 1999).

k=friction loss coefficient inside the solar chimney

$k= k_{in}+ 4\cdot C_d \cdot H/d$ where, for the operation range of Reynolds numbers inside the solar chimney, the drag friction factor C_d is approximately equal to 0.003, see (White 1999) and for no available data k_{in}it is estimated to 0.15.

η_T= turbo generators overall efficiency, if not available data estimated to 0.8.

T_0= ambient air temperature

$T_{02}= T_0+0.5$

p_0= ambient atmospheric pressure on ground level at the place of installation of the SAEP, if not available data it is assumed as equal to 101300 Pa.

p_4= ambient atmospheric pressure on top exit at height H, estimated by the formula:

$$p_4=p_0\cdot(1-\frac{g\cdot H}{c_p\cdot T_0})^{3.5} \tag{6}$$

g= gravity constant 9.81

c_p= specific heat of air approximately equal to 1005
R= air constant approximately equal to 287

υ_{ex}= average air speed at the top exit of the solar chimney $\upsilon_{ex}=\frac{R\cdot T_4\cdot \dot{m}}{P_4\cdot A_{ch}}$

and: $T_4'=T_{03}\cdot\frac{T_0\text{-}C_1}{T_0}, C_1=\frac{g\cdot H}{c_p}, C_2=\frac{a}{2\cdot c_p}\cdot\left(\frac{R\cdot\dot{m}}{A_{ch}\cdot p_4}\right)^2, C_3=T_{03}\cdot(\eta_T\text{-}1)+\frac{g\cdot H}{c_p}$

The system of the previous equations can been simplified (see Papageorgiou, 2004),leading to a forth order polynomial equation for T_4 given by:

$$w_1\cdot T_4^4+w_2\cdot T_4^3+w_3\cdot T_4^2+w_4\cdot T_4+w_5=0 \tag{7}$$

Where the coefficients w_1, w_2, w_3, w_4 and w_5 are given by the relations:

$$w_1=C_2^2\cdot(1\text{-}k), w_2=C_2\cdot(2\text{-}k\text{-}\eta_T\cdot C_2\cdot T_4'), w_3=(1\text{-}k)\cdot C_2\cdot C_3+1\text{-}2\cdot\eta_T\cdot C_2\cdot T_4'$$
$$w_4=C_3\text{-}\eta_T\cdot T_4'\cdot(1\text{-}C_1\cdot C_2), w_5=\text{-}\eta_T\cdot T_4'\cdot C_1$$

The proper root of the previous polynomial equation is the temperature T_4.
It is easy using the previous relations to calculate T_{03te} by the formula:

$$T_{03te}=T_4+C_2\cdot T_4^2+\frac{g\cdot H}{c_p} \tag{8}$$

Thus the overall electrical power of the generators is given by the relation:

$$P=\dot{m}\cdot c_p\cdot(T_{03}-T_{03te})=\dot{m}\cdot c_p\cdot(T_{03}\text{-}T_4\text{-}C_2\cdot T_4^2-\frac{g\cdot H}{c_p}) \tag{9}$$

As a final result we can say that the air mass flow $\dot{m}$ and the exit temperature T_{03} of the moving air mass through solar collector can define, through the previous analytical procedure, based on the thermodynamic cycle analysis, the electrical power output P of the SAEP.
The proposed thermodynamic analysis, though it looks more complicated than the analysis based on the buoyancy of warm air inside the chimney and the relevant pressure drop to the air turbine used by Bernades M.A. dos S., Vob A., Weinrebe G. and Pretorius J.P., Kroger D.G., it is an equivalent thermodynamic analysis that takes into consideration all necessary and non negligible effects and parameters of the process in the SAEP.
An approximate procedure for T_{03} calculation is given by Shlaigh in his relative book.
The approximate average equation relating the average exit solar collector air temperature T_{03} to its input air temperature T_{02} near the point of optimal operation of the SAEP can be written as follows:

$$ta\cdot G_{av}\cdot A_c=\dot{m}\cdot C_p\cdot(T_{03}-T_{02})+\beta\cdot A_c\cdot(T_{03}\text{-}T_{02}) \tag{10}$$

where:

- β is the approximate thermal power losses coefficient of the Solar Collector (to the ambient and ground) per m^2 of its surface area and °C of the temperature difference (T_{03}-T_{02}) . An average value of β for double glazing solar collectors is ~3.8÷4 $W/m^2/°C$.

- Gav is the annual average horizontal irradiance on the surface of the solar collector.
- The annual average solar horizontal irradiance G_{av} is given by the formula: Wy/8760hours, where W_y is the annual horizontal irradiation of the place of installation of the SAEPP, (in KWh/m^2)
- ta is the average value of the product: {roof transmission coefficient for solar irradiation X soil absorption coefficient for solar irradiation}.An average value of the coefficient ta for a double glazing roof is ~ 0.70 .
- and Ac is the Solar Collector's surface area.

Using in the equation an approximation for the function T_{03} ($\dot{m}$), it gives as:

$$T_{03}(\dot{m}) = [\, ta \cdot G \,/\, (\beta + \dot{m} \cdot C_p / A_c)\,] - T_{02} \tag{11}$$

Where T_{02} is, approximately, equal to the ambient temperature (T_0 in ^0K), plus 0.5 degrees of Celsius. The increase is due mainly to ground thermal storage around the Solar Collector. The inlet ambient air temperature as passing above it is increasing entering to the solar collector.

The proper value of β, giving the average solar collector thermal losses, has been calculated by the heat transfer analysis of the solar collector. An introduction on this analysis is given on the next paragraph. The heat transfer analysis uses time Fourier series in order to take into account the ground thermal storage phenomena during a daily cycle of operation.

The instantaneous efficiency of the SAEP is given by the formula:

$$\eta = P \,/\, (A_C \cdot G) \tag{12}$$

where $A_C \cdot G$ is the solar irradiation power arriving on the horizontal solar collector surface area A_c and P is the maximum generated electric power. This efficiency is for a given value of horizontal solar irradiance G. However we can prove that for an almost constant mass flow near the point of maximum power output, the maximum electric power P and the horizontal irradiance G are almost proportional, thus the previous formula is giving also the annual efficiency of the SAEP defined as the annual generated electricity in KWh divided by the annual horizontal irradiation arriving on top of the roof of the greenhouse of the SAEP i.e

$$\eta = P_{av} \,/\, (A_C \cdot G_{av}) = E_y(KWh) \,/\, W_Y \tag{13}$$

As an example let us consider that a SAEP has the following dimensions and constants: $A_c = 10^6 m^2$ (DD=1000m), H = 800 m, d=40 m, k = 0.49, α = 1.1058, η_T = 0.8, the average ambient temperature is T_o = 296.2 ^oK and the ambient pressure is P_o = 101300 Pa. Let us assume that the horizontal solar irradiance G is varying between 100 W/m^2 to 500 W/m^2 ($G_{av} \approx$ 240W/m^2). In following figure the effect of the G on the power output as function of mass flow of this SAEP is shown.

If the maximum (daily average during summer operation) G_{av} is 500 W/m^2 the maximum power output of this SAEP, achieved for $\dot{m}_M$ = ~10000 Kg/sec is 5 MW. Thus its efficiency is approximately 1%. Let us assume that the rated power output P_R of a SAEP is the maximum power output for the maximum average solar irradiance. As we can observe on the above figure, the maximum power output point of operation ($\dot{m}_M$) is approximately the same for any horizontal solar irradiance G.

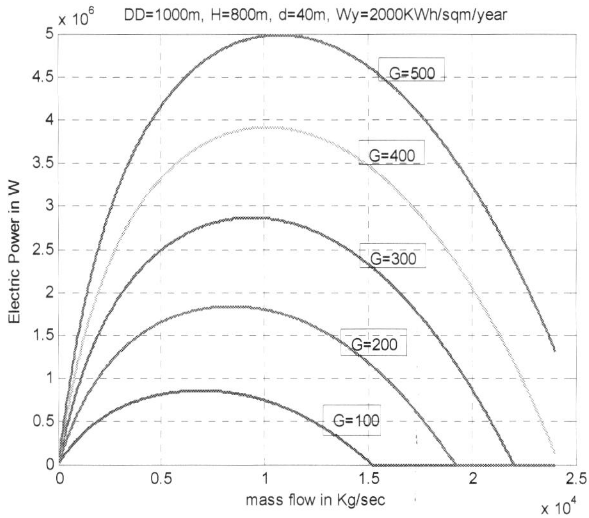

Fig. 9. Electrical Power as function of mass flow for various values of G

Thus if we can control the operation of the SAEP to operate with the proper constant mass flow, close to $\dot{m}_M$, we should achieve almost the possible maximum electric power output by the SAEP for any horizontal solar irradiance. This is referred to as an optimal operation of the SAEP.

As we see later this can be achieved by using induction generators and gear boxes of proper transmission rate.

As a rule of thumb we can state that $\dot{m}_M$ for optimal operation of the SAEP can be calculated approximately by the formula $\dot{m}_M = \rho \cdot \upsilon \cdot (\pi \cdot d^2/4)$, where air speed is υ it is estimated to 7-8 m/sec, the air density is given by $\rho = p_0/(287 \cdot 307.15)$ and d is the internal solar chimney diameter.

A more accurate calculation can be done if we work out on the mass flow for maximum electric power output per annual average horizontal solar irradiance $G_{av,annual}=Wy/8760$. This can be done using the thermodynamic cycle analysis for variable mass flow $\dot{m}$ and G_{av}. The calculated efficiency for the annual average horizontal solar irradiance $G_{av}=2100000/8760 \approx 240 W/m^2$, of the previously defined SAEP, is 0.94 % (i.e. 6% lower than the calculated efficiency of 1% for the maximum summer average horizontal solar irradiance of $500 W/m^2$).

4.2 Maximum exit warm air speed without air turbines

Using the thermodynamic cycle diagram, the maximum top exit warm air speed of the solar collector plus the FSC alone (i.e. without the air turbines) can be calculated.

In the previous set of equations we should assume that $n_T=0$. Thus:

T_{03}=T_{03te} and $T'_{04}=T''_{04}$. If we consider that the kinetic losses are approximately equal to $T'_{04}-T'_4 \approx \frac{a \cdot \upsilon^2}{2 \cdot c_p}$, the friction losses are equal to $T''_{04}-T'_{04} = k \cdot \frac{a \cdot \upsilon^2}{2 \cdot c_p}$ and taking into consideration that the equations $T'_4 = T_{03} \cdot \frac{T_0\text{-}C_1}{T_0}, C_1 = \frac{g \cdot H}{c_p}$ the following relation is derived:

$$2 \cdot g \cdot H \cdot \frac{\Delta T}{T_0} = (k+1) \cdot a \cdot \upsilon^2 \tag{14}$$

Where $\Delta T = T_{03} - T_0$ (we can approximately consider that $T_0 \approx T_{02}$).

Thus the maximum exit top air speed in a free passage solar chimney (without air turbines) is given by the formula:

$$\upsilon = \sqrt{2 \cdot g \cdot H \cdot \frac{\Delta T}{T_0} / [(k+1) \cdot a]}\,. \tag{15}$$

For example the exit top speed of the up-drafting air inside the FSC of H=800m height, with ordinary values for coefficients a=1.1058 and k=0.49 and ambient air temperature T_0=296.2 ^{0}K (23 ^{0}C) as function of ΔT is given in the next figure:

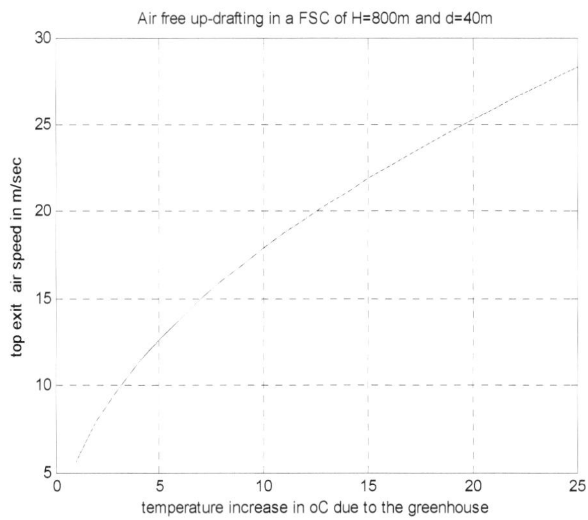

Fig. 10. Free air speed as a function of temperature increase

The temperature increase ΔT as a function of the greenhouse surface area A_c is given by the approximate formula $\Delta T \approx$ ta$\cdot$G / (β + $\dot{m}$ $\cdot c_p/A_c$) where ta≈0.7, $\beta \approx 4$, c_p=1005, and $\dot{m}_M = \rho \cdot \upsilon \cdot (\pi \cdot d^2/4)$ where $\rho \approx 1.17 Kg/m^3$, and d=40m. Thus The approximate double glazing solar collector area, generating the free up-drafting air speed υ can be defined by ΔT, υ and G by the equation $A_c \approx \dot{m} \cdot c_p / [(\text{ta} \cdot G)/\Delta T - \beta]$.

The approximate solar collector area A_c as a function of the temperature increase ΔT for various values of equivalent horizontal solar irradiance G=250,300,350,400 and 450 W/m^2, is shown in the following figure.

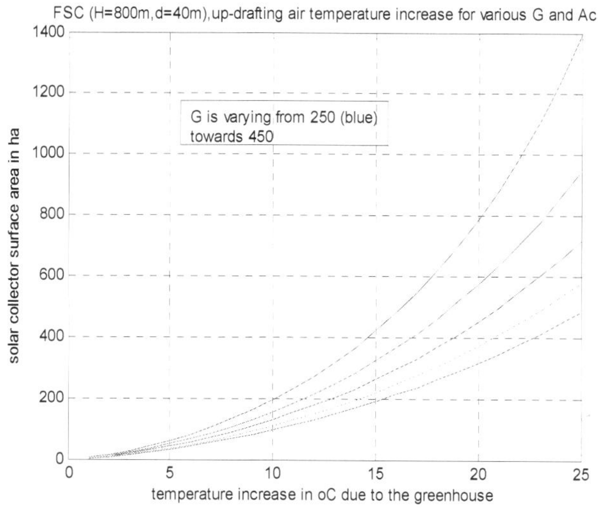

Fig. 11. The solar collector area as a function of its generating temperature increase

Example: for a solar collector of surface area A_c=400Ha (i.e.400000m^2), with a diameter $D_c \approx 715m$, for an equivalent horizontal solar irradiation G of 250W/m^2, the created temperature difference ΔT is ~14.5^{0}C and the free up-drafting air speed υ inside the FSC of H=800m height and d=40m internal diameter will be ~21m/sec, while for G=450W/m^2, ΔT is ~22.5^{0}C and υ is ~27m/sec.

For one dimensional analysis $a \approx 1$ and if the friction losses are negligible, i.e. $k \approx 0$, we have:

$$\upsilon \approx \sqrt{2 \cdot g \cdot H \cdot \frac{\Delta T}{T_0}} \tag{17}$$

Therefore free up-drafting warm air top speed formula, in an adiabatic and free friction FSC, due to its buoyancy, is similar to free falling water speed due to gravity given by:

$$\upsilon_{water} \approx \sqrt{2 \cdot g \cdot H}$$

4.3 The thermal heat transfer model of the SAEP

In order to use the previous thermodynamic cycle analysis of the SAEP we should calculate the warm air temperature T_{03} at the entrance of the air turbine or at the exit of the solar collector. The calculation of this average temperature can be done by using the previously proposed approximate analysis. However the temperature T_{03} is varying during the 24 hours daily cycle.

In order for the daily variation to be calculated and consequently the electric power daily variation using the previously proposed thermodynamic cycle analysis, we should make a

heat transfer model and use it for the calculation of the exit temperature as function mainly of daily horizontal irradiance profile and ambient temperature daily profile.
The SAEP heat transfer model with a circular collector is shown in the indicative diagram of the previous figure.
The circular solar collector of this SAEP is divided into a series of M circular sectors of equal width Δr as shown in the next figure.
In this figure the cut of a circular sector of the solar collector of the SAEP is shown with the heat transfer coefficients of the process (radiation and convection) and the temperatures of ground (T_s), moving air (T), inner curtain (T_c), outer glazing (T_w), ambient air (T_0) and sky (T_{sk}). The ground absorbs a part of the transmitted irradiation power due to the horizontal solar irradiance G (ta ·G).
The wind is moving with a speed υ_w and on the ground it is a thin sheet of water inside a dark plastic film. The ground is characterized by its density ρ_{gr}, its specific heat capacity c_{gr} and its thermal conductivity k_{gr}.

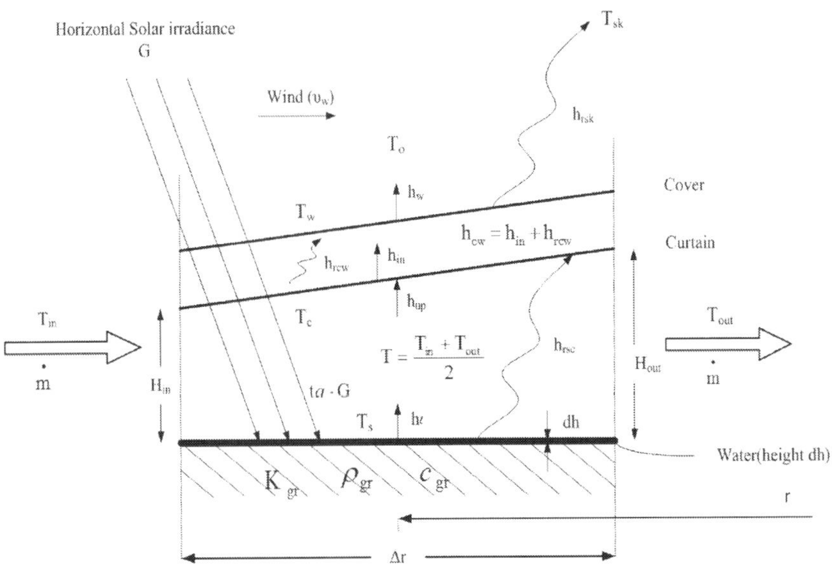

Fig. 12. The cut of a circular sector of a double glazing circular solar collector

The mth circular sector (m=1up to M) will have a width $\Delta r = (D_c - D_{in})/M$, an average radius $r_m = Dc/2 - \Delta r \cdot (m1 - 1/2)$ and an average height $H_m = (H_{in,m} + H_{ex,m})/2$.
For a linear variation of the roof height $H_m = Hin + (Hout - Hin) \cdot (m - 1/2)/M$, where D_c=solar collector diameter and D_{in}=Final internal diameter of the solar collector.
These consecutive circular sectors, for the moving air stream of mass flow $\dot{m}$, are special tubes of nearly parallel flat surfaces and therefore they have equivalent average diameters $d_{e,m} = 2 * H_m$.
As the ambient air moves towards the entrance of the first circular sector it is assumed that its temperature T_0 increases to $T_0 + dT$ due to the ground heat transfer convection to inlet air, around the solar collector. As an approximation dT is estimated to 0.5 °K.

The exit temperature of the first sector is the inlet temperature for the second etc. and finally the exit temperature of the final M[th] sector is the T_{03}, i.e. the inlet stagnation temperature to the air turbines.

The solar chimney heat transfer analysis during a daily 24 hours cycle, is too complicated to be presented analytically in this text however we can use the results of this analysis in order to have a clear picture of the operational characteristics of the SAEPs. Using the code of the heat transfer analysis for moving mass flow $\dot{m}_M$, the daily variation of the exit temperature T_{03} can be calculated. Using these calculated daily values of the T_{03} and by the thermodynamic cycle analysis for the optimal mass flow $\dot{m}_M$ the daily power profile of the electricity generation can be calculated.

With this procedure the 24 hour electricity generation power profile of a SAEP with a solar collector of surface area $A_c=10^6 m^2$ and a FSC of H=800m height and d=40m internal diameter for an average day of the year has been calculated. The SAEP is installed in a place with annual horizontal solar irradiation W_y=1700 KWh/m^2.

In the following figure three electric power profiles are shown with or without artificial thermal storage.

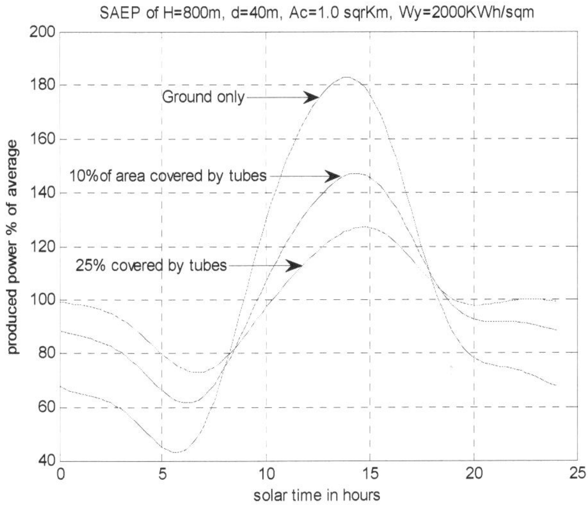

Fig. 13. The average daily SAEP's electricity generating profiles

The relatively smooth profile shows the electric power generation when only the ground acts as a thermal storage means. While the smoother profiles are achieved when the greenhouse is partly covered (~10% or ~25% of its area) by plastic black tubes of 35cm of diameter filled with water, i.e. there is also additional thermal storage of an equivalent water sheet of 35·π/4=27.5 cm on a small part of the solar collector.

The daily profiles show that the SAEP operates 24hours/day, due to the greenhouse ground (and artificial) thermal storage. That is a considerable benefit of the FSC technology compared to the rest solar technologies and the wind technology which if they are not equipped with energy mass storage systems they can not operate continuously.

As shown in the produced curves on the previous figure, with a limited (~10%) of the greenhouse ground covered by plastic tubes (35 cm) filled with water, the maximum daily power is approximately 140% of its daily average, or the daily average is 70 % of its maximum power.

Taking into consideration the seasonal power alteration and assuming that the average annual daily irradiation at a typical place is approximately 70% of the average summer daily irradiation, the annual average power can be estimated as a percentage of the maximum power production (at noon of summertime) as the product of 0.77·0.70=0.49.

The maximum power is equal to the rating of the power units of the SAEP (Air turbine, electric generator, electric transformer etc.), while the average power multiplied by 8760 hours of the year defines the annual electricity generation. Therefore the capacity factor of a SAEP equipped with a moderate artificial thermal storage can be as high as ~49%.

Without any artificial thermal storage the average daily power is approximately 0.55 of its maximum thus the capacity factor is ~37% (0.55·0.70≈0.385).

This means that in order to find the annual energy production by the SAEP we should multiply its rating power by ~3250÷4300 hours. However we should take into consideration that the SAEPs are operating continuously (24x365) following a daily and seasonal varying profile.

5. The major parts and engines of Floating Solar Chimney technology

5.1 The solar collector (Greenhouse)

The solar collector can be an ordinary circular greenhouse with a double glazing transparent roof supported a few meters above the ground. The periphery of the circular greenhouse should be open to the ambient air. The outer height of the greenhouse should be at least 2 meters tall in order to permit the entrance of maintenance personnel inside the greenhouse. The height of the solar collector should be increased as we approach its centre where the FSC is placed. As a general rule the height of the transparent roof should be inversely proportional to the local diameter of the circular solar collector in order to keep relatively constant the moving air speed. The circular greenhouse periphery open surface can be equal or bigger than the FSC cut area.

Another proposal with a simpler structure and shape the greenhouse can be of a rectangular shape of side DD. The transparent roof could be made of four equal triangular transparent roofs, elevating from their open sides towards the centre of the rectangle, where the FSC is placed. Thus the greenhouse forms a rectangular pyramid.

The previous analysis is approximately correct and can be figured out by using an equivalent circular greenhouse external diameter $D_c \approx DD \cdot \sqrt{4/\pi}$.

The local height of each inclined triangular roof is almost inversely proportional to the local side of the triangle in order to secure constant air speed.

Both solar collector structures are typical copies of ordinary agriculture greenhouses although they are used mainly for warming the moving stream of air from their periphery towards the centre where the FSC of the SAEP is standing. Such greenhouses are appropriate for FSC technology application combined with special agriculture inside them.

In desert application of the FSC technology the solar collectors are used exclusively for air warming. Also in desert or semi desert areas the dust on top of the transparent roofs of the conventional greenhouses could be a major problem. The dust can deteriorate the transparency of the upper glazing and furthermore can add unpredictable weight burden on the roof structure. The cleaning of the roof with water or air is a difficult task that can eliminate the desert potential of the FSC technology.

Furthermore in desert or semi-desert areas the construction cost of the conventional solar collector (a conventional greenhouse) could be unpredictably expensive due to the unfavourable working conditions on desert sites.

For all above reasons another patented design of the solar collectors has been proposed by the author.The proposed modular solar collector, as has been named by the author, will be evident by its description that it is a low cost alternative solar collector of the circular or rectangular conventional greenhouse which can minimize the works of its construction and maintenance cost on site.

We can also use and follow the ground elevation on site, and put the FSC on the upper part of the land-field therefore the works on site for initial land preparation will be minimized.

The greenhouse will be constructed as a set of parallel reverse-V transparent tunnels made of glass panels as shown in the next figure (14). The maximum height of the air tunnel should be at least 190cm in order to facilitate the necessary works inside the tunnel, as it is for example the hanging of the inner crystal clear curtains.

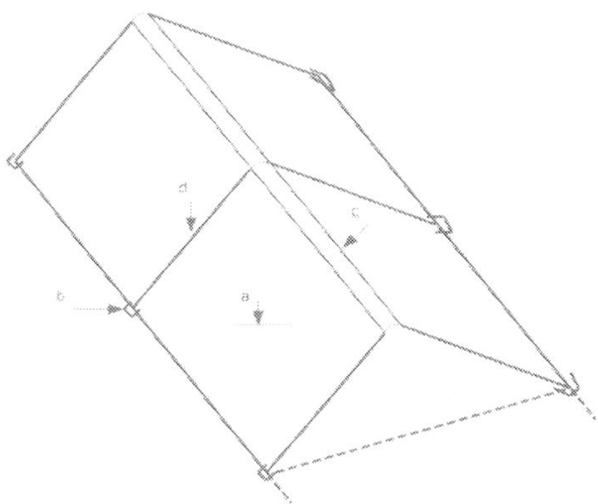

Fig. 14. A part of the triangular tunnel of two panels (a)glass panel, (b)ground support, (c)glass panel connector (d)glass plastic separator

An indicative figure of a greenhouse made of ten air tunnels is shown in next figure. Among the parallel air tunnels it is advisable that room should be made for a corridor of 30-40cm of width for maintenance purposes.

By above description it is evident that the modular solar collector is a low cost alternative of a conventional circular greenhouse for the FSC technology in desert or semi-desert areas that minimize the works on site and lower the construction costs of the solar collector and its SAEP. Furthermore the dust problem is not in existence because the dust slips down on the inclined triangular glass panels.

The average annual efficiency of the modular solar collector made by a series of triangular warming air tunnels with double glazing transparent roofs is estimated to be even higher than 50%. Thus its annual efficiency will follow the usual diagram of efficiency (or it will be even higher).

The total cut area of all the triangular air tunnels should be approximately equal to the cut area of the FSC for constant air speed. The central air collecting corridor cut should also follow the constant air speed rule for optimum operation and minimum construction cost.

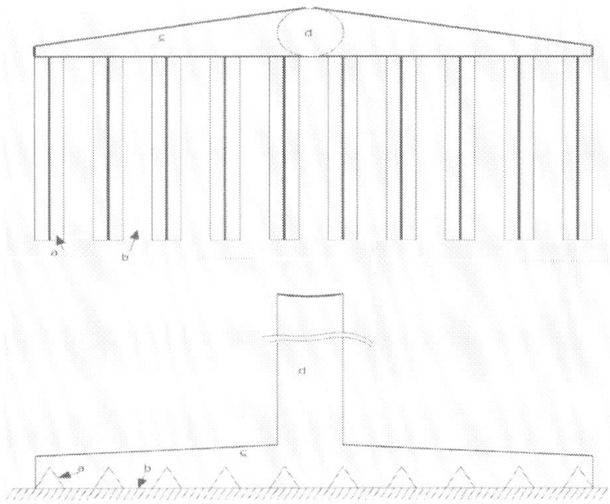

Fig. 15. Modular solar collector with ten air tunnels (a)Triangular tunnel, (b)Maintenance corridor (c)Central air collecting tube, (d)FSC

5.2 The Floating Solar Chimney (FSC)

A small part of a typical version of the FSC on its seat is taking place in the figure(16) below.

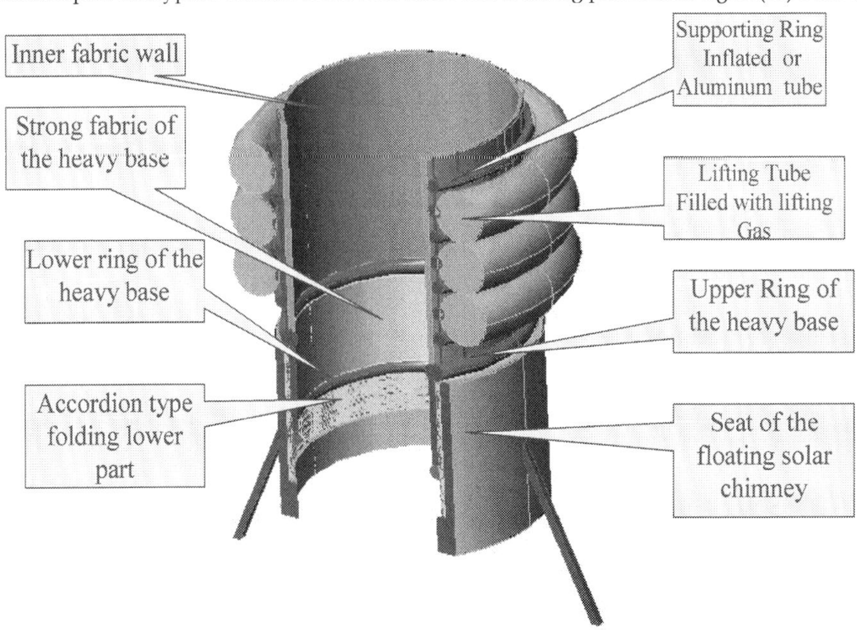

Fig. 16. A small part of a typical version of the FSC on its seat

The over-pressed air tubes of the fabric structure retain its cylindrical shape. While the lifting tubes (usually filled with NH_3) supply the structure with buoyancy in order to take its upright position without external winds. Both tubes can be placed outside the fabric wall as they are shown in the figure or inside the fabric wall. When the tubes are inside the fabric core they are protected by the UV radiation and the structure has a more compact form for the encountering of the external winds unpredictable behavior. But inside the warm air friction losses are increased and in order to have the same internal diameter the external diameter of the fabric core should be greater. In the first demonstration project both shapes could be tested in order that the best option is chosen.
Therefore the FSCs of the SAEPs are free standing fabric structures and due to their inclining ability they can encounter the external winds. See the next indicative figure (17) describing its tilting operation under external winds.

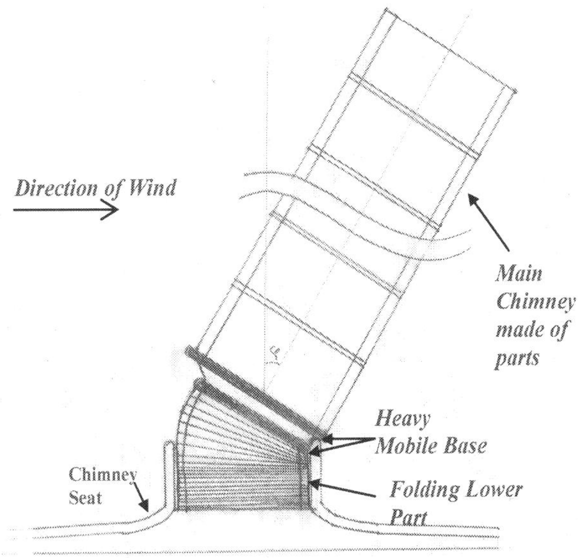

Fig. 17. Tilting operation of the FSC under external winds

However in areas with annual average strong winds the operating heights of the inclining fabric structures are decreasing. The following figure (18) presents the operating height loss of the FSCs as function of the average annual wind speed, for Weibull average constant k≈2.0. The net buoyancy of the FSC is such that will decline 60^0 degrees when a wind speed of 10 m/sec appears.
For example using the diagram in figure (18), for an average wind speed of 3 m/sec and a net lift force assuring a 50% bending for a wind speed of 10 m/sec, the average operating height decrease is only 3.7%.
As a result we can state that the best places for FSC technology application are the places of high average horizontal solar irradiation, low average winds and limited strong winds. The mid-latitude desert and semi-desert areas, that exist in all continents, combine all these properties and are excellent places for large scale FSC technology application.

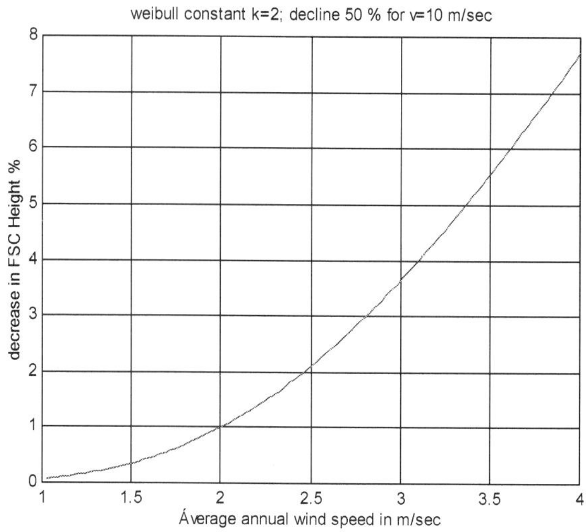

Fig. 18. FSC's operating height average decrease under external winds.

5.3 The air turbines

The air turbines of the SAEPs are either of horizontal axis placed in a circular pattern around their FSCs or with normal axis placed inside the FSCs (near the bottom). The later case with only one air turbine is most appropriate for the FSC technology, while the former is more advisable for concrete solar chimney technology applications.

The air turbines of the solar chimney technology are caged (or ducted) air turbines. These air turbines are not similar to wind turbines that transform the air kinetic energy to rotational energy, therefore their rotational power output depends on the wind speed or the air mass flow. The caged air turbines transform the dynamic energy of the warm air, due to their buoyancy, to rotational. Therefore their rotational power output does not depend on the mass flow only but on the product of the mass flow and the pressure drop on the air turbine. Therefore the warm air mass flow, as we have noticed already, is possible to remain approximately constant during the daily operation (in order that an optimal operation is achieved) while its rotational power and its relative electric power output vary during the daily cycle. The varying quantity is the pressure drop of the air turbine. This pressure drop depends on the warm air temperature i.e. the warm air proportional buoyancy and the FSC height.

The air turbines are classified according to the relation between their mass flows and their pressure drops. The wind turbines are class A turbines (large mass flow small pressure drop). The useful classes for solar chimney application are the class B and C. The class B are the caged air turbines with lower pressure drop and relatively higher mass flow and made without inlet guiding vanes, while the class C air turbines are with higher pressure drops and relatively lower mass flows and should be made of inlet guiding vanes in order that optimal efficiency is achieved.

Considering that the floating or concrete solar chimney SAEPs can have the same heights (between 500m÷1000m) the defining factor for air turbines with or without inlet guiding vanes is the solar collector diameter.

For the expensive concrete solar chimney the respective solar collectors are made with high diameters in order to minimize the construction cost of their SAEPs. While the low cost floating solar chimneys can be designed with smaller solar collectors for minimal cost and optimal operation.

The diameters of the solar collectors are proportional to the increase of the warm air temperatures $\Delta T=T_{03}-T_0$, thus proportional also to the buoyancies and to the pressure drops on the air turbines.

Therefore the Floating Solar Chimney SAEPs can be designed with air turbines of class B (i.e. without inlet guiding vanes). These caged air turbines are lower cost units per generated electricity KWh in comparison with class C air turbines which are appropriate for concrete solar chimney SAEPs.

5.4 The electric generators

There are two types or electric generators which can be used in SAEPs, the synchronous and the induction or asynchronous electric generators.

The synchronous electric generators for FSC technology should have a large number of pole-pairs pp. The frequency of the generated electricity by the multi-pole synchronous electric generator should be equal to the grid frequency f.

The generated electricity frequency of the synchronous generators f_{el} is proportional to its rotational frequency f_g i.e. $f_{el} = pp \cdot f_g$. Thus in case of varying f_g an electronic drive is necessary, for adjusting the generated electric frequency f_{el} to the grid electric frequency f.

A multi-pole (high value of pp) synchronous electric generator combined with an electronic drive can be a reasonable solution in order to avoid the adjusting gear box.

In order to control the set to operate the whole SAEP under optimal conditions we either control its electronic drive unit or its air turbine blade pitch.

The induction generators are of two types. The squirrel cage and the double fed or wound rotor induction generators. The squirrel cage induction generators rotate with frequencies close to their synchronous respective frequencies f/pp defined by the grid frequency and their pole-pairs. For given pole-pairs (for example for four pole caged induction generators pp=2) the induction generator should engage itself to the air turbine through an appropriate gear box that is multiplying its rotational frequency in order that the generator rotational speed matches to the frequency $(f/pp)\cdot(1+s)$, where s is the absolute value of the slip and it is a small quantity in the range of 0.01 for large generators.

The electric power output of the squirrel cage induction generator is approximately proportional to the absolute value of the slip s near their operating point. Thus even high power variations can be absorbed with small rotational frequency variations. Therefore the squirrel cage induction generators engaged to the air turbines with proper gear boxes are supplying the grid always with the proper electric frequency and voltage without any electronic control. The only disadvantage of the squirrel cage induction generators is that they always produce an inductive reactive power. This reactive power should be compensated using a parallel set of capacitors creating a capacitive reactive power.

The wound rotor or doubly fed induction generators are characterized by the fact that their rotors are supplied with a low frequency electric current. With proper control of the voltage and frequency of the rotor supply we can make them operate as zero reactive power units. The electronic system supplying the rotor with low frequency current is a power electronic unit of small power output (~3% of the power output of the generator). However the doubly fed induction generators with these small electronic supplies of their rotors are more

expensive than the squirrel cage induction generators with reactive power compensating capacitors.
The SAEPs with normal axis air turbines have enough space underneath the air turbine to accommodate a large diameter multi-pole generator with a large number of pole pairs in order to avoid the rotation frequency adjusting gear box.
I believe that the large scale application of the FSC technology will boost the research and production of large diameter multi-pole squirrel caged or wound rotor induction generators in order to avoid the sensitive and expensive adjusting gear boxes and to lower the cost of large electronic drives of multi-pole synchronous generators.

5.5 The gear boxes

The gear box is a essential device for adjusting the frequency of the rotation of the air turbines f_T to the electric frequency f of the grid through the relation
$f = pp \cdot f_T \cdot rt$. The rt is the rate of transmission of the gear box i.e the generator rotates with frequency $f_g = f_T \cdot rt$.
When conventional electric generators with a few pole pairs (low pp) are used, as electricity generating units, gear boxes with a proper rate of transmission rt are necessary. However if multi-pole electric generators are used with high pole-pair values (pp_h) then the gear boxes can be avoided (if $pp_h = pp \cdot rt$).
The gear boxes are mechanical devices made of gears of various diameters and combinations in order to transform their the mechanical rotation incoming and out-coming characteristics (i.e.the frequency of rotation f_{in} , f_{out} and the torque Tq_{in} and Tq_{out}) by the relations $f_{in}/f_{out} = Tq_{out}/Tq_{in} = rt$ =rate of transmission.
The gears demand a continuous oil supply and have a limited life cycle. Thus the gear boxes being huge and heavy devices of high maintenance and sensitivity, if possible they should not be preferred.
The electric power production by the SAEPs, is calculated as a function of the inlet air speed υ (i.e. the air mass $\dot{m}$) in the air turbines by a relation of the form:

$$P = \dot{m} \cdot c_p \cdot (T_{03} - T_{03te}) = \dot{m} \cdot c_p \cdot (T_{03} - T_4 - C_2 \cdot T_4^2 - \frac{g \cdot H}{c_p}) \quad (9)$$

Where T_{O3}, T_{O3te} are functions of mass flow $\dot{m}$ and FSC top exit temperature T_4.
We have shown that T_4 is the (appropriate) root of a fourth order polynomial equation:

$$w_1 \cdot T_4^4 + w_2 \cdot T_4^3 + w_3 \cdot T_4^2 + w_4 \cdot T_4 + w_5 = 0 \quad (7)$$

where w_1, w_2, w_3, w_4 and w_5 are functions of the geometrical, the thermal and ambient parameters of the SAEP, the air turbine efficiency η_T and the equivalent horizontal solar irradiance G.
The mass flow $\dot{m}$ and the warm air speed υ are proportional ($\dot{m} = \rho \cdot A_t \cdot \upsilon$) Thus:

P=Function (υ)

The efficiency of the air turbine is in general a function of the ratio $\upsilon / \upsilon_{tip}$
i.e. $\eta_T(\upsilon / \upsilon_{tip})$ where υ_{tip} is the blades' end rotational speed.
The air turbines of the SAEPs with their geared electric generators are generating electric power following the air turbine characteristics given by the two operating functions P (υ),

and η_T (υ / υ_{tip}). Considering that $\upsilon_{tip} = \pi \cdot f_T \cdot d_T$, where f_T is the air turbine frequency of rotation and d_T the turbine diameter.

The electric frequency for the geared electric generators is equal to f_n where: $f_n = f_t \cdot rt \cdot pp$, rt is the gear box transmission ratio and pp the number of their pole pairs. Hence:

$$\upsilon_{tip} = \frac{\pi \cdot d_T \cdot f_n}{rt \cdot pp} \tag{18}$$

For optimal power production by a SAEP, for an average solar irradiance G, the maximum point of operation of P(υ) should be reached for an air speed υ for which the efficiency η_T (υ / υ_{tip}) is also maximum.

The value of υ_m for maximum electric power can be defined by the SAEP operating function for η_T=constant (usually equal to 0.8) and a given solar irradiance G.

The value of the ratio $(\upsilon / \upsilon_{tip})_m$ for maximum air turbine efficiency can be defined by the turbine efficiency function $\eta_T(\upsilon / \upsilon_{tip})$.

Thus the appropriate υ_{tip} is defined by the relation:

$$\upsilon_{tip,m} = \frac{\upsilon_m}{\left(\upsilon / \upsilon_{tip} \right)_m} \tag{19}$$

Where the index m means maximum power or efficiency.

Thus for $\upsilon_{tip,m}$ the maximum power production under the given horizontal solar irradiance G is generated. Taking into account that υ_{tip} and f_n are proportional, f_n should vary with the horizontal solar irradiance G.

However as we have stated the mass flow for maximum power output by the SAEP is slightly varying with varying G, thus we can arrange the optimum control of the SAEP for the average value of G.

A good choice for this average G is a value of 5÷10% higher than the annual average $G_{y,av}$, defined by the relation $G_{y,av}=W_y/8760$.

Following the previous procedure for the proposed G, if the air turbine efficiency function $\eta_T(\upsilon / \upsilon_{tip})$ is known or can be estimated, the value of $\upsilon_{tip,m}$ can be calculated.

The frequency f of the produced A.C. will follow f_n by the relation $f = (1+s) \cdot f_n$, where s is the absolute value of the operating slip. Taking into consideration that the absolute value of slip s, for large induction generators, is less than 1%, $f \approx f_n$.

Thus the gear box transmission ratio will be defined by the approximate relation:

$$rt \approx \frac{\pi \cdot d_T \cdot f}{\upsilon_{tip,m} \cdot pp} \tag{20}$$

If the air turbine efficiency function $\eta_T(\upsilon / \upsilon_{tip})$ is not known we can assume that for caged air turbines without inlet guiding vanes their maximum efficiency is achieved for $\upsilon_{tip,m}=(6 \div 8) \cdot \upsilon$.

Thus:

$$rt \approx \frac{\pi \cdot d_T \cdot f}{(6 \cdots 8) \cdot \upsilon_m \cdot pp} \tag{21}$$

Where: υ_m= the air speed for maximum efficiency of the SAEP (derived by the SAEP basic equation for the chosen value of G), d_T= the caged air turbine diameter (smaller by 10% of the FSC diameter usually), f=the grid frequency (usually 50 sec^{-1}), pp=2 (usually the generators are four pole machines).

6. Dimensioning and construction cost of the Floating Solar Chimney SAEPs

6.1 Initial dimensioning of Floating Solar Chimney SAEPs

The floating solar chimneys are fabric structures free standing due to their lifting balloon tube rings filled with a lighter than air gas. The inexpensive NH_3 is the best choice as lifting gas for the FSCs. As we will see later the FSCs are low cost structures, in comparison with the respective concrete solar chimneys.

The annual electricity generation by the SAEPs (E) is proportional to their FSC's height (H), their solar collector surface area (A_c) and the annual horizontal irradiation at the place of their installation W_y i.e. $E=c \cdot H \cdot A_c \cdot W_y$.

As for the concrete solar chimney SAEPs, due to their concrete solar chimneys high cost, it is obvious that in order to minimize their overall construction cost per produced KWh, it is preferable to use one solar chimney, of height H and internal diameter d, and a large solar collector of surface area A_c.

In case of the floating solar chimney SAEPs, generating the same annual amount of electricity, a farm of N similar SAEPs should be used. Their FSCs will have the same height (H) and their solar collectors a surface area A_c/N. If the internal diameters of these FSCs are $d_{FSC} \approx d/\sqrt{N}$ then both Power Plants they will have the same efficiency and power production. Usually $d_{FSC} > d/\sqrt{N}$ therefore the FSC farm has higher efficiency and generates more electricity than the concrete solar chimney SAEP for the same solar collector area.

We have several benefits by using farms of FSC technology as for example:

- The handling of FSC lighter than air fabric structures is easy if their diameters are smaller. The diameter d_{FSC} should not be less than 1/20 of FSC height H.
- This choice will give us the benefit of using existing equipment (electric generators, gear-boxes, etc.) already developed for the wind industry.
- The smaller surface areas of the solar collectors will decrease the average temperature increase ΔT of the moving air mass, and consequently it is advisable that simpler and lower cost air turbines should be used (class B instead of class C air turbines i.e. caged air turbines without inlet guiding vanes).

The following restrictions are prerequisite for a proper dimensioning of the Floating Solar Chimney SAEPs.

- The FSC height H should be less than 800m.
- Their internal diameter should be less than 40m
- The solar collector active area should be less than 100 Ha (i.e. $10^6 m^2$)

If the solar collectors are equipped with artificial thermal storage the SAEP will have a rating power of $P_r = W_y \cdot \eta \cdot A_c/4300$. For maximum height 800m, and d=40m the SAEP annual efficiency is $\eta \approx 1\%$. In desert places W_y can be as high as 2300 KWh/m^2. Thus P_r for the maximum solar collector surface area of $10^6 m^2$ is less than 5MW.Generators and respective gear-boxes up to 5MW are already in use for wind technology. Furthermore if we choose an internal diameter of 40m for the FSC, it can be proven that for rating power less than 5MW, the optimal air turbine should be of class B, i.e. without the inlet guiding vanes. The air turbine will be placed onto the normal axis inside the bottom of the FSC. A useful notice concerning the dimensioning of the SAEPs is that for constant FSC height H, rating power and annual horizontal irradiation the solar collector equivalent diameter D_c and the FSC internal diameter d are nearly proportional.Let us apply the dimensioning rules in the case of desert SAEPs, considering for example that the annual horizontal irradiation is not less than 2100 KWh/m^2.Let us consider that the FSC height H is varying, while the solar collector area is remaining constant to1.0Km^2 and the FSC internal diameter is also constant and equal to 40m. The rating power of the respective SAEPs, with artificial thermal storage, is shown on the following table(2).

Solar collector area in Km^2	FSC internal diameter d in m	FSC height H in m	Rating power P_r in MW
1.0	40	180	1.0
1.0	40	360	2.0
1.0	40	540	3.0
1.0	40	720	4.0
1.0	40	800	4.5

Table 2. Dimensions and rating of SAEPs of 1Km^2 with artificial thermal storage

In the following table (3) initial dimensions of the SAEPs of FSC height 720m installed on the same area for rating power 1MW, 2MW, 3MW and 4 MW are shown.

Solar collector area in Km^2	Minimum FSC internal diameter d in m	FSC height H in m	Rating power P_r in MW
0.25	36	720	1.0
0.50	36	720	2.0
0.75	36	720	3.0
1.0	36	720	4.0

Table 3. Dimensions and rating of SAEPs of 720m height with artificial thermal storage

6.2 Estimating the direct construction cost of Floating Solar Chimney SAEPs

The direct construction cost of a Floating Solar Chimney SAEP with given dimensions is the sum of the costs of its three major parts, the solar collector cost (C_{SC}), the FSC cost (C_{FSC}) and the Air turbines gear boxes and generators cost (C_{TG}).The construction cost of the solar collector is proportional to its surface area. A reasonable rough estimate of modular solar collectors including the cost of their collecting corridors is:

$$C_{SC}=6.0\cdot A_c \text{ in EURO } (A_c \text{ in } m^2) \tag{22}$$

The construction cost of the FSC is the sum of the cost of its fabric lighter than air cylinder, and the cost of the heavy base, the folding accordion and the seat. A reasonable rough estimation of above costs is:

$$C_{FSC}=60 \cdot H \cdot d+ 300 \cdot d^2 \text{ in EURO (H, d in m)} \quad (23)$$

The construction cost of the Turbo-Generators is proportional to the rating power P_r of the SAEP a reasonable rough estimation for this cost is:

$$C_{TG}=300 \cdot P_r \text{ in EURO } (P_r \text{ in KW}) \quad (24)$$

The estimating rough figures are reasonable for SAEPs of rating power of 1÷5 MW. Any demonstration SAEP and maybe the first few operating SAEPs possible will give us a construction cost up to ~100% higher than the estimated by the previous rough formulae but gradually the direct construction cost of the SAEPs should have even lower construction costs than estimated by the given rough formulae. In the following tables (4,5) the construction costs of the previously dimensioned SAEPs are given.

Taking into consideration that the rating power multiplied by 4300 hours (for solar collectors reinforced with artificial thermal storage) will give the annual electricity generation, the construction cost per produced KWh/year is also presented in the tables (4,5).

Solar collector area in Km²	FSC internal diameter d in m	FSC height H in m	Rating power P_r in MW	Construction cost in million EURO	Construction cost in EURO per produced KWh/year
1.0	40	180	1.0	7.2	1.54
1.0	40	360	2.0	8.0	0.85
1.0	40	540	3.0	8.7	0.62
1.0	40	720	4.0	9.4	0.50
1.0	40	800	4.5	9.8	0.47

Table 4. Direct construction cost of various SAEPs

Solar collector area in Km²	Minimum FSC internal diameter d in m	FSC height H in m	Rating power P_r in MW	Construction cost in million EURO	Construction cost in EURO per produced KWh/year
0.25	36	720	1.0	2.75	0.64
0.50	36	720	2.0	5.45	0.63
0.75	36	720	3.0	7.35	0.57
1.0	36	720	4.0	9.15	0.53

Table 5. Direct construction cost of various SAEPs

7. Floating Solar Chimney versus concrete chimney SAEPs

The optimum dimensions and power ratings of the concrete solar chimney SAEPs are far higher than the Floating Solar Chimney dimensions and rating. In order for them to be compared we should consider a concrete solar chimney SAEP with given dimensions and construction cost and a Floating Solar Chimney SAEP farm generating annually the same electricity and having the same solar chimney height.

In a paper presented in 2005 (Shlaigh et al., 2005) it was mentioned the estimates on the construction cost of large SAEPs of concrete solar chimneys (Solar Updrafts Towers as they name them). According to these estimates concerning a 30 MW SAEP with a concrete solar chimney of 750 m height and 70 m of internal diameter and a solar collector of 2900m diameter(i.e. 6.6 Km^2 of surface area) the SAEP will generate 99 million KWh/year and will have a construction cost of 145 million EURO (2005 prices). Prof Jorg Schlaigh in a recent speech was estimating the construction cost of a similar concrete solar chimney SAEP of a solar chimney of 750m height and 3Km diameter to be 250÷300 million EURO (prices 2010).

Let us compare this concrete chimney SAEP with a farm of 9 Floating Solar Chimney SAEPs each one with a solar collector of surface area 740000m^2 (all of them together will cover approximately the same land area of the concrete solar chimney SAEP of 6.6Km^2). Furthermore let as assume that all of them have the same FSC of ~750m height and an internal diameter of ~40m. Let us also assume that the power rating of each FSC SAEP is ~3MW.

Although it is reasonable to assume that with these assumptions both electricity generating power plants will generate the same KWh of electricity per year (~99million KWh/year), the FSC farm could generate30% more electricity. This is the result of having a higher overall solar chimney cut in the farm of nine SAEPs, or equivalently the FSC farm will have an equivalent solar chimney diameter of 120m ($120m = 40m \cdot \sqrt{9(SAEPs)}$). Thus the warm air speed, in the FSCs, is lower than the air speed within the concrete chimney, therefore the kinetic energy losses of the exit air are lower in the FSCs and the efficiency of the FSC farm is higher.

Using the previous construction cost relations the estimated construction cost of each Floating Solar Chimney SAEP of the farm is ~6million EURO (2010 prices). Thus the whole FSC farm will have a construction cost of 54 million EURO.

The final result is that the capital expenditure for the Floating Solar Chimney farm, for similar electricity generation with the concrete solar chimney solar updraft tower, is 3 to 5 times smaller.

8. Direct production cost of electricity KWh of the FSC technology

8.1 Direct production cost analysis

The direct production cost of MWh of any electricity generating power plant is the sum of three costs:

- The capital cost related to the capital expenditure (CapEx) on investment
- The operation and maintenance cost
- The fuel cost
- The CO_2 emission cost

For renewable technology PPs the fuel and the carbon dioxide emission costs are zero.

The base load continuous operating technologies are dominating the electricity generation and their average estimated direct production cost per MWh is, without any carbon emission penalty within the range of 55÷60 EURO (EU area 2009).

The onshore wind turbine farms have succeeded to generate electricity almost with the same cost in average. However it is generating intermittent electricity thus it can enter to the grid up to 45% in power and cover the 15÷20 of the electricity demand.

Let us calculate the direct production cost of the solar chimney technology.

The assumptions we use are the following for FSC and concrete solar chimney SAEPs:

- The life cycle of both SAEPs is high (minimum 40 years)

- The CapEx is a long term loan repaid in 40 equal installments
- The interest rate of above loans is 6% (2009)
- The fabric FSCs should be replaced every 6÷10 years. This cost goes along with the maintenance cost.
- The initial construction period of the concrete chimney SAEPs is 3÷5 years while the period for FSC SAEPs is 1÷2 years. The repayments will start after those periods.
- Thus the annual repayment installment will be equal to 7% for the FSC farm and 7.5% for the concrete solar chimney PP (with the cost of initial grace period to be included)
- The rest operation and maintenance cost of both SAEPs is in the range of 5.0 EURO per generated MWh.
- The land lease is not included in the calculation because it is a negligible cost for desert or semi desert installation

In order to calculate the FSC technology average direct production cost we can use the figures of the previous paragraph for the SAEP farm of 9 similar units. The dimensions of which are H=750m, d=40m and A_c=740000m^2. Each one of these SAEPs will have a rating power of 3MW and an annual generating ability of ~12.9GWh/year. Thus their construction cost was estimated to 6 million. The Annual repayment amount for each FSC SAEP will be 420000 EURO or a capital cost of 32.3 EURO per produced MWh/year.

For the concrete SAEP we consider as a moderate estimation the amount of 200 million EURO construction cost with an annual generation of ~100 GWh/year. Thus the annual repayment cost will be 15 million EURO or a capital cost of ~150EURO per MWh/year.

The fabric structure of the FSC should be replaced every 6÷10 years. Its replacement cost is estimated to be 50·H·d=1.5 million EURO (present value) or a maximum of 250000 EURO/year i.e. 19.2EURO MWh/year (for 6 year replacement period).

The rest operation and maintenance cost for both SAEPs is ~5 EURO per produced MWh. Thus the direct production cost of MWh/year by the two technologies is:

- FSC technology ~56.5 EURO/MWh
- Concrete solar chimney technology ~155 EURO/MWh

Both SAEP technologies operate 24 hours/day year round and they can replace the base load fossil fueled power plants (Coal, Natural Gas and Nuclear).

8.2 Direct production cost comparison

The following table (6) gives the comparison of the major electricity generating technologies. The figures for the rest technologies are average values of collected official data, released by EU authorities in various publications.

The conventional base load electricity generating technologies are the coal and the natural gas fueled technologies of combined cycle and the nuclear fission technology.The first two technologies are emitting greenhouse gases and should sooner or later be replaced by alternative zero emission technologies, while the third-one although it is of zero emission technology it is considered to be dangerous and health hazardous technology. A necessary condition for the replacement of the base load electricity generating technologies by alternative renewable technologies is that these alternative technologies should operate continuously and their sources should be unlimited. The nuclear fusion technology is an alternative but its progress is slow, while the global warming threat demands urgent actions. That goes too for the promising carbon capture and storage technology, besides the problems related to carbon dioxide safe sequestration

Fuel or Method of Electricity Generation	MWh Direct Production Cost in EURO	Investment in EURO per produced MWh/year	Mode of operation and Capacity factor
Coal fired (not including carbon emission penalties)	55-60	200	Combined cycle base load 85%
Coal fired with CCS (Carbon capture and storage)	80-100	300-400	Combined cycle base load 85%
Natural Gas fired (not including carbon emission penalties)	60-65	150	Combined cycle 85%
Nuclear Fission	65-75	400÷450	Base load 95%
Wind parks onshore	60	500	Intermittent 30%
Wind parks offshore	75	650	Intermittent 30%
Concentrating Solar CSP	180	2000	Continuous with thermal storage 30%
Photo Voltaic PV	280	3000	Intermittent 15-17%
Solar Chimney concrete	155	~2000	Continuous ~50%
Floating Solar Chimney	~60	~500	Continuous ~50%
Biomass	55-75	500-÷700	Continuous 85%
Geothermal	50-70	500-÷800	Continuous 90% (limited resource)
Hydroelectric	50-60	500÷800	Continuous (load following, limited resource)

Table 6. A cost comparison of electricity generating technologies

The wind and solar technologies are appropriate technologies if they are equipped with massive energy storage systems for continuous operation. With today's technology only the solar concentrating power plants (CSP) can be equipped with cost effective thermal energy storage systems and generate continuous electricity. However their MWh direct production cost is three times higher in comparison with the respective cost of the existing base load technologies. The FSC technology is by nature equipped with ground thermal storage and operates continuously. Due to its low investment cost and its almost equal direct production cost to the conventional base load electricity technologies it is an ideal candidate to replace the fossil fueled base load technologies.

9. Large scale application of the FSC technology in deserts

9.1 Desert solar technologies

The mid-latitude desert or semi desert areas of our planet are more than enough in order to cover the present and any future demand for solar electricity. According to most conservative estimations, a 3% of these areas with only 1% efficiency for solar electricity generation can supply 50% of our future electricity demand. Also these kinds of lands exist in all continents and near the major carbon emitting countries (USA, China, EU and India).

The desert solar technologies for continuous electricity generation are the following:

- The photo voltaic (PV) large scale farms equipped with batteries
- The concentrating solar power plants (CSP) equipped with thermal storage tanks
- The concrete solar chimney SAEPs or Solar Up-draft Towers
- The floating solar chimney (FSC) farms

The following table (7) is giving us a comprehensive comparison of these desert solar technologies (OM means operation and maintenance).

Desert Technology of continuous operation	Major benefits	Major problems	MWh Direct production cost in EURO	Investment per produced MWh/year
PV with energy storage batteries	-Demands no water -Low OM care and cost	-The replacement cost of the batteries	Very high 280	Very high >3000
CSP with thermal storage	-Low cost thermal storage	-Demands water for its operation -Demands OM personnel on site	High 180	High >2000
Solar up-draft Tower (concrete solar chimney)	- No water demand -High operating life -Low OM care and cost	-High initial cost -High construction period on site	High 155	High >2000
Floating Solar Chimney	-No water demand -Easy and fast deployment on site -Low OM care	-Periodic replacement of the FSC fabric parts	Low 60	Low 500

Table 7. Comparison of desert solar technologies

9.2 The Desertec project

The Desertec project is a proposal to EU for using the desert or semi desert areas in MENA area (Middle East and North Africa) in order to generate solar electricity. Using an appropriate area of 300KmX300Km in MENA with only 1% efficiency up to 50% of its present and future electricity demand can be generated.

The transmission of the generated electricity to the EU can be achieved by using UHVDC (Ultra High Voltage Direct Curent) lines. Using the existing technology up to 6.4 GW of electricity power can be transmitted by only one UHVDC line of two conductors (±800KV and 4000A).

The UHVDC lines can be overhead, underground or undersea lines with different construction costs but the same safety and reliability.

The farm of desert power plants generates AC electricity (up to 6.4 GW). This AC electricity is converted to DC electricity, at a special power station near the farm. Through a UHVDC line the DC electricity is transmitted to the chosen place of EU, where a reverse converter power station is transforming the DC to AC electricity with the suitable characteristics for the EU local grid.

The losses of the UHVDC transmission (including the losses of two converting power stations) are not more than ~5% per 2000 Km of transmission distance. Their construction cost for 2000Km average distance between MENA and EU areas, depends on the mode of the UHVDC line and will range between 1÷2 Billion EURO.

The following table shows a comparison cost for an electricity generation system of 6.4GW installed in MENA area and transmitting its electricity power to a EU grid for a distance of 2000Km. It is assumed that due to the energy storage systems of all the desert power plants their capacity factor is more or less similar (~50%). This practically means that the desert solar farms would generate electricity of ~6.4GW X (8760/2)hours≈28000GWh/year, of which ~95% or ~26500 GWh/year (or 26.5 TWh/year) will be transmitted to the EU chosen place.

In order to cover 40÷50% of the present and future EU electricity demand i.e. 1060÷1500 TWh/year we should build a set of 40 to 56 independent solar farms of 6.4GW that can be installed in appropriate MENA areas and connected through UHVDC lines to the proper places of EU countries. In order to build 40-56 farms we should invest capital of the amounts as shown in the next table (8) for respective technologies.

Desert Technology of continuous operation	Investment cost (including UHVDC lines cost of 1.5 billion EURO) for the solar farm of 6.4GW in billion EURO	Investment cost for building 40÷56 similar solar farms in billion EURO	MWh direct production cost in EURO (26.5 TWh supplied to EU)
PV with energy storage batteries	>85.5	3420 4778	>285
CSP (parabolic through or tower) with thermal storage	57.5	2300 3220	185
Solar up-draft Towers	57.5	2300 3220	160
Floating Solar Chimney	15.5	620 868	65

Table 8. Cost comparison of solar desert farms of 6.4 GW

The maximum desert or semi desert area for the installation of one solar farm of 6.4GW is not more than 1600 Km^2 or a square area ~(40Km X 40Km). Thus the maximum neaded area in order to cover the 40÷50% of the present and future EU electricity demand, with zero emission solar electricity, is 64000÷90000Km^2 (i.e. a square area of 250Km X 250Km up to 300Km X 300Km)

This maximum area is indispensable for solar chimney farms (concrete or floating) of 1% efficiency. As for the rest solar technologies a much smaller desert area is adequate. However the maximum area needed is not more than 2% of proper desert or semi desert area in MENA territory.

By the presented data it is evident that the FSC technology has tremendous benefits in comparison with its solar competitors for desert application.

Its major benefits are:

- Low investment cost
- Low KWh direct production cost (almost the same with the fuel consuming base load electricity generating technologies)
- 24hours/day uninterrupted operation due to the ground thermal storage
- The daily power profile can be as smooth as necessary using low cost additional thermal storage
- Demands no water for its operation and maintenance
- Easy and fast deployment on site
- It uses recycling and low energy production materials (mainly plastic and glass)
- Minimum personnel on site during its construction and operation

Large scale desert application of the Floating Solar Chimney technology can be one of the major tools for global warming elimination and sustainable development.

10. Climate change warning

Climate change indications due to the global warming threat are accelerating. Climate change policies should be agreed upon and urgent measures should be taken. Global warming due to greenhouse gases emissions (CO_2, CH_4 etc.) is a reality scientifically documented.

Intergovernmental Panel on Climate Change (IPCC) is a Nobel Priced UN committee studying carefully and objectively the global warming due to greenhouse gases produced by human activity on earth. The major producer is the fossil fuels used in residential, industrial, and transportation activities, of which the major-one is the electricity generation of fossil fueled power plants. According to IPCC estimations the global average temperature increase on earth will follow the pattern shown in the next figure (19) depending on our future model of energy use, electricity generation scenarios and greenhouse gases concentration. According to mentioned estimations, pertaining the existing technology and applying an internationally agreed upon strict policy on greenhouse gas emissions, the scenario most likely to come up is an eventuality between I and II.

According to mentioned scientifically documented estimations, global temperatures in excess of 1.9 to 4.6 ^{0}C warmer than pre-industrial would appear and it will be possibly sustained for centuries.

The major global warming effects on our planet, according to IPCC are:

- Anthropogenic warming and sea level rise would continue for centuries even if the greenhouse gas concentrations were to be stabilized
- Eventual melting of the Greenland ice sheet, would raise the sea level by 7 m compared to 125,000 years ago
- Due to precipitation changes fertile land devastation is possible to appear in many areas
- The existing atmospheric models can not exclude the appearance of extreme catastrophic atmospheric phenomena such as: very strong typhoons, tornados, snow or hail storms etc.

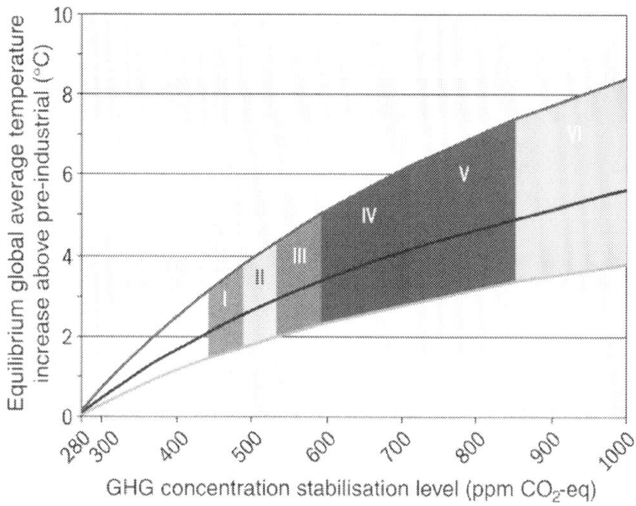

Fig. 19. IPCC scenarios of global temperature increase

The energy sector is the major source of the greenhouse gases due to its fossil fuelled technologies of electricity generation, transportation, industrial activities etc. For the year of 2010 an estimated quantity of 29,000 Mt of carbon dioxide will be spread all over the environment from fossil fuel combustion of which:

- 36.4 % from electricity generation
- 20.8 % from the industry
- 18.8 % from transport and
- 14.2 % from household, service and agriculture and
- 9.8 % from international bunkers

The mechanism of Kyoto protocol aims to create an "objective" over the external cost at least for the threatening carbon dioxide (CO_2) emissions through trading their rights.

The cost of the emitted CO_2, sooner or later it will reach at prices 20-30 EURO per ton of CO_2 and after the year 2012 for EU the fossil fuelled PPs should pay for each ton of CO_2 emitted by them. Taking into consideration that 1 Kg of coal has a thermal energy of ~8.14 KWh, thus a modern coal fired power plant with efficiency ~45% will generate by this ~ 3.66 KWh and will emit to the environment 3.667 Kg of CO_2. Thus in a modern coal fired plant approximately 1.0 Kg of CO_2 is emitted per generated KWh. For the lignite coal fired power plants this figure is 50% higher and for modern combined cycle natural gas power plants could be 50% smaller.

11. Conclusion

Although electricity generation is a major carbon dioxide producer we should notice that electricity can replace all the energy activities related to fossil fuelled technologies. Thus a solution to the global warming is possible if we succeed to generate zero emission clean electricity.

The renewable electricity generating technologies is a major tool, some believe that it should be the exclusive technology, towards the aim of eliminating the greenhouse emissions threatening the future on our planet.
It is possible to mitigate global warming if the world-wide consumption of fossil fuels can be drastically reduced within the next 10 to 15 years. I believe that the only viable scenario that could lead to a successful and real reduction of fossil fuels is the large scale application of the FSC technology in desert or semi desert areas. This means that we should start building, for the next 30 years, Floating Solar Chimney SAEP desert farms of overall rating power ~160 GW/year, that could generate ~720 TWh/year.
Thus for the next 30 years we will build SAEP desert farms generating more than 21600 TWh/year solar electricity that could replace fossil fuelled generated electricity. The global investment cost for this choice will not exceed the amount of 360 billion EURO/year or 11.5 trillion EURO for the next 30 years. These investments in electricity generation are reasonable taking into consideration that the future electricity demand could reach the 45000 TWh. The necessary land for the 30 years FSC power plants is 1.000.000 Km^2 (1000 Km X 1000 Km)

12. References

[1] Bernades M.A. dos S., Vob A., Weinrebe G., 2003 *"Thermal and technical analyses of solar chimneys"* Solar Energy 75 ELSEVIER, pp. 511-52.

[2] Backstrom T, Gannon A. 2000, *"Compressible Flow Through Solar Power Plant Chimneys"*. August vol 122/ pp.138-145.

[3] Gannon A. , Von Backstrom T 2000, *"Solar Chimney Cycle Analysis with System loss and solar Collector Performance"*, Journal of Solar Energy Engineering, August Vol 122/pp.133-137.

[4] Papageorgiou C. 2004 *"Solar Turbine Power Stations with Floating Solar Chimneys"*. IASTED proceedings of Power and Energy Systems, EuroPES 2004. Rhodes Greece, july 2004 pp,151-158

[5] Papageorgiou C. 2004, *"External Wind Effects on Floating Solar Chimney"* IASTED Proceedings of Power and Energy Systems, EuroPES 2004, Conference, Rhodes Greece ,July 2004 2004 pp.159-163

[6] Papageorgiou C. 2004, *"Efficiency of solar air turbine power stations with floating solar chimneys"* IASTED Proceedings of Power and Energy Systems Conference Florida, November 2004, pp. 127-134.

[7] Papageorgiou C. "Floating Solar Chimney" E.U. Patent 1618302 April. 29, 2009.

[8] Pretorius J.P., Kroger D.G. 2006,*"Solar Chimney Power Plant Performance"*, Journal of Solar Energy Engineering, August 2006, Vol 128 pp.302-311

[9] Pretorius J., "*Optimization and Control of a Large-scale Solar Chimney Power Plant*" Ph.D. dissertation, Dept. Mechanical Eng., Univ. Stellenbosch 7602 Matieland, South Africa 2007.

[10] Schlaich J. 1995, *"The Solar Chimney: Electricity from the sun"* Axel Mengers Edition, Stutgart

[11] J. Schlaich J. e.al 2005, *"Design of commercial Solar Updraft Tower Systems-Utilization of Solar Induced Convective Flows for Power Generation"* Journal of Solar Energy Engineering Feb. 2005 vol 127, pp. 117-124R.

[12] White F. *"Fluid Mechanics"* 4th Edition McGraw-Hill N.York 1999

11

Organic Solar Cells Performances Improvement Induced by Interface Buffer Layers

J. C. Bernède[1], A. Godoy[2], L. Cattin[1], F. R. Diaz[3], M. Morsli[1] and M. A. del Valle[3]
[1]*Université de Nantes, Nantes Atlantique Universités, LAMP, EA 3825, Faculté des Sciences et des Techniques, 2 rue de la Houssinière, BP 92208, Nantes, F-44000*
[2]*Facultad Ciencias de la Salud, Universidad Diego Portales. Ejército 141. Santiago de Chile*
[3]*Facultatd de Quimica, PUCC, Casilla 306, Correo 22, Santiago,*
[1]*France*
[2,3]*Chile*

1. Introduction

The energy sector has a constrained future, since increasing demand coincides with "prise de conscience" of the negative implications of fossil energy use. Global warming is finally a clear evidence of the fundamental idea of the "old" Newtonian physics: there is no action without reaction. Fundamental principle neglected by the occidental world during the last century. That is to say, we cannot continue to emit continuously carbon dioxide, nitrogen dioxide... and others pollutants produced from the burning of fossil energies into our environment without suffering the consequences. Some environmental scientists have highlighted this problem for some time [Lüthi et al., *Nature*, 2008], but only now are some governments giving the issue the attention that it deserves. Man-made climate change is one of the greatest threats our world faces. Renewable energies issued from our natural environment, such as wind power, solar thermal, photovoltaic, geothermal heat, marine and hydro power..., can help reduce our dependence on fossil energies. The present review is dedicated to photovoltaic energy and more precisely to some specific photovoltaic devices based on organic materials.

Photovoltaic cells belong to the family of the optoelectronic devices. As evidenced by their denomination, such devices use the optical and electronic transport properties of different materials to either produce electromagnetic radiation (light emitting diodes) or to generate electricity (photovoltaic cells -PV cells). Photovoltaic cells also called solar cells are used to generate electrical power. A PV cell is a device based on the photoconductive properties of semiconductor materials -for carriers generation- coupled with the ability of these semiconductors to form junctions -for carriers separation. The photoconductivity is the process in which electromagnetic energy is absorbed by a material and converted to excitation energy of electric charge carriers so that the material becomes quite conductor. When irradiated by a light, PV cells produce electrical energy across any connected external load. When irradiated without load a PV cell produces a maximum photogenerated voltage V_{oc}, the open-circuit voltage. When shorted, the PV cell produces the maximum short circuit current I_{sc}. When connected to a load the power output of the cell is given by the voltage current product VxI. The maximal power generated possible is $V_{oc}xI_{sc}$. In fact the maximum power a PV cell is able generating depends on the dark I-V characteristics, that is to say on the diode properties of the junction constituting the device. When the load value is optimised, the maximum power provided by the cell is $Pm = V_m x I_m$. A figure of merit called the fill factor, FF, for the PV cells is given by:

$$FF = V_m x I_m / V_{oc} x I_{sc} \quad (1).$$

Up to now, inorganic materials are used in photovoltaic cells. Crystalline, polycrystalline and amorphous silicon represent more than 95 % of the world production, while CdTe and $Cu(In,Ga)Se_2$ (CIGS) are now emerging in the market. Crystalline (or polycrystalline) devices allow achieving efficiencies up to 25%. However, efficient crystalline (or polycrystalline) devices are difficult and expensive to produce and the pay-back time of such modules is around three years. Traditionally, optoelectronic devices were grown using inorganic compounds. However, some years ago, research devoted to organic light emitting diodes (OLEDs) encounter an unexpected success [Jain et al., *Semiconductors and semimetals*, 2007] and they are now available on the market. Moreover it has been shown that the quantum efficiency of the electron transfer from an excited polymer to fullerene (C_{60}) is very high [Xiong Gong et al. *Sciences*, 1992]. So, since the pioneering work of Tang [Tang, *Appl. Phys. Lett.*, 1986] the interest devoted to organic solar cells has been raising very fast, which has undergone a gradual evolution of the energy conversion efficiency, η, from less than 1% to more than 5% [Kim et al., *Sciences*, 2007, Xue et al., *J. Appl. Phys.*, 2005]. These significant progresses demonstrate that organic solar cells are a potential avenue to low cost next generation solar cells. However, some efforts are still necessary to improve the cell efficiency and lifetime. To overcome the quite narrow absorbance domain of the organic photoactive layer, several approaches such as low band gap organic material, incorporation of metal nanostructures, use of inorganic optical spacer between the active layer and the electrode can be used. It is also well known that carriers exchange at interfaces organic material/electrode can greatly influence device performance. In the present review, based on our recent studies, we will discuss more specifically possible device improvement through interface optimisation. The plan of the manuscript is as follow, after recalling some generality on organic solar cells and the classical interface theory in semiconductors, impact of electrode/organic interface properties on cells performances will be discussed using different published results, and more specifically studied from our last results. All the results will be critically discussed in the context of how to improve the fundamental understanding of interface behavior to enhance solar cells performance.

2. A short comparison with organic light emitting diodes

As said above, the development of efficient organic displays based on organic light emitting devices (OLEDs) has shown that organic electronic components are viable. Those displays are now developed using low cost technology and these new technologies development for OLED can be tested for PV solar cells realisation [Bernède et al., *Current Trends in polymer Sciences*, 2001]. Basically the underlying principle of a photovoltaic solar cell is the reverse of the principle of OLED Figure 1.

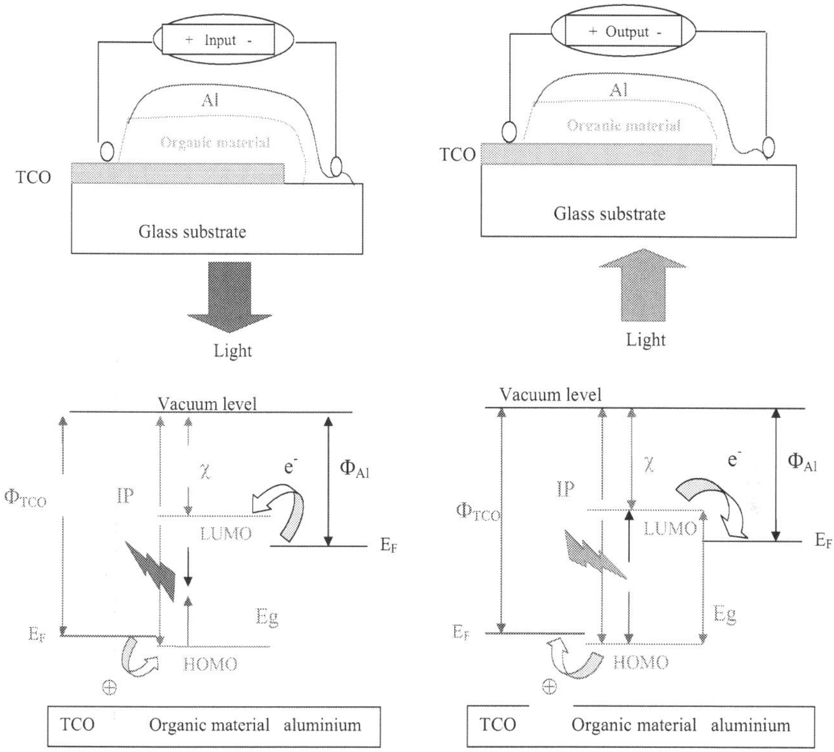

Fig. 1. Principle of an OLED (left) and a solar cell (right) (Band scheme without contact)

In OLED, electrons are injected at the low work function electrode (cathode), while holes are injected at the high work function electrode (anode). At some point in the organic, the electron and hole meet and recombine with light emission. The reverse happens in a PV cell, when light is absorbed an exciton forms. After exciton dissociation, the electron must reach the low work function electrode and the hole the high work function electrode.

In fact, when the organic material is put into contact with electrode, the shape of the band scheme depends on the conductance of the organic material Figure 2.

When the cells are short circuited, the Fermi levels of the electrodes align. If the organic is an insulator, the field profile changes linearly through the cell (fig. 2b). If the organic is a p-type semiconductor a depletion layer forms on the side of the metal with small work function, we have Schottky contact (fig. 2c). Usually the former scheme is used in OLEDs, the organic films used being quite insulating and the latter scheme is often used in solar cells, the organic active layers being semiconducting.

Almost all organic optoelectronic devices have a planar layered structure, where the organic active layer(s) is (are) sandwiched between two different electrodes. One of them must be transparent. A transparent conductive oxide (TCO) is used, usually indium tin oxide (ITO) because it allows achieving better results. The other electrode is very often aluminium, even if calcium has a better work function, because Al is stable in air while Ca is not. From the above comparison it can be concluded that a device which exhibits high electro-luminescent properties will be a poor solar cell and vice versa. However in both devices families the properties of the contact electrode/organic material are determinant to the efficiency of the devices, and the progress in that field for one device family is very helpful for the other family.

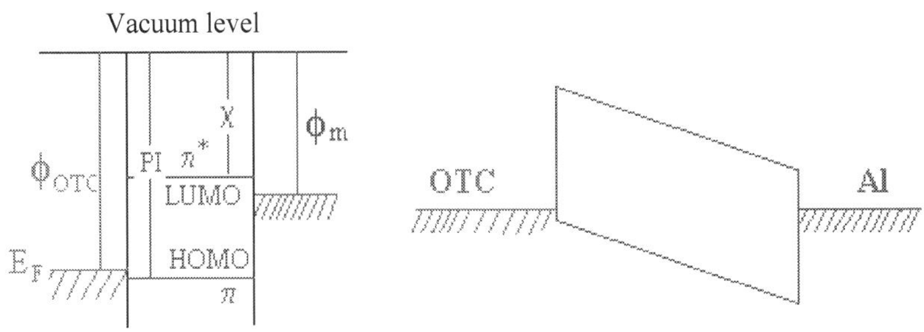

a: before contact b: after contact, with an insulating organic material

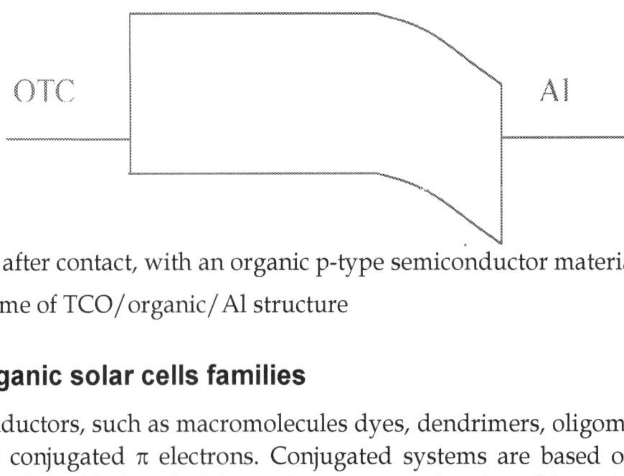

c: after contact, with an organic p-type semiconductor material

Fig. 2. Band scheme of TCO/organic/Al structure

3. Different organic solar cells families

Organic semiconductors, such as macromolecules dyes, dendrimers, oligomers, polymers…, are all based on conjugated π electrons. Conjugated systems are based on an alternation between single and double bonds. The main property related to this conjugation is that π electrons are more mobile than σ electrons. Therefore the π electrons can move by hopping. These π electrons allow light absorption, in the case of solar cells, and emission, in the case of OLEDs. Molecular π-π* orbitals correspond respectively to the Highest Occupied Molecular Orbital (HOMO) and Lowest Unoccupied Molecular Orbital (LUMO). For sake of simplicity, such organic material can be regarded as a semiconductor-like material, where the band gap corresponds to the difference between the LUMO and the HOMO.
Photons absorption by inorganic semiconductors produces free electrons and holes, the charge separation is more difficult in organic semiconductors. When a photon of

appropriate energy is incident upon organic semiconductor it can be absorbed to produce an excited state called exciton, that is to say an electron-hole pair in a bound state which is transported as a quasi-particle. In organic materials excitons are strongly bounded as a consequence of their low dielectric constant. Organic solar cells belong to the class of photovoltaic cells known as excitonic solar cells [Thompson, Fréchet, *Angnew. Chem. Int. Ed.*, 2008]. The excitons can have appreciable life-time before recombination. To produce photocurrent the electron-hole pair of the exciton must be separated. If not, they can recombine either radiatively (luminescence is a loss mechanism in photovoltaic cells) or non-radiatively with heat production. Therefore after light absorption and exciton formation, the carriers should be separated. Even if not well understand the dissociation occurs at defects, impurities, contacts or any other inhomogeneities. The separation occurs in the electric field induced around the inhomogeneity. If the ionisation takes place at a random defect in a region without an overall electric field, the generated carriers will be lost. To avoid such loss, exciton dissociation should occur in high electric field region associated with a contact or a junction. To produce an internal electric field which occupy a substantial volume of the device, the usual method is to juxtapose two materials with different appropriate properties. One of these materials is an electron donor and the other one is called electron acceptor. The interface between these two materials is called heterojunction. Therefore it is clear that the active donor-acceptor pair governs the separation mechanism. While in the case of inorganic materials the both materials of the heterojunction are clearly identified, the electron acceptor is the n-type material related to its electron excess and the electron donor is the p-type material related to its hole excess, it is not so simple in the case of organic materials. The donor or acceptor nature of an organic semiconductor depends on its carrier mobility which is determined by intrinsic properties of this material. Moreover, it is known that the donor or acceptor character of a material in an organic couple depends also on their relative HOMO and LUMO values. For instance, CuPc, which is a usually an electron donor, has been also used as electron acceptor with a triphenylamine derivative used as donor [Chen et al., *Sol. Energy Mater. Sol. Cells*, 2006]. Therefore an organic material with intermediary HOMO and LUMO values can be used as an electron donor for one organic material and as an electron acceptor for another organic material. In figure 3 we can

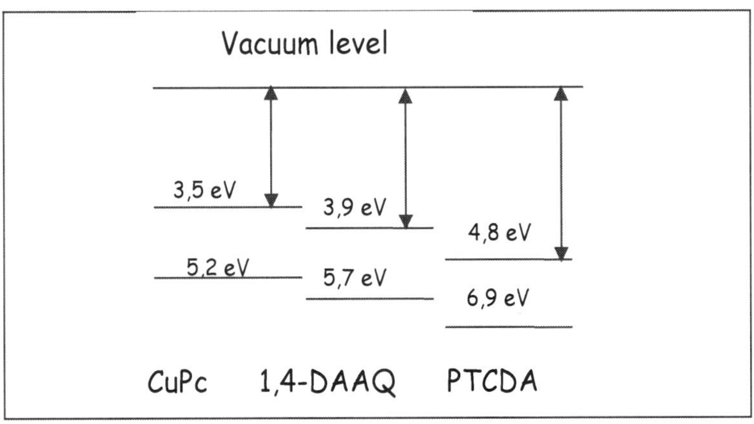

Fig. 3. Relative position of the HOMO and LUMO of CuPc/1,4-DAAQ/PTCDA.

see the example of the 1,4-diaminoanthraquinone (1,4-DAAQ), which is an acceptor relatively to the copper phthalocyanine (CuPc) and a donor relatively to the perylene 3,4,9,10-tetracarboxylic dianhydride (PTCDA) [Berredjem et al., *Dyes and Pigments,* 2008].
It is well known that exciton dissociation is efficient at the interface between materials with different electron affinity EA (i.e. LUMO) and ionisation potential IP (i.e. HOMO) (Figure 4). The difference in electron affinities creates a driving force at the interface between the two materials that is strong enough to separate charge carriers of photogenerated excitons. EA and IP of the electron acceptor should be higher than those of the donor.

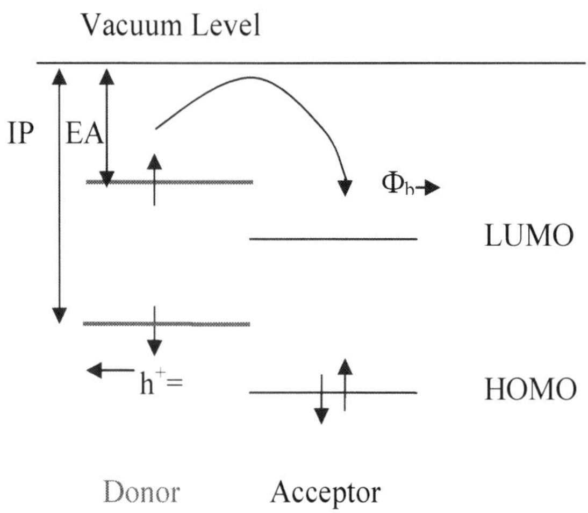

Fig. 4. Charge separation of an exciton into a free electron/hole pair at a donor acceptor interface.

Then the separated electron and hole should be collected at the cathode and anode electrodes respectively. In a crystalline inorganic semiconductor the LUMOs and HOMOs form the conduction and valence band respectively, and carriers transport within a band. It is usually different in the case of the organic semiconductors. In that case intermolecular forces are Van der Waals forces and no large conduction band and valence band are formed. Therefore the charge transport proceeds by hopping between localized states. This means that charge carrier mobility in organic materials is smaller than that in inorganic semiconductors. Mobility is improved with the molecular ordering in the films, it is reduced by impurities as well as by oxygen traps. Therefore charge transport is improved by improving order, purification and no oxygen contamination.
So, the operations of an organic solar cell can be summarized in five fundamental steps:

1. Photon absorption by the active organic layer to create excitons.
2. Diffusion of the excitons to a donor/acceptor interface.
3. Excitons dissociation at the interface and formation of electron-hole carrier pairs.
4. These free carriers should move toward the cathode (electrons) and the anode (holes).
5. The electrons (holes) should be extracted to the external circuit by the cathode (anode).

Up to 1986 very low power conversion efficiencies were achieved (#10^{-6}%) because of low concentration and mobility of free charge carriers. In 1986 Tang reported on a bilayer heterojunction solar cell based on the copper-phthalocyanine (CuPc)/perylene tetracarboxylic derivative with 1% energy conversion efficiency [Tang, *Appl. Phys. Lett.*, 1986]. In spite of numerous efforts to improve the cells performances, there was no breakthrough and it remained around 1%. However, during the last ten years a new and strong interest has been devoted to organic photovoltaic cells, motivated by two recent developments in the organic semi-conductor field. First it has been shown that the quantum efficiency of the electron transfer from an excited polymer to C_{60} is very high and the transfer is very fast, [Sariciftci et al., *Sciences*, 1992; Sariciftci et al., *Appl. Phys. Lett.*, 1993] , which is promising for change carrier separation in PV cells. Secondly the development of efficient organic displays based on organic light emitting devices (OLEDs) has shown that organic electronic components are viable. Those displays are now developed using low cost technology and these new technologies development for OLED can be tested for PV solar cells realisation [Bernède et al., *Current Trends in polymer Sciences*, 2001]. Therefore OLED technology allows drawing a research guideline in organic PV cells. Since the beginning of the 21st century, power conversion efficiency is increasing rapidly and the organic photovoltaic devices are getting closer to commercialisation [Kim, Lee, Chin, *Sol. Energy Mater. Sol. Cells.*, 2009].

Two technological routes are mainly used: the deposition of polymers and nanoparticles from solution by spin coating technique and the vacuum sublimation of small molecules. The former route has given the concept of bulk heterojunction solar cells (BHJ), while the later gives the multi-heterojunction solar cells family. Development of BHJ interpenetrating network structures using polymer blended with a fullerene derivative allows overcoming inherent disadvantageous very short exciton diffusion length, which improves significantly cells performances. Controlled morphology of poly-paraphenylene-vinylene/fullerene blend derivatives increased conversion efficiency up to 2.5% in 2001 [Shaheen et al., *Appl. Phys. Lett.*, 2001]. A post-annealing treatment improved efficiency to 3.5 [Padinger et al., *Adv. Funct. Mater.*, 2003]. To day, the best energy conversion efficiencies (>6%) have been reach with poly-3(hexyl-thiophene) (P3HT) serving as electron donor and [6,6]-phenyl-C_{61}-butyric acid methyl ester (PCBM), which is a fullerene derivative as electron acceptor [Kim et al., *Science*, 2007].

In the multi-heterojunction solar cell family the simplest efficient configuration is based on a bilayer system between two electrodes. About these electrodes, one should be transparent and the other highly reflexive. The bilayer is composed of, at least, one absorbing layer, which is usually an electron donor such as a conjugated polymer or a dye like a metal phthalocyanine and an electron acceptor. Up to now, fullerenes are the best acceptors for organic solar cells [Hoppe and Sariciftci, *J. Mater. Res.*, 2004]. As a matter of fact, they have an energetically deep LUMO, which induces a very high electron affinity relative to numerous organic donors. They exhibit relatively high electron mobility, up to 6 cm^2/Vs [Anthopoulos et al., *Appl. Phys. Lett.*, 2006, Zhang, Domercq, and Kippelen, *Appl. Phys. Lett.*, 2007]. In the bilayer cells, charge separation predominantly occurs at the organic heterojunction. The built-in potential is determined by the HOMO/LUMO gap energy difference between the two materials contacting to form the heterojunction.

In order to improve the efficiency of planar herojunction structures, either p-i-n or stacked junctions have been probed. The p-i-n heterojunction structure consists of an active intrinsic region sandwiched between a n and a p-doped layer. [Niggemann et al., *Phys. Stat. Sol. (a)*, 2008]. As said above, usually a metal phthalocyanine is the p-type (electron donor) and

fullerene the n-type (electron acceptor). The i photoactive layer can be a combination of both [Drechsel et al., *Organic electronics,* 2004].
As a conclusion of this chapter, in order to facilitate the discussion of the interface behaviour it is helpful to use an equivalent circuit model, which could be helpful in understanding of organic solar cells by providing a quantitative estimation for losses in the cells. The equivalent circuit commonly used to interpret the I-V characteristics of solar cells consists of a photogenerator connected in parallel with a diode, which represents the I-V characteristics in the dark. This corresponds to an ideal model in absence of parasitic resistances.

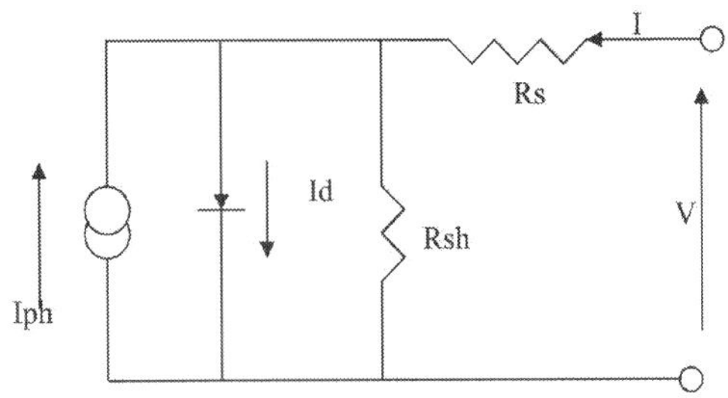

Figure 5 a

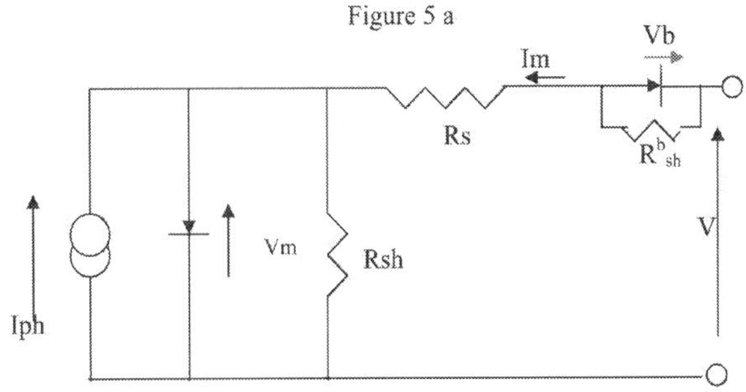

Figure 5 b

Fig. 5. Equivalent circuit of: a-solar cell with ohmic contacts, b-solar cell with rectifying back-contact.

However, in real organic solar cells, it is necessary to introduce a series resistance, Rs, and a shunt resistance, Rsh (Figure 5 a). For such solar cells the mathematical description of this circuit is given by the following equation:

$$I = I_0\left[\exp\left(\frac{V - I \times Rs}{nkT}\right) - 1\right] + \frac{V - I \times Rs}{Rsh} - Iph \tag{2}$$

As discussed below, the values of Rs and Rsh depend on the properties of the electrode/organic interfaces.

4. Impact of electrode/organic interface properties on cells performances

As shown above organic optoelectronic devices are formed by sandwiching one or more semiconducting organic layers between conducting electrodes. The properties of the interface between the organic layers and the electrodes are determinant for the efficiency of the devices. In photovoltaic devices the contact must have good carrier extracting properties. Also electrodes must generate a built-in electric field that is sufficient to collect separated charges. Therefore a good understanding of the interface between the organic layers and the electrodes is crucial to achieve good organic optoelectronic devices.

When a π-conjugated organic material is put into contact with another material such as electrodes in optoelectronic devices the contact may result in different effects depending on the electrode properties. With a metal which surface is passivated the interaction results in a physisorption with possible integer electron charge transfer. If the metal is non reactive there is possibility of weak chimisorption and possible partial charge transfer. In the case of reactive metal, there is chemisorption with iono-covalent bonding between the metal and the organic material and corresponding partial charge transfert.

Often the surface work function of an electrode depends on the electrode history. The surface work function of a clean metal or TCO can be measured *in situ* by UPS for exemple (Φ_M = (5.2 eV for clean gold). However, after air, or only classical vacuum (10^{-3} Pa) exposure there is some physisorbtion at the metal surface which decreases its work function, for instance in place of the 5.2 eV expected, 4.5 eV are often obtained. Therefore air ambient exposure of gold results in a reduction by 0.7 eV of its work function [Braun, Salaneck, and Fahlman, *Adv. Mater.*, 2009]. Similar results have been obtained with TCO. For instance, Φ_{ITO} varies from 5 to 4.5 eV depending on the sample history [Li et al., *Thins Solid Films,* 2005; Johnev, et al., *Thins Solid Films* 2005]. So, the effective work function of the electrode when put in contact to an organic material may be significantly lower than that expected, which induces unexpected energy level alignment at the organic material/electrode interface. Often, in order to study the energy level alignment at the organic material/electrode interface UPS is used, and samples are grown *in-situ* to avoid contamination, that is to say all the process takes place under ultra-high vacuum. The result can fail in the attempt to describe the interface band alignment of optoelectronic devices contact, since they are processed under moderate vacuum, neutral gas or even room air conditions. That fact should be kept in mind when one attempts to understand the optoelctronic devices behaviour through UPS studies done in ultra high vacuum.

Energy level alignment at organic material/electrode interfaces is one of the main fundamental issues about the optoelectronic devices. In the light of inorganic semiconductor/metal interfaces the simple Schottky-Mott model has been often applied to organic contacts. The model is presented in Figure 6. When an organic semiconductor is put in contact with an electrode the Schottky-Mott model assumes an alignment of the vacuum level and a band bending in the space charge layer (SCL) to achieve alignment of the bulk Fermi levels. Therefore the barrier height for hole extraction Φ_B corresponds to:

$$\Phi_B = \Phi_M - \Phi_S \tag{3}$$

Φ_S being the work function of the organic semiconductor (Figure 6).

More generally, if ϕ_M is the work function of the metal, i.e. its ionisation potential IP, and ϕ_S the work function of the semiconductor, i.e. the energy difference between the Fermi level and the vacuum level, for the semiconductor we will have:

- -in the case of n-type semiconductor
  - a Schottky-Mott contact if $\phi_M > \phi_S$, electrons diffuse from the semiconductor to the metal owing to small carrier density in semiconductors and to high carrier density

in metals, a depletion layer appears in the semiconductor: there is a rectifying contact.

- an ohmic contact if $\phi_M < \phi_S$, electrons diffuse from the metal to the semiconductor. There is a negative accumulation in the n-type semiconductor, no barrier forms, the contact is ohmic.

- -in the case of p-type semiconductor
  - a Schottky-Mott contact if $\phi_M < \phi_S$, holes diffuse from the semiconductor to the metal. Owing to small carrier density in semiconductors and to high carrier density in metals, a depletion layer appears in the semiconductor: there is a rectifying contact.
  - an ohmic contact if $\phi_M > \phi_S$, electrons diffuse from the semiconductor to the metal. There is a positive accumulation in the p-type semiconductor, no barrier forms, the contact is ohmic.

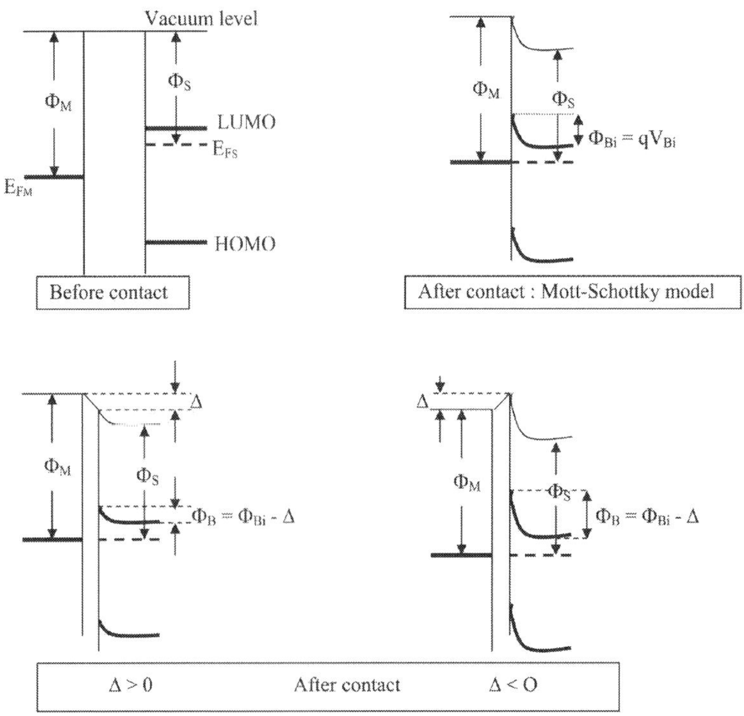

Fig. 6. Interface band alignment, before contact, after contact, without and with an interface dipole

Even if many studies have experimentally demonstrated a strong correlation between the metal work function and the barrier Φ_B for carriers exchange at electrode/organic material interfaces there is still matter of controversy.

The described Schottky-Mott model is an ideal and simple model, however, real energy level alignment should often consider a vacuum level discontinuity associated with an

interface dipole, ID, resulting from charge rearrangement upon interface formation [Lee et al., *Appl. Phys. Lett.*, 2009].
In the case of inorganic metal/semiconductor contacts two limit models have been proposed. The Schottky-Mott model where the vacuum level of the organic and metal aligned, forming a region of net space charge at the interface and the Bardeen model, where a large density of surface states induces a pining effect of the Fermi level and the presence at the interface of a barrier independent of the metal work function. The Cowley-Sze model is an intermediate model, where interface states would be induced in the original band gap of the semiconductor upon contact with a metal giving the interfacial dipole Δ'. The effective barrier height for hole exchange $\Phi_{b,eff}$ is therefore given by :

$$\Phi_{B,eff} = \Phi_B - \Delta' \quad (4)$$

Δ' is proportional to the amount of charge transferred due to energy difference between the metal Fermi level and the charge neutrality level (CNL). If we assume a uniform distribution of metal-induced interface state, it can be shown that $\Phi_{B,eff}$ varies linearly with the metal work function with a slope, S, smaller than one [Lee et al., *Appl. Phys. Lett,*. 2009]. In the absence of metal-induced interface state, the injection barrier follows the Schottky-Mott limit with S = 1. The other limit corresponds to S = 0, the interface dipole reaches a saturated value with the organic CNL aligned to the metal's Fermi level. There is Fermi level pining and the variation of the metal work function is fully compensated by the metal-induced interface state dipole.
By analogy with inorganic metal/semiconductor contacts two limit models have been proposed when an organic semiconductor is deposited onto a conducting material. The first is the above described Schottky-Mott simple model. The second proposed that a charge dipole forms on the interface due to effect such as chemical interaction and/or formation of interface states, in that case the vacuum level does not align at the interface. This interface dipole (ID) induces vacuum level shift Δ. Therefore the Mott-Schottky barrier height should be modified by the amount of Δ:

$$\Phi_B = \Phi_M - \Phi_S - \Delta \quad (5)$$

The sign of Δ depends on the nature of the contact (Figure 6) and it will be discussed below.
Moreover, another question is, does band bending occur in organic semiconductors? Following S. Braun and W.R. Salaneck, M. Fahlman [Braun, Salaneck, and Fahlman, *Adv. Mater.*, (2009)] band bending should not be expected for organic semiconductors, as they do not have band structure but localized state featuring hopping transport. Charge can be exchanged at the interface but only organic material in close vicinity to the metal surface takes part in the charge exchange. Yet, they admit that band-bending like behaviour has been demonstrated for π-conjugated organic thin films deposited on metal substrates. It has been shown that localized energy levels of the organic material are shifted depending on the distance to the metal interface, until depletion region thickness is reached [Nishi et al., Chem. Phys. Lett., (2005); Ishii et al., *Phys. Stat. Sol* (a), 2004]. Also, J. C. Blakesley and N. C. Greenham [Blakesley and Greenham, *J. Appl. Phys.*, 2009] have shown that there is a good agreement between UPS measurements and theoretical band bending calculations. UPS measurements of thin organic layers on conducting substrates have shown the presence of band bending within a few nanometers [Hwang et al., *J. Phys. Chem. C*, 2007]. It has been proposed that this band bending effect is due to transfert of carriers from the substrate into

the organic film. Such integer charge transfer (ICT) at organic/passivated conducting substrate interface has been proposed by Salaneck group [Tengstedt et al., *Appl. Phys. Lett.* (2006); Fahlman et al., *J. Phys.: Condens. Matter,* (2009)]. The ICT model proposes that electron transfer via tunnelling through the passivating surface layer, which implies the transfer of an integer amount of charge, one electron at a time. Tunnelling occurs when the substrate work function is greater (smaller) than the formation energy of positively (negatively) charged states in the organic material. The energy of a positive integer charge transfer state E_{ICT+} is defined as the energy required to take away one electron from the organic material and, in the case of negative integer, the charge transfer state, E_{ICT-} is defined as the energy gained when one electron is added to the organic material. In the case of a positive integer charge transfer, the organic material at the interface becomes positively charged, while the substrate becomes negatively charged, creating an interface dipole Δ that down-shift the vacuum level. The electron transfer begins when the organic is put into contact with the substrate, and it goes on up until equilibrium is reached, i.e. when E_{ICT+} Δ is equal to the substrate work function (Figure 7).

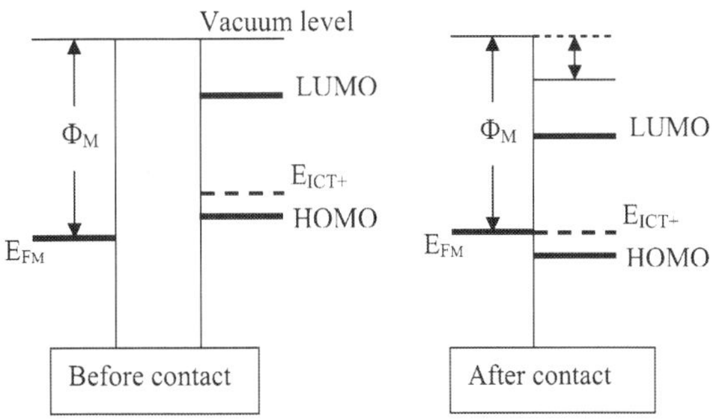

Fig. 7. Integer charge transfer model.

Here also there is some controversy about the formation, or not, of a band bending. However the model predicts the Fermi level pinning experimentally encountered when $\Phi_M < E_{ICT-}$ and $\Phi_M < E_{ICT+}$, while it varies linearly with Φ_M between these two values [Tanaka et al., *Organic Electronics,* 2009].

In addition, Fermi level alignment is a critical problem. However in practical situation of organic solar cells, band bending coupled with interface dipole formation have demonstrated their potentiality to account for experimental results.

If the ICT model, with or without band bending, is efficient for passivated surface substrates other models should be used when there is some chemical interaction between the organic and the substrate.

In the case of strong chemisorption, for instance when the metal electrode is deposited by evaporation onto the organic material there is diffusion of metal atoms into the organic film and the situation is quite complicated, since often the organic material may offer different feasible bonding sites for the metal. Chemisorption can be used voluntarily to modify the properties of the substrate surface, typically by using self-assembled monolayers (SAM). SAM will be discussed in the paragraph dedicated to the contact anode/electron donor.

More generally, the chemical bonding between the metal and the organic molecule may involve a transfer of charge which up-shift, when there is an electronic charge transfer to the molecule, or down-shift, when there is an electronic charge transfer to the metal, the vacuum level by introducing a dipole-induced potential step at the interface (Figure 8). Therefore here also there is a shift Δ of the vacuum level at the interface.

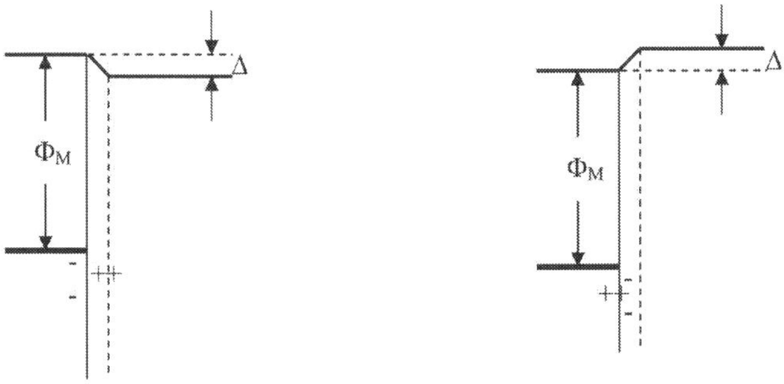

Fig. 8. Interface dipole involved by chemisorption's

As a conclusion it can be said that, whatever its origin, an interface dipole is often present at the interface electrode/organic. Following its sign, this dipole can increase or decrease the potential barrier present at the interface. However, this dipole is only one contribution to the interface barrier, the difference between the work function of the electrode (anode-cathode) and the energy level (HOMO-LUMO) of the organic material is another significant contribution, which allows predicting, at least roughly, the behaviour of the contact.

5. Interface characterisation techniques

One key issue for organic optoelectronic is the understanding of the energy-level alignment at organic material/electrode interfaces, which induces, a fortiori, the knowledge of the electrode work function and ionisation potential (HOMO) and electron affinity (LUMO) of organic semiconductors. For the investigation of the chemistry and electronic properties of interfaces X-ray photoelectron spectroscopy (XPS) and ultraviolet photoelectron spectroscopy (UPS) are often used [Braun, Salaneck, and Fahlman, *Adv. Mater.*, 2009]. Energy level alignment at organic/electrode interfaces can be also carefully studied with Kelvin probe [Ishii et al., .*Phys. Stat. Sol* (a) (2004)]. Cyclic voltammetry is also a valuable tool to estimate the HOMO and LUMO of the organic materials [Cervini et al., *Synthetic Metals*, 1997; Brovelli et al., *Poly. Bull.*, 2007].

5.1 Electron spectroscopy for chemical analysis (ESCA): X-ray photoelectron spectroscopy (XPS) and ultraviolet photoelectron spectroscopy (UPS)

ESCA is a widely used technique for studying chemical and electronic structure of organic materials. More precisely, the method is very useful for the study of surfaces and interfaces. In the case of UPS, the photoelectron inelastic mean free path is less than ten Angstroms.
The well known basic equation used in interpreting photoelectron spectra is:

$$E_B = h\nu - E_{kin} - \Phi_{SP} \qquad (6)$$

Where E_B is the binding energy, $h\nu$ is the photon energy, ϕ_{SP} spectrometer specific constant (the work function of the spectrometer). Assuming that due to the removal of an electron from orbital i the rest of the electron system is not affected (frozen approximation), E_B corresponds to orbital energies $-\varepsilon(i)$. However, the remaining electrons in the environment can screen the photohole, which induces an additional relaxation contribution and impacts the measured E_B value. Changes in the valence electron density induces small, but significant, shift of the core level binding energy, called chemical shift. Hence, charge transfer and chemical bond formation can be probed using XPS. UPS is used for valence electronic study because the photoionisation cross-section for electrons is orders of magnitude higher in the valence band region for UPS and the photon source (He lamps) has high resolution. The source of photons is either HeI ($h\nu$ = 21.2 eV) or HeII radiation ($h\nu$ = 40.8 eV). These energies allow for mapping the valence electronic states of organic materials. The UPS spectra give information about the electronic structure of the material and its work function. It also measures the change Δ of the work function after coverage (Figure 9).

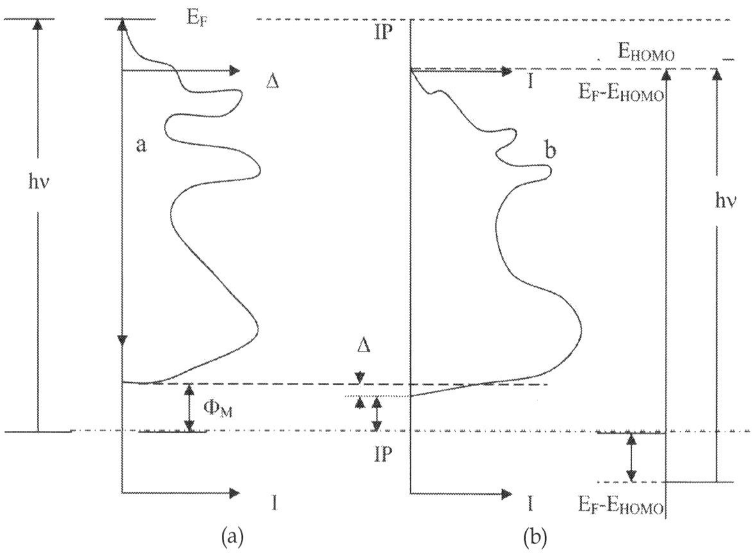

Fig. 9. Shows the principle of UPS for the study of an interface:
a- clean metal, b- metal covered with an organic monolayer.

The UPS spectrum of a clean metal substrate can be seen in Figure 9a. Electrons below the Fermi level are excited by the uv light and emitted into vacuum. The kinetic energy E_{kin} distribution of the emitted electrons is called the UPS spectrum and reflects the density of the occupied states of the solid.

Only photoelectrons whose kinetic energy is higher than the work function ϕ_M of a sample can escape from the surface, consequently ϕ_M can be determined by the difference between the photon energy and the width of the spectrum (Figure 9 a). The width of the spectrum is given by the energy separation of the high binding energy cutoff (E_{cutoff}) and the Fermi energy ($E_b = 0$):

$$\phi_M = h\nu - E_{cutoff} \tag{7}$$

A change in work function, Δ, then can be tracked by remeasuring the E_{cutoff} after deposition of an organic monolayer.
Possible shift of the cutoff and thus of the vacuum level suggests the formation of an interfacial dipole layer Δ [Crispin, *Solar Energy Materials & Solar Cells*, 2004; Kugler et al., *Chem. Phys. Lett.*, 1999; Seki, Ito and Ishii, *Synthetic Metals*, 1997] (Figure 9 b).
In this case the small binding energy onset corresponds to the emission from the highest occupied molecular orbital (HOMO) and the high binding energy (low kinetic energy) cutoff corresponds to the vacuum level at the surface of the organic layer.
Therefore as said above we can visualise the relative position of the energy levels at the interface, and examine the difference of the vacuum level between the metal and organic layer which corresponds to Δ (Figure 10).
UPS is a very powerful tool to detect the presence-or not- and to measure the interface dipole and therefore to understanding of the energy-level alignment at interfaces organic material/electrode.

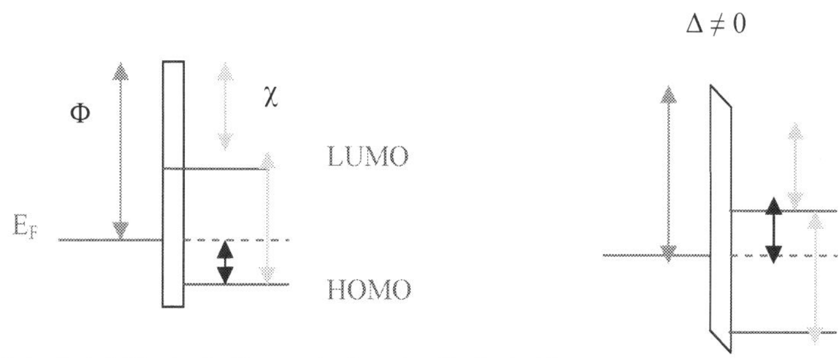

Fig. 10. Interfacial dipole Δ after contact: a: Δ = 0, b: Δ ≠ 0.

4.2 Kelvin probe

The principle of Kelvin probe was put in evidence by Lord Kelvin in 1898 [Phil. Mag., 1898]. The principle was first applied, using a vibrating capacitor by Zisman [Zisman, *Rev. Sci. Instrum.*, 1932]. Nowadays, the Kelvin probe method (KPM) is used to measure the work function of various surfaces. The sample and a metallic vibrating reference electrode constitute a capacitor. The vibration of the reference electrode induces an alternative current, this current is zero when the voltage applied to the reference electrode is equal to the contact potential difference between the reference and the sample. When the sample is conductor, there is no difficulty, the surface of the sample works as a plate of the capacitor and charges are accumulated at the surface. It is more complicated when the sample is a semiconductor or an insulating material. Some part of the charge is into the sample, this situation has been discussed by different authors [Ishii et al., *Phys. Stat. Sol* (a), 2004; Pfeiffer, Leo and Karl, *J. Appl. Phys*, 1996]. They conclude that the vacuum level of the reference electrode exactly coincides with that of the sample, in the case of null-detection condition. Therefore it can be said that KPM probes the surface potential of the sample with precision.
For instance, the energy level alignment at CuPc/metal interfaces has been studied using KPM [Tanaka et al., *Organic Electronics*, 2009]. In order to study the vacuum level (VL) shift at CuPc/metal interfaces different metals presenting a wide range of Φ_M have been probed. Moreover, the deposition of the CuPc onto the metal was performed in a stepwise manner

with Kelvin probe measurement at each step to follow the VL shift as a function of the CuPc film thickness. The study showed that the organic layer onto the metal surface plays two important roles in the energy level alignment: formation of an interfacial dipole (ID) and passivation of the metal surface. The deposition of the first nanometers (<2 nm) induces a large VL shift indicating a charge redistribution at the interface related to the interface dipole (ID) formation. For thicker thickness the VL variation depends on the Φ_M value. When Φ_M is higher than $LUMO_{CuPc}$ very little VL shift occurs for thicker films, the energy level alignment is determined by Δ_{ID} and Φ_M. Therefore the barrier height at the interface varies with Φ_M. When Φ_M is smaller than $LUMO_{CuPc}$, VL varies up to 5nm of CuPc, there is a spontaneous charge transfer (CT) from metal to the CuPc until $LUMO_{CuPc}$ is located above the Fermi level. There is a pinning of the Fermi level and the barrier height at the interface does not vary with Φ_M. This example shows the KPM could be an efficient tool for studying the interfaces organic materials/electrodes.

5.2 Cyclic voltammetry

Electrochemistry is a simple technique, which allows estimating the HOMO and LUMO of organic material [Li et al., *Synthetic Metals*, 1999].

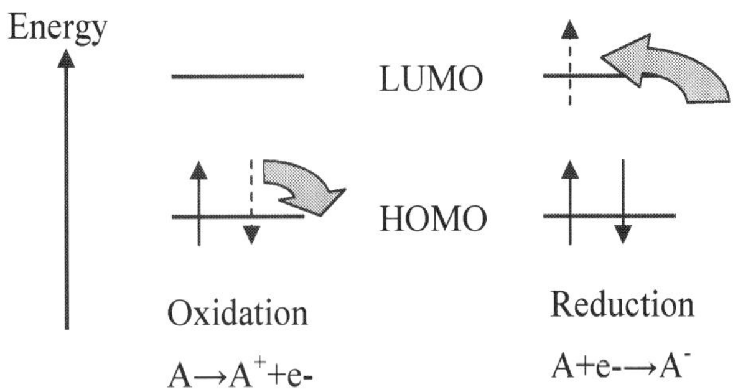

Fig. 11. Oxidation and reduction of an organic molecule.

When the organic material shows an electron reversible reduction and oxidation wave, cyclic voltammetry (CV) is recognised as an important technique for measuring band gaps, electron affinities (LUMO) and potential ionisations (HOMO). The oxidation process corresponds to removal of charge from the HOMO energy level whereas the reduction cycle corresponds to electron addition to the LUMO (Figure 11).

The experimental method is based on cyclic voltammetry [Cervini et al., *Synthetic Metals*, 1997; Li et al., *Synthetic Metals*, 1999.]. The electrochemical set up was based on classical three electrodes cells. The reference electrode was Ag/AgCl.

The electrochemical reduction and oxidation potentials of the organic material are measured by cyclic voltammetry (CV). When the CV curves showed a one electron reversible reduction and oxidation wave, the HOMO and LUMO energy can be determined from the first oxidation and reduction potential respectively. The potential difference Eg = LUMO – HOMO can be used to estimate the energy gap of the dye. The energy level of the normal hydrogen electrode (NHE) is situated 4.5 eV below the zero vacuum energy level [Brovelli et al., *Poly. Bull.,* 200)]. From this energy level of the normal hydrogen and the reduction potential of the reference electrode used, for example Ag/AgCl i.e. 0.197 V versus NHE, a simple relation can be written which allows estimating the both energy values (7):

$$LUMO = [(-4.5)-(0.197-Ered)]eV.$$

$$HOMO = [(-4.5)-(0.197-Eox)]\ eV. \quad (8)$$

As an example the curve corresponding to N,N'-diheptyl-3,4,9,10-perylenetetracarboxylicdiimide (PTCDI-C7) is presented Figure 12.

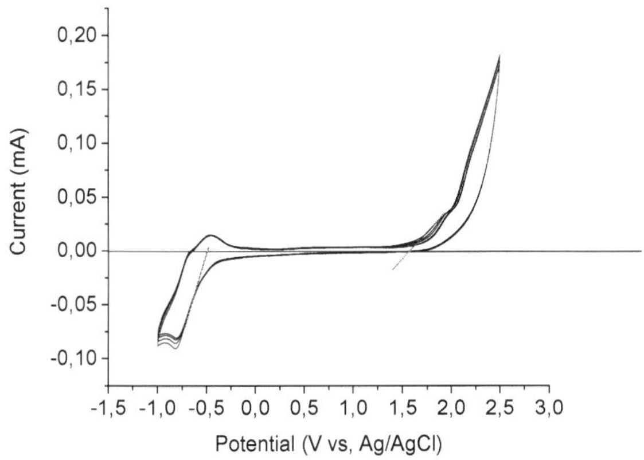

Fig. 12. Cyclic voltammogram of PTCDI-C7 on Pt disc electrode in medium of anhydride dichloromethane.

As working electrode, a polycrystalline platinum disc was used. The reference electrode was Ag/AgCl in solution of tetraethylammonium chloride (Et_4NCl). The potential was adjusted to 0.199 mV with respect to the normal hydrogen electrode (NHE) [East and del Valle, *J. Chem. Educ.,* 2000]. As counter-electrode, spiral platinum was used in a separated compartment of work electrode by fritted glass

The electrochemical reduction and oxidation potentials of the PTCDI-C7 were measured by cyclic voltammetry (CV) (see Figure 12). From CV curves, PTCDI-C7 in dichloromethane showed a one electron reversible reduction and oxidation waves.

The HOMO and LUMO energy of PTCDI-C7 can be determined from the first oxidation and reduction potential respectively. The potential difference Eg = LUMO-HOMO can be used to estimate the energy gap of the dye. The energy level of the normal hydrogen electrode (NHE) is situated 4.5 eV below the zero vacuum energy level [Bard and Faulkner, *Fundamentals and Applications, Wiley* 1984]. From this energy level of the normal hydrogen and the reduction potential of the reference electrode used in the present work Ag/AgCl i.e. 0.199 V versus NHE, a simple relation allows us to estimate the both energy values:

$$LUMO = [(-4.5)-(0.199-Ered)]\ eV$$

$$HOMO = [(-4.5)- (0.199-Eox)]\ eV \quad (9).$$

The values of oxidation and reduction potential are 1.57 V and -0.38 V respectively. Relatively to the vacuum level the energy values of HOMO and LUMO levels are -6.30 eV and -4.30 eV respectively. Therefore the band gap estimated from the electrochemical measurements is 2.0 eV. This value is only slightly higher than the optical band gap of a PTCDI-C7 thin film (1.95 eV). So, the energy gap calculated from the difference between the LUMO and HUMO energies is quite close to the optical band gap, which testifies that the cyclic voltammetry provides a useful rough estimate for the location of the LUMO and the HOMO of the organic materials.

6. Interface organic acceptor/cathode

For electron injection (OLED) or collection (solar cells) it is necessary to incorporate a low work function as cathode. However low work function metals such as Mg, Li, Ca... are not suitable because they have high reactivity in air. Historically works on OLEDs have shown that aluminium coupled with LiF is a very efficient cathode. Hung et al. [Hung, Tang and Mason, *Appl. Phys. Lett.* 2008] have shown that when an ultra thin (1 nm) LiF layer is deposited onto the organic material before Al, this LiF/Al bilayer cathode greatly improved the electron injection and reduced the threshold voltage.

The increase in luminance and efficiency is attributed to enhancement of the electron injection from the aluminium into the organic acceptor. The LiF/Al cathode improves injection by raising the Fermi energy and shifting the effective injection interface deeper into the organic film [Baldo and Forrest, *Phys. Rev.*, 2001.]. Effectively there is Li doping of the organic layer during Al deposition.

In the case of solar cells, insertion of a thin LiF layer (< 1.5 nm) at the organic/aluminium interface allows improving the power conversion efficiency of the cells. An increase in the forward current and in the fill factor is observed upon reducing the serial resistivity across the contact. The optimum LiF thin film thickness is around 1 nm. For higher values the high resistivity of the LiF decreases its beneficial influence. From (I-V) curves it has been estimated that the insertion of a thin LiF layer decreases the serial resistivity of the diodes by a factor 3-4, while the shunt resistivity is stable [Brabec et al., *Appl. Phys. Lett.*, (2002).]. The precise mechanism of LiF on the interface properties is still under discussion. Moreover, it should be highlighted that, in the case of solar cells, LiF is not as successful as in the case of OLEDs. Therefore a lot of works have been dedicated at the improvement of the organic acceptor/cathode interface. Different buffer layers have been probed and the main results are summarized below.

We have seen that the maximum value of Voc is Voc $\leq$ $LUMO_A$ - $HOMO_D$. The same dependence of Voc with $LUMO_A$ - $HOMO_D$ has been encountered whatever the structure used, bulk heterojunction or multiheterojunction structures. The same controversy on the dependence of Voc with the cathode work function [Chan et al., *Appl. Phys. Lett.*, 2007; Rand, Burk and Forrest, *Phys. Rev.*, 2007] is present for both structure families. Indeed, if the Voc value is effectively related to $\Delta(LUMO_A - HOMO_D)$, it depends also of others parameters such as the dark current (leakage current), Voc decreases when this current increases, that is to say when the shunt resistance, Rsh, is faint (Figure 5). In order to check the variation of Voc with $\Delta(LUMO_A - HOMO_D)$ and Rsh, we have studied a cell family with the structure ITO/Donor/Acceptor/Al/P, with donor = ZnPc or CuPc, acceptor = C_{60}, PTCDA, PTCDI-C7 and 1,4-DAAQ and P a protective layer from oxygen and humidity contamination, which allows keeping the device in room air after assembling. P_I corresponds to an encapsulation before breaking the vacuum and P_A an encapsulation after 5 min of room air exposure [Karst and Bernède, *Phys. Stat. Sol. (a)*, 2006]. While in the former case there is no aluminium post depot oxidation, at least during the first hours of air exposure, in the latter case, 5 min of air exposure induces air diffusion at the grain boundaries of the polycrystalline Al layer and formation of a thin Al_2O_3 between the anode and the organic material.

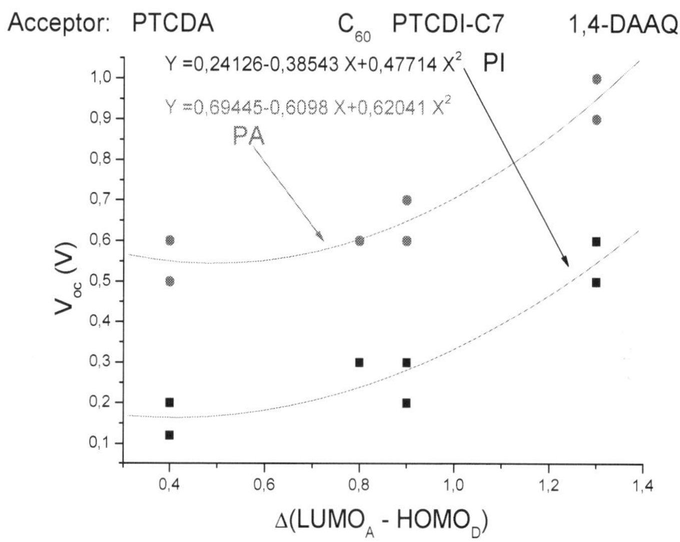

Fig. 13. Voc variation with $\Delta(LUMO_A - HOMO_D)$.

The results are summarized in Figure 13. It can be seen that, as expected, the Voc value increases with the $\Delta(LUMO_A - HOMO_D)$. However, it can be seen also that two curve families are clearly visible. One with small Voc values, which corresponds to cell encapsulated without breaking the vacuum and another with higher Voc values, which corresponds to cells encapsulated after 5 min of air exposure. The two curves are nearly parallel, which demonstrates that the same phenomenon is at the origin of the Voc increase. Since the only difference between these two families is the contact or not with room air, the translation of the curve should be attributed to the presence of the thin natural Al_2O_3 layer at the electron acceptor/aluminium interface. This natural oxide does not depend on the organic material but only on the alumium electrode air exposure, which is in good agreement with the translation effect of the two curves. Such ultra thin Al_2O_3 layer (1nm) increases the shunt resistance value, which justifies the Voc value increase. Such effect of aluminium oxidation on the open circuit voltage has already been proposed by Singh and coll.[Singh et al., *Appl. Phys. Lett.*, 2005; Singh et al., *Sol. Energy Mater. Sol. Cells,* 2006], thanks to our in situ encapsulation process we have directly put this effect in evidence. However, if the increase of the shunt resistance of the cells through insulating oxide formation at the interface cathode/organic materiel, allows increasing the open circuit voltage and therefore the solar cells efficiency, the limit of the positive effect of such oxide layer is rapidly achieved. Indeed, it is only efficient when electrons can tunnel through the oxide layer. Beyond 2.5 nm, not only the shunt resistance increases but also the series resistance and therefore the current and cell efficiency.

Moreover other limitation at the interface organic/cathode has been highlighted through the experiments described below. It has been shown that one way for circumventing the diffusion length limitation is to use cells with multiple interfaces. Peumans et al. [Peumans et al., *Appl. Phys. Lett.*, 2000] have shown that the introduction of a thin large band gap organic material allows improving significantly the device performances. He called electron blocking layer (EBL) this thin film, because its bandgap was substantially larger than those of the organic donor and acceptor, which block excitons in the organic semiconducting layer far from the cathode avoiding any quenching effect at the cathode/organic interface. Will see more precisely the effect of this "EBL", but first we will conclude on the effectiveness of the very thin oxide layer between the cathode and the organic electron acceptor. In order to discriminate between the effect of an EBL and an oxide layer deposited before the cathode we have worked with ITO/CuPc/C_{60}/Alq_3/Al/P, Alq_3 being used as EBL layer. It is shown in Table 1, that, as expected, the EBL improve significantly the cells performances, while the encapsulation process does not modify the strongly the I-V characteristics.

Devices	Jsc (mA/cm^2)	Voc (V)	Rsh (Ω)
ITO/CuPc/C_{60} /Al/PI	4.75	0.24	90
ITO/CuPc/C_{60} /Al/PA	4.40	0.41	1650
ITO/CuPc/C_{60} /Alq_3/Al/PI	7.75	0.45	1800
ITO/CuPc/C_{60} /Alq_3/Al/PA	7.45	0.48	1850

Table 1. Jsc and Voc values of the different devices under AM1.5 conditions.

In fact, the Voc value in the presence of Alq_3 does not depend strongly on the encapsulation process, while it does when simple CuPc/C_{60} junction is used. This difference can be explained by the variation of the value of the shunt resistance, Rsh. Without Alq_3, a thin Al_2O_3 layer is necessary to improve Rsh and therefore Voc, with Alq_3, Rsh is sufficient and the alumina is not necessary to optimise the Voc value (Table 1).

Accordingly to the present discussion, the EBL is sufficient to confine the photogenerated excitons to the domain near the interface where the dissociation takes place and prevents parasitic exciton quenching at the photosensitive organic/electrode interface. Also it limits the volume over which excitons may diffuse. For vapor deposited multilayer structures, a significant increase in efficiency occurs upon the insertion of the exciton blocking interfacial layer, interfacial layer, between the cathode and the electron acceptor film. Bathocuproine (BCP) is often used as exciton blocking buffer layer [Peumans et al., *Appl. Phys. Lett.*, 2000; Huang et al., *J.Appl. Phys.*, 2009]. However, with time, BCP tends to crystallize, which induces some OSCs performance degradation [Song et al., *Chem. Phys. Lett.*, 2005]. Consequently, either other more conductive [Refs] or more stable, e.g, aluminium tris(8-hydroxyquinoline) (Alq_3), materials have been tested as EBL [Song et al., *Chem. Phys. Lett.*, 2005; Hong, Huang and Zeng, *Chem. Phys. Lett,.*, 2006; Bernède and al., *Appl. Phys. Lett.*, 2008]. Therefore, many organic materials with quite different HOMO and LUMO values can be used as EBL. Indeed, it appears that EBL can also protect the electron accepting film from atoms diffusion during deposition of the electrode. Also it is thick enough and sufficiently homogeneous to fill pinholes and others shorting effect which increases Rsh and therefore Voc and the cell efficiency. Therefore the EBL protects the fragile organic films from damage produced during electrode deposition onto the organic material. The large band gap of the EBL, larger than that of the adjacent organic film, allows blocking the excitons in this film. If the EBL blocks the excitons it should not block all charge carriers. Therefore the EBL should be chosen so that it allows electrons collection at the cathode. However the offset energy of the highest occupied molecular orbital (HOMO) of the electron donor (often the fullerene) and the EBL (such as the bathocuproine) is large. Moreover, the optimum EBL thickness is around 8 nm, which is too thick to allow high tunnelling current. So, even if the EBL is an electron conducting material, the difference of the LUMO levels of C_{60} and Alq_3 implies that electrons must overcome a large energy barrier to reach the Al cathode in case of electron transport via LUMO levels (Figure 14-1). More probably, the charge transport in the EBL is due to damage induced during deposition of the cathode, which introduces conducting levels below its LUMO (Figure 14-2) and explains the reason why the transport of electron is not weakened. [Rand et al., *Adv. Mater.*, 2005]. As a conclusion, the EBL, not only block the excitons far from the cathode where they can be quenched, but also prevents damage of the electron acceptor film during cathode deposition. It should be transparent to the solar spectrum to act as a spacer between the photoactive region and the metallic cathode and it must transport electrons to avoid high series resistance. The EBL is also important for fabricating large-area devices with a low density of electrical shorts.

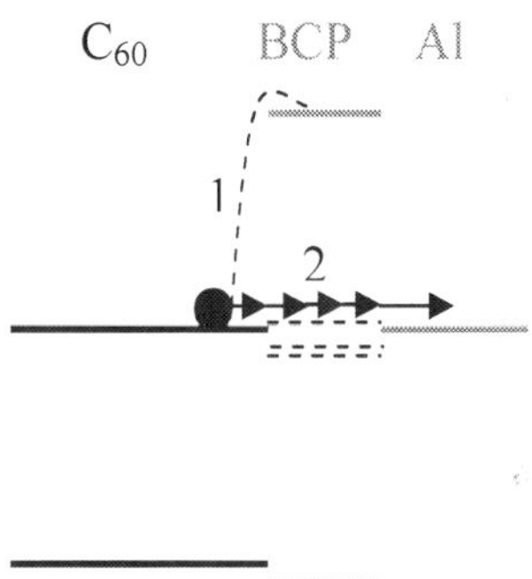

Fig. 14. Band schemes of organic films and cathode contact.

7. Anode/organic donor interface

Globally, the electrodes in contact with the organic materials have great importance in the device behaviour. Of course, in optoelectronic devices it is necessary to allow the maximum amount of photons of the solar spectrum to enter the active part of the device. Therefore it is necessary that at least one of the electrodes should exhibit high transparency and should not be reflecting. In classical configuration the transparent electrode, a transparent conductive oxide thin film (TCO), is the anode. Typically, glass coated with the degenerate semiconductor indium-tin oxide (ITO), is used as anode electrode. ITO, which commonly serves as anode in organic optoelectronic devices, attracted considerable interest due to its unique characteristics of high conductivity, good transparency in the visible domain and easy patterning ability, moreover it is commercially available.

A crucial point in organic devices is the interface between the inorganic electrodes and the organic materials. The key parameter at the anode interface is the hole collection from the organic semiconductor to the anode. A barrier for carrier transport is often present at the interface. It is usually determined by the electrode/organic band offset, that is to say, in the case of holes, the difference between the work function of the anode and the highest occupied molecular orbital (HOMO) of the electron donor, even if, as discussed above, the barrier height depends also of the presence, or not, of an interface dipole. About the influence of the barrier height at the interface anode/organic donor Kang et al [Kang, Tan and Silva, *Organic Electronics,* 2009] have shown a clear relation between the work function of the anode and the devices performances. They show that the energy conversion efficiency of the cells follows the variation of the value of the anode work function. The work function was measured by Kelvin probe, the anode, ITO/PTFE (polytetrafluoroethylene), was treated with different UV exposure time. The work function increases during the first five minutes and then decreases, also the devices performances and mainly the short circuit current Jsc. High Jsc in organic solar cells are mainly due to small barrier height between the anode and the organic and subsequently improved carriers extraction process. Therefore, the influence of the barrier height at the contact anode/electron donor being well established, it is necessary to control the work function of the anode to achieve good band alignment and ohmic contact. High work function anode is desirable to decrease the series resistance. High and reproducible work functions are difficult to obtain for ITO [Bruner et al., *J. Am. Chem. Soc.,* 2002]. Many processes have been proposed to achieve this goal. First, as discussed in paragraph 3, it should be underlined that ITO work function depends strongly of the thin film history and it is quite difficult to predict. It has been shown that ITO surface chemistry is difficult to control, because its surface is covered by hydrolysed oxides [Armstrong et al., *Thin Solid Films,* 2004), Donley et al., *Langmuir* (2002); Kim, Friend & Cacialli, *J. Appl. Phys.,* 1999]. In fact, he surface chemical functionality of ITO is not well understood [Katkova et al., *Appl. Surf. Sciences,* 2008]. Authors propose the presence of hydroxyl [Purvis et al., *J. Am. Chem. Soc,.* (2000)] others not [Chaney & Pehrsson, *Appl. Surf. Sci.,* 2003]. What is clear for all experimenters in the field of organic optoelectronic devices is that cleanliness of the ITO surface is critically important for efficient hole exchange at the organic material/ITO interface. Devices performances, not only depend on the surface treatment but also on the deposition batches [Berredjem et al., *The European Physical Journal: Applied Physics,* 2008].

Moreover, it is well known that crystals in polycrystalline ITO thin films have pyramidal shape, which induces a significant surface roughness (some nm) of these films. This surface roughness is often evocated as a source of leakage current and lifetime limitation in optoelectronic devices. Also, ITO electrodes were reported to interact chemically, which contributes to the degradation of optoelectronic devices performances [Kugleret al., *Synthetic Metals,* 1997)]. For instance, even in the absence of oxygen and moisture, oxidation of organic material in contact with ITO has been reported [Scott et al., *J. Appl. Phys.*1997]. It appears that ITO anode serves as source of oxygen. At least, it should be underlined that ITO, is not ideal due to the scarcity of its main component: indium. Indeed, to day, ITO is widely used as electrode in optoelectronic devices and demand for indium is expected to outstrip supply these years, making devices based on ITO expensive. All that justifies, not only the different works dedicated to ITO surface treatment itself, but also original works on different TCOs and transparent anodes.

First of all different surface treatments of ITO have been probed. Hydrogen peroxide treatment improves the devices performance through work function increase ($4.7< \Phi_M < 4.8$

eV), however, even with similar Φ_M, different turn-on voltage are measured. Obviously, additional factors should be considered such as surface roughness performance [Kugleret al., *Synthetic Metals,* 1997]. Different acidic solutions have been probed (HCl, H_3PO_4).

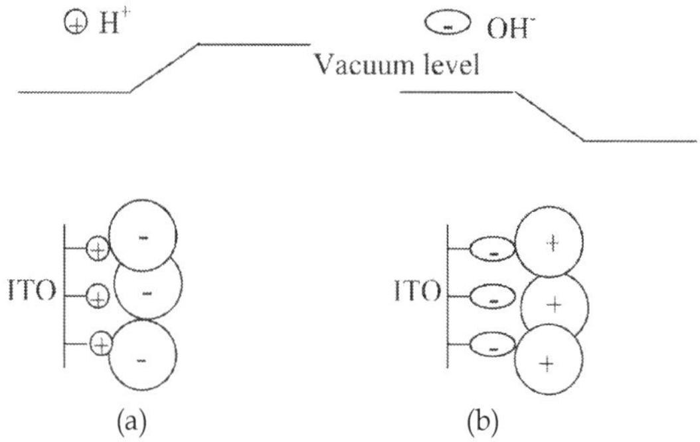

Fig. 15. Surface structure of passivated ITO for acid (a) and base (b) treatment.

Treatments with phosphoric acid lead to an increase in work function of about 0.7 eV (4.5 to 5.2 eV) with good homogeneity [Johnev et al., *Thin Solid Films,* 2005)].
Such effect is induced by monolayer adsorption (Figure 15 a), which allows improving the solar cells efficiency from 1.2 to 1.5 %.
However this efficiency remains smaller than that obtained with conducting polymer buffer layer, which will be discussed below. When treated by a base a decrease of Φ_M is obtained, which means that Φ_M can be shifted of 1 eV [Nüesch et al., *Appl. Phys. Lett,.* 74 1999]. Moreover, it is necessary to use an appropriate plasma treatment before chemical adsorption. As a matter of fact, plasma treatments are often used to increase the ITO work function. In the case of plasma treatment, after chemical pre-cleaning, the sample was treated by RF plasma, usually the ambient gas is Ar or O_2, with better results achieved with O_2. Not only the plasma treatment cleans the ITO surface, increases Φ_M, but also smoothes the film surface, whole things resulting in performance improvement of devices [Lu & Yokoyama, *Journ. Crys. Growth,* 2004, Zhong & Jiang, *Phys. Stat. Sol.* (a), 2006]. Another well known technique used to tune the ITO surface work function, is the deposition of self-assembled monolayers (SAM) onto the ITO film surface. A SAM consists of a molecular backbone terminated by an anchoring group and, at the other extremity, by an end group that may induce a dipole, Δ_{SAM}. Δ_{SAM} is defined as positive if it up-shifts the vacuum level on the organic/SAM side. Different organic material families can be used as SAM, tin phenoxides [Bruner et al., J. Am. Chem. Soc., 2002], thiophene phosphonates , phosphonic acids [Hansson et al., J. Am. Chem. Soc., 2005; Sharma et al., *J. Appl. Phys.,* 2009], also polymeric (LBL) assembly has been used for anode modification layer by layer [Kato, J. Am. Chem. Soc., (2005].
If these chemical techniques allow tuning efficiently the work function of the ITO thin films, physical techniques such as spin coating, vapor deposition can be used also with success.
The conducting polymer the most widely used to help the charge transporting at the interface ITO/organic is the poly(ethylene dioxythiophene) doped with polystyrene sulfonic acid (PEDOT:PSS) [Hoppe and Sariciftci, *J. Mater. Res.,* 2004]. PEDOT:PSS is a p-type semiconductor, a good hole transport material, it is soluble in water and easy to depose by spin coating. Its work function is 5.2 eV.

The initial solution of PEDOT-PSS is 3 wt. % in water. It is spun at 2000-5000 rpm to form a 50-100 nm thick layer. After deposition, to prevent the presence of water in the device, PEDOT:PSS coated ITO is annealed for half to an hour at 100-150°C. Then the different organic constituents and the cathode of the optoelectronic device are deposited. The PEDOT:PSS buffer layer allows the device performance to be significantly improved, OLEDs and solar cells. It is admitted that the high value of its work function allows a good band alignment with the HOMO of the electron donor, which decreases the barrier height at the interface and therefore assures a better hole collection from the polymer into the ITO electrode. Also it is supposed that the PEDOT:PSS spin coated onto the ITO surface smoothes its surface and, therefore, any possible short circuiting due to the spiky roughness of the ITO surface is prevented. It improves the contact between the polymer and the ITO. It is admitted that this buffer layer enhances adhesion to the organic layer. Also it prevents direct contact between the oxygen of the ITO and the organic material.

However, PEDOT:PSS is problematic since its poor conductivity is a major limiting factor for device performance and it degrades under UV illumination [Chang & Chen, *Appl. Phys. Lett,*. 2007; Kang et al., *J. Phys. D: Appl. Phys.*, 2008]. Even after baking, due to its hygroscopic nature some amount of water always appears in PEDOT:PSS, which introduces water into the active layer, it is also slightly acidic [Van de Lagemaat et al., *Appl. Phys. Lett.*, 2006, Johnev et al., *Thin Solid Films,* 2005]. Moreover, not only the depositing process from aqueous solution introduces impurities but the reproducibility is in need of improvement [Johnev et al., *Thin Solid Films,* 2005].

Therefore, other solutions have been proposed, each one based on original buffer layers such as metal or oxides. Some attempts using thin metal buffer layers have been done during the last years, however the results were quite disappointing, the metal thin film used being thick of some nanometers, the transmission of the visible light decreases significantly (Figure 16) and also the devices efficiency [Yoo et al., *Synthetic Metals,* 2005].

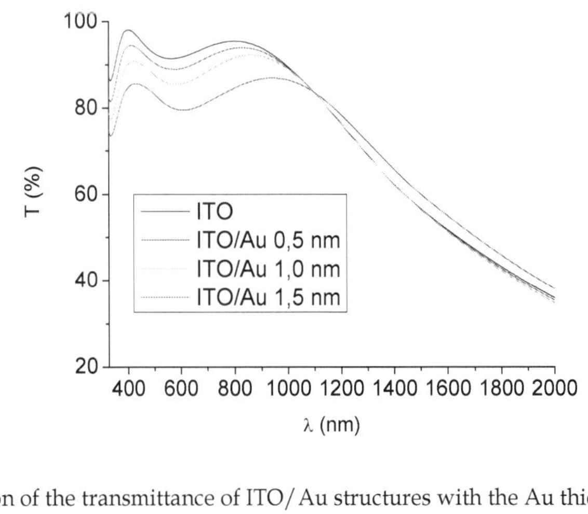

Fig. 16. Variation of the transmittance of ITO/Au structures with the Au thickness (0 to 1.5 nm).

We have shown that this difficulty could be overcome by using an ultra thin (0.5 nm) gold film. The introduction of this ultra-thin metal layer at the interface anode/electron donor allows improving significantly the energy conversion efficiency of the organic solar cells [Bernède et al., *Appl. Phys. Lett.*, 2008; Bernède et al., *Sol. Energy Mater. Sol. Cells*, 2008].

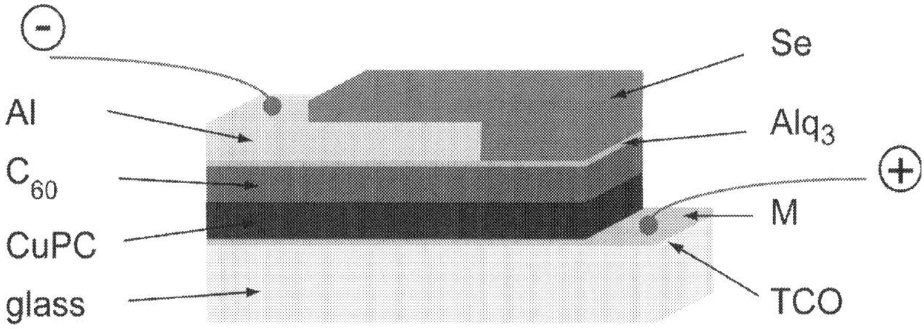

Fig. 17. Schematic structure of the fabricated solar cells with the ultra-thin gold layer onto the TCO.

The efficiency improvement is even more remarkable as regards to TCO initial quality. The effect of this ultra-thin metal buffer layer has been probed on multi-heterojunction organic solar cells (Figure 17) and we present, with more details, this example of efficient buffer layer at the interface anode/electron donor. The electron donor used was copper phthalocyanine (CuPc) (some attempts have been done using pentacene and similar behaviour has been obtained), the electron acceptor was fullerene (C_{60}) and the electron blocking layer was the tris(8-hydroxyquinoline) (Alq_3) [Kim et al., *Sciences*, 2007, Berredjem et al., *Eur. Phys. Journ.: App. Phys.* 2007]. CuPc, C_{60} and Alq_3 have been deposited in a vacuum of 10^{-4} Pa. The thin film deposition rate and thickness were estimated in situ with a quartz monitor. The deposition rate and final thickness were 0.05 nm/s and 35 nm in the case of CuPc, 0.05 nm/s and.40 nm in the case of C_{60} and 0.1 and 9 nm for Alq_3. These thicknesses have been chosen after optimisation.

After organic thin film deposition, the aluminium upper electrodes were thermally evaporated, without breaking the vacuum, through a mask with 2 mm x 8 mm active area. This Al film behaves as the cathode, while the ITO is the anode. Some ITO anodes have been covered with an ultra thin metal film deposited by vacuum evaporation, the metal being Au, Cu, Ni. The thickness of these ultra thin metal films, M, has been varied from 0.3 to 1.2 nm. Finally, the structures used were: glass/ITO(100nm)/M ($0 \leq x \leq 1.2$ nm)/CuPc(35nm)/C_{60}(40nm)/Alq_3(9 nm)/Al(120nm). It can be seen in figure 18 and table 2 that the presence of the ultra-thin gold layer improves significantly the solar cells performances. When different batches of ITO were used, without Au buffer layer, the solar cells performance vary strongly, while they were of the same order of magnitude when an ultra-thin gold layer was deposited onto ITO (Table 2).

Similar results have been obtained when AZO and FTO are used. The performances of organic solar cells using this ultra thin metal layer, are nearly similar, whatever the TCO used [Bernède et al., *Appl. Phys. Lett.*, 2008, Bernède et al., *Sol. Energy Mater. Sol. Cells*, 2008]. This suggests that indium free organic devices with high-efficiency can be achieved, which can contribute to the sustainable development.

Batch	Anode	Jsc (mA/cm2)	Vco (V)	FF	η %
a	ITO	7.31	0.45	0.44	1.45
	ITO/Au (0.5nm)	8.09	0.49	0.56	2.25
b	ITO	6.80	0.40	0.25	0.67
	ITO/Au (0.5nm)	8.34	0.45	0.50	1.86

Table 2. Photovoltaic performance data of devices achieved with batches a and b of ITO, under AM1.5 conditions.

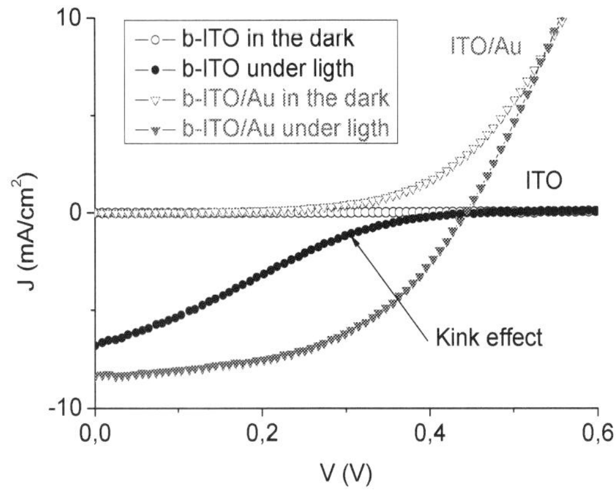

Fig. 18. Typical J-V characteristics of solar cells, with an anode of ITO (batch b) covered (▼) or not (●) with 0.5 nm of Au, in the dark (open symbol) and under illumination of AM1.5 solar simulation (100 mW/cm²) (full symbol).

As said above, others metal such as Cu, Ni, Ag, Pt have been probed, however, up to now, the best results have been obtained with a gold ultra thin film. It should be noted that, roughly, the organic solar cells performance increases with the metal work function, which means that, Ag which have the smallest Φ_M gives the worst performances.
In order to understand the ultra-thin gold layer effect, TCO covered with such gold layers have been characterized by scanning electron microscopy (SEM), atomic force microscope (AFP), X-ray photo-electron spectroscopy, optical transmission measurement (Figures 16, 19). It is shown that the ultra-thin film is discontinuous, while the roughness of the TCO/Au electrode is not different from that of bare TCO (0.8 nm). However it can be seen through the XPS study that the CuPc has grown more homogeneously when deposited onto gold covered TCO. Moreover as shown in Figure 18 the shape of the J-V characteristic depends on the anode configuration, in the case of bare TCO a "kink" effect is clearly visible, while classical diode characteristics are obtained when the TCO is covered by the ultra-thin gold layer.
In order to discuss the effect of the ultra thin gold film on solar cells performances, we recall shortly supposed beneficial effect of the classical buffer layer, the PEDOT:PSS.
As said above, up to now, the most common buffer layer inserted at this interface is the PEDOT:PSS, its contribution to the improvement of solar cells performance has been attributed to:

a-ITO (rms = 0.80 nm)

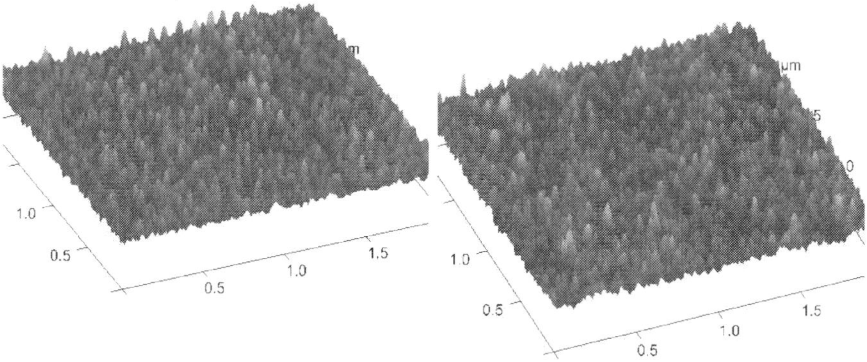

b-ITO/Au (rms = 0.75 nm)

Fig. 19. AFM images of Au (0.5 nm) covered ITO (a) and bare ITO (b).

- the smoothing effect of the quite rough TCO surface, therefore, any possible short circuiting due to the spiky roughness of the TCO surface is prevented.
- the physical separation, which avoids direct contact between the oxygen of the TCO and the organic material.
- its work function ($\phi_{M(PEDOT:PSS)}$ = 5.1 eV), which allows to decrease the barrier height at the interface, since the work function of the TCO is smaller (4.5-4.7 eV for ITO), while the HOMO of CuPc is 5.2 eV.
- the improvement of the contact between the polymer and the TCO. It is admitted that this buffer layer enhances adhesion to the organic layer.

It can be seen immediately that the two first contributions can be excluded when an ultra-thin gold film is substituted by PEDOT:PSS, since the roughness of the modified anode is similar to that of the initial TCO and the gold film is discontinuous. The two last contributions seem more probable. The work function of gold is 5.1 eV and therefore the ultra-thin gold layer can improve the matching between the work function of the anode and the highest occupied molecular orbital (HOMO) of the organic electron donor (Figure 5a, b, c). Such contribution will be discussed more carefully below. The fourth contribution is in good agreement with the XPS study, which shows that the CuPc films are more homogeneous when deposited onto gold modified TCO.

Therefore from the examples presented above it can be concluded that the two main contributions to the interface improvement by inserting an ultra-thin metal buffer layer between the TCO anode and the CuPc are a better matching of the band structure (Figure 20) and a higher homogeneity of the organic film.

The equivalent circuit model (figure 5 a) could be helpful in understanding of organic solar cells by providing a quantitative estimation for losses in the cells. As said above, the equivalent circuit commonly used to interpret the I-V characteristics of real solar cells consists of a photogenerator connected in parallel with a diode and a shunt resistance, and a series resistance. For such solar cells the mathematical description of this circuit is given by the equation (1).

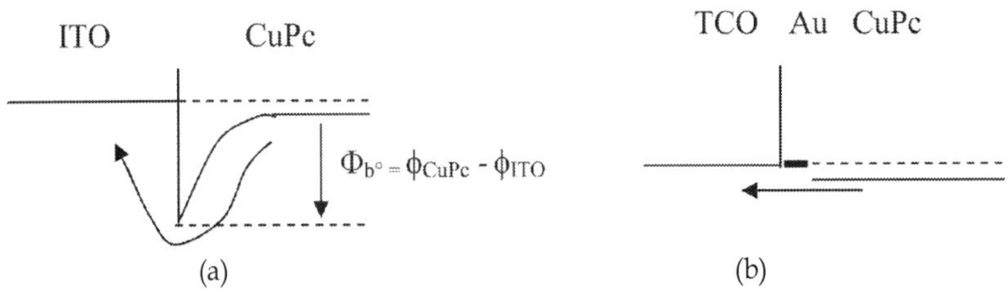

Fig. 20. Contact anode (ITO)/ electron donor (CuPc)
a-without buffer layer; b-with a gold buffer layer.

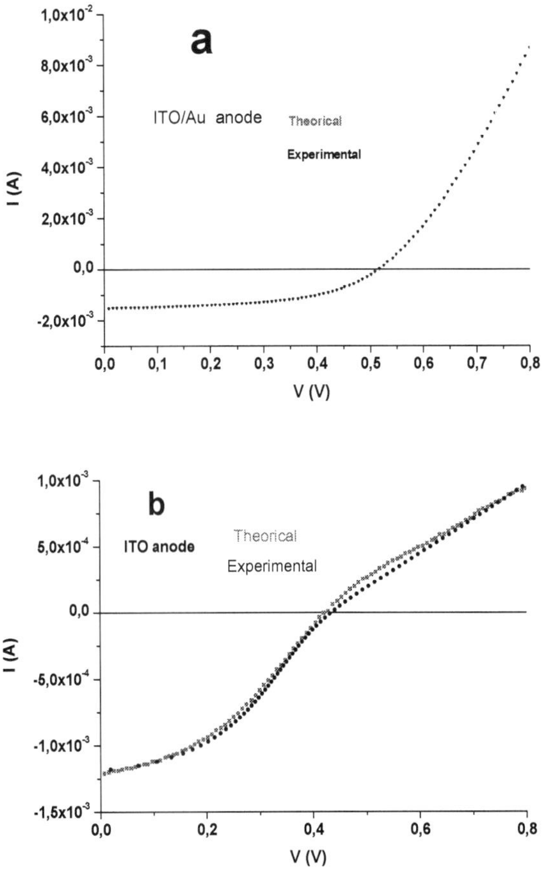

Fig. 21. I-V characteristics under AM1.5 illumination of a solar cell using a ITO/Au anode (a) and a ITO anode (b), (•) experimental and (▪) theoretical curves.

In a recent contribution [Kouskoussa et al., *Phys. Stat. Sol.* (a), 2009], we have shown that the Lambert W-function method can be used to determinate Rs, the series resistance, Rsh the shunt resistance, n the ideality factor of the diode and Iph the photo-generated current.
The problem to be solved is the evaluation of a set of five parameters Rs, Rsh, n, Iph and Is in order to fit a given experimental I-Vcharacteristics using a simple diode circuit.
A good agreement between the experimental and theoretical fitted curves is achieved with ITO/Au anode (Figure 21 a), while it is not in the case of bare ITO anode.
As said above no agreement could be achieved in the case of a bare ITO or Ag covered anode, whatever the series and shunt resistance proposed. Such impossibility shows that the simple equivalent scheme used in this theoretical study cannot explain the experimental results obtained with a bare ITO anode.
In the case of bare and Ag covered ITO anode, it is necessary to assume the presence of a back-contact barrier at the ITO/CuPc interface (figure 5 b), to achieve a good fit between experimental and theoretical results. Assuming a thermoionic current at this interface, the hole current is:

$$I_b = - I_{b0} (\exp(-qV_b/kT)-1) \quad (10)$$

With: - I_{b0} saturation current,
- V_b voltage across the back contact,
- k Boltzmann constant,
- T temperature.

Therefore the current-limiting effect, "rollover", is due to the back-contact barrier height. It occurs because the total current saturate at a value J_{b0} [Demtsu and Sites, *Thin Solid Films*, 2006]. The value of J_{b0} is the current value where the J-V curve starts to show rollover.
Demtsu and Sites have treated the main junction and the back-contact junction (Figure 6 b) as independent circuit element. Here, when a forward bias V is applied to the circuit, the voltage is divided between V_m across the main CuPc/C_{60} junction, V_b across the back-junction TCO/CuPc and IRs across the series resistance:

$$V = V_m + V_b + IRs.$$

Under illumination the current across the main junction is:

$$I_m = I_{m0} (\exp(qV_m/nkT)-1) - Iph + Vm/Rsh \quad (11)$$

And through the back contact it is:

$$I_b = - I_{b0} (\exp(-qV_b/kT)-1) + Vb/ R^b_{sh} \quad (12)$$

Equating equations (10) and (11):

$$I_{m0} (\exp(qV_m/nkT)-1) - Iph + Vm/Rsh + I_{b0} (\exp(-qV_b/kT)-1) - Vb/ R^b_{sh} = 0 \quad (13).$$

The parameters Rs and I_{m0}, n, Rsh of the main diode are calculated in the region far from the saturation current Ib. As said above, I_{b0} is the current value where the J-V curve starts to show rollover. Then equation (13) can be solved.
A good agreement can be achieved between experimental and theoretical curves, (Figure 21 b), which validates the hypothesis of the presence of a rectifying effect at the ITO/CuPc contact.

The parameters extracted are reported in Table 3. It should be noted that when ITO is covered with Ag, two diodes are necessary to obtain a good fit between the experimental and theoretical curves. Since the work function of Ag is only 4.3 eV, this result is in good agreement with the present discussion. One can see in Table 3 that, when the two diodes model is used, the estimated Rs values are of the same order of magnitude as those of ITO/Au anode. Therefore the introduction of a back junction diode is a good interpretation of the interface ITO/CuPc. It justifies the small fill factor and short circuit current values when bare ITO is used as anode in the heterojunction solar cells. The Voc value can be improved thought an increase of the shunt resistance value. It can be seen in Table 3 that the shunt resistance value of the cells with a ITO/Au anode is significantly higher.

Anode	Main junction CuPc/C_{60}				Back-contact junction	
	n	Rsh	Rs	I_{mo} (A)	$R^b_{sh}(\Omega)$)	I_{bo}(A)
ITO	2.55	330	20	$2\ 10^{-6}$	15000	$2.58\ 10^{-6}$
ITO/Ag	2.6	400	40	$1\ 10^{-6}$	13200	$2.24\ 10^{-6}$
ITO/Au	1.5	6900	25	$2.23\ 10^{-9}$	-	-

Table 3. Parameters calculated using a main diode and (or not) a back contact diode.

Table 3 shows that the ideality factor, n, decreases, while Rsh increases, when the ITO anode is covered by an ultra-thin Au layer, which corresponds to a significant improvement of cell performance. When Ag is used the improvement is not significant. The increase of Rsh can be related to the growth of pinhole free organic films when deposited onto Au coated TCO [Kim et al., *Sol. Energy Mater. Sol. Cells,* 2009].

When thicker metal films are introduced at the anode/organic interface, surface plasmon effect has been proposed as alternative approach toward enhanced light absorption without the need for thick films. [Chen, et al., *Appl. Phys. Lett.,* 2009]. A surface plasmon is an optically generated wave, which propagates along a metal/dielectric interface. In tuning the light excitation, a resonance can occur when the frequency of the incident photon equals the collective oscillation frequency of conduction electrons of metallic particles. These properties can be used in the photovoltaic domain in order to improve the light absorption. Theoretical calculations have been performed to determine optimal plasmonic materials to optimise light absorption [Duche, et al. *Sol. Energy Mater. Sol. Cells,* 2009]. Enhanced absorbance up to 50% has been experimentally obtained in a 50 nm thick blend film including silver nanospheres with a diameter of 40 nm. Devices based on BHJ structures using ITO anode covered with Ag nanodots permits 20% improvement of the efficiency [Kim et al., *Appl. Phys. Lett.,* 2008]. The plasmon resonance wavelength depends on the metal nanodots properties, it can be tuned by annealing the silver film or using other metal [Morfa,. *Appl. Phys. Lett.,* 2008]. Even if the efficiency of the surface plasmon effect in the field of solar cells is always under discussion, it is often proposed to improve the light absorption.

After metals, different oxides (MoO_3, WO_3, ZnO) have been probed as buffer layer at the TCO anode /CuPc interface.

In the case of OLEDs [Im et al., *Thin Solid Film,* 2007; Hsu & Wu, *Appl. Phys. Lett.,* 2004; Shi, Ma, and Peng, *Eur. Phys. J. Appl. Phys.,* 2007; Qiu et al., *J. Appl. Phys.,* 2003;

Matsushima et al., *Appl. Phys. Lett.,* 2007; Jiang et al., *J. Phys. D: Appl. Phys.,* 2007; You et al., J. *Appl. Phys.,* 2007] and organic thin film transistors [Chu et al., *Appl. Phys. Lett.,* 2005;

Park, Noh and Lee, *Appl. Phys. Lett.,* 2006], different attempts have been done, with some success, using different oxide thin films as buffer layer between the TCO and the organic material. Also some attempts have been probed in the case of organic solar cells [Yoo et al., *Synthetic Metals*, 2005; Chan et al., *J. Appl. Phys.*, 2006]. The use of such buffer layer is based on the idea that the potential barrier at the ITO/organic donor interface can be reduced by elevation of ITO surface work function, giving high hole transport at this interface. In order to check this hypothesis, using the same multiheterojunctions as above, we have probed different buffer layers at the ITO/CuPc interface. The buffer layers used were MoO_3, WO_3 and ZnO. The resulting different I-V characteristics are compared to those of a reference cell built on untreated anode. The different energy band diagrams are proposed and discussed. Finally, we demonstrate that the reduction process of the barrier at the ITO/organic donor interface depends on the type of material: oxide or metal.

During this study, the majority of the ITO anodes have been covered with an oxide buffer layer. These thin buffer layers were deposited by vacuum evaporation. The MoO_3 oxide thin films were thermally evaporated onto the ITO coated glass. The thickness of these ultra thin oxide films ranged between 1 and 7.

Finally, the structures used were: glass/ITO(100nm)/bufferlayer/CuPc(40nm)/C_{60}(40nm)/Alq_3(9nm)/Al(120nm)/P_{Se}.

The optimum efficiencies were achieved for thickness of around 3.5 ± 1 nm, that means that, as in the case of gold, the oxide film does not completely cover the anode surface since it has been shown that 5 nm are necessary to obtain continuous thin MoO_3 film, using thermal evaporation [Song et al., *Chem. Phys. Lett.*,2005]. For thicker oxide films, if the fill factor (FF) stays far higher than the value obtained with bare ITO, the short circuit current (Jsc) is slightly smaller and it progressively decreases as the oxide thickness increases. With thinner oxide films the inflection point, in the J-V characteristics, typical of the bare ITO anode (Figure 22), is still slightly present.

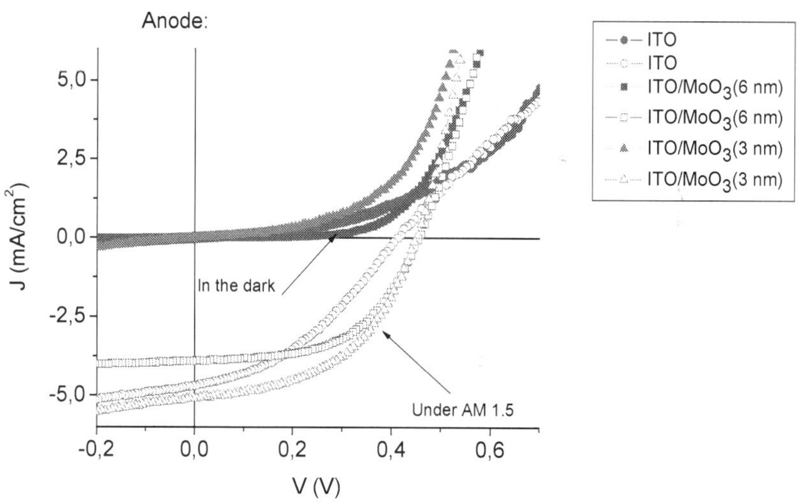

Fig. 22. Typical J-V characteristics of Anode/CuPc/C_{60}/Alq_3/Al structure, with Anode = ITO (●), ITO/MoO_3 (6 nm) (■) and ITO/MoO_3 (3 nm) (▲), in the dark (full symbol) and under illumination of AM1.5 solar simulation (100 mW/cm^2) (open symbol).

It should be highlighted that the main improvement induced by the oxide is related to the fill factor (Table 4). Here also the improvement induced by the oxide does not depend on the TCO used, since the best result achieved by the cells deposited during the run corresponds to FTO. Here also the presence, or not, of the kink effect testifies of the efficiency of oxide.
A similar approach to the one used in the case of gold buffer layer can be used in the case of oxide buffer layer. Often the positive effect on the properties of the organic optoelectronic devices of the anode buffer layer is mainly attributed to the reduction of the barrier energy between the ITO, which is usually the anode, and the organic layer (electron donor for solar cells, hole transporting layer for OLEDs...).

Anode	ITO	ITO/MoO_3 (6 nm)	ITO/MoO_3 (3 nm)
Jsc (mA/cm^2)	4.69	3.90	5.05
Voc (V)	0.41	0.46	0.45
η (%)	0.73	0.97	1.13
FF (%)	38	54	49.7

Table 4. Photovoltaic performance data under AM1.5 conditions of devices using ITO, ITO/MoO_3 (3 nm) and ITO/MoO_3 (6 nm) anodes.

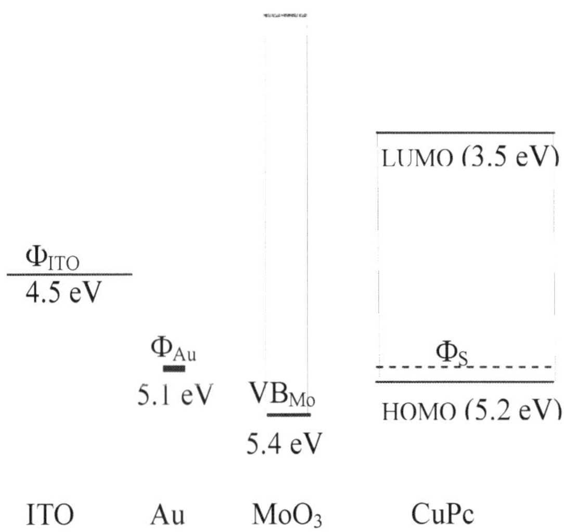

Fig. 23. Band scheme before contact

If the buffer layer material exhibits a high work function value (Figure 23), i.e., a work function in good agreement with the HOMO of the organic layer, it is often suggested that there is a simple alignment of the energetic levels of the anode and the organic material, without any discussion of the electrical properties of the buffer layer. However the electrical behavior of conductive, semi conducting and insulating materials is very different and therefore the nature of their interfaces will depend on the their properties.

If we introduce an insulating film, (I), between the metal, (M) (or degenerated semiconductor), and the semiconductor, (S), the contact behaves like a MIS structure. This means that there is not band alignment between the TCO, the insulating layer and the organic semiconductor but, qualitatively, the band diagram of Figure 24. This results from the fact that the Fermi level must be constant throughout the metal/insulating layer/semiconductor and therefore [Demtsu et Sites, *Thin Solid Films,* 2006]:

$$\Phi_{b°}=\Phi_b + \Delta \tag{15}$$

The distribution between Φ_b and Δ depends on the insulating properties of the oxide and the semiconductor properties. The space charge, which forms in the depletion layer of the semiconductor, can be expressed as an equivalent surface state density Qsc. In the ideal case, that is to say in the absence of any space charge effect in the interfacial insulating layer, an exactly equal and opposite charge Qm develops on the metal surface. In this case of an ideal insulating layer the potential Δ across the interfacial layer can be obtained, following the application of Gauss law to the surface charge on the metal and semiconductor:

$$\Delta = \delta(Qm/\varepsilon_i) \tag{16}$$

ε_i being the dielectric constant of the insulating layer and δ its thickness [Sze, Physics of Semiconductor Devices, John Wiley Editor, 1981]. Therefore, even if there is not a simple alignment of the work function of the insulating film with those of the conductive anode and of the semiconductor, the presence of this layer can significantly modifies the barrier value at the interface through the potential Δ [Park, Noh and Lee, *Appl. Phys. Lett,* 2006]. As a summary, in the case of the present work, since we have (Figure 23): Φ_{ITO} = 4.5 eV, Φ_{Au} = 5.1 eV, $HOMO_{CuPc}$ = 5.2 eV, VB_{MoO3} = 5.4 eV.

- the ITO/CuPc contact induces a pseudo Schottky contact, $V_{bi} = \phi_M - \phi_S$ between 0.5 and 0.7 eV, CuPc being an electron donor (p-type semiconductor), ϕ_S is slightly smaller than $HOMO_{CuPc}$ (Figure 20 a), while the Au/CuPc, and therefore ITO/Au/CuPc, contacts induce a nearly ohmic contact (Figure 20 b).
- the ITO/MoO_3/CuPc behaves like a MIS structure, that is to say the insulating layer induces a decrease of the barrier height with $\Phi_b = \Phi_{b°} - \Delta$ [Cowley and Sze, *J. Appl. Phys.,* 1965]. As discussed above the efficiency of the insulating layer in the band alignment depends on the insulating properties and the thickness of the MoO_3 layer (Figure 24).

We have now all the keys necessary to explain the behaviour of the different cells studied in the present work. It is well known that a barrier is present at the interface ITO/organic material (electron donor in the case of solar cells and hole transporting layer in the case of OLEDs). The introduction of an interfacial layer between the ITO and the organic material allows to decrease the barrier height at the interface, which facilitate the hole collection (solar cells) or hole injection (OLEDs). It is clear from the discussion above that an oxide or a metal with sufficiently high work function value should be used.

MoO_3 has already been used to increase the hole injection in OLEDs [Matsushima et al., *Appl. Phys. Lett.,* 2007; Jiang et al., *J. Phys. D: Appl. Phys.,* 2007]. The decrease in the barrier height, allowed by the introduction of a MoO_3 layer, increases with its thickness (equation 16). However, as shown by the experimental study, there is an optimum thickness value.

When the film is too thin (≤ 1.5 nm) the Δ value is small and moreover the covering efficiency of the ITO by the MoO_3 is not complete. For MoO_3 layers thicker than 6 nm, the short circuit current significantly decreases (Figure 22).
In Figure 24 it is supposed that the carriers cross the insulating layer by tunnel effect. The interfacial buffer layer is assumed to be a few angstroms thick and transparent to carriers whose energy is greater than the potential barrier of the semiconductor. However, when the thickness of the insulating layer increases, the probability of tunnel effect decreases. It is admitted that up to 3 nm the efficiency of the tunnel effect is maximum, while it decreases progressively for thicker films.
We have shown that the optimum MoO_3 thickness is around 3.5± 1 nm, which means that the insulating layer is efficient even for thickness higher than the theoretical optimum value. However evaporated molybdenum oxide films are strongly oxygen deficient and Rozzi et al have shown, in a theoretical study, that, when some oxygen is removed from the MoO_3 crystal, some Mo4d antibonding orbitals located in the gap are filled up [Rozzi et al., *Phys. Rev.*, 2003]. Therefore holes can cross the insulating film by multiple tunnelling steps [A.G. Milnes, D.L. Feucht, "Heterojunctions and metal semi-conductor junctions.", Academic Press Editor, 1972.] effect through these gap states introduced by the oxygen vacancies. As a result, the probability of tunnelling though the barrier increases. Finally the optimum MoO_3 thin film thickness, 3.5± 1 nm, corresponds to a compromise between an optimum ITO coverage and a sufficient transparency to charge carriers.

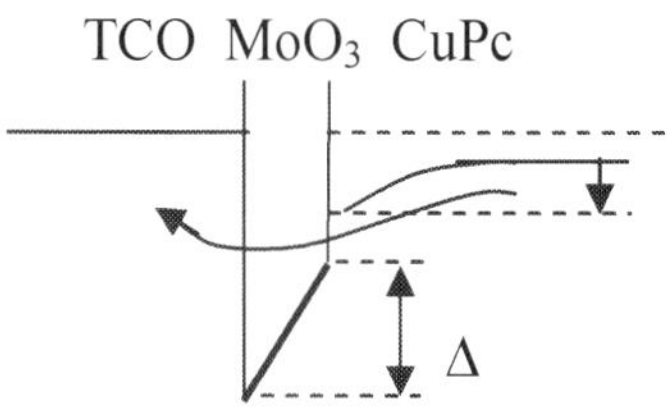

Fig. 24. Band scheme after contact: Hole transfer from the organic electron donor to the anode for different anode configurations after contact.

8. Example of the effect of an organic buffer layer used in different configurations

Factors which limit organic solar cells performance include limited spectral sensitivity, carriers separation at the interface acceptor/donor, low carrier mobility, energy step at the organic material/electrode. In devices based on ED/EA heterojunction, the theoretically obtainable open circuit potential (Voc) is given by the difference between the highest occupied molecular orbital of the electron donor ($HOMO_D$) and the lowest unoccupied molecular orbital of the electron acceptor ($LUMO_A$): $\Delta(HOMO_D\text{-}LUMO_A)$ [Zimmermann et al., *Thin Solid Films* 2005]. The introduction, at the electron donor /electron acceptor interface, of an ultra-thin organic layer could allows controlling the Voc value by modifying this difference as shown by Kinoshita and col. [Kinoshita et al., Appl. Phys. Lett., 2007]. They use with success this multicharge separation interface concept to increase the Voc value. They introduce an ultra-thin CuPc layer, which is usually used as electron donor, at the pentacene/fullerene interface. CuPc has also been proved to be an efficient buffer layer [Hong, Huang, and Zeng, *Chem. Phys. Lett.*, 2006]. In order to improve carrier mobility, structural templating of CuPc has been obtained using an ultra-thin 3,4,9,10-perylenetetracarboxyl dianhydride interlayer [Sullivan, Jones and Ferguson, *Appl. Phys. Lett.*, 2007]. In the present example, similar experiments have been done with a perylene derivative:N,N'-diheptyl-3,4,9,10-perylenebiscarboximide (PTCDI-C7).

8.1 Synthesis of the N,N'–diheptyl-3,4,9,10-perylenebiscarboximide (PTCDI-C7)

PTCDI-C7 (Figure 25) was synthesized by condensation of PTCDA with heptylamine [Demmig and Langhals, *Chem. Ber.*, 1988]. This compound is sufficiently soluble in chloroform to allow a chromatographic method for purification. Yield: 85 %. The elemental microanalyses results are in good agreement with the expected ones (weight %: C = 77.76; H = 6.74; N = 5.11).

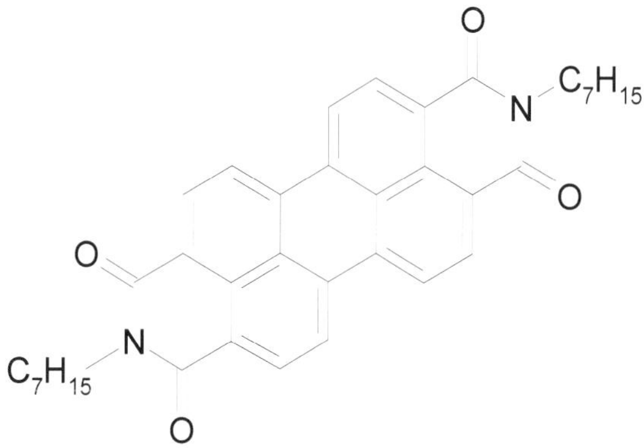

Fig. 25. PTCDI-C7 molecule.

8.2 Organic solar cells realisation

In the present series of experiments, the TCO electrode was a layer of indium tin oxide (ITO) on glass substrate. CuPc, PTCDI-C7, C_{60} and Alq_3 have been deposited in a vacuum of 10^{-4} Pa. The thin film deposition rate and thickness were estimated *in situ* with a quartz monitor. The deposition rate and final thickness were 0.05 nm/s, 35 nm for CuPc, 0.05 nm/s and 40 nm in the case of PTCDI-C7 and or C_{60} and 0.05 nm/s and 8 nm for Alq_3. The thicknesses have been chosen after optimisation.

We have shown earlier that the presence of an ultra-thin metallic film at the TCO/organic material interface allows improving significantly the power conversion efficiency of organic solar cells [Bernède et al., *Appl. Phys. Lett.*, 2008]. So, a very thin Au film (0.5 nm) has been introduced at the interface ITO/CuPc in the most cases. The thickness of the organic films has been checked with a scanning electron microscope (SEM). After organic thin film deposition, aluminium upper electrodes were thermally evaporated (150 nm thick), without breaking the vacuum, through a mask with 1.5 mm x 6 mm active area.

8.3 Experimental tests and discussion of the effect of PTCDI-C7 buffer layers

As justified above, PTCDI-C7 has been probed as a buffer layer at the CuPc/C_{60} and anode/electron donor interfaces. Following the multicharge separation interfaces concept, in order to improve Voc, an ultra-thin PTCDI-C7 has been introduced at the heterojunction CuPc/C_{60} interface. The PTCDI-C7 thickness was 4 nm. However, while there was a strong decrease of the short circuit current Jsc, there was not any Voc increase (Figure 26). It has already been shown [Zhou et al., *Sol. Energy Mater. Sol. Cells*, 2007] that, in the case of bulkheterojunctions, charge separation yield achieved with perylene derivatives is smaller than the one achieved with C_{60}. The same effect can explain the strong Jsc decrease in the presence of PTCDI-C7 at the CuPc/C_{60} interface.

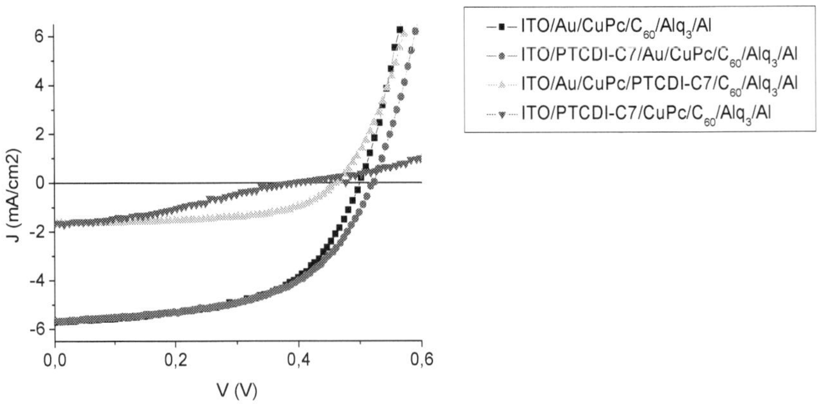

Fig. 26. Typical J-V characteristics under illumination of AM1.5 solar simulation (100 mW/cm^2) with different anodes.

Then PTCDI-C7 has been probed as buffer layer at the anode/electron donor interface. A thin PTCDI-C7 ultra-thin layer has been introduced at the anode/CuPc interface. The results are summarized in Figure 26 and table 5. It can be seen that, if, as expected, the presence of Au improves significantly the devices performances, it is not the case when PTCDI-C7 is substituted to Au. However, when a PTCDI-C7 ultra-thin film (0.2nm) is deposited onto ITO and then covered with Au (0.5 nm) there is a significant increase of Voc. Such increase is reproducible. A systematic study of the thickness of this buffer layer has shown that 0.2 nm is the optimum value, for thicker PTCDI-C7 film there is a fast decrease of Jsc, FF and therefore of the efficiency of the cells (Table 5).

Au thickness (nm)	0.5	0.5	0.5	0.5	0.5	0	0
PTCDI-C7 thickness (nm)	0	1	2	3	4	2	0
Voc (V)	0.50	0.50	0.52	0.48	0.47	0.1	0.44
Jsc (mA/cm^2)	5.63	5.62	5.64	3.75	2.88	0.35	3.21
FF %	57	56	56	34	26	22	0.24
η %	1.60	1.58	1.63	0.61	0.35	0.008	0.33

Table 5. Photovoltaic characteristics, under light AM1.5 (100 mWcm^{-2}), for different cells using PTCDI-C7 as buffer layer deposited onto the ITO

Sullivan and col. [Sullivan, Jones and Ferguson, *Appl. Phys. Lett.*, 2007] have shown that, in the case of PTCDA buffer layer, there is some structural templating of CuPc which results in a higher carrier mobility and therefore Jsc increase. It can be observed in Figure 27, that the CuPc film grown on PTCDI-C7 exhibits a slightly higher ordering than those grown onto ITO or ITO/Au. The tendency of CuPc to grow with higher degree of order is expected to improve the carrier mobility in the films.

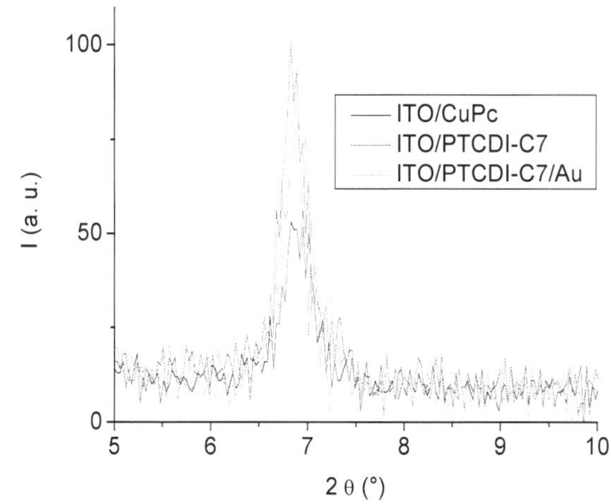

Fig. 27. Difractograms of CuPc thin films, 35 nm thick, deposited onto different cathodes

However, in our study, the more significant effect of the PTCDI-C7 buffer layer is not an increase of Jsc but of Voc. Probably, since the presence of the PTCDI-C7 buffer layer induces higher degree of order of CuPC films it should avoid the presence of pinholes and other defects growth, which increases the shunt resistance value of the solar cells. We have already shown that such diode improvement allows increasing the Voc value [Karst and Bernède, *Phys. Stat Sol. (a)*, 2006]. The study of the PTCDI-C7 films has shown that it has small conductivity. When the film thickness is only 2 nm, the carriers can cross the buffer layer by tunnel effect and the photocurrent is not decreased, for thicker buffer layer, the tunnel effect efficiency decreases and does the current.

In conclusion, the example of PTCDI-C7 developed above is a good illustration of the very broad possibility of original buffer layers. Here, a first gold buffer layer allows, as shown above to improve significantly the solar cells performances mainly through a decrease of the barrier height at the interface organic donor/TCO anode. Besides, the gold buffer layer increases the shunt resistance by decreasing the density of pinholes present in the CuPc films. We have shown here that this last effect could be improved by inserting a thin PTCDI-C7 at the interface just onto the TCO. This layer induces structural templating of CuPc as shown by the XRD study. Therefore, while PTCDI-C7 is clearly not an ideal surface modification of the anode in term of energy level alignment for hole collection (cf 4.3), the effect of shunt resistance increase allows improving the open circuit voltage.

9. Conclusion, toward the future.

In the last 22 years that have elapsed since the pioneering work of Tang [Tang, *Appl. Phys. Lett.,* 1986], significant improvement in the fundamental understanding and cells construction have led to efficiencies higher than 6%. The new concept of polymer:fullerene BHJ solar cells has allowed dramatic improvements in devices efficiency. It has induced a healthy competition with the multi-heterojunction devices base on small organic molecules, which induces significant progress in both cells families. A common aspect of both families is the great difficulty to control the electrode/organic material interface. This aspect is largely developed in the present work. One key issue for organic optoelectronic is the understanding of the energy-level alignment at organic material/electrode interfaces and the knowledge of the influence of the bottom electrode on the structural properties of the organic films. Even if not easy to predict, since dipole effect related to different possible origin are often present, the behaviour of an electrode/organic material contact can be roughly estimated using quite classical theory such as the Schottky-Mott model. Therefore the knowledge of the HOMO and LUMO values of the organic materials and that of the work functions of the electrodes are very helpful in the choice of new materials and structures configuration.

The challenges for the next organic solar cells improvements focus not only on new efficient absorbing organic materials, but developing new buffer layers, new cells architecture, new transparent conductive electrodes….

In the case of BHJ devices, the use of a TiO_x buffer layer inserted between the blend and the cathode gives considerable enhancement in the photocurrent attributed to the optical spacer effect of the TiO_x film [Kim et al., *Adv. Mater.,* 2006]. In the same order of idea, as discussed above, the introduction of an ultra-thin gold film (0.5 nm) (or a 3 nm thick MoO_3 film) at the anode/electron donor interface, allows strong solar cells improvement, whatever the transparent conductive oxide used, which demonstrates that well fitted buffer layer could allow in the new future to grow organic solar cells indium free.

The extent to which subtle interfacial buffer layers modification can change devices performance indicates that much room exists for improvement of organic solar cells.

Nowadays, two manufacturing techniques are mainly used, vacuum processing for the small organic molecules and wet processing for the polymers. In a first approach, wet processing, if transposable to simple ink jet process, has the advantage of low cost technique, however vacuum processing presents the advantages of easy fabrication of high quality thin films from highly purified materials, well controlled films thickness and therefore simplicity to construct complex multilayers structures. Even if, in a first approach, vacuum process appears more costly, its reproducibility can made it competitive for large scale massive production.

A cost factor of both cell families is oxygen and water protection. To day packaging is obtained using additive multilayer structures [Kim et al. *Appl. Phys. Lett.* (2009)], However, packaging techniques must be compatible with the substrate, which could be flexible, the active material and the final cost of the panel. One possible opportunity could be to use buffer layer which gives devices not only with high efficiency but also with long life structures.

The future of organic solar cells is bounded to the concept of "electronics everywhere", where light and flexible devices such as organic solar cells could play a major role in combination with thin films batteries…

10. References

T.D. Anthopoulos, B. Singh, N. Marjanovic, N.S. Sariciftci (2006). High performance n-channel organic field-effect transistors and ring oscillators based on C_{60} fullerene films, *Appl. Phys. Lett.*, 89, 213504.

N.R. Armstrong, C. Carter, C. Donley, A. Simmonds, P. Lee, M. Brumbach, B. Kippelen, B. Domercq, S. Yoo, (2003). Interface modification of ITO thin films: organic photovoltaic cells, *Thin Solid Films*, 445, 342-352.

M.A. Baldo, S.R. Forrest, (2001). Interface-limited injection in amorphous organic semiconductors, *Phys. Rev.* B 64, 085201.

A. J. Bard, L. R. Faulkner (1984). Electrochemical Methods. Fundamentals and Applications, Wiley, New York, p 634.

J.C Bernède, L. Cattin, M. Morsli, Y. Berredjem (2008). Ultra thin metal layer passivation of the transparent conductive anode in organic solar cells. *Solar Energy Materials and Solar Cells*, 92, 1508-1515.

J. C. Bernède, V. Jousseaume, M. A. Del Valle, F. R. Diaz (2001), Current Trends in polymer Sciences, From the organic electroluminescent diodes to the new organic photovoltaic cells, Vol 6, pp.135-155.

J.C. Bernede, Y. Berredjem, L. Cattin, M. Morsli (2008). Improvement of organic solar cells performances using a zinc oxide anode coated by an ultra thin metallic layer. *Appl. Phys. Lett.*, 92, 083304.

Y. Berredjem, J.C. Bernede, S. Ouro Djobo, L. Cattin, M. Morsli, A. Boulmokh (2008). On the improvement of the efficiency of organic photovoltaic cells by the presence of an ultra-thin metal layer at the interface organic/ITO. *The European Physical Journal: Applied Physics*, 44, 223-228.

Y. Berredjem, N. Karst, A. Boulmokh, A. Drici, J.C. Bernede (2007) Optimisation of the interface "organic material/aluminium" of CuPc/C_{60}based photovoltaic cells, the *European Physical Journal: Applied Physics*, 40, 163-167.

Y. Berredjem, N. Karst, L. Cattin, A. Lakhdar-Toumi, A. Godoy, G. Soto, F. Diaz, M.A. del Valle, M. Morsli, A. Drici, A. Boulmkh, A.H. Gheid, A. Khelil, J.C. Bernède (2008). Plastic photovoltaic cells encapsulation, effect on the open circuit voltage. *Dyes and Pigments*, 78, 148-156.

J.C. Blakesley and N.C. Greenham (2009). Charge transfert at polymer-electrode interfaces: The effect of energetic disorder and thermal injection on band bending and open-circuit voltage disorder and thermal injection on band bending and open-circuit voltage, *J. Appl. Phys.*, 106, 034507.

C.J. Brabec., J. E. Shaheen, C. Winder, N.S. Sariciftci, P. Denk (2002). Effect of LiFÕmetal electrodes on the performance of plastic solar cells, *Appl. Phys. Lett.*, 80, 1.

S.S. Braun, W.R. Salaneck, M. Fahlman (2009). Energy-level alignment at organic/metal and organic/organic interfaces, *Adv. Mater.*, 21, 1450.

F. Brovelli, B.L. Rivas, J.C. Bernède, M. A. Del Valle, F. R. Diaz (2007). Electrochemical and optical studies of 1,4-diaminoanthraquine for solar cells applications, *Polym. Bull.*, 58, 521-527.

E.L. Bruner, N. Koch, A.R. Span, S.L. Bernasek, A. Kahn, J. Schwartz (2002). Controlling the Work Function of Indium Tin Oxide: Differentiating Dipolar from Local Surface Effects, *J. Am. Chem. Soc.*, 124, 3192-3193.

R. Cervini, X. C. Li, G. W. C. Spences, A. B. Holmes, S. C. Moratti, R. H. Friend. (1997). Electrochemical and optical studies of PPV derivatives and poly(aromatic oxadiazoles), *Synthetic Metals*, 84, 359-360.

M.Y. Chan, S.L. Lai, M.K. Fung, C.S. Lee, S.T. Lee, (2007). Doping-induced efficiency enhancement in organic photovoltaic devices, *Appl. Phys. Lett.*, 90, 023504.

J.A. Chaney, S.E. Koh, C.S. Dulcey, P.E. Pehrsson (2003). Surface chemistry of carbon removal from indium tin oxide by base and plasma treatment, with implications on hydroxyl termination, *Appl. Surf. Sci.*, 218, 259-267.

C-H. Chang, S.-A. Chen (2007). Effect of ionization potential change in poly(3,4-ethylenedioxythiophene):poly(styrenesulfonic acid) on the performance of polymer light emitting diodes due to its reaction with indium tin oxide, *Appl. Phys. Lett.*, 91, 103514.

F.-C. Chen, J.-L. Wu, C.-L. Lee, Y. Hong, C.-H. Kuo, M.H. Huang, (2009). Plasmonic-enhanced polymer photovoltaic devices incorporating solution-processable metal nanoparticles, Appl. Phys. Lett., 95, 013305.

L.L. Chen, W.L. Li, H.Z. Wei, B. Chu, B. Li (2006). Organic ultraviolet photovoltaic diodes based on copper phthalocyanine as an electron acceptor, *Sol. Energy Material*, 90, 1788-1796.

A.M. Cowley, S. M. Sze (1965). Surface States and Barrier Height of Metal-Semiconductor Systems, *J. Appl. Phys.*, 36, 3212.

X. Crispin. (2004). Interface dipole at organic/metal interfaces and organic solar cells, *Solar Energy Materials & Solar Cells* 83, 147-168.

S. Demmig, H. Langhals (1988). Leichtlösliche, lichtechte Perylen-Fluoreszenzfarbstoffe, *Chem. Ber.*, 121, 225-230.

S.H. Demtsu, J.R. Sites (2006). Effect of back-contact barrier on thin-film CdTe solar cells, *Thin Solid Films*, 510, 320-324.

C. Donley, D. Dunphy, D. Paine, C. Carter, K. Nebesny, P. Lee, D. Alloway, N.R. Amstrong (2002). Characterization of Indium–Tin Oxide Interfaces Using X-ray Photoelectron Spectroscopy and Redox Processes of a Chemisorbed Probe Molecule: Effect of Surface Pretreatment Conditions, Langmuir 18, 450-457.

J. Drechsel, B. Männig, D. Gebeyehu, M. Pfeiffer, K. Leo, H. Hoppe (2004). MIP-type organic solar cells incorporating phthalocyanine/fullerene mixed layers and doped wide-gap transport layers, *Organic electronics* 5 175-186.

D. Duche, P. Torchio, L. Escubas, F. Monestier, J.-J. Simon, F. Flory, G. Mathan, (2009). Improving light absorption in organic solar cells by plasmonic contribution, *Sol. Energy Mater. Sol. Cells*, 93, 1377-1382.

G. East, M. A. del Valle (2000). Easy to make Ag/AgCl reference electrode, *J. Chem. Educ.* 77(1) 97.

M. Fahlman, A. Crispin, X. Crispin, S.K.M. Henze, M.P. de Jong, W. Osikowicz, C. Tengstedt, W.R. Salaneck (2007). Electronic structure of hybrid interfaces for polymer-based electronics, *J. Phys.: Condens. Matter*, 19 183202.

Z.R. Hong, Z.H. Huang, X.T. Zeng, (2006). Investigation into effects of electron transporting materials on organic solar cells with copper phthalocyanine/C60 heterojunctions, *Chem. Phys. Lett.*, 425, 62-65.

H. Hoppe, N.S. Sariciftci (2004). Organic solar cells: An overview, *J. Mater. Res.*, 7, 1924-1945.

C.M. Hsu, W.-T. Wu, (2004). Improved characteristics of organic light-emitting devices by surface modification of nickel-doped indium tin oxide anode, *Appl. Phys. Lett.*, 85, 840.

J. Huang, J. Yu, H. Lin, Y. Jiang (2009). Detailed analysis of bathocuproine layer for organic solar cells based on copper phthalocyanine and C60, *J. Appl. Phys.*, 105, 073105.

H.C. Im, D.C. Choo, T.W. Kim, J.H. Kim, J.H. Seo, Y.K. Seo, Y.K. Kim (2007). Highly efficient organic light-emitting diodes fabricated utilizing nickel-oxide buffer layers between the anodes and the hole transport layers, *Thin Solid Films*, 515, 5099-5101.

H. Ishii, N. Hayashi, E. Ito, Y. Washizu, L. Sugi, Y. Kimura, M. Niwano, Y. Ouchi, K. Seki (2004). Kelvin probe study of band bending at organic semiconductor/metal interfaces: examination of Fermi level alignment, *Phys. Stat. Sol* (a) 201 1075_1094.

C. Jain, M. Willander, V. Kumar, (2007). Conducting Organic Materials and Devices, High-Detectivity Polymer Photodetectors with Spectral Response from 300 nm to 1450 nm *Semiconductors and Semimetals*, 81, 1-188.

X-Y. Jiang, Z.-L. Zhang, J. Cao, M.A. Khan, K.-ul-Haq, W.-Q. Zhu (2007). White OLED with high stability and low driving voltage based on a novel buffer layer MoOx, *J. Phys. D: Appl. Phys.*, 40, 5553-5557.

B. Johnev, M. Vogel, K. Fostoropulos, B. Mertesacker, M. Rusu, M.-C. Lux-Steiner, A. Weidinger (2005). Monolayer passivation of the transparent electrode in organic solar cells, Thins Solid Films, 488, 270-273.

B. Kang, L.W. Tan, S.R.P. Silva (2009). Ultraviolet-illuminated fluoropolymer indium-tin-oxide buffer layers for improved power conversion in organic photovoltaics, *Organic Electronics* 10 1178-1181.

K.S. Kang, H.K. Lim, K.J. Han, J. Kim (2008). Durability of PEDOT : PSS-pentacene Schottky diode, *J. Phys. D: Appl. Phys.*, 41, 012003.

N. Karst, J.C. Bernède (2006). On the improvement of the open circuit voltage of plastic solar cells by the presence of a thin aluminium oxide layer at the interface organic aluminium. *Physica Status Solidi (a)*, 203 R70-R72.

M.A. Katkova, V.A. Ilichev, A.N. Konev, M.A. Batenkin, L.L. Pestova, A.G. Vitukhnovsky, Bochkarev M.N. (2008). Modification of anode surface in organic light-emitting diodes by chalcogenes, *Appl. Surf. Sciences* 254 2216-2219.

S. Kato (2005). Designing Interfaces That Function to Facilitate Charge Injection in Organic Light-Emitting Diodes, J. Am. Chem. Soc., 127 11538-11539.

J.S. Kim, R.H. Friend, F. Cacialli (1999). Surface energy and polarity of treated indium-tin-oxide anodes for polymer light-emitting diodes studied by contact-angle measurements, *J. Appl. Phys.* 86, 2774.

J.Y. Kim, K. Lee, N.E. Coates, D. Moses, T.Q. Nguyen, M. Dante, A.J. Heeger, (2007) Efficient Tandem Polymer Solar Cells Fabricated by All-Solution Processing, *Science*, 317, 222

J.Y. Kim, S.H. Kim, H.-H. Lee, K. Lee, W. Lee, W. Ma, X. Gong, A.J. Heeger, (2006). New Architecture for High-Efficiency Polymer Photovoltaic Cells Using Solution-Based Titanium Oxide as an Optical Spacer, *Adv. Mater.*, 18 572-576.

N. Kim, W.J. Potscavage Jr., B. Domercq, B. Kippelen, S. Graham, (2009). A hybrid encapsulation method for organic electronics, *Appl. Phys. Lett.*, 94, 163308.

S.-S Kim, S.-I. Na, J. Jo, D.-Y. Kim, Y.-C. Nah, (2008). Plasmon enhanced perfomance of organic solar cells using electrodeposited Ag nanoparticles, *Appl. Phys. Lett.*, 93, 073307.

S.Y. Kim, K. H. Lee, B.D. Chin, J.-W. Yu (2009). Network structure organic photovoltaic devices prepared by electrochemical copolymerization, *Solar Energy Materials & Solar Cells* 93, 129-135.

Y. Kinoshita, T. Hasobe, H. Murata (2007). Control of open-circuit voltage in organic photovoltaic cells by inserting an ultrathin metal-phthalocyanine layer, Appl. Phys. Lett., 91, 083518.

Th. Kugler, W. R. Salaneck, H. Rast, A. B. Holmes (1999). Polymer band alignment at the interface with indium tin oxide: consequences for light emitting devices, *Chem. Phys. Lett.*, 310, 391-386.

B. Kouskoussa, M. Morsli. K. Benchouk, G. Louarn, L. Cattin, A. Khelil, J.C. Bernede, (2009). On the improvement of the anode/organic material interface in organic solar cells by the presence of an ultra-thin gold layer. *Physica Status Solidi (a)* 206, 311-315.

Th. Kugler, A. Johansson, I. Dalsegg, U. Gelius, W.R. Salanek (1997). Electronic and chemical structure of conjugated polymer surfaces and interfaces: applications in polymer-based light-emitting devices, *Synthetic Metals*, 91, 143-146.

Th. Kugler, W. R. Salaneck, H. Rast, A. B. Holmes (1999). Polymer band alignment at the interface with indium tin oxide: consequences for light emitting devices, *Chem. Phys. Lett.* 310, 391-396.

A. Latef, J. C. Bernède (1991). Study of the thin film interface Aluminium -Tellurium, *Phys. Stat. Sol. (a)*, 124, 243.

C.S. Lee, J.X. Tang, Y.C. Zhou, S.T. Lee, Interface dipole et metal-organic interfaces: contribution of metal induced interface states (2009). *Appl. Phys. Lett.* 94 113304.

C.N. Li, C.Y. Kwong, A.B. Djurisic, P.T. Lai, P.C. Chui, W.K. Chan, S.Y. Liu (2005). Improved performance of OLEDs with ITO surface treatments, *Thins Solid Films* 477 57-62.

Y. Li, Y. Cao, D. Wang, G. Yu, A. S. Heeger (1999). Electrochemical properties of luminescent polymers and polymer light-emitting electrochemical cells, *Synthetic Metals*, 99, 243-248.

S. G. Liu, G. Sui, R. A. Cormier, R. M. Leblanc, B. A. Gregg (2002). Self-Organizing Liquid Crystal Perylene Diimide Thin Films: Spectroscopy, Crystallinity, and Molecular Orientation, *J. Phys. Chem. B*, 106, 1307-1315.

Lord Kelvin (1898). *Philos. Mag.*, 46, 82.

H-T. Lu, M. Yokoyama (2004). Plasma preparation on indium-tin-oxide anode surface for organic light emitting diodes, *Journ. Crys. Growth*, 260, 186-190.

D. Lüthi, , M. Le Floch, B. Bereiter, T. Blunier, J.-M. Barnola, U. Siegenthaler, D. Raynaud, J. Jouzel, H. Fischer, K. Kawamura, and T.F. Stocker (2008). High-resolution Carbon dioxide concentration record 650,000–800,000 years before present, Nature. 453 379-382.

T. Matsushima, Y. Kinoshita, H. Murata (2007). Formation of Ohmic hole injection by inserting an ultrathin layer of molybdenum trioxide between indium tin oxide and organic hole-transporting layers, *Appl. Phys. Lett.*, 91, 253504.

A.G. Milnes, D.L. Feucht 1972. Heterojunctions and metal semi-conductor junctions. Academic Press Editor, New York,

A.J. Morfa,. K.L. Rowlen, T.H. Reilly III, M.J. Romero, J. van de Lagemaat, (2008). Plasmon-enhanced solar energy conversion in organic bulk heterojunction photovoltaics, *Appl. Phys. Lett.*, 92, 013504.

M. Niggemann, M. Riede, A. Gombert, K. Leo (2008). Light trapping in organic solar cells, *Phys. Stat. Sol. (a,)* 205, 2862-2874.

T. Nishi, K. Kanai, Y. Ouchi, M.R. Willis, K. Seki (2005). Evidence for the atmospheric *p*-type doping of titanyl phthalocyanine thin film by oxygen observed as the change of interfacial electronic structure, *Chem. Phys. Lett.*, 414, 479-482.

F. Nüesch, L.J. Rothberg, E.W. Forsythe, Q. Toan Le, Y. Gao, (1999). A photoelectron spectroscopy study on the indium tin oxide treatment by acids and bases, *Appl. Phys. Lett.*, 74, 880.

F. Padinger, R.S. Rittenberger, N.S. Sariciftci (2003). Effects of Postproduction Treatment on Plastic Solar Cells, *Adv. Funct. Mater.*, 13, 85-88.

S. Y. Park, Y. H. Noh, H. H. Lee (2006). Introduction of an interlayer between metal and semiconductor for organic thin-film transistors, *Appl. Phys. Lett.*, 88, 113503.

P. Peumans, V. Bulovic, S.R. Forrest (2000). Efficient photon harvesting at high optical intensities in ultrathin organic double-heterostructure photovoltaic diodes, *Appl. Phys. Lett.* 76, 2650.

P. Peumans, S.R. Forrest (2001). Very-high-efficiency double-heterostructure copper phthalocyanine/C_{60} photovoltaic cells, *Appl. Phys. Lett.* 79, 126-128

M. Pfeiffer, K. Leo, N. Karl (1996). Fermi level determination in organic thin films by the Kelvin probe method, *J. Appl. Phys.* 80, 6880,

K.L. Purvis, G. Lu, J. Schwartz, S.L. Bernasek (2000). Surface Characterization and Modification of Indium Tin Oxide in Ultrahigh Vacuum, J. Am. Chem. Soc., 122, 1808-1809.

C. Qiu, Z. Xie, H. Chen, M. Wong, H.S. Kwok (2003). Comparative study of metal or oxide capped indium-tin oxide anodes for organic light-emitting diodes, *J. Appl. Phys.*, 93, 3253.

B.P. Rand, D.P. Burk, S.R. Forrest (2007). Offset energies at organic semiconductor heterojunctions and their influence on the open-circuit voltage of thin-film solar cells, *Phys. Rev.* B 75, 115327,

B.P. Rand, J. Li, J. Xue, R.J. Holmes, M.E. Thompson, S.R. Forrest (2005). Organic double-heterostructure photovoltaic cells employing thick tris(acetylacto)ruthenium(III) exciton-blocking layers, *Adv. Mater.*, 17, 2714-2718.

C.A. Rozzi, F. Mangi, F. Parmigiani (2003). *Ab initio* Fermi surface and conduction-band calculations in oxygen-reduced MoO_3, *Phys. Rev.*, B 68, 075106.

N.S. Sariciftci, D. Braun, C. Zhang, V. I. Srdanov, A. S. Heeger, G. Stucky, F. Wuld (1993). Semiconducting polymer-buckminsterfullerene heterojunctions: Diodes, photodiodes, and photovoltaic cells, *Appl. Phys. Lett.*, 62, 585.

N. S. Sariciftci, L. Smilowitz, A. J. Heeger, and F. Wudl, (1992). Photoinduced Electron Transfer from a Conducting Polymer to Buckminsterfullerene, *Science*, 258, 5087, 1474-1476.

J.C. Scott, J.H. Kaufman, P.J. Brock, R. DiPietro, J. Salem, J.A. Goitia (1996). Degradation and failure of MEH-PPV light-emitting diodes, *J. Appl. Phys.*, 79, 2745.

K. Seki, E. Ito, H. Ishii (1997). Energy level alignment at organic/metal interfaces studied by UV photoemission, *Synthetic Metals*, 91, 137-142,

S.E. Shaheen, C.J. Brabec, N.S. Sariciftci, F. Padinger, T. Fromherz, J.C. Hummelen, (2001). 2.5% efficient organic plastic solar cells, *Appl. Phys. Lett.*, 78, 841-843.

A. Sharma, P. J. Hotchkiss, S.R. Marder, B. Kippelen, (2009). Tailoring the work function of indium tin oxide electrodes in electrophosphorescent organic light-emitting diodes, *J. Appl. Phys.*, 105, 084507.

S.W. Shi, D.G. Ma, J.B. Peng (2007). Effect of NaCl buffer layer on the performance of organic light-emitting devices (OLEDs), Eur. Phys. J. Appl. Phys., 40, 141-144.

V.P. Singh, B. Parthasarathy, R.S. Singh, A Aguilera, J. Antony, M. Payne (2006). Characterization of high-photovoltage CuPc-based solar cell structures *Solar Energy Materials and Solar Cells*, 90, 798-812.

V.P. Singh, R.S. Singh, B. Parthasarathy, A Aguilera, J. Antony, M. Payne (2005). Copper-phthalocyanine-based organic solar cells with high open-circuit voltage, *Appl. Phys. Lett.*, 86, 082106,

Q.L. Song, F.Y. Li, H. Yang, H.R. Wu, X.Z. Wang, W. Zhou, J.M. Zhao, X.M. Ding, C.H. Huang, X.Y. Hou (2005). Small-molecule organic solar cells with improved stability, Chem. Phys. Lett., 416, 42.

P. Sullivan, T.S. Jones, A.J. Ferguson (2007). Structural templating as a route to improved photovoltaic performance in copper phthalocyanine/fullerene (C_{60}) heterojunctions, *Appl. Phys. Lett.*, 91, 233114.

S.M. Sze 1981"Physics of Semiconductor Devices" 2nd Edition, John Wiley Editor, New York.

Y. Tanaka, K. Kanai, Y. Ouchi, K. Seki (2009). Role of interfacial dipole layer for energy-level alignment at organic/metal interfaces, *Organic Electronics*, 10, 990-993.

C.W. Tang (1986). Two-layer organic photovoltaic cell, *Appl. Phys. Lett.*, 48, 183.

C. Tengstedt, W. Osikowioz, W.R. Salaneck, I.D. Parker, C.-H. Hsu, M. Fahlman (2006). Fermi-level pinning at conjugated polymer interfaces, *Appl. Phys. Lett.*, 88, 053502.

B.C Thompson., J.M.J. Fréchet (2008). Polymer-Fullerene Composite Solar Cells, *Angew. Chem. Int. Ed.*, 47, 58-77.

J. Van de Lagemaat, T.M. Barnes, G. Rumbles, S. E. Shaheen, T.J. Coutts, C. Weeks, I. Levitsky, J. Peltola, P. Glatkowski (2006). Organic solar cells with carbon nanotubes replacing In_2O_3:Sn as the transparent electrode, *Appl. Phys. Lett.*, 88, 233503.

J. Xue, B.P. Rand, S. Uchida, S.R. Forrest, (2005). Mixed donor-acceptor molecular heterojunctions for photovoltaic applications. II. Device performance, *J. Appl. Phys.*, 98, 124903.

I. Yoo, M. Lee, C. Lee, D.-W. Kim, I.S. Moon, D.-H. Hwang (2005). The effect of a buffer layer on the photovoltaic properties of solar cells with P3OT:fullerene composites, Synthetic Metals, 153, 97-100.

H. You, Y. Dai, Z. Zhang, D. Ma (2007). Improved performances of organic light-emitting diodes with metal oxide as anode buffer, J. Appl. Phys., 101, 26105.

X-H. Zhang, B. Domercq, B. Kippelen (2007). High-performance and electrically stable C60 organic field-effect transistors, *Appl. Phys. Lett.* 91 092114.

D. Zhang, Y. Li, Guohui Zhang, Y. Gao, L. Duan, L. Wang, Y. Qiu Lithium cobalt oxide as electron injection material for high performance organic light-emitting diodes (2008). *Appl. Phys. Lett.*, 92, 073301.

Z.Y. Zhong, Y.D. Jiang (2006). Surface treatments of indium-tin oxide substrates for polymer electroluminescent devices, *Phys. Stat. Sol., (a)*, 203, 3882-3892.

Y. Zhou, Y. Wang, W. Wu, H. Wang, l. Han, W. Tian, H. Bässler (2007). Spectrally dependent photocurrent generation in aggregated MEH-PPV:PPDI donor–acceptor blends, *Sol. Energy Mat. Solar cells*, 91, 1842-1848.

B. Zimmermann, M. Glatthaar, M. Niggermann, M. Riede, A. Hinsch (2005). Electroabsorption studies of organic bulk-heterojunction solar cells, *Thin Solid Films*, 493, 170-174.

W.A. Zisman. (1932). A new method of measuring contact potential differences in metals, Rev. Sci. Instrum. 3, 367-370.

12

New Trends in Designing Parabolic trough Solar Concentrators and Heat Storage Concrete Systems in Solar Power Plants

Valentina A. Salomoni[1], Carmelo E. Majorana[1], Giuseppe M. Giannuzzi[2], Adio Miliozzi[2] and Daniele Nicolini[2]
[1]*University of Padua*
[2]*ENEA – Agency for New Technologies, Energy and Environment*
Italy

1. Introduction

Energy availability has always been an essential component of human civilization and the energetic consumption is directly linked to the produced wealth. In many depressed countries the level of solar radiation is considerably high and it could be the primary energy source under conditions that low cost, simple-to-be-used technologies are employed. Then, it is responsibility of the most advanced countries to develop new equipments to allow this progress for taking place. A large part of the energetic forecast, based on economic projection for the next decades, ensure us that fossil fuel supplies will be largely enough to cover the demand. The predicted and consistent increase in the energetic demand will be more and more covered by a larger use of fossil fuels, without great technology innovations. A series of worrying consequences are involved in the above scenario: important climatic changes are linked to strong CO_2 emissions; sustainable development is hindered by some problems linked to certainty of oil and natural gas supply; problems of global poverty are not solved but amplified by the unavoidable increase in fossil fuel prices caused by an increase in demand. These negative aspects can be avoided only if a really innovative and more acceptable technology will be available in the next decades at a suitable level to impress a substantial effect on the society. Solar energy is the ideal candidate to break this vicious circle between economic progress and consequent greenhouse effect. The low penetration on the market shown today by the existent renewable technologies, solar energy included, is explained by well-known reasons: the still high costs of the produced energy and the "discontinuity" of both solar and wind energies. These limitations must be removed in reasonable short times, with the support of innovative technologies, in view of such an urgent scenario.

On this purpose ENEA, on the basis of the Italian law n. 388/2000, has started an R&D program addressed to the development of CSP (*Concentrated Solar Power*) systems able to take advantage of solar energy as heat source at high temperature. One of the most relevant objectives of this research program (Rubbia, 2001) is the study of CSP systems operating in the field of medium temperatures (about 550°C), directed towards the development of a new and low-cost technology to concentrate the direct radiation and efficiently convert solar energy into high temperature heat; another aspect is focused on the production of hydrogen by means of thermo-chemical processes at temperatures above 800°C.

As well as cost reductions, the current innovative ENEA conception aims to introduce a set of innovations, concerning: i) *The parabolic-trough solar collector*: an innovative design to reduce production costs, installation and maintenance and to improve thermal efficiency is defined in collaboration with some Italian industries; ii) *The heat transfer fluid*: the synthetic hydrocarbon oil, which is flammable, expensive and unusable beyond 400°C, is substituted by a mixture of molten salts (sodium and potassium nitrate), widely used in the industrial field and chemically stable up to 600°C; iii) *The thermal storage* (TES): it allows for the storage of solar energy, which is then used when energy is not directly available from the sun (night and covered sky) (Pilkington, 2000). After some years of R&D activities, ENEA has built an experimental facility (defined within the Italian context as PCS, "*Prova Collettori Solari*") at the Research Centre of Casaccia in Rome (ENEA, 2003), which incorporates the main proposed innovative elements (Figure 1). The next step is to test these innovations at full scale by means of a demonstration plant, as envisioned by the "Archimede" ENEA/ENEL Project in Sicily. Such a project is designed to upgrade the ENEL thermo-electrical combined-cycle power plant by about 5 MW, using solar thermal energy from concentrating parabolic-trough collectors.

Fig. 1. PCS tool solar collectors at ENEA Centre (Casaccia, Rome).

Particularly, the Chapter will focus on points i) and iii) above:

- loads, actions, and more generally, the whole design procedure for steel components of parabolic-trough solar concentrators will be considered in agreement with the Limit State method, as well as a new approach will be critically and carefully proposed to use this method in designing and testing "special structures" such as the one considered here;
- concrete tanks durability under prolonged thermal loads and temperature variations will be estimated by means of an upgraded F.E. coupled model for heat and mass transport (plus mechanical balance). The presence of a surrounding soil volume will be additionally accounted for to evaluate environmental risk scenarios.

Specific technological innovations will be considered, such as:

- higher structural safety related to the reduced settlements coming from the chosen shape of the tank (a below-grade cone shape storage);
- employment of HPC containment structures and foundations characterized by lower costs with respect to stainless steel structures;
- substitution of highly expensive corrugated steel liners with plane liners taking advantage of the geometric compensation of thermal dilations due to the conical shape of the tank;
- possibility of employing freezing passive systems for the concrete basement made of HPC, able to sustain temperature levels higher than those for OPC;
- fewer problems when the tank is located on low-strength soils.

2. Description of parabolic-trough solar concentrators

The parabolic-trough solar concentrators are one of the basic elements of a concentrating solar power plant. The functional thermodynamic process of a solar plant is shown in (Herrmann et al., 2004). The main elements of the plant are: the solar field, the storage system, the steam generator and the auxiliary systems for starting and controlling the plant. The solar field is the heart of the plant; the solar radiation replaces the fuel in conventional plants and the solar concentrators absorb and concentrate it. The field is made up of several collector elements composed in series to create the single collector line. The collected thermal energy is determined by the total number of collector elements which are characterized by a reflecting parabolic section (the concentrator), collecting and continuously concentrating the direct solar radiation by means of a sun-tracking control system to a linear receiver located on the focus of the parabolas. A circulating fluid flows inside a linear receiver to transport the absorbed heat.

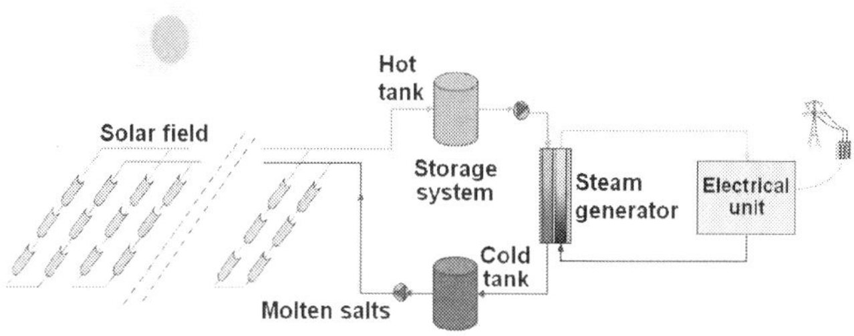

Fig. 2. Functional thermodynamic process flow of a solar plant.

A solar parabolic-trough collector line is divided into two parts from a central pylon supporting the hydraulic drive system (Antonaia et al., 2001). Each part is composed by an equal number of identical collector elements, connected mechanically in series. Each collector element consists of a support structure for the reflecting surfaces, the parabolic mirrors, the receiver line and the pylons connecting the whole system to a solid foundation by means of anchor bolts. The configuration of a solar parabolic-trough collector is that of a cylindrical-parabolic reflecting surface with a receiver tube co-axial with the focus-line, as a first approximation. The reflecting surface must be able to rotate around an axis parallel to the receiver tube, to constantly ensure that the incident radiation and the plane containing the parabolic sections' axes are parallel. In this way the incident solar light on the reflecting

surfaces is concentrated and continuously intercepted by the receiver tube in any assumed position of the sun during its apparent motion. The parabolic-trough collector is then constituted by a rotating "mobile part" to orientate the concentrator reflecting surfaces and by a "fixed part" guaranteeing support and connection to the ground of the mobile part.
The solar collector performances, in terms both of mechanical strength and optical precision, are related to one side to the structural stiffness and on the other to the applied load level. The main load for a solar collector is that coming from the wind action on the structure and it is applied as a pressure distributed on the collector surfaces.
From a structural point of view, it must be emphasized that the parabolic-trough concentrator is composed mainly by three systems: the *concentration*, the *torque* and the *support system*. Other fundamental elements, not treated in this document for sake of brevity, are the *foundation* and the *motion systems*. In Table 1 the subsystems and basic elements characterizing the structure of the concentrator developed by ENEA are shown. All elements should be considered when designing a parabolic-trough concentrator and verified for *"operational"* and *"survival"* load conditions. Corrosion risks and safe-life (about 25-30 years) must be taken into account as well.
The following basic operational conditions, listed in Table 2, can be considered valid for a parabolic-trough concentrator; they define different performance levels under wind conditions. *"Design conditions"* can be fixed consequently.
Finally, on the basis of what described above, the main requirements when designing a parabolic-trough concentrator can be summarized as follows:

- *Safety*: the collector structures exposed to static loads must guarantee adequate safety levels to ensure public protection, according (in our case) to the Italian Law 1086/71. This is translated into a suitable strength level or more generally in safety factors for the construction within the *Limit State Analysis*.
- *Optical performance*: the structure must guarantee a suitable stiffness in order to obtain, under operational conditions, limited displacements and rotations, the optical performance level being related to the capacity of the mirrors concentrating the reflected radiation on the receiver tube.
- *Mechanical functionality*: the structural adaptation to loads must not produce interference among mobile and fixed parts of the structure under certain load conditions.
- *Low cost*: the structure has to respond to typical economic requirements for solar plant fields (e.g. known from experiences abroad): unlimited plant costs lead to non-competitive sources employments. This can lead to tolerate fixed damage levels of the structure under extreme conditions (i.e. collapse of not-bearing elements, local yield, etc.), but still respecting the above mentioned requirements of public protection.

3. Codes of practice and rules

The parabolic-trough concentrator, on the basis of its structural shape and use and further considering available National and European recommendations, is classifiable as a *"special structure"* (Majorana & Salomoni, 2004 (a); Giannuzzi et al., 2007): it is not a machine or a standard construction. The definition "special" comes directly from a subdivision in classes and categories according to the criterion of the "Rates for professional services" as it results from the Italian law n. 143/1949; this law places "Metallic structures of special type, notable constructive importance and requiring ad-hoc calculations" into class IX e subclass b.

Systems	Subsystems	Elements
Concentration system	Reflecting surfaces	Mirrors, Mirror-structure connection
	Mirrors support structures	Girders, Girder-framed structure connection Framed structure, Framed structure-torque tube connection
Torque system	Torque tube, plate, hinge	Torque tube, Torque tube-plate connection, Plate, Plate-hinge connection, Hinge
Module supports	Intermediate / final pylons	Cylindrical pin joint, Pin joint-support connection, Framed structure, Plate, Anchor bolts
	Central pylon	Cylindrical pin joint, Pin joint-support connection, Framed structure, Engine support structure, Plate, Anchor bolts
Other	Foundations	Piles and/or plinths, Anchor bolts
	Drive system	Hydraulic drive/pistons, etc.

Table 1. Example of structural elements of a parabolic-trough concentrator.

Level	***Condition***
W1	Response under normal operational conditions with light winds. The concentration efficiency must be as high as possible under wind velocity less than a value v_1 characterizing this level.
W2	Response under normal operational conditions with medium winds. The concentration efficiency is gradually diminishing under wind velocity comprised between v_1 and v_2. The wind velocity v_2 characterizes this level.
W3	Transition between normal operating conditions and survival positions under medium-to-strong or strong winds. The survival must be ensured in any position under medium-strong winds. The drive must be able to take the collector to safe positions for any wind velocity comprised between v_2 and v_3. The wind velocity v_3 characterizes this level.
W4	Survival under strong winds in "rest" positions. The survival wind velocity must be adapted to the requests of the site according to recommendations. The wind velocity v_4 characterizes this level.

Table 2. Operational conditions.

From the functional analysis of the structure its special typology clearly emerges, according to its design, technical arrangements and innovation. When the parabolas are stopped in an assigned angular configuration, the nature of the structure can be determined: steel structure of mixed type founded on simple or reinforced concrete placed on a foundation soil having characteristics closely correlated to a chosen site, also under the seismic profile. From the structural point of view, the dynamic characteristics play a major role, with the response deeply influenced not only by the drive-induced oscillations, but also by dominant winds or seismic actions. Taking into account the above considerations, it is then possible to state that the examined structure is "*special*".

Moreover, such a structure requires appropriate calculations since some parts are mobile, even if with a slow rotation; at the same time the structure is subjected to wind actions, especially relevant due to the parabolas dimension. The simultaneous thermal and seismic actions, acting as self-equilibrated stresses in an externally hyperstatic structure, are equally important. Special steel made structures are e.g. cranes: they are designed using specific recommendations; in our case the reference to existing codes of practice is necessary, even if with the aim of adapting them and/or proposing new ones for CSP systems. Hence it clearly appears that such structures, built within the European countries, are currently designed and verified out of standards; the only two Italian recommendations acting as guidelines are:

- Law 5/11/71, n.1086, Norms to discipline the structures made by plain and pre-stressed reinforced concrete and by metallic materials.
- Law 2/2/74, n.64, Procedures devoted to structures with special prescriptions for seismic zones.

Moreover, several "technical norms" are related to the above ones, in form of "Minister (of Public Works) Decrees", or "explanation documents", or other documents giving rise to a certain amount of duplications and repetitions; however, a progressive compulsory use of Eurocodes is being introduced to push Italian engineers more properly into the European environment. In this case, Eurocodes 3 and 8 are of interest for the structural design of solar concentrators, also in view of their seismic performance. It is important to make an advanced choice regarding the body of recommendations to be followed in the design and checking phases and to proceed further with them, avoiding the common mistake of some designers to take parts from one norm (i.e. Italian) and mix it with parts of another norm (i.e. Eurocodes). The main problems in the so-called harmonization of rules within Europe reside in finding safety coefficients to be applied for considering special conditions (e.g. environmental) in each country, as well as those for materials. This is a source of difficulty in the creation of a unique body of rules valid in the whole European territory. The last product of recommendations recently emitted by the actual Ministry of Public Works in Italy is a 438 pages document (plus two Annexes) named "*Testo Unico per le Costruzioni*". It is compulsory in the Italian territory from July 1st 2009. The aim of this decree was also to unify a series of previous decrees into a single document. As already stated, it has been here chosen to follow the current Italian laws, and Eurocodes for comparison, in view of the possible application of solar concentrators at Priolo Gargallo (near Syracuse, Sicily). In principle, with a few changes, it is possible to apply the technology in other sites, as well as outside Italy or even Europe: slight changes in dimensioning could occur.

Hence, to take into account the specificity of the investigated structures, it was necessary to combine together *operational states* (OSs) (Table 2), *characteristic positions* and *load actions*, reaching to the interpretation of Table 3 within the context of a *limit state* (LS) analysis (Salomoni et al., 2006). Additionally, within the *serviceability limit states* (SLSs) the conditions of maximum rotation (W_1 operational state) and maximum deformation (W_2) must be verified; W_3 requires the collector operability within an elastic *ultimate limit state* (ULS), i.e. absence of permanent deformations. Differently, such deformations can be present within W_4 but without leading to a structural collapse.

Characteristic positions →								
Operational states			-120°	-30°	0°	30°	60°	75°
v_{ref} [m/sec]	↓	Limit state						
7	W_1	SLS		Y				
14	W_2	SLS					Y	
21	W_3	ULS	Y	Y	Y	Y	Y	Y
28	W_4	ULS	Y		Y			

Table 3. Example of combinations among characteristic positions, operational states and load actions to study CSPs in the context of LS analyses.

4. Materials

The solar concentrator supporting structure is made of hot-laminated steel. Hence, according to Eurocode 3 and UNI EN 10025, steels in form of bars, plates or tubes must be of the types shown in Table 4.

However recommendations allow for using different types of steel once the ensured safety level remains the same, justifying this through appropriate theoretical and experimental documentations. Under uniaxial stress states, their design strengths can be deduced from tables; in case of multiaxial states, suitable combinations are additionally given. In our calculations, the following material characteristics are considered: elastic modulus E = 210000 N/mm², Poisson's coefficient ν = 0.3, thermal expansion coefficient $\alpha = 12 \cdot 10^{-6}$ °C^{-1} and density ρ = 7850 kg/m³. If welding is used for connecting elements, the behaviour of steel types S235 and S275 is distinguished from that of S360.

Nominal steel type	thickness t [mm]			
	t ≤ 40		40 < t ≤ 100	
	f_y [N/mm²]	f_u [N/mm²]	f_y [N/mm²]	f_u [N/mm²]
Fe360 / S235 (EN 10025)	235	360	215	340
Fe430 / S275 (EN 10025)	275	430	255	410
Fe510 / S360 (EN 10025)	360	510	335	490

Table 4. Strengths and failure stresses (nominal values) for structural steels.

5. Loads

Given the design loads, subdivided in *permanent* and *variable* ones, wind conditions are here examined more in detail, whose effects on the structure are connected to the parabolas aerodynamics in their different characteristic positions (see below). The role of the snow has been additionally considered.

5.1 Variable loads

5.1.1 Wind action on the parabolas

The mean value of wind velocity, as a function of the distance from soil $V_m(z)$, is expressed by

$$V_m(z) = C_r(z) \cdot C_t(z) \cdot V_{ref} \tag{1}$$

where V_{ref} is the reference wind velocity, $C_r(z)$ the roughness coefficient and $C_t(z)$ the topographic coefficient.

The reference wind velocity V_{ref} is defined as the mean wind speed over a time period of 10 min, at 10 m height on a second category soil, with a 50 years "return period". The reference wind speeds for each Italian area is given by recommendations; e.g. a site located near the sea in Southern Italy has a reference wind speed of about 28 m/s. An important wind speed value is the peak wind speed which can be seen as the superposition of the mean wind speed plus its variation due to turbulence conditions on site. It can be evaluated as

$$V_{peak}(z) = G(z) \cdot V_m(z) \tag{2}$$

where G(z) is the "peak factor", that is,

$$G = \sqrt{1 + \frac{7}{C_t(z) \cdot \ln(z / z_0)}} \tag{3}$$

Usually G is comprised between 1.5 and 1.6. It should be emphasized that the check under failure loads must be necessarily performed on the basis of the peak velocity, since this gives an overload capable of making the material reach its strength limit, even if its duration is short. As far as the operational performance is concerned, it is more feasible to use the mean velocity. The roughness coefficient $C_r(z)$ takes into account the variability of the mean wind speed and the site characteristics by considering the height over the soil and the soil roughness as functions of the wind direction. The roughness coefficient at height z is defined by the logarithmic profile

$$C_r(z) = k_r \ln(z / z_0) \tag{4}$$

where k_r is the soil factor and z_0 is the roughness length, both related to the soil exposure category on its turn linked to the geographic location of the investigated area within Italy and on the basis of the soil roughness. In case of an open country, k_r is 0.19 and z_0 is 0.05 m. The topographic coefficient $C_t(z)$ takes into account the increment in the mean wind speed on escarpments and isolated hills; in our case $C_t = 1$ can be taken.

The solar concentrator shape is taken into account by means of aerodynamic coefficients. The different aerodynamic shape coefficients have been identified by means of a CFD analysis carried out in (Miliozzi et al., 2007). These coefficients have been determined starting from wind actions exerted on the linear parabolic collector as functions of its angular position (Figure 3). Such coefficients have been calculated for the most (external) and the least (internal) stressed collectors (Giannuzzi, 2007), see e.g. Figure 4. An external collector is one of those belonging to the first line without any artificial barrier against wind actions, whereas an internal collector is one on the sixth line, taken as representative of all the others.

Full tables for shape coefficients in case of "external" parabolas as well as "internal" ones are reported in (Majorana & Salomoni, 2005 (a)) and used in (Majorana & Salomoni, 2005 (b)) for structural assessment within the Limit State Design. Shape coefficients have been used to evaluate drag (C_{fx}), lift (C_{fy}), torsion (C_{Mz}) and mean pressure (C_{pm}), each of them being function of the concentrator rotation angle, where the allowed rotation is in the range +/- 120°. Then, shape coefficients for mean pressures have been calculated as functions of the aperture angle for "external" or "internal" parabolas. By analyzing the above coefficients it is possible to identify the parabolas' characteristic positions listed in Table 5.

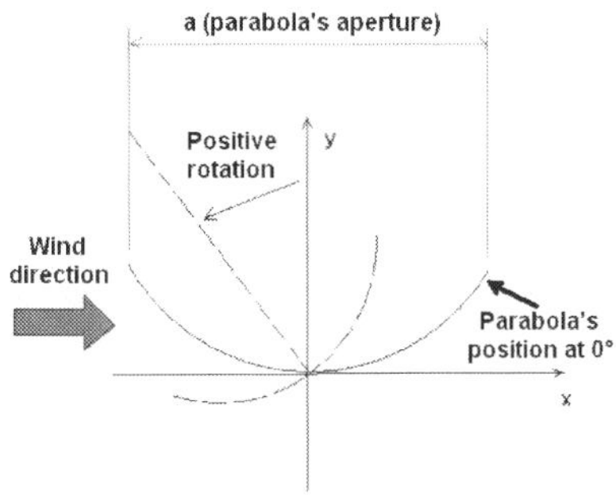

Fig. 3. Parabolic concentrator scheme at different angular positions.

Starting from the calculated shape coefficients, the corresponding effects referring to drag and lift force, torsion, mean pressure and pressure distribution have been determined.
By analyzing the results of the CFD analysis, it has been evidenced that aerodynamic coefficients and associated loads are largely reduced at the internal collectors. The main reason resides in the shielding effect produced by the first collectors' rows. This remark leads to the necessity of designing "strong" collectors along the external rows (Figure 4) and "light" collectors along the internal ones. Alternatively, it is possible to choose a different

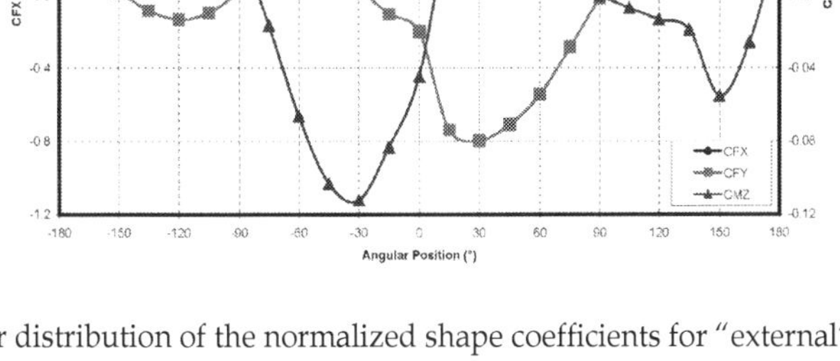

Fig. 4. Angular distribution of the normalized shape coefficients for "external" parabolas.

design strategy, based on the introduction of opportune windbreak barriers and on the realization of "light" collectors only. The position characterized by smaller loads is at 180°. This is only a theoretical, unattainable position because of the interferences between receivers and pylons. The safety position to be really taken in consideration is at about -120°. The waiting position (at 0°) does not guarantee an adequate level of protection for the mirrors. All the positions shown in Table 5 must be taken into account during the design phase but the most relevant position is, without doubt, the one associated to the maximum torque action. This is consequence of the fact that torque effects are accumulated along all the line, producing the maximum stresses on the structural elements close to the central pylon. This can be considered the key action in the parabolic-trough solar concentrators wind design.

5.1.2 Snow

The snow load is usually evaluated on the roofs (here parabolas), by means of the following expression

$$q_S = \mu_i \cdot q_{Sk} \tag{5}$$

where q_s is the snow load on the roof, μ_i the roof shape coefficient and q_{sk} the reference value of the snow load on the ground.

Characteristic Effect	**Angular position (degrees)**	
	"External" collector	**"Internal" collector**
Safety position	-120	-120
Waiting position	0	0
Maximum torque effect	-30	-15
Maximum bending action on the torque tube	+60	+30
Maximum drag force	+75	-45
Maximum lift force	+120	-45
Maximum crush force	+30	+30

Table 5. Wind effect: characteristic positions.

The load acts along the vertical direction and it is referred to the horizontal projection of the covering surface. The snow load on the ground depends on local environmental and exposure conditions, where the variability of the snowfall from region to region is taken into account. The reference snow load in locations at heights less than 1500 m over the mean sea level (m.s.l.) has to be evaluated on the basis of given expressions (whose values correspond to a "return period" of about 200 years). In case of a region like Sicily and a site located at a reference height less then 200 m m.s.l., q_{sk} is about 0.75 kN/m^2. The shape coefficients to be used for the snow load are those indicated in Table 6, being α (degrees) the angle between cover and the horizontal plane.

The shape coefficients μ_1, μ_2, μ_3, μ_{1^*} refer to roofs having one or more slopes, and they should be evaluated as functions of α, as indicated by the codes. For given parabolas positions, other coefficients can be used, as e.g. those related to cylindrical covers. In absence of rifting inhibiting snow sliding, for cylindrical covers of any shape and single curvature of constant sign, the worst uniform and not-symmetric load distribution is there considered.

	$0° <= \alpha <= 15°$	$15° < \alpha <= 30°$	$30° < \alpha <= 60°$	$\alpha > 60°$
μ_1	0.8	0.8	0.8(60-α)/30	0.0
μ_2	0.8	0.8+0.4(α-15)/30	(60- α)/30	0.0
μ_3	0.8+0.8α/30	0.8+0.8α/30	1.6	-
μ_1^*	0.8	0.8(60- α)/45		0

Table 6. Shape coefficient for the snow load (Eurocode1-Part 2.3).

In our case, to determine the shape coefficients μ_i for the parabolas, it is possible to approximatively evaluate the maximum slope of the parabolic collector with respect to the horizontal line, if it is rotated with its concavity upwards, being the element profile defined by means of the equation

$$y = x^2 / 4f \tag{6}$$

where -2950 < x < 2950 (mm), f = 1810 (mm); and the slope by

$$y' = x / 2f \tag{7}$$

with maximum value equal to 0.815, corresponding to an angle α such that tgα = 0.815, i.e. α ≈ 39°. On the other side, taking into account the value corresponding to x/2, then α = 22°. Hence, assuming α = 22° as a mean value, it is possible to calculate the shape coefficients as indicated in the recommendations. The corresponding load conditions are shown in Figure 5.

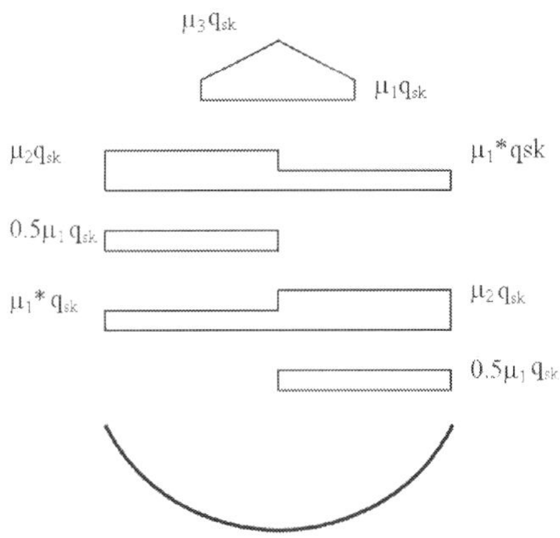

Fig. 5. Snow conditions for the parabolas when the solar collector is rotated to the waiting position.

As demonstrated in (Majorana & Salomoni, 2005 (b)), snow effects are fundamental when verifying the structure in the safety position (Table 5) or when seismic effects are included.

Discussing about the real significance of such an effect when considering desert locations (as those typical for CSP systems) is reasonable: this should be another example of the necessity for ad-hoc codes of practice when studying *special* structures in possibly *special* sites.
The parabola's configuration with its concavity upwards (Figure 5) is not the only possible one when evaluating the effects of the snow; being the snowy phenomenon largely predictable, so that a rotation of the collector towards the safety position is expected, an additional investigated angular position for analyzing snow effects refers to $\alpha = \pm 120°$. When e.g. $\alpha = +120°$, the situation is the one of Figure 6; the remaining characteristic positions, even associable to different OSs, can be considered as characterized by a null snow action: in fact, in case of snow, the collector would be evidently moved to its safety position with no tracking. Additionally, being L_1 (distance between the point, on the rotated parabola, with null tangent and the origin of the vertical axis) > L (= 2950 mm in our case), it is precautionarily assumed $L_1 = L$ and hence from Eurocodes $\mu_1 = 0.8$, $\mu_2 = 2.0$ ($\mu_3 = 1.0$), with loads as those of Figure 7.

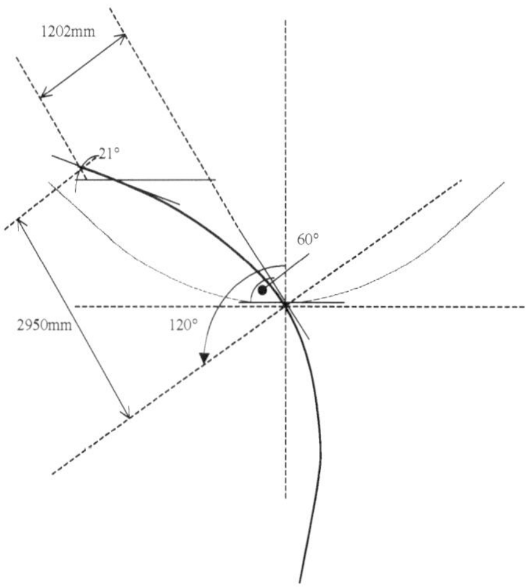

Fig. 6. Parabola's position for evaluating snow effects.

Two changes have been essentially introduced to what indicated by the codes: being, as already stated, $L_1 > L$, the point of null load amplified by μ_2 goes outside the effective parabola's projected dimension (consequently, the effect of μ_3 is zero; anyway, we are still in favour of safety being $\mu_2 > \mu_3$) and this explains the chosen trapezoidal shape for the load of Figure 7; the load cusp (from Eurocodes falling on the point whose slope on the curve is of 30°, i.e. at $L_1/4$), considered the not-symmetric parabolic profile, is moved with respect to $L_1/4$.
Then, when combining the loads, among the various indicated load conditions only those revealed as heaviest for the structural system have been adopted.
Hence, the main load combinations are reported in Table 7, where the multiplicative coefficients related to each basic action (permanent, G_k, and variable, Q_k) and to strength (f_y) are additionally indicated, for the OSs of Table 3.

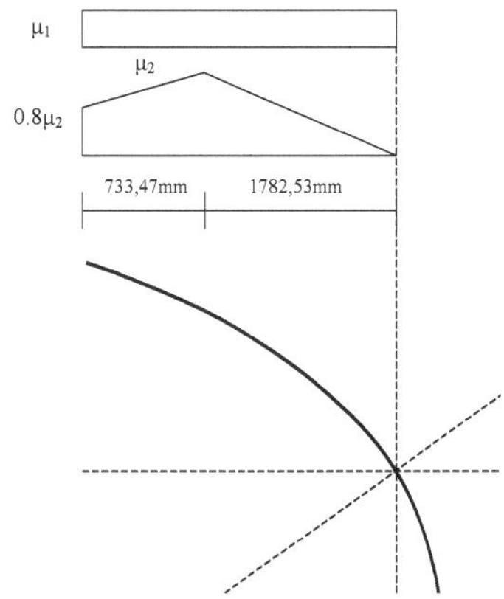

Fig. 7. Snow condition for the parabola when the solar collector is rotated to the safety position.

Combinations		G_k	Q_{1k} (wind)	Q_{2k} (snow)	f_y
$W_1 30$	$W_1 30R$	1.	1.	0.	1.
$W_2 60$	$W_2 60R$	1.	1.	0.	1.
$W_3\alpha$ -120° ≤ α ≤ 75° (Table 3)	$W_3\alpha E$	1.4*	1.5	0.	1.
$W_4\beta$; β = 0° (Table 3)	$W_{4}\beta E$	1.4*	1.5	0.	1.
$W_4\beta$; β = -120° (Table 3)	$W_{4}\beta E$	1.4*	1.5	0.	1.
		1.4*	1.5	1.05	1.
		1.4*	1.05	1.5	1.
		1.4*	0.	1.5	1.
	$W_{4}\beta P$	1.4*	1.5	1.05	0.83
		1.4*	1.05	1.5	0.83

Table 7. Main load combinations and corresponding multiplicative coefficients (*: if not acting in favour of safety; R: rare; E: elastic limit state; P: plastic collapse).

Particularly, for combinations related to states $W_1 30$ (W_1, 30°) and $W_2 60$, just *rare* ones are considered, being *frequent* and *quasi-permanent* combinations already included.
In the following, the main results related only to the concentration system are reported, being the conducted design and analysis methodology repeatable in the same way to the other macro-systems, i.e. the torque system and the module's support.

6. Analysis and verification of the concentration system

The concentration system is composed by three main elements: centering, stringers and reflecting mirrors (Figures 8 and 9). The system has been analysed considering a single modulus of 12 m, reproducing also the torque tube to which the centerings are linked.

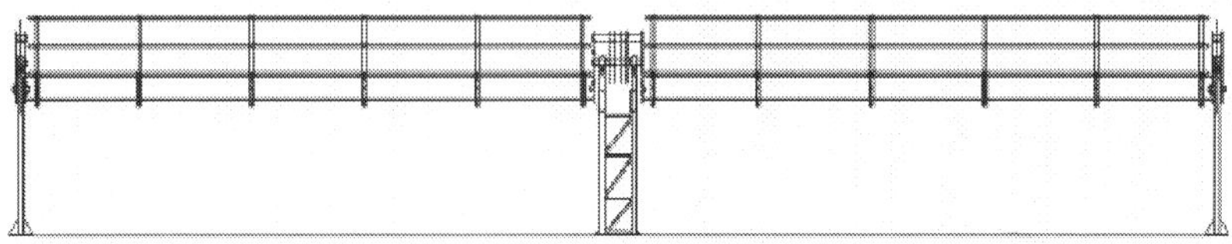

Fig. 8. Sketch of the solar collector (portion).

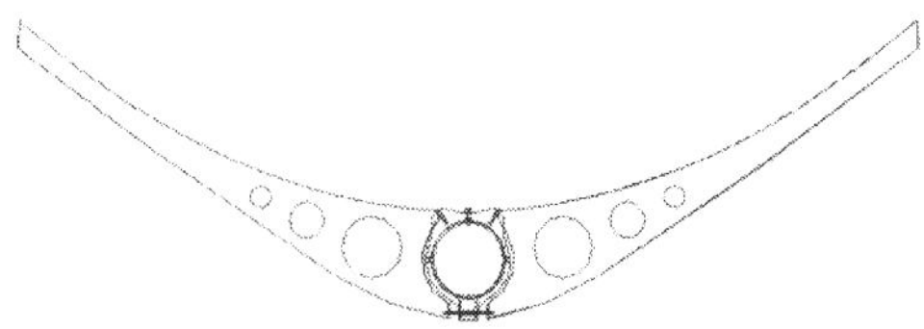

Fig. 9. Sketch of a typical centering (first proposal).

6.1 Limit states and load combinations

As already reported in the previous Section, the considered limit states and load combinations are summarized in Table 8. Correspondingly, OSs W_1 and W_2 are associated to SLSs for which the wind loads refer to a medium velocity and the serviceability limits referring to maximum torsion and maximum deformation, respectively, must be verified. Differently, in the ULSs W_3 and W_4 the structural permanence within the elastic state as well as tightness under loads corresponding to a characteristic peak wind must be verified. Particularly, in the W4 state the possible presence of snow has to be additionally accounted for. For both ULSs, a structural instability verification has to be conducted.

Operational states	V_{ref} (m/s) @10m	Limit state	Reference velocity
W_1	7	Serviceability	Medium
W_2	14	Serviceability	Medium
W_3	21	Ultimate, Elastic	Peak
W_4	28	Ultimate, Collapse	Peak

Table 8. Summary of adopted limit states and load combinations for the concentration system.

In Table 9 all the possible load combinations are shown which have been considered for developing the above-mentioned verifications. It is to be noticed that, for the collapse ULS, in addition to the combinations required by the Recommendations, two other combinations have been evaluated in which the snow and the wind alone are present: this was necessarily done due to the fact that the concurrent presence of the two loads, if from one side it increases the acting forces, from the other it reduces the magnitude of the torque bending, so reducing the stress state in some fundamental structural components. All the analyses have been performed in an elastic state and just in those cases, corresponding to a collapse ULS, in which the structure is particularly stressed, a tightness evaluation within a plastic state has been conducted.

LS	*OS*	*Angle* (°)	*G_receiver*	*G_conc*	*Q_ wind*	*Q_ snow*	*c_G*	*c_Qv*	*c_Qn*	f_y	*ID*
ELS	W_1	30	yes	yes	yes	no	1.00	1.00	0.00	1.00	*w1p030c1*
ELS	W_2	60	yes	yes	yes	no	1.00	1.00	0.00	1.00	*w2p060c1*
ULS elastic	W_3	75	yes	yes	yes	no	1.40	1.50	0.00	1.00	*w3p075c1*
							1.00	1.50	0.00	1.00	*w3p075c2*
		60	yes	yes	yes	no	1.40	1.50	0.00	1.00	*w3p060c1*
							1.00	1.50	0.00	1.00	*w3p060c2*
		30	yes	yes	yes	no	1.40	1.50	0.00	1.00	*w3p030c1*
							1.00	1.50	0.00	1.00	*w3p030c2*
		0	yes	yes	yes	no	1.40	1.50	0.00	1.00	*w3p000c1*
							1.00	1.50	0.00	1.00	*w3p000c2*
		-30	yes	yes	yes	no	1.40	1.50	0.00	1.00	*w3m030c1*
							1.00	1.50	0.00	1.00	*w3m030c2*
		-120	yes	yes	yes	no	1.40	1.50	0.00	1.00	*w3m120c1*
							1.00	1.50	0.00	1.00	*w3m120c2*
ULS collapse	W_4	-120	yes	yes	yes	load1	1.40	1.50	1.05	0.83	*w4m120c1*
							1.00	1.50	1.05	0.83	*w4m120c2*
							1.40	1.05	1.50	0.83	*w4m120c3*
							1.00	1.05	1.50	0.83	*w4m120c4*
							1.40	1.50	0.00	0.83	*w4m120c5*
							1.00	1.50	0.00	0.83	*w4m120c6*
							1.40	0.00	1.50	0.83	*w4m120c7*
							1.00	0.00	1.50	0.83	*w4m120c8*
		-120	yes	yes	yes	load2	1.40	1.50	1.05	0.83	*w4m120c9*
							1.00	1.50	1.05	0.83	*w4m120c10*
							1.40	1.05	1.50	0.83	*w4m120c11*
							1.00	1.05	1.50	0.83	*w4m120c12*
							1.40	1.50	0.00	0.83	*w4m120c5*
							1.00	1.50	0.00	0.83	*w4m120c6*
							1.40	0.00	1.50	0.83	*w4m120c13*
							1.00	0.00	1.50	0.83	*w4m120c14*
		0	yes	yes	yes	no	1.40	1.50	0.00	0.83	*w4p000c1*
							1.00	1.50	0.00	0.83	*w4p000c2*

Table 9. Details of the load combinations for the concentration system.

6.2 Analysis methodologies.

The structural element has been studied through the F.E. Cast3M code, realizing a 3D model of the 12 m concentration system (Figure 10). Reflecting mirrors, centerings, stringers, torque tube and edge plates. Apart from the plates, which have been modelled through infinitely-rigid beams, all the other components, being made by thin plates, have been modelled through 2D shell elements, able to take into account membrane as well as bending and shear stresses.

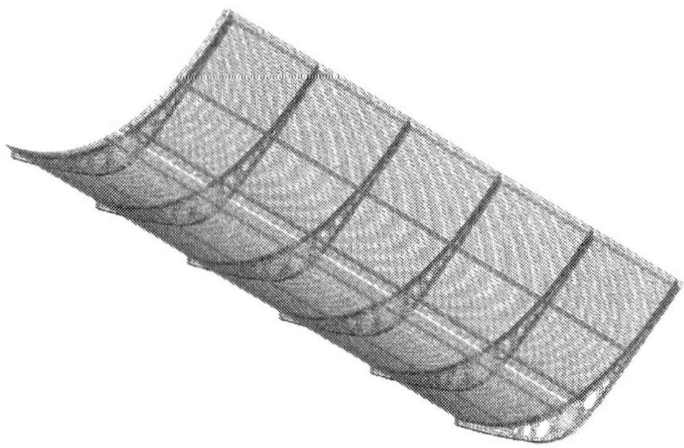

Fig. 10. 3D F.E. model of the concentration system.

The global structural constraints, applied to the edges of the connecting plates, have been applied such to create an isolated and isostatic system, so the stress state doesn't change due to possible loads transmitted by the adjacent moduli.

6.3 Discussion of the main numerical results.

The main results referring to SLSs for weak and medium wind, as well as to ULSs (elastic and collapse) are depicted in Table 10; stresses are calculated as the maximum equivalent Tresca stress, F_{saf} is the safety factor obtained by dividing the material yield limit (reduced in case of ULS, see Tables 7 and 9) by the above stress. The medium value of the parabola's deformation is additionally reported (which is always lower than ± 5 mm, the assumed limit within SLSs).

It is hence evidenced that in both elastic and collapse ULSs the safety factors are generally lower than one; by examining the results in more detail, it has been found that local yielding occur in the higher and middle part of the centering and in some zones connecting the centering with the stringers.

Anyway, it is to be said that the model has been developed to study the global stress level in the various components and not to locally analyse the connecting constructive details which need specific 3D models; a possible local overcome in the stress yield limit and/or consequent re-distributions of stresses can't be caught by such an approach, as explained below.

For sake of brevity, the contour maps of stresses have been included (Figures 11 and 12) referring only to w3p060c1 and w4m120c9 combinations: it is here evidenced the much localized nature of plasticization, as explained above.

	Component	σ_{max}	F_{sat}	Δ (mm)
w1p030c1	Stringers	87	3,17	2,33
	Centerings	108	2,54	
w2p030c0	Stringers	120	2,29	3,66
	Centerings	148	1,86	
w2p060c1	Stringers	117	2,36	3,05
	Centerings	142	1,94	
w3p075c1	Stringers	545	0,51	
	Centerings	597	0,46	
w3p075c2	Stringers	542	0,51	
	Centerings	578	0,48	
w3p060c1	Stringers	571	0,48	
	Centerings	622	0,44	
w3p060c2	Stringers	563	0,49	
	Centerings	602	0,46	
w3p030c1	Stringers	549	0,50	
	Centerings	604	0,46	
w3p030c2	Stringers	534	0,51	
	Centerings	566	0,49	
w3p000c1	Stringers	239	1,15	
	Centerings	262	1,05	
w3p000c2	Stringers	214	1,29	
	Centerings	232	1,19	
w3m030c1	Stringers	218	1,26	
	Centerings	221	1,24	
w3m030c2	Stringers	217	1,27	
	Centerings	183	1,50	
w3m120c1	Stringers	316	0,87	
	Centerings	299	0,92	
w3m120c2	Stringers	305	0,90	
	Centerings	285	0,97	
w4m120c1	Stringers	580	0,47	
	Centerings	564	0,49	
w4m120c2	Stringers	570	0,48	
	Centerings	540	0,51	
w4m120c3	Stringers	453	0,61	
	Centerings	494	0,56	

	Component	σ_{max}	F_{sat}
w4m120c1	Stringers	580	0,47
	Centerings	564	0,49
w4m120c2	Stringers	570	0,48
	Centerings	540	0,51
w4m120c3	Stringers	453	0,61
	Centerings	494	0,56
w4m120c4	Stringers	442	0,62
	Centerings	470	0,59
w4m120c5	Stringers	532	0,52
	Centerings	500	0,55
w4m120c6	Stringers	522	0,53
	Centerings	505	0,54
w4m120c7	Stringers	237	1,16
	Centerings	333	0,83
w4m120c8	Stringers	210	1,31
	Centerings	298	0,92
w4m120c9	Stringers	615	0,45
	Centerings	639	0,43
w4m120c10	Stringers	605	0,45
	Centerings	615	0,45
w4m120c11	Stringers	531	0,52
	Centerings	612	0,45
w4m120c12	Stringers	524	0,53
	Centerings	577	0,48
w4m120c13	Stringers	392	0,70
	Centerings	490	0,56
w4m120c14	Stringers	386	0,71
	Centerings	454	0,61
w4p000c1	Stringers	357	0,77
	Centerings	383	0,72
w4p000c2	Stringers	332	0,83
	Centerings	353	0,78

Table 10. Numerical results (static analyses) for the concentration system.

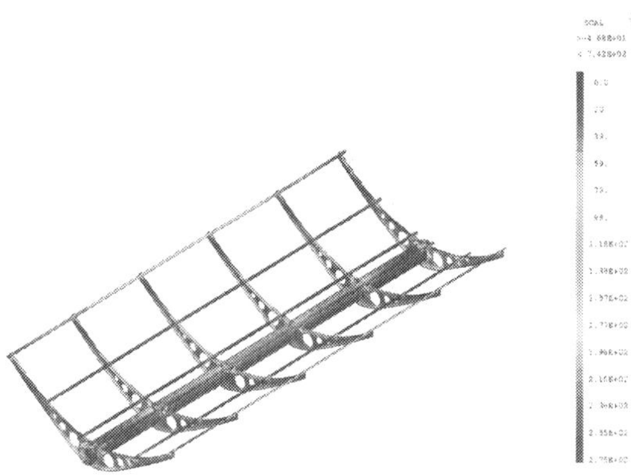

Fig. 11. Contour map of maximum equivalent Tresca stresses for w3p060c1.

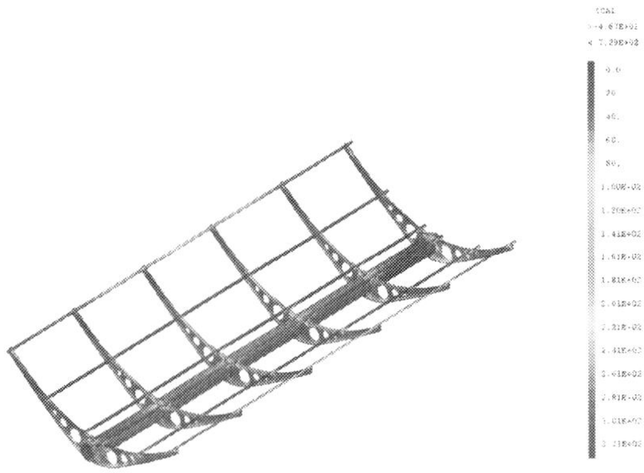

Fig. 12. Contour map of maximum equivalent Tresca stresses for w4m120c9.

The 3D static analyses revealed an appropriate response of the structure under a variety of actions and once, for example, the material strength had been locally overcome, appropriate design procedures have been updated and nonlinear (for material and geometry) analyses performed (see e.g. Figure 13).

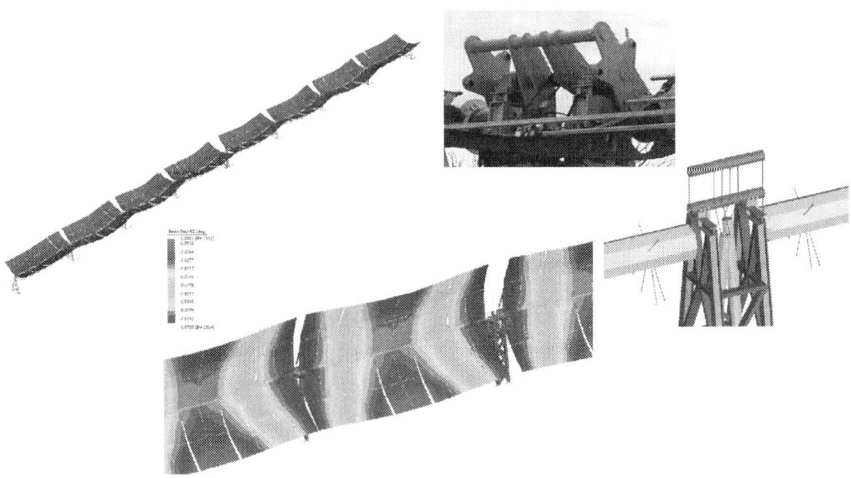

Fig. 13. Typical results from modal and seismic analyses and scheme for a modelled joint.

Additional modal, spectral and generally dynamic analyses have been conducted for the whole CSP system (see Figure 14) to understand the global structural behaviour and to newly upgrade the first design sketch.

The discussion about such results and the corresponding structural response can't be reported here for sake of brevity.

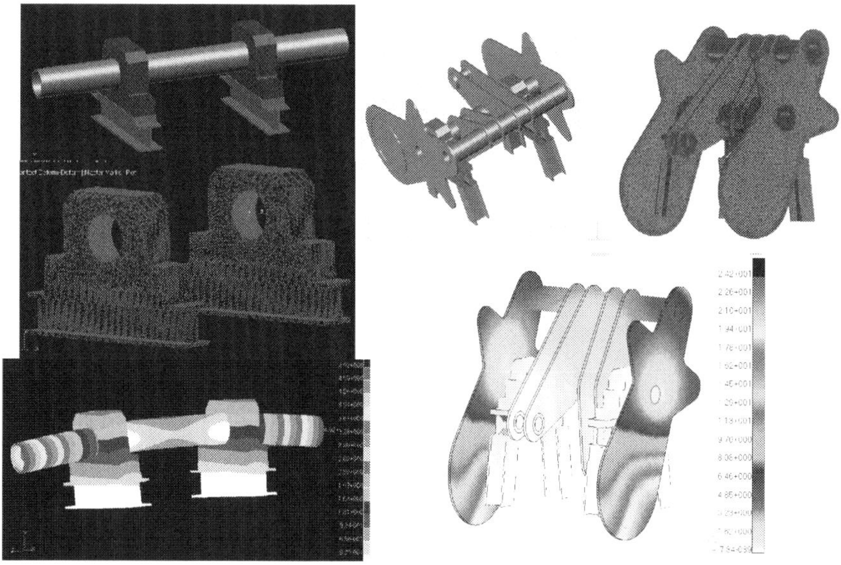

Fig. 14. Joints, pins and specific nodes studied through 3D nonlinear analyses for material and geometry to test their effective structural response and to verify the requirements of the different operational states.

7. Description of heat storage concrete systems

The main advantage of thermal solar power plants is the possibility to use relatively economical storage systems, if compared to other renewable energies (i.e. photo-voltaic and wind). Storing electricity is much more expensive than storing thermal energy itself. Thermal Energy Storage (TES) option can collect energy in order to shift its use to later times, or to smooth out the plant output during irregularly cloudy weather conditions. Hence, the functional operativeness of a solar thermal power plant can be extended beyond periods of no solar radiation without the need of burning fossil fuel. Periods of mismatch among energy supplied by the sun and energy demand can be reduced. Economic thermal storage is a technological key issue for the future success of solar thermal technologies.

In our days, among eight thermal storage systems in thermo-electric solar plants, seven have been of experimental or prototypal nature and only one has been a commercial unit (Salomoni et al., 2008). All the considered systems are "at sensible heat storage": two single-tanks oil thermo-cline systems, four two-tanks single medium systems (one oil- and three molten salt-) and two single-tanks double medium systems. Actually the most advanced technology for heat storage in solar towers and through collector plants considers the use of a two-tanks molten salt system (Ives et al., 1985).

Generally, the hot and cold tanks are located on the ground and they are characterized by an internal circumferential and longitudinally-wrinkled liner, appropriately thermally insulated. The cost of the liner is the primary cost of such a tank. In recent studies it has been shown that an increase in the hourly capacity accumulation reduces sensibly the cost of the

produced electrical energy (LEC); this leads to increase the reservoir dimensions from the 11.6 m diameter and 8.5 m height of the Solar Two power plant to the larger 18.9 m diameter and 2.5 height calculated in the Solar Tres power plant design phase.
Already in 1985, the Solar Energy Research Institute (SERI) commissioned the conceptual design of a below-grade cone shape storage (Figure 15) with 900°C molten carbonate salts (Copeland et al., 1984). This solution, even though interesting because of the use of low cost structural materials, showed some limits connected to the high level of corrosion induced by carbonate and high temperature.

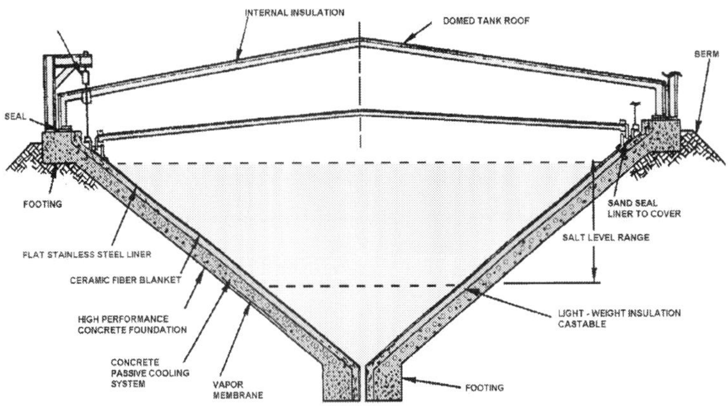

Fig. 15. Conical storage partially buried in the ground.

Such a type of storage is here reconsidered in combination with nitrate molten salts at a maximum temperature of 565°C, using an innovative high performance concrete (HPC) for the tanks. From the technological point of view, the innovations rely in:

- higher structural safety related to the reduced settlements;
- employment of HPC containment structures and foundations characterised by lower costs with respect to stainless steel structures;
- substitution of highly expensive corrugated steel liners with plane liners taking advantage of the geometric compensation of thermal dilations due to the conical shape of the tank;
- possibility of employing freezing passive systems for the concrete basement made of HPC, able to sustain temperature levels higher than those for OPC;
- fewer problems when the tank is located on low-strength soils.

The planned research activities required the upgrade of a F.E. coupled model for heat and mass transport (plus mechanical balance) to estimate concrete tanks durability under prolonged thermal loads and cyclic temperature variations due to changes in the salts level. The presence of a surrounding soil volume is additionally accounted for to evaluate environmental risk scenarios.

7.1 Mathematical-numerical modeling of concrete

Concrete is treated as a multiphase system where the voids of the skeleton are partly filled with liquid and partly with a gas phase (Baggio et al., 1995; Gawin et al., 1999). The liquid phase consists of bound water (or adsorbed water), which is present in the whole range of water contents of the medium, and capillary water (or free water), which appears when water content exceeds so-called solid saturation point S_{ssp} (Couture et al., 1996), i.e. the upper limit of the hygroscopic region of moisture content. The gas phase, i.e. moist air, is a mixture of dry air (non-condensable constituent) and water vapour (condensable gas), and is assumed to behave as an ideal gas.

The approach here is to start from a phenomenological model (Schrefler et al., 1989; Majorana et al., 1997; Majorana et al., 1998; Majorana & Salomoni, 2004 (b); Salomoni et al., 2007 (a)), originally developed by Bažant and co-authors, e.g. (Bažant, 1975; Bažant & Thonguthai, 1978; Bažant & Thonguthai, 1979; Bažant et al., 1988), in which mass diffusion and heat convection-conduction equations are written in terms of relative humidity, to an upgraded version in which its non-linear diffusive nature is maintained as well as the substitution of the linear momentum balance equations of the fluids with a constitutive equation for fluxes, but new calculations of thermodynamic properties for humid gases are implemented too to take into account different fluid phases as well as high ranges of both pressure and temperature. Additionally, Darcy's law is abandoned when describing gas flow through concrete.

The proposed model couples non-linear geometric relations with empirical relations; to enhance its predictive capabilities, a predictor-corrector procedure is supplemented to check the exactness of the solution. For additional details the reader is referred to (Salomoni et al., 2007 (b); Salomoni et al., 2008; Salomoni et al., 2009).

7.2 Numerical analyses

A conical tank for storing hot salts has been modelled through the F.E. research code NEWCON3D (Figure 16) using 330 8-node isoparametric elements (axis-symmetric condition). In agreement with the design criteria, it is proposed to employ a High Performance Concrete (HPC), particularly a C90 for this analysis, to increase both the operational temperature up to 120°C -against the usual 90°C for ordinary concretes- and concrete durability. The whole tank is composed by a flat stainless steel liner in contact with the salts and a ceramic fibre blanket (not modelled here) close to the concrete main structure (Figure 15). An additional passive cooling system is supposed to be added within the concrete thickness to reach such operational temperature on concrete surfaces. Geometric details have not been included for privacy reasons.

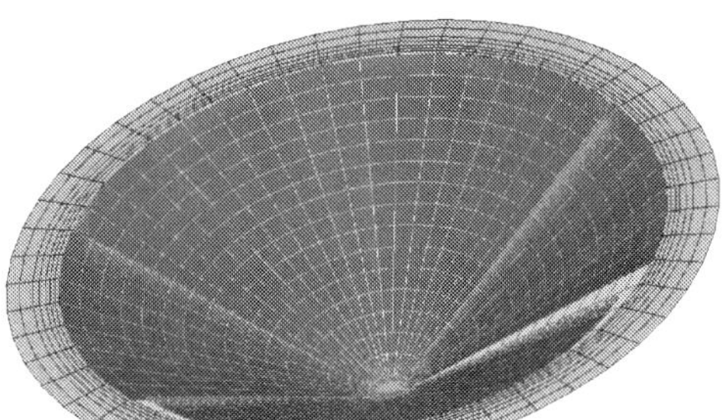

Fig. 16. F.E. discretization for the thermal storage concrete tank.

The adopted material properties are listed in Table 11.

Water/cement ratio	0.29
Elastic modulus [MPa]	$0.367 \cdot 10^5$
Poisson's ratio	0.18
Reference diffusivity along x/y directions [mm²/day]	$0.1 \cdot 10^2$
Intrinsic liquid permeability [mm²]	$2.0 \cdot 10^{-19}$
Unrestrained shrinkage for h = 0 (ε_{sh})	$-0.4 \cdot 10^{-2}$
Thermal expansion coefficient of solid	$0.12 \cdot 10^{-4}$
Hygro-thermal coefficient	$0.5 \cdot 10^{-2}$
Thermal capacity [N/(mm² K)]	2.0
Heat conductivity along x/y directions [N/(day K)]	$0.18144 \cdot 10^6$
Coefficient α_0 for diffusivity	$0.5 \cdot 10^{-1}$

Table 11. Material parameters for concrete C90.

The concrete tank is subjected to transient heating from the internal side assuming to reach the maximum temperature of 100°C in 8 days; the concrete tank has initially a relative humidity of 60% and a temperature of 30°C. In the first analyses (pushed up to about 4 months) the tank is supposed to be simply supported on its basement only.

The results in terms of R.H. (a) and temperature (b) are presented in Figure 17 (3D plot): the development of the R.H. bowl in time (along a typical tank section) is clearly evident; the peaks in R.H. for the zone closest to the heated surface are not referable to the phenomenon of "moisture-clog" (Majorana et al., 1998; Chung et al., 2006), because of the limited value of the temperature gradient, but anyway it is driven by the coupling between humidity and temperature fields and it is connected to the low intrinsic liquid permeability of the adopted HPC. Once 100°C has been reached, concrete starts depleting itself of water, thereby making the relative humidity values tend towards zero (but very slowly: in fact, after about 4 months, a concrete thickness of about 255 mm is still in saturated conditions).

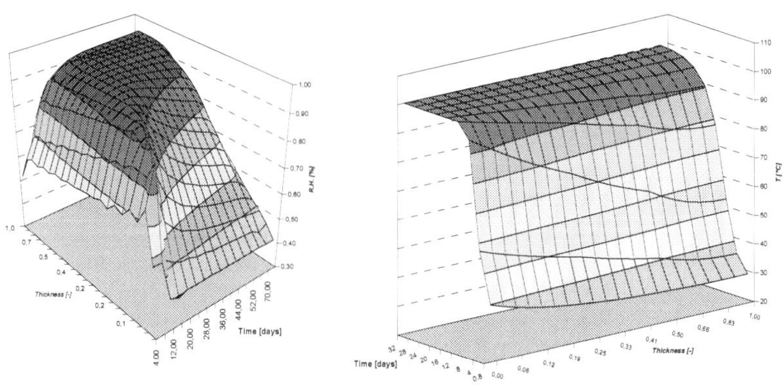

(a) (b)

Fig. 17. R.H. (a) and temperature (b) time-history along a typical tank section.

The desaturation occurring within the concrete tank is mainly caused by the evaporation of water, resulting in formation of a zone of increased vapour pressure. Vapour pressure gradients cause vapour flow towards both the heated surface and the external side of the tank. Moreover, the existing temperature gradient causes thermo-diffusion of water vapour towards the colder layer of the wall. These vapour flows result in an increase in R.H. above its initial value as well as in condensation of vapour in the colder layers and subsequent slight increase in saturation.

The effect of a surrounding ground volume (dry sand) has been additionally evaluated on the development of the thermal front up to 2 years; the main results are here recalled (Salomoni et al., 2008): if considering a sand with a specific heat of 781.25 J/(kg K) and a thermal conductivity of 0.35 W/(m K), a shift the maximum temperature to farther times is revealed with respect to the results obtained from the analysis for the tank only. Additionally, the peak of 100°C seems to be not reachable even after several months.

The situation is clearly preferable when considering the durability performances of the concrete tank, but anyway the heat level is again not admissible for the soil: under such temperatures chemical reactions can take place if organic materials are present, even if a soil treatment is usually performed; moreover, a desiccation of zones around the tank combined with possible re-wettings due to rainfalls could induce relative displacements and consequently tank movements. Hence, being the hypothesis of an additional cooling system (or an additional foundation) too expensive in the design of such structures, it should be planned to use gravelly soils when preparing the surrounding embankment.

8. Conclusions

This Chapter gives a general view related to the last experiences R&D in the field of new technologies for solar energy exploitation within the Italian context, directly exportable abroad due to the followed design and analysis methodologies. The main structures and elements characterizing a solar power plant with parabolic-trough solar concentrators and a double-tank below ground system are studied, evidencing the fundamental design and modelling results.

From the process of analysis and verification of a 100 m parabolic trough solar concentrator, with reference to each system, subsystem and construction details, some remarks can be evidenced. Within this design approach, four operational states corresponding to different design winds, i.e. W_1 = 7 km/h, W_2 = 14 km/h, W_3 = 21 km/h e W_4 = 28 km/h, have been defined, to which two SLSs (for medium wind) and two ULSs (for peak wind) are respectively associated. Then, considering that the snow load acts simultaneously with the self-weight and with some probability contemporaneously with the wind load, the following conclusions can be stated:

- *Concentration system*: the adoption of a 3D FE model, corresponding to a 12 m module, has been considered as adequate to perform such a design and verification procedure. Safety or functioning problems have not been evidenced in the SLSs, linked to the operational states with W_1 and W_2 wind types. With reference to the elastic ULS (ELS), the maximum value for the σ_{eq} has been locally overcome (higher and middle part of the centering and some zones connecting the centering with the frames); such trend has been confirmed in the collapse ULSs.
- *Torque system* (*omissis*): the adoption of a simplified 3D FE model, in which the structure is characterized by a series of beams with sections of appropriate bending and torsion

inertias, has been considered as adequate to perform such a design and verification procedure. Safety, functioning and instability problems have not been evidenced in the SLSs, linked to the operational states with W_1 and W_2 wind types. In fact, with reference to the SLS verifications for weak wind, the only imposed constraint is that the maximum rotation of the torque tube is lower than 2 mrad and this constraint is largely satisfied, as well as the displacements requirements (limit of 20 mm for the camber). In relation to the ELS, the only constraint is the maintenance of the structural behaviour within the elastic limits in any condition, as proved by the obtained high safety factors. When the structure is verified under the collapse ULS, collapse must not occur in the standard safety position: two load cases exist for which such requirement is not satisfied and the critical action is given by the snow. Differently, the stability analyses referring to the heaviest load conditions (W_3 and W_4) show safety factors always higher than one. As regards the constructive details, the requirement of minimum distances between the connectors and the plates' edges, as indicated by the Eurocode 3, is not satisfied so that appropriate design prescriptions have been planned by the authors.

- *Module's support system* (*omissis*): once again a simplified 3D FE model, in which the structure is characterized by a series of beams with sections of appropriate bending and torsion inertias, has been adopted. Both the intermediate and final supports, under SLSs for weak wind, are characterized by low stresses, as well as under SLSs for medium wind. The constraint of a structural behaviour within the elastic range in the elastic ULS is fully respected, as well as the absence of collapse in the standard safety position (collapse ULS). The stability analyses referring to the heaviest load conditions (W_3 and W_4) show again safety factors always higher than one.

A series of linear and non-linear static 3D analyses have been followed by modal, spectral and generally dynamic analyses to understand the structural behaviour of the whole CSP system and the first design sketch has been consequently updated. The analyses revealed an appropriate response of the whole structure under a variety of actions and once, e.g., the material strength had been locally overcome, local non-linear analyses have been conducted and the constructive details eventually re-designed.

Additionally, within the medium temperature field, an innovative approach has been presented for the conceptual design of liquid salts concrete storage systems. A multi-tank sensible-heat storage system has been proposed for storing thermal energy, with a two-tanks molten salt system. The hygro-thermal behaviour of a HPC tank has been assessed through a coupled F.E. code based on the theory by Bažant and enhanced by additional constitutive and thermodynamic relationships. A predictor-corrector procedure has been included to check the exactness of the solution. The study allows for estimating the durability performances of the tank: after about one month, all the structure is fully heated, possibly inducing thermal damage within concrete; such a result is slightly modified when modelling the domain more in detail, i.e. tank plus surrounding ground, or when changes in the salts level are considered. Even if at present some geometric and mechanical characteristics are still to be fixed, so that they can consequently induce an unavoidable uncertainty on the numerical results, the generality of the approach is not affected by such restrictions, and the results themselves can be evaluated as first guidelines in defining design criteria for liquid salts concrete systems. In fact, this study is the first step in a new research field and is being extended within the Italian Research Project "Elioslab - Research Laboratory for Solar Technologies at High Temperatures" started at the end of 2007.

Again, independently on the specificity and interest of the application, it has been shown here that the fully coupled mathematical-numerical model proposed (whose details have been reported for the hygro-thermal field only and whose predictive abilities have already been demonstrated in (Salomoni et al., 2007) even if in its not-upgraded form) has been enhanced through additional thermodynamic and constitutive relationships, allowing for obtaining more complete results in terms of water vapour pressure, gas pressure and capillary pressure which become fundamental variables mainly when higher temperature regimes are to be considered.

9. References

Antonaia, A.; Avitabile, M.; Calchetti, G.; Crescenzi, T.; Cara, G.; Giannuzzi, G.M.; Maccari, A.; Miliozzi, A.; Rufoloni, M.; Prischich, D.; Vignolini, M. (2001). Design sketch of the parabolic trough collector for solar plants. *ENEA/TM/PRES/2001_09*, Rome, Italy (technical report, in Italian).

Baggio, P.; Majorana, C.E.; Schrefler, B.A. (1995). Thermo-hygro-mechanical analysis of concrete. *International Journal for Numerical Methods in Fluids*, Vol. 20, 573-595.

Bažant, Z.P. (1975). Pore pressure, uplift, and failure analysis of concrete dams. *Int. Commission on Large Dams*, Swansee, UK.

Bažant, Z.P.; Thonguthai, W. (1978). Pore pressure and drying of concrete at high temperature. *Journal of the Engineering Materials Division*, ASME, Vol. 104, 1058-1080.

Bažant, Z.P.; Thonguthai, W. (1979). Pore pressure in heated concrete walls: theoretical predictions. *Magazine of Concrete Research*, Vol. 31, No.107, 67-76.

Bažant, Z.P.; Chern, J.C.; Rosenberg, A.M.; Gaidis, J.M. (1988). Mathematical Model for Freeze-Thaw Durability of Concrete. *Journal of the American Ceramic Society*, Vol. 71, No. 9, 776-83.

Chung, J.H.; Consolazio, G.R.; McVay, MC. (2006). Finite element stress analysis of a reinforced high-strength concrete column in severe fires. *Computers and Structures*, Vol. 84, 1338-1352.

Copeland, R.J.; West, R.E.; Kreith, F. (1984). Thermal Energy Storage at 900°C. *Proc. 19th Ann. Intersoc. Energy Conversion Engrg. Conf.*, San Francisco, Aug. 19-24, 1171-1175.

Couture, F.; Jomaa, W.; Ruiggali, J.R. (1996). Relative permeability relations: a key factor for a drying model. *Transport in Porous Media*, Vol. 23, 303-335.

ENEA (2003). http://www.enea.it/com/ingl/solarframe.htm.

Gawin, D.; Majorana, C.E.: Schrefler, B.A. (1999). Numerical analysis of hygro-thermal behaviour and damage of concrete at high temperature. *Mechanics of Cohesive-Frictional Materials*, Vol. 4, 37-74.

Giannuzzi, G.M.; Majorana, C.E.; Miliozzi, A.; Salomoni, V.A.L.; Nicolini, D. (2007) Structural design criteria for steel components of parabolic-trough solar concentrators. *Journal of Solar Energy Engineering*, Vol. 129, 382-390.

Herrmann, U.; Kelly, B.; Price, H. (2004). Two-tank molten salt storage for parabolic trough solar power plants. *Energy*, Vol. 29, No. 5-6, 883-893.

Ives, J.; Newcomb, J.C.; Pard, A.G. (1985). High Temperature Molten Salt Storage. *SERI/STR-231-2836* (technical paper).

Majorana, C.E.; Salomoni, V.; Secchi, S. (1997). Effects of mass growing on mechanical and hygrothermic response of three-dimensional bodies. *Journal of Materials Processing Technology*, PROO64/1-3, 277-286.

Majorana, C.E.; Salomoni, V; Schrefler, B.A. (1998). Hygrothermal and mechanical model of concrete at high temperature. *Materials and Structures*, Vol. 31, 378-386.

Majorana, C.; Salomoni, V. (2004) (a). Selection, elaboration and application of recommendations for designing parabolic trough solar concentrators. Functional description, classification and selection of design codes for their structural elements. *Report ENEA-TRASTEC*, Rome-Padua, Italy (in Italian).

Majorana, C.; Salomoni, V. (2004) (b). Parametric analyses of diffusion of activated sources in disposal forms. *Journal of Hazardous Materials*, A113,. 45-56.

Majorana, C.; Salomoni, V. (2005) (a). Design guide for parabolic trough solar concentrators. *Report ENEA-TRASTEC*, Rome-Padua, Italy (in Italian).

Majorana, C.; Salomoni, V. (2005) (b). Analyses and structural verifications for a 100 m-parabolic trough solar concentrator. *Report ENEA-TRASTEC*, Rome-Padua, Italy (in Italian).

Miliozzi, A.; Nicolini, D.; Giannuzzi, G.M. (2007). Evaluation of wind action on parabolic trough concentrators for a high temperature solar plant. *Enea Report RT/2007/13/TER* (in Italian).

Pilkington Solar International GmbH (2000). Survey of Thermal Storage for Parabolic-Trough Power Plants, *NREL/SR-550-27925* (technical report).

Rubbia, C.; and ENEA Working Group (2001). Solar thermal energy production: guidelines and future programmes of ENEA. *ENEA/TM/PRES/2001_7*, Rome, Italy (technical report).

Salomoni, V.A.; Giannuzzi, G.M.; Majorana, C.E.; Miliozzi, A.; Nicolini, D. (2006). Structural design of parabolic-trough solar concentrators' steel components against wind and natural hazards, *Proceedings of the 3^ Int. Conf. on Protection of Structures Against Hazards* (Majorana, Salomoni, Lok Eds.), Venice, Italy, sept. 28-29 (ISBN 981-05-5561-X) (keynote lecture).

Salomoni, V.A.; Mazzucco, G.; Majorana, C.E. (2007) (a). Mechanical and durability behaviour of growing concrete structures. *Engineering Computations*, Vol. 24, No. 5, 536-561.

Salomoni, V.A.; Majorana, C.E.; Khoury, G.A. (2007) (b). Stress-strain experimental-based modeling of concrete under high temperature conditions. In: B.H.V. Topping (Ed.), *Civil Engineering Computations: Tools and Techniques*, Ch. 14, Saxe-Coburg Publications, 319-346.

Salomoni, V.A.; Majorana, C.E.; Giannuzzi, G.M.; Miliozzi, A. (2008) Thermal-fluid flow within innovative heat storage concrete systems for solar power plants. *International Journal of Numerical Methods for Heat and Fluid Flow* (Special Issue), Vol. 18(7/8), 969-999.

Salomoni, V.A.; Majorana, C.E.; Mazzucco, G.; Xotta, G.; Khoury, G.A. (2009). Multiscale Modelling of Concrete as a Fully Coupled Porous Medium. In: J.T. Sentowski (Ed.), *Concrete Materials: Properties, Performance and Applications*, Ch. 3, NOVA Publishers, 2009 (in press).

Schrefler, B.A.; Simoni, L.; Majorana, C.E. (1989). A general model for the mechanics of saturated-unsaturated porous materials. *Materials and Structures*. Vol. 22, 323-334.

13

Charge Carrier Recombination in Bulk Heterojunction Organic Solar Cells

Gytis Juška and Kęstutis Arlauskas
Vilnius University
Lithuania

1. Introduction

Photovoltaic phenomenon was first observed by E. Becquerel (Becquerel) in 1839. He observed the electric current-lit silver electrode, immersed in the electrolyte. In 1894, taking advantage of the observed photoconductivity phenomenon in amorphous selenium the semiconductor solar cell was developed.

The very first silicon p-n junction solar cell was made in 1954, energy conversion efficiency of which was 6% and the energy price $200/W did not seem promising for wide application. Later, the development of satellites needed to provide sustainable energy sources and the cadmium sulfid, cadmium telluride, gallium arsenide and more efficient solar cells of other materials were created.

The first solar cell breakthrough was something like of the 1970 year, feeling the lack of oil, which oncreased interest in alternative energy sources. The basic raw materials, in addition to crystalline silicon, a polycrystalline silicon, were also amorphous silicon and other, suitable for thin solar cells, materials. Although, due to the high cost of these energy sources, extracted energy was only a small part of total energy production, but the lending spread as energy sources in various areas of small devices such as mobile phone, calculators, meteorogical instruments, watches and so on. A solar powered cars and even solar powered aircraft were constructed. Major Solar cells used for the purification of salt water, as well as supply power to isolated objects: mountains, islands or jungle living population.

The second and much greater solar energy use breakthrough occurred in the first decade of the twenty-first century. This is caused by the earth's climate warming due to the increasing threat of thermal energy and the increasing CO_2 in the atmosphere. Many governments in many ways stimulated the solar energy lending. Germany in the decade from 1994 to 2004, installed as much as 70 times more solar energy equipment, and now is installed more than 1GW: produced over 3TWh energy, which cost around 0.5 €/kWh. In Japan solar power energy is less costly than the heat. The main price of solar energy is caused by the installation consts - ~ 1€/W. Till 2004 there have already been installed over 1GW, while in 2006, the world's installed 6.5 GW. In 2007, the European Union in the fight against climate warming threat committed by 2030 to achieve that 25% of the total energy from alternative sources, mainly from the Sun. It should be around 1200 GW, the cost should not exceed 0.1 €/kWh. Another reason for the needed alternative energy sources is projected oil and gas resource depletion.

Crystalline silicon still remains the unrivaled leader in the development of solar cells. However, the demant of renewable energy sources stimulated a search for a new, low-cost technologies and materials. Hydrogenated amorphous silicon (a-Si:H) has long been regarded as one of the most promising materials for development of cheap, lightweight and technologicall solar cells. However, a Si:H solar cells degraded in high intensity-light. Thus, forward-looking, more efficient microcrystalline (μc-Si:H) and nanocrystalline silicon (nc-Si:H) solar cells began to compete successfully with a-Si:H.
The first organic materials were investigated for more than a hundred years ago and for a long time the widest application, in scope of optoelectronics, was electrography. However, in 1977 A. J. Heeger, A. G. MacDiarmid and H. Shirakawa showed that the π-conjugated polymers can be doped, and change the properties of substances. This work demonstrated the possibility use polymers to create optoelectrical devices, resulted in huge interest and in 2000 was awarded the Nobel Prize. During the period from 1977 on the base of π-conjugated polymers has been built a number of electronic and optoelectronic devices: diodes, field effect transistors, sensors, photodiodes, etc. On 1993 - 2003 years π-conjugated polymers have been investigated in order to create a light-emitting diodes (OLED) and their systems, and these studies culminated in the creation of a colour OLED matrix, which is adapted to different types of displays. Recently, organic polymers mainly involved studies of organic solar cells and other organic electronics appliances, effectiveness of which is determined by the drift and recombination of charge carriers.
In order to develop efficient solar cells it is necessary the maximum possible the light absorption, the carrier photogeneration quantum efficiency, and that all photogenerated carriers be collected in a solar cell electrodes. The collection of charge carriers depends on their mobility and recombination. Thus, the investigations of carrier mobility and their density dependencies on the electric field, temperature and material structure are essential for the formation of understanding of charge carrier transport in these materials, which is essential to find effective new inorganic and organic materials and to development of new optoelectronic structures.
One of the main factors limiting efficiency of organic solar cells (OSC) is charge carrier recombination. In crystals, where the carrier location uncertain, recombination is caused by the probability to transfere energy: or emit photon - radiation recombination, or to another electron - Auger recombination, or induction phonons through the deep states. The latter depends on the density of deep states. In disordered structures, with a lot of localized states, should be very rapid recombination, but there recombinationis caused by the meeting probability of electron and hole in space, as the only their meeting at a distance closer than the Coulomb radius causes their recombination (named Langevin), likely as gemini recombination. It is valid only if the energy dissipation or jump distance is less than the Coulomb radius. Thus, the Langevin bimolecular recombination is ordained by the mutual Coulomb attraction drift time, because under this attraction electron is moving toward the nearest hole, while at the same time, due to diffusion, with equal probability in any direction. The Langevin recombination time can be expressed as:

$$\tau_{\mathrm{L}} = \int_0^r \frac{\mathrm{d}x}{(\mu_{\mathrm{n}} + \mu_{\mathrm{p}})F} = \frac{\varepsilon\varepsilon_0}{e(\mu_{\mathrm{n}} + \mu_{\mathrm{p}})n} \tag{1}$$

Here μ_n, μ_p are electron and hole mobility, respectively; $F = e / 4\pi\varepsilon\varepsilon_0 x^2$ is strength of Coulomb electric field; n is density of charge carriers $(1/n = 4\pi r^3/3)$, and r is a mean distance between electrone and hole. Thus, from the expression of Langevin bimolecular recombination coefficient $B_L = e(\mu_n + \mu_p) / \varepsilon\varepsilon_0$ it is clearly seen that recombination is caused by the features of charge carrier transport. In bulk heterojunction organic solar cells the reduced Langevin recombination is observed.

In this work we describe methods of investigation of charge carrier recombination in disordered structures, where stochastic transport of charge carriers complicates interpretation of experimental results: integral time of flight (i-TOF) (Juška et, 1995), using of which allows easily estimate the temperature dependence of recombination coefficient; charge carriers extraction by linearly increasing voltage (CELIV) (Juška et, 2000, a), which allows independently measure relaxation of density and mobility of photoexcited charge carriers; double injection current transient (DoI) (Juška et, 2005; Juška et, 2007), which is additional method of investigation of charge carrier recombination and, which allows to measure dependence of recombination coefficient on electric field.

In this study we represent how using current transient methods may be cleared up the features of charge carrier transport and recombination in disordered inorganic and organic materials. The microcrystalline silicon and π-conjugated polymers have been investigated as a typical inorganic and organic material.

2. Investigation methods

The disordered structure of material causes that mobility of charge carriers is low, because their motion is slowed down by the interaction with spectrum of the local states. Thus, the classical investigation methods: the Hall and magnetoresistance measurements are invalid. The carrier transport in disordered inorganic and organic materials, conductivity (σ) of which is low, is studied using time-of-flight (TOF) method. However, the conductivity of many π-conjugated polymers is high and does not fulfill the latter condition. Thus, for their investigation has been adapted and refined microcrystalline hydrogenated silicon (μc-Si:H) used the extraction of charge carriers by linearly increasing voltage (CELIV) method. The latter method allows to investigate the transport properties of charge carriers both in conductive and low conductivity materials. For investigation of charge carrier transport and recombination the double injection current (DoI) transient method is promising as well.

2.1 TOF method

Time-of-flight method is widely used for investigation of transport, trapping-retrapping and recombination of charge carriers in disordered materials and structures. This method is applicable only for investigation of low conductivity materials, i.e. where the Maxwell relaxation time exceeds the duration of transit time (t_{tr}) of charge carriers through the interelectrode distance (d):

$$\tau_\sigma = \frac{\varepsilon\varepsilon_0}{\sigma} >> t_{tr} = \frac{d^2}{\mu U} \tag{2}$$

TOF method is based on the current transient measurement when photogenerated of the same sign charge carriers is moving in the electric field (E) created in the interelctrode distance (d) of the sample and during a drift time (t_{tr})the package achieves an opposite electrode. The simplicity and efficiency of method meant that it is a widely used for study of mobility (μ), trapping (τ_t) and lifetime of charge carriers (τ) in low conductivity ($\tau_\sigma > t_{tr}$)

materials. Low conductivity of material ensures that during the drift of photogenerated charge carriers through interelectrod distance the density equilibrium charge carriers will be too low to redistribute the electric field inside the sample, and the electric field will be steady at the moment of charge carrier photogeneration, i.e. $RC < t_{del} < \sqrt{\tau_\sigma t_{tr}}$ (here R is total resistance of measurement system and sample elctrodes, C is geometric capacitance of sample). TOF method, dependently on amount of initial injected charge (Q_0) and, also, on characteristic time RC of measurement system, is devided into a number of regimes.

Small charge drift currents (SCDC). This regime is ensured when an amount of photogenerated charge is much less than an amount of charge on sample electrodes at given voltage (U_0), i.e. $eL = Q_0 << CU_0$. Here L amount of charge carriers photogenerated by pulse of light. In this regime there are a few cases:

a. ***current (diferencial) regime* ($t_{tr} > RC$).** In case of strong absorption of light ($\alpha d >> 1$, α is absorption coefficient) and nondispersive transport, the shape of pulse of photocurrent transient is close to rectangular, duration of which is t_{tr} (Fig. 1a, $L = 0{,}3$), and form the area of current transient the Q_0 can be estimated. In case of weak absorption of light ($\alpha d << 1$), charge carriers are photogenerated in the bulk of sample, thus, the shape of photocurrent pulse is triangular, which's duration is t_{tr}, and area is equal $Q_0/2$. The dispersive transport of charge carriers, due to dependence of charge carrier mobility on time, causes that pulse of current transient did not demonstrate obvious break points, form which will be possible to estimate t_{tr} (even if $\alpha d >> 1$). In this case, if the current transient is represented by a double-log scale ($\lg j = f(\lg t)$), the turning point corresponds the t_{tr}.

b. ***charge (integral) regime* ($t_{tr} < RC$).** Even in case of strong absorption of light and nondispersive transport of charge carriers the shape of photocurrent pulse is not so informative as in current regime (Fig. 1a, $L = 0{,}3$): the drift time of charge carrier package is estimated as halftime ($t_{1/2}$) of rise time of photocurrent pulse, i.e. $t_{tr} = 2\, t_{1/2}$. The magnitude of photocurrent pulse is equal to amount of charge (Q) collected onto the sample electrodes during the charge carrier drift time. In case of bulk absorption of light, the magnitude of photocurrent pulse is equal $Q/2$, and $t_{tr} = 3{,}41\, t_{1/2}$

If the voltage of backward direction is applied onto solar cell electrodes and, by short pulse of light the charge carrier pairs are photogenerated, the photocurrent pulse of their drift is observed, from which's duration (t_{tr}) the mobility of the charge carriers of the same polarity as illuminated electrode is estimated. An amount of drifting charge carriers is estimated from the area of photocurrent pulse, from which, when amount of absorbed quanta of light is known, the quantum efficiency is evaluated (Fig.1).

In case of trapping with characterstic trapping time τ or in case of stochastic transport, after photogeneration, the shape of photocurrent pulse is decreasing, and, from the area of photocurrent pulse, estimated dependence of amount of photogenerated charge carriers on voltage follows Hecht's dependence (Eg. (3)). From the latter dependence the $\mu\tau$-product, which determines both the diffusion and drift lengths of charge carries, and causes effectiveness of solar cell, is estimated.

$$\frac{N}{N_0} = \frac{\mu\tau E}{d}\left(1 - \exp\left(-\frac{d}{\mu\tau E}\right)\right) \tag{3}$$

Space charge limited photocurrent (SCLP). In this case an amount of phtogenerated charge is higher than charge on sample electrodes at U_0, i.e. $Q_0 >> CU_0$. The shape of photocurent pulse depends on Q_0 (Fig. 1a), and strongly absorbed light ($\alpha d >> 1$) creates reservoir of charge carriers at the illuminated elektrode, from which not more than CU_0 charge package can drift to the opposite elektrode. This package is moving in growing electric field, thus, in case of nondispersive transport and when $t_{tr} > RC$, drift time is $t_{SCLC} = 0{,}78\, t_{tr}$, which is estimated from the spike of current transient (Fig. 1a). When $t > t_{SCLC}$, current flows until the whole charge is extracted from reservoir and the second turning point, at extraction time (t_e), appears on the pulse of photocurrent. An amount of charge extracted from the reservoir (Q_e), as well as t_e, depend on recombination speed of charge carriers in reservoir.

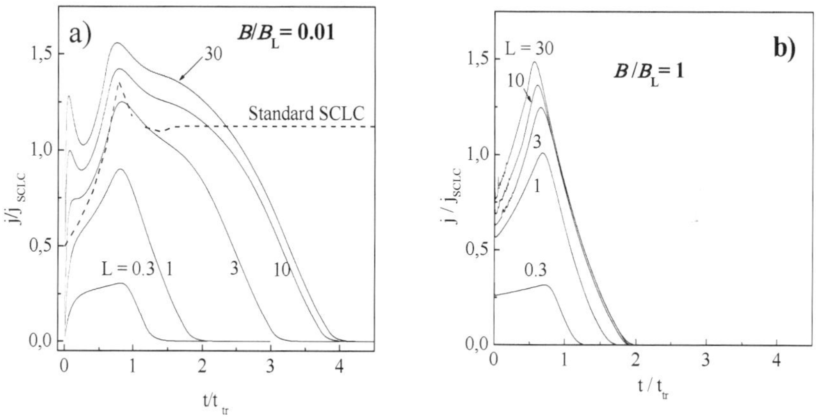

Fig. 1. Numerically modelled photocurrent transients of charge carrier drift dependence on exciting light intensity in case when $B/B_L = 0.01$ (a), and when Langevin recombination prevails (b). Density of photogenerated charge carriers is normalised to amount of charge on sample electrodes in SCLC regime

For investigation of charge carrier recombination by photocurrent transient methods the dependence of collected onto sample electrodes charge on intensity of photoexciting light pulse is measured (Pivrikas et, 2005). When, due to increasing intensity of light pulse, the amount of photogenerated charge achieves amount of charge carriers on sample electrodes ($Q_0 = CU_0$), the TOF regime changes from small charge drift current (SCDC) to space charge limited current (SCLC) (Fig. 2a). Further increase of light pulse intensity not follows by increase of photocurrent, but increases the duration ($t_e \geq t_{tr}$) of photocurrent pulse, which is caused by the extraction of charge carriers from reservoir. The faster charge carrier recombination in reservoir, the shorter extraction time (t_e), and, when recombination is very fast, $t_e \rightarrow t_{tr}$. Thus, the dependence of t_e on intensity of exciting light pulse L gives information about recombination process in charge carrier resrvoir: dependence as $t_e(L) \approx \ln L$ indicates that monomolecular recombination prevails; if, at high intensity of light pulse, t_e saturates with L, than the bimolecular or of higher order of charge carrier recombination prevails.

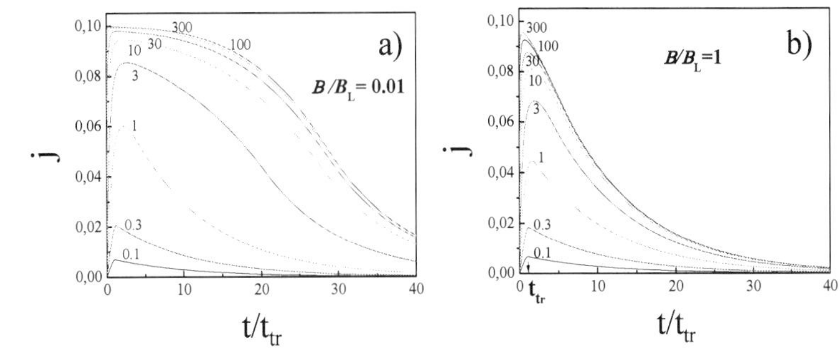

Fig. 2. Numerically modelled integral TOF current transients ($RC = 10\ t_{tr}$)

In organic polymers the bimolecular recombination typically is of Langevine-type. The photocurrent transients of this case are shown in Fig. 2b, and the maximal amount of extracted charge is estimated as:

$$\frac{Q}{CU} = 1 - \exp\left(-\frac{eL}{CU}\right) \tag{4}$$

The maximal amount of extracted charge $Q = CU$.
When the bimolecular recombination is weaker than Langevin's one, from the saturation of extraction time, which is estimated as difference of photocurrent pulse halwidths at space charge limited and at small charge regimes, i.e. $t_e = t_{1/2}\ (L>1) - t_{1/2}(L<<1)$, the ratio of bimolecular recombination coefficient (B) with Langevin's one according to expression:

$$\frac{B}{B_L} = \frac{t_{tr}^2}{t_e(t_e + t_{tr})}\frac{1}{\alpha d} \tag{5}$$

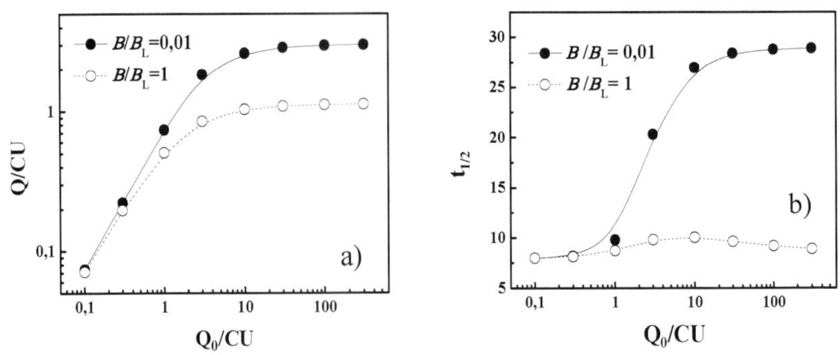

Fig. 3. Dependencies of amount of extracted charge Q (a) and of photocurrent halfwidth (b) on amount of photogenerated charge in case of Langevin and reduced bimolecular recombination

Here α is absorption coefficient. However, it is easer to measure the recombination coefficient using integral TOF when $RC > t_{tr}$ (Pivrikas et al, 2005). The examples of numerically modelled transients are demonstrated in Fig.2. Using this method the coefficient of bimolecular recombination is estimated as (Fig. 3):

$$B = \frac{edS}{t_e Q} \quad (6)$$

2.2 Charge carrier extraction by linearly increasing voltage (CELIV) method.

Method has the advantage that it is suitable for investigation of both high and low conductivity materials (Juška et al, 2000 a; Juška et al, 2004). After the triangular voltage pulse is connected to the sample electrodes in backward direction, the current caused by geometric capacitance of sample ($j(0)$) and conductivity current Δj are observed (Fig. 4).

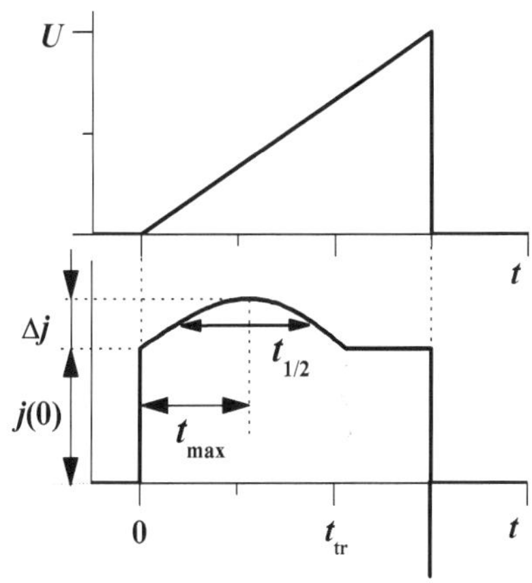

Fig. 4. Voltage pulse and current density observed by CELIV method

The measuring device is very simple: triangular pulse voltage generator and oscilloscope. Another advantage is that after triangular pulse of voltage is applied onto sample electrode, there is no initial, caused by capacitance, current peak, which disturb to monitor drift current in conductive materials.

The current transients were calculated by using standard solution method from continuity, current and Poisson equations in case when one of electrodes is blocking: Schottky or *p-i* barrier, or even special structure with isolating sublayer. From Poisson equation when density of equilibrium carriers is n_0 the extraction depth l ($0 \le l \le d$) is estimated:

$$\frac{Q(t)}{\varepsilon\varepsilon_0} = \frac{en_0 l(t)}{\varepsilon\varepsilon_0} = E(0,t) - E(d,t), \quad (7)$$

here Q is the amount of extracted charge, $E(0,t)$ and $E(d,t)$ are the magnitudes of electric field at the front and back electrodes correspondingly.

From the continuity equation:

$$\frac{\mathrm{d}Q}{\mathrm{d}t} = j_{\mathrm{d}} = \sigma_0 E(d,t)\,, \tag{8}$$

In case of linearly increasing voltage $U = A\,t$ and

$$\int_0^d E\mathrm{d}x = At = E(d,t)\cdot d + \frac{E(0,t)-E(d,t)}{2}\cdot l(t)\,,$$

the Rikati equation is obtained for $l(t)$

$$\frac{\mathrm{d}l(t)}{\mathrm{d}t} + \frac{\sigma}{2\varepsilon\varepsilon_0 d} l^2(t) = \frac{\mu At}{d}\,,$$

Then the curent transient is

$$j(t) = \frac{\varepsilon\varepsilon_o A}{d} + \frac{\sigma}{\mu}\left(1-\frac{l(t)}{d}\right)\left(\frac{\mu At}{d} - \frac{\sigma}{2\varepsilon\varepsilon_o d}\cdot l^2(t)\right), \tag{9}$$

The first component is caused by capacitance, and second one by conductivity.
When $\tau_\sigma = \varepsilon\varepsilon_0/\sigma >> t_{\mathrm{tr}}$

$$j(t) = \frac{A}{d}\left[\varepsilon\varepsilon_o + \sigma t\left(1-\frac{\mu At^2}{2d^2}\right)\right], \text{ when } t < d\sqrt{\frac{2}{\mu A}} = t_{tr}\,, \tag{10a}$$

$$j(t) = \frac{A}{d}\cdot\varepsilon\varepsilon_o = j(0)\,, \text{ when } t > t_{tr} \tag{10b}$$

From experimentally observed current transient the thickness of sample and/or dielectric permitivity may be estimated:

$$\frac{\varepsilon\varepsilon_0}{d} = \frac{j(0)}{A}\,, \tag{11}$$

The dielectric relaxation time may be estimated as

$$\tau_\sigma = \frac{2}{3}\cdot t_{\max}\frac{j(0)}{\Delta j}\,, \tag{12}$$

The mobility of equilibrium charge carriers can be estimated as

$$\begin{aligned} \mu &= \frac{2d^2}{3At_{\max}^2} \quad \text{if } \Delta j \le j(0), \text{ i.e. } \tau_\sigma \ge t_{\mathrm{tr}} \\ \mu &= \frac{\tau_\sigma d^2}{At_{\max}^3} \quad \text{if } \Delta j >> j(0), \text{ i.e. } \tau_\sigma << t_{\mathrm{tr}} \end{aligned} \tag{13}$$

The bulk conductivity of sample follows from:

$$\sigma_{bulk} = \varepsilon\varepsilon_0 \left.\frac{dj}{dt}\right|_{t=0} / j_0\,. \tag{14}$$

The density of charge may be calculated from:

$$p_0 = \frac{2}{ed}\int_0^{\infty} \Delta j dt \quad . \tag{15}$$

In Fig. 5 there are demonstrated the results of modelling without trapping and with single trap level (Juška et al, 2000, b). For high and low A the modelling very well reproduces $t_{max}(A) \approx A^{-0.5}$ and $A^{-0.33}$, predicted by Eq. (13). When trapping is accounted then in both limiting cases the same expressions for t_{max} and Δj like without trapping are obtained, if one substitute μ by(μf), where f is the trapping factor or for single trap $f = \tau_C/(\tau_C + \tau_R)$, where τ_R is the release time.

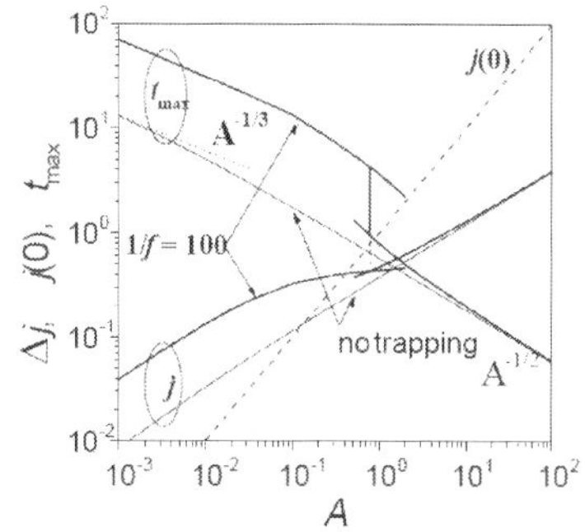

Fig. 5. Numerical modelling results of dependencies of Δj, t_{max}, $j(0)$ on A. $A = 1$, when $t_{tr} = \tau_\sigma$. Bold lines demonstrate dependencies when shallow trapping is accounted ($\tau_C = 1$, $\tau_R = 100$); lines correspond case when shallow trapping is absent. Density of current is normalized to magnitude of $j(0)$, when $A = 1$, and time is normalized to τ_σ

The basic measurable parameters of CELIV Δj, t_{max} depend on charge carrier interaction with trapping states, and this is reflected in dependencies of t_{max} and Δj (Fig. 6). Numerical modelling (Juška et al, 2000 b), taking into account energy distribution of trapping states as $N(E) \sim \exp(-E^2/2\delta^2)$, demonstrate that measurements of $\Delta j(A) \sim A^{\beta}$ and $t_{max}(A) \sim A^{\gamma}$ dependencies in various temperatures and electric fields, while choosing such A that $\Delta j \cong j(0)$, and estimating the rates of change as coefficients $\beta = \left.\frac{d(\ln j)}{d(\ln A)}\right|_{\Delta j = j(0)}$ and $\gamma = \left.\frac{d(\ln t_{max})}{d(\ln A)}\right|_{\Delta j = j(0)}$, the nature of charge carrier interaction with trapping states can be cleared up, i.e., which charge carrier transport model is prevailing (Fig. 6):

1. if $\mu(F)$ dependence is caused by stochastic transport, then $(\beta - \gamma) = 1$, $(\beta + \gamma) < 0$;
2. if $\mu(E)$ dependence is caused by Poole-Frenkel type dependence of micromobility on electric field ($\mu \sim \exp(a\sqrt{E})$) then $(\beta - \gamma) > 1$, and $(\beta + \gamma) < 1$, and the latter is independent or decreases with increasing a (T decreases);
3. if the characteristic release from trapping states time τ_R depends on electric field, i.e. $\tau_R \sim \exp(-b\sqrt{E})$ then, when b increases, $(\beta - \gamma) > 0$, and $(\beta + \gamma)$ increases or even changes the sign.

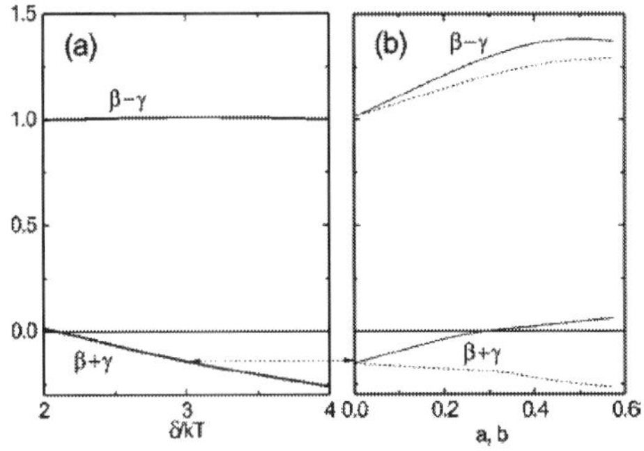

Fig. 6. Numerical modelling results of ($\beta + \gamma$) and ($\beta - \gamma$) dependencies on: (a) parameter δ/kT of Gaussian distribution of localized states; (b) Poole-Frenkel parameters a (doted line) and b (line) when $\delta/kT = 3$

2.3 Photo-CELIV method

Photo-CELIV method demonstrate even more opportunities where, by short pulse of light, photogenerated charge carriers are extracted by delayed (delay time t_{dU}) triangular pulse of voltage (Fig. 7) (Österbacka et al, 2004). Measurements of amount of extracted charge dependence on the delay time t_{dU} allow investigation of the relaxation of charge carrier density and mobility, independently. The latter are important in case of stochastic transport.

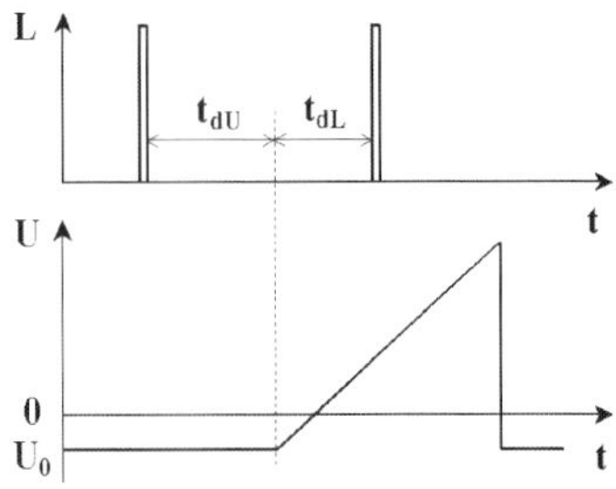

Fig. 7. Time chart of photo-CELIV method

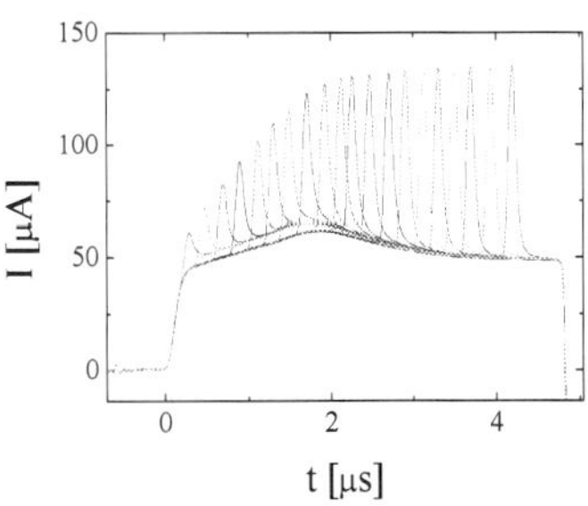

Fig. 8. Photocurrent transients of photo – CELIV for different t_{dL}

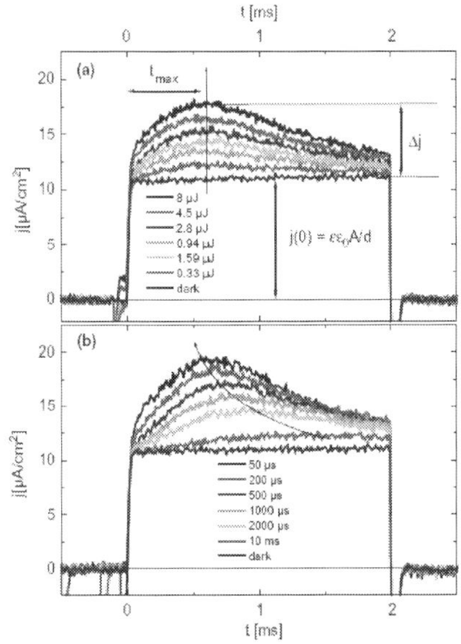

Fig. 9. Photocurrent transients of photo-CELIV at different intensity of pulse of light and fixed delay time (a), and at different delay time t_{dU} and fixed intensity of pulse of light (b) in RRa-PHT layer

2.4 Double injection current transient (DoI) method.

After the voltage is applied onto solar cell's electrodes in forward direction, the double injection current is observed. When the dielectric relaxation time is longer than charge carrier drift time ($\tau_\sigma = \varepsilon\varepsilon_0/\sigma >> t_{tr}=d/\mu E$), in case of bimolecular Langevin recombination, the whole injected charge carriers recombine while moving through interelectrode distance, and the observed current transient matches space charge limited current transient in case of sum of mobilities of both sign carriers.

When the recombination is weaker, then, after the drift time (t_{sl}) of slower charge carriers, an amount of injected charge carriers and, at the same time, current increases till saturates, due to recombination. Thus, the dependence of saturated density of current on voltage is:

$$j = 2\sqrt{\frac{e\varepsilon\varepsilon_0\mu_n\mu_p\left(\mu_n + \mu_p\right)}{B}}\frac{U^2}{d^3} = 2\varepsilon\varepsilon_0\sqrt{\frac{B_L}{B}\mu_n\mu_p}\frac{U^2}{d^3}\text{., when } \tau_\sigma >> t_{tr}; \quad (16)$$

From the shape of current transient pulse it is possible evaluate whether recombination is of Langevin-type or weaker. In Fig. 10 there are shown measurable parameters, from which the transport and recombination values is estimated. The sum of mobilities of both sign charge carriers as

$$\mu_n + \mu_p = 0.8\frac{d^2}{Ut_{sc}} \quad (17)$$

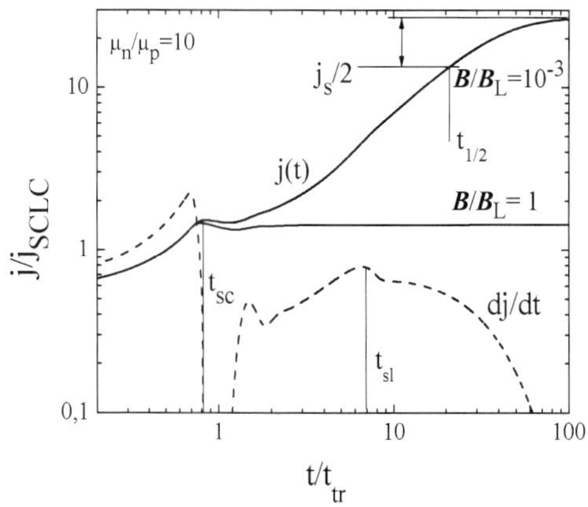

Fig. 10. Transients of double injection currents into dielectric in cases of Langevin and reduced bimolecular recombination

or in other case

$$\mu_n + \mu_p \cong \frac{j_{SCLC} d^3}{\varepsilon\varepsilon_0 U^2} . \tag{18}$$

The mobility of slower charge carriers

$$\mu_{sl} = 0.8 \frac{d^2}{U t_{sl}} . \tag{19}$$

The coefficient of bimolecular recombination

$$B = \frac{\ln 3}{2} e\left(\mu_n + \mu_p\right) \frac{U}{j_s t_{1/2} d} . \tag{20}$$

In case of plasma injection into semiconductor, when dielectric relaxation time is shorter than charge carrier transit time ($\tau_\sigma = \varepsilon\varepsilon_0/\sigma >> t_{tr}=d/\mu E$), after ambipolar transit time of charge carriers ($t_a = t_m$), an amount of injected plasma, and, thereby, the current, increases till, due to recombination, saturates.

$$j(t) = \begin{cases} \sigma E \left(1 - \frac{2}{3}\frac{t}{t_a}\right)^{-\frac{1}{2}}, & t < \frac{5}{6} t_a \\ \frac{3}{2}\sigma E + \left(j_S - \frac{3}{2}\sigma E\right) \tanh\left(B \Delta n_S \left(t - \frac{5}{6} t_a\right)\right), & t > \frac{5}{6} t_a \end{cases}$$

$$j_s = \frac{8}{9} e \sqrt{\frac{(\mu_p + \mu_n)\mu_p \mu_n (n_0 - p_0)}{B}} \cdot \frac{U^{3/2}}{d^2} = \frac{8}{9} \cdot \frac{U}{d} \varepsilon\varepsilon_0 \sqrt{\frac{B_L}{B} \cdot \frac{1}{\tau_\sigma t_a}}, \text{ when } \tau_\sigma << t_a. \quad (21)$$

From the maximum of differential of current transient, using Eq. (22), the ambipolar mobility (μ_a) is estimated:

$$t_m \left(\left. \frac{dj}{dt} \right|_{max} \right) = t_a = \frac{5}{6} \cdot \frac{d^2}{\mu_a U}. \quad (22)$$

Using Eq. (23), the coefficient of bimolecular recombination is:

$$B / B_L = 0.45 \frac{\tau_\sigma t_m}{t_{1/2}^2}. \quad (23)$$

From shape of double injection current pulse the information about charge carrier trapping is obtained (Fig. 11) (Juška et al, 2008). During the trapping of the slower charges, after the transit time, through the interelectrode distance, of faster charge carriers, the space charge limited current is flowing through the sample till the whole trapping states are filled in by slower charge carriers ("hole trapping" in Fig. 11). When the trapping of faster charge carriers is dominating, the current is decreasing and begins to increase after trapping states are filled in ("electron trapping" in Fig. 11). Thus, the integration of current until time when the current starts to rise, allows evaluate density of trapping states.

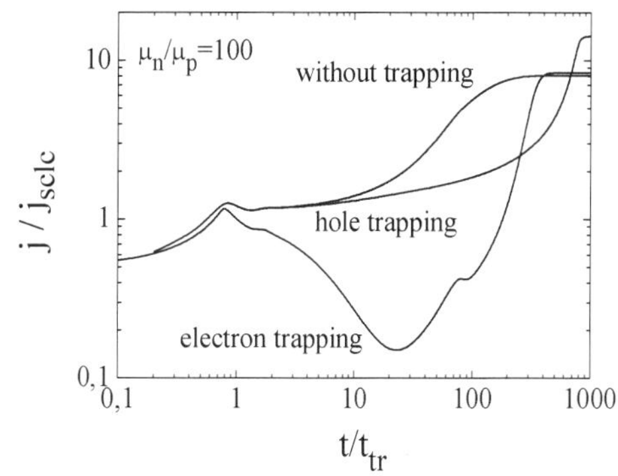

Fig. 11. Numerical modelling of double injection current transients when trapping is absent and when faster or slower charge carriers are trapped

The high capacitance of thin solar cells, immediately after application onto electrodes of rectangular pulse of voltage, causes high initial spikes of current, which complicates measurement and analysis of double injection current transients. To around the latter problem is possible by modification of DoI method, i.e. immediately after forward voltage pulse to apply the pulse of backward direction, and to measure the extraction current transient (Fig. 12) (Juška et al, 2006).

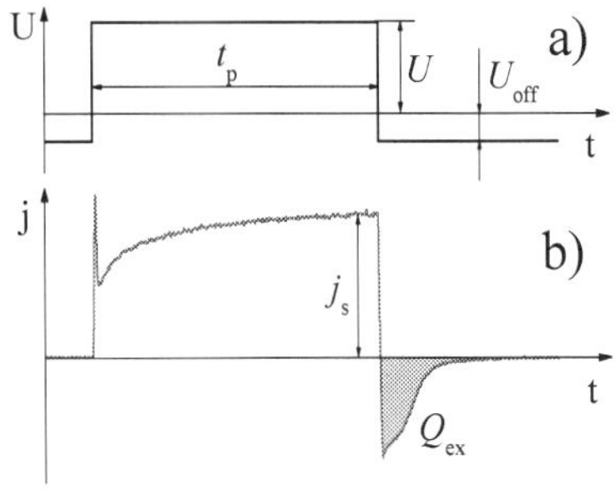

Fig. 12. Time charts of voltage pulse (a) and transients of double injection and extraction currents

The integral of extraction current gives an amount of extracted charge, from which's dependence on duration of injecting voltage (Fig. 13), the charge carrier mobility and bimolecular recombination coefficient can be estimated as:

$$B = \frac{\ln 3}{2} \frac{edS}{t_{1/2} Q_s}. \tag{24}$$

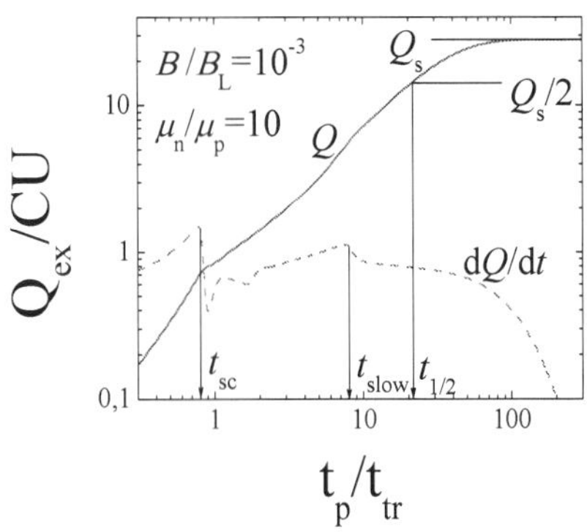

Fig. 13. Dependence of extracted charge on duration of injecting voltage pulse

3. Experimental results

3.1 Amorphous and microcrystalline silicon

Light-induced degradation of amorphous hydrogenated silicon (a-Si:H) is a serious problem of a-Si:H based photovoltaic solar cells. The most probable driving force for a-Si:H degradation is the energy (more than 1 eV) released during nonradiative bimolecular recombination of electron-hole pairs (which prevails at high light intensity) and that is why the discovery of mechanism of this recombination is of great importance.

For the study of bimolecular recombination coefficient (B) we have proposed the photoelectrical method (Juška et al, 1995), which is based on the measurement of extraction time (t_e) of the charge carrier reservoir using the space-charge limited photocurrent (SCLP) transient method. This method gives a possibility to estimate the monomolecular recombination time from the shape of the t_e dependence on the light intensity (L) and the bimolecular recombination coefficient B from the saturated value of t_e. These photoelectrical measurements demonstrated that the bimolecular recombination begins to prevail if charge carrier density is approximately 10^{17} cm^{-3}, and $B \cong 10^{-9}$ cm^3/s.

In a-Si:H layers it was observed the reduced bimolecular recombination, which, possibly, is reduced because electron and hole, immediately after photogeneration, are separated by internal random potential field. Fig. 14 demonstrates that the bimolecular recombination coefficient is lower in a-Si:H layers, which are deposited at high grow speeds (internal random field is greater), and that temperature dependence of B is stronger than that of high-quality amorphous silicon layers.

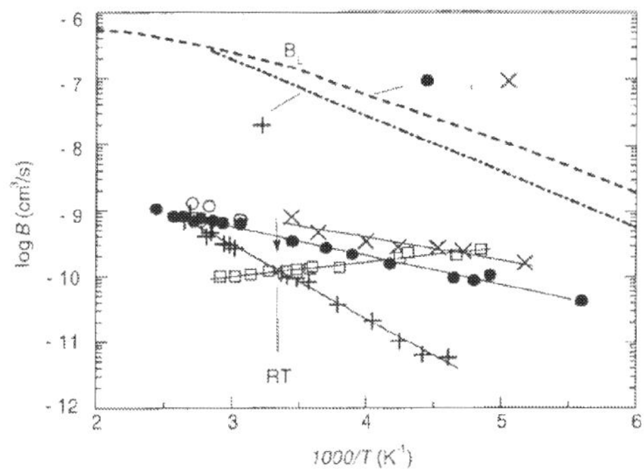

Fig. 14. Dependence of bimolecular recombination coefficient of electrons (•) and holes (○) on temperature in: high grade a-Si:H (×, •, ○), high deposition rate a-Si:H (+) and μc-Si:H (□) layers. Temperature dependencies of Langevin recombination coefficient (B_L)

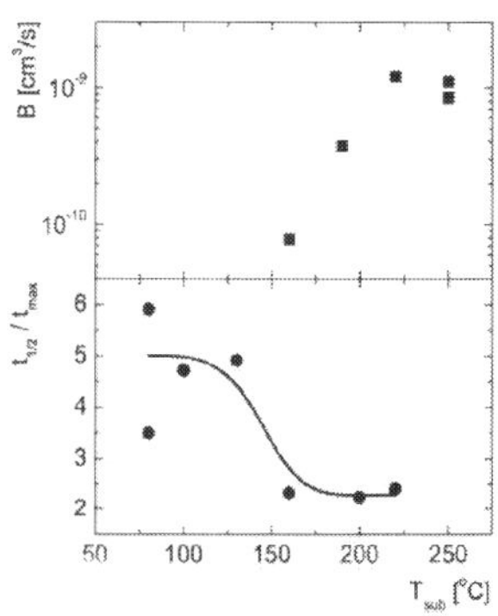

Fig. 15. Dependencies of bimolecular recombination coefficient B and dispersion parameter, estimated as $t_{1/2}/t_{max}$ from CELIV (Fig. 4), on substrate temperature during deposition of μc-Si:H layer

In similar way the internal random potential influences bimolecular recombination in microcrystalline hydrogenated silicon (μc-Si:H). The temperature of substrate during deposition of μc-Si:H strongly influences the magnitude of internal random potential, and, through the latter, influences dispersion of charge carrier transport. Thus, decreasing of the substrate temperature leads to increase of dispersion of charge carrier transport, but decreases coefficient of bimolecular recombination (Fig. 15).

3.2 π-conjugated polymers

Recently, the great opportunity to create enough effective, large area and low-cost organic solar cells (OSC) increased interest in π-conjugated polymers, but also has raised several problems. First of all, in disordered materials, which include π-conjugated polymers, the mobility of charge carriers, due to hopping, is low ($\mu << 1$ cm²/Vs) in comparison with crystalline materials, and the mean hopping distance of charge carriers is shorter than Coulomb radius. In turn, the low charge carriers jump distance results in a low photogeneration quantum efficiency and conditions the diffusion-controlled, Langevin type charge carrier recombination. Latter is caused by, the diffusion and Coulomb inter-traction field controlled, meeting probability of electron and hole in space. On the other hand, in order to OSC current density would be comparable with the crystalline semiconductors the density of photogenerated charge carriers should be much higher than in the crystalline solar cells.

The density of charge carriers, due to bimolecular recombination, causes small their lifetime because $\tau = (Bn_f)^{-1}$. However, for effective OSC, it is necessary that the lifetime of charge carriers should be higher than their drift time through the interelectrode distance in intrinsic electric field, i.e. $\tau > t_{tr}^{i} = d^2 / \mu U_i$ (U_i is intrinsic potential). Thus, for higher than 5% efficiency of OSC, when open circuit voltage is ~ 0.5 V, thickness of sample is 300 nm, it is necessary that density of photocurrent will be higher than 15 mA/cm², and $\mu B_L / B > 5\times10^{-3}$ cm²/Vs. Thus, the bimolecular recombination limits efficiency of organic solar cell in region of high intensity light, and ratio of bimolecular recombination coefficient with Langevin's one allows evaluate effectiveness of materials and structures.

As a model material for investigation of features of bimolecular recombination was chosen π-conjugated polymer RRa PHT. In RRa PHT layer the TOF current transients were nondispersive at low intensity of light pulses. With increase of intensity of the light pulse, the shape of photocurrent transient changes to the classic SCLC kinetics, and, at a very high intensity of light, its shape stopped to change (Fig. 16). An amount of extracted charge linearly increased with intensity of light and saturated in the region of high light intensity, when $Q_e / CU_0 = 1$. Such saturation of the $Q_e(L)$ dependence is the consequence of Langevin recombination (See Eq. (4)).

To assess the coefficient of Langevin recombination, it is necessary to know the charge carriers mobility, when the electric field is zero, because in the depth of photogeneration the electric field is shielded by the carriers. By measuring the hole mobility dependence on electric field strength and by extrapolation to $E = 0$, the hole mobility was evaluated as $\mu_p(E = 0) = 6.5\times10^{-6}$ cm²/Vs, and, considering that the mobility of electrons at least by one order lower than of the holes, $B_L = 4\times10^{-12}$ cm³/s was estimated.

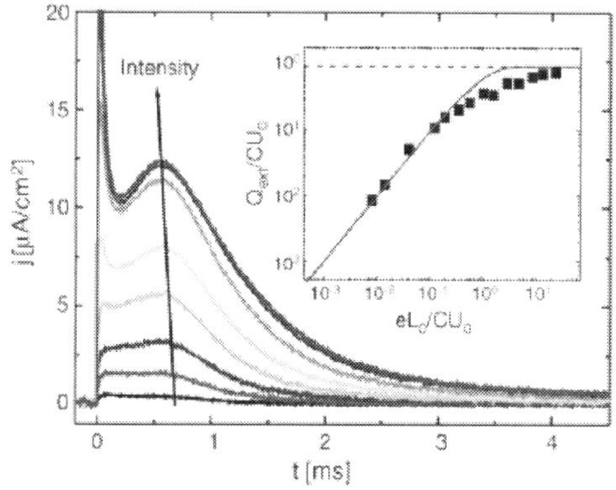

Fig. 16. TOF transients of photocurrent for different intensity pulse of light. Thickness of RRa PHT layer is 8 μm, $E = 10^5$ V/cm

For direct measurements of photogenerated charge carrier density and mobility relaxation on time, the photo-CELIV method was used. From the duration t_{max} (see Fig. 4) the mobility of charge carriers and from integral of conductivity current ($\frac{1}{e}\int_0^\infty \Delta j dt$, here Δj is density of conductivity current) the density of charge carriers (p) at given t_{dU} (see Fig. 7) are estimated. The presence in the structure of intrinsic electric field has been compensated by offset voltage. Possible inaccuracy of this method can be caused by spatially distributed intrinsic electric field, which separates photogenerated charge carriers, thereby, decreasing recombination.

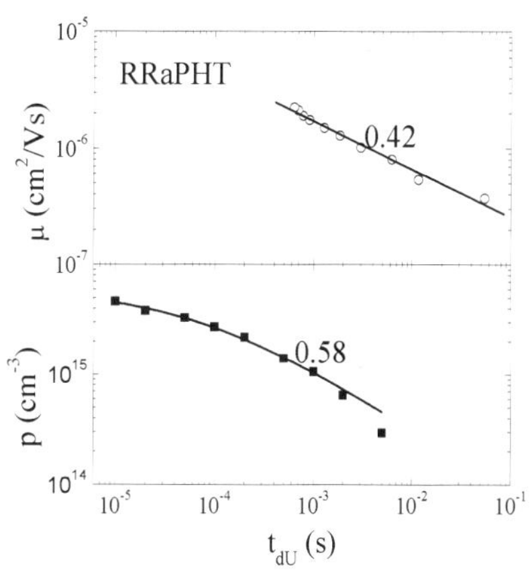

Fig. 17. Charge carrier mobility and density dependencies on t_{dU}. Solid line is $p(t)$ results according $B(t) = e\mu(t)/\varepsilon\varepsilon_0$ in case of Langevin recombination

In Fig. 17 the typical for RRa PHT mobility and density dependencies of hole on time are demonstrated. Fitting of mobility relaxation as $\mu = at^{-0,42}$ gives that Langevin recombination coefficient changes with time too, i.e. $B(t) = e\mu(t)/\varepsilon\varepsilon_0$. Thus, charge carrier density follows expression

$$p(t) = \frac{p(0)}{1 + p(0)B\int_0^t \mu(t)dt}. \quad (25)$$

Therefore

$$p(t) = \left(\frac{1}{p(0)} + \frac{e}{\varepsilon\varepsilon_0} \cdot \frac{at^{0.58}}{0.58}\right)^{-1}. \quad (26)$$

which is shown by solid line in Fig. 17. The coincidence of experimental results and theory confirms that, in low mobility organic material, bimolecular recombination is of Langevin-type. So, the same result has been obtained from the saturation of SCLC transients with intensity of light (Pivrikas et al, 2005 [5]).

3.3 Recombination in conjugated polymer/fullerene bulk heterojunction solar cells

One of possibilities reduce bimolecular recombination is to make junction of two organic material layers, in one of which are mobile the electrons and in another one the holes. The excitons, immediately after photoexcitation, are destroyed by electric field of heterojunction and separated electrons and holes are moving each of its transport material to sample electrodes. However, in organic polymers the diffusion distance not exceed 100 nm. Thus, an efficiency of such solar cell would be low because the thickness of solar cell would be approximately 100 nm and absorption of light weak. From Fig. 18a follows, that, oppositely to MEH PPV (poly(2-methoxy-5-(2′-ethylhexykoxy)-1,4-phenylenevinylene) layer, the heterojunction of MEH PPV/perilene more effectively separates photogenerated pairs, i.e. the charge carrier reservoir is created and an amount of collected charge approximately twice exceeds CU_0 (Fig. 18b). Supporting the latter experimental result, the numerical modelling, taking into account the Langevin recombination, demonstrates that, in case of bulk absorption of light and of small resistor, causing extraction current, an amount of extracted charge can exceed CU a few times, too. Thus, the obtained experimental results do not deny the Langevin recombination in heterojunction.

Another charge carriers bimolecular recombination reduction method has been identified investigating a-Si:H and μc-Si:H layers: separate photogenerated charge carriers by an internal random field in space, so, that they move towards the electrodes in different ways. This method has been used for organic semiconductor structures: layer cast mixing transporting materials of holes and electrons. Such a bulk heterojunction blends allow to expect a significant reduction of bimolecular recombination, as in the bulk of samples created excitons are in the vicinity to heterojunctions. When they disintegrate, resulting electrons and holes moving towards each of its material to the contrary of the electrodes, i.e. separated in space.

Experimentally there were investigated bulk heterojunctions of various organic polymers with PCBM.

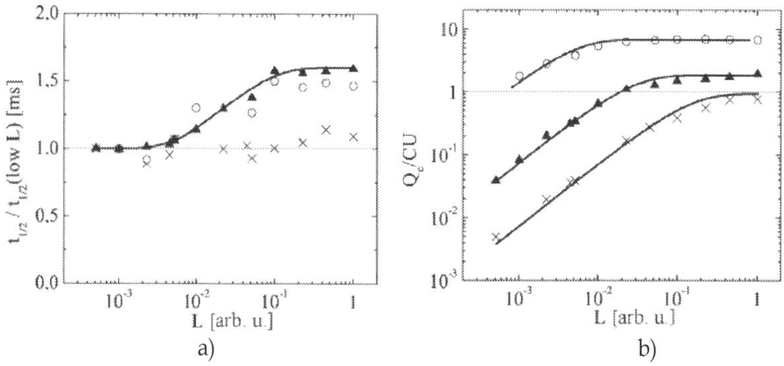

Fig. 18. Dependencies of TOF current transient halfwidth (a) and of collected charge (b) on the intensity of light pulse: × - MEH PPV, ▲ – perilene/MEH PPV junction, O - MEH PPV:PCBM blend

Time-dependent mobility and recombination in the blend of poly[2-methoxy-5-(3,7-dimethyloctyloxy)-phenylene vinylene] (MDMO-PPV) and 1-(3-methoxycarbonyl)propyl-1-phenyl-(6,6)-C_{61} (PCBM) is studied simultaneously using the photoinduced charge carrier extraction by linearly increasing voltage technique (Mozer et al, 2005).

Photo-CELIV transient at various delay times, light intensities and applied voltages have been recorded, and the charge carrier mobility and lifetime simultaneously studied. It is found that, shortly after photoexcitation, both the charge mobility and the recombination are time-dependent (dispersive) processes, which is attributed to the initial relaxation of the charge carriers towards the tails states of the density of states distribution. The results confirm that the recombination dynamics within the studied μs - ms time scale is a thermally activated process rather than a temperature independent tunneling. The obtained time-dependent mobility values are used to directly describe the recombination dynamics (see Fig. 19 and Fig. 20). Density decay of charge carriers fitted according to Eq. (25).

Therefore results suggest that the recombination dynamics is nearly Langevin-type, i.e. controlled by diffusion of the charge carriers towards each other.

3.4 Recombination in P3HT:PCBM Bulk Heterojunction Organic Solar Cells

In poly(3-hexylthiophene): 1-(3-methoxycarbonyl) propyl-1-phenyl[6,6]C_{61} (P3HT:PCBM) bulk heterojunction solar cells, a reduction of the Langevin recombination is commonly observed after thermal treatment. This treatment has been shown to modify significantly the nanomorphology of the photoactive composite, inducing a crystallization of both the donor and the acceptor phases (Pivrikas et al, 2007). In Fig. 21 the experimentally measured results using integral TOF SCLC regime are presented. By comparing experimentally measured bimolecular recombination coefficient B with calculated Langevin recombination coefficient B_L, it was shown that $B/B_L \cong 10^{-3}$.

According to (Adriaenssens et al, 1997), if the reduction of bimolecular recombination is caused by random potential, the bimolecular recombination has follow $B \cong B_L \exp(-\Delta E/kT)$ dependence on temperature (here ΔE is mean random potential energy). Thus, the activation energy of B has to be higher than one of B_L, while experimentally it is obtained an opposite result. In case if bimolecular recombination is caused by tunnelling, the B/B_L ratio will

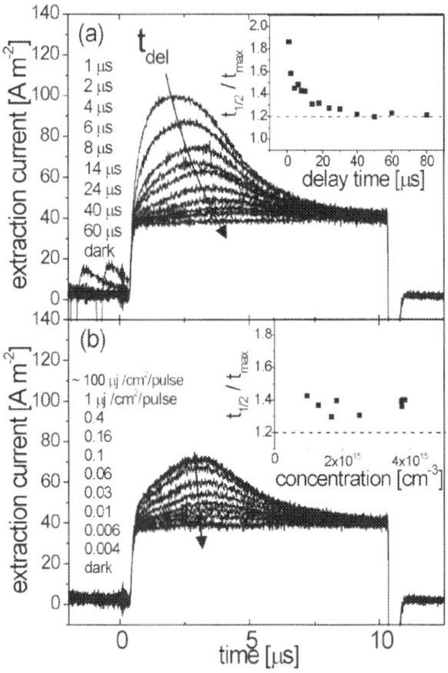

Fig. 19. Photo-CELIV transients recorded at 300 K at (a) various delay times at fixed light intensity; (b) varying illumination intensities attenuated using optical density filters at fixed 5 µs delay time. The voltage rise speed *A* was 4 V/10 µs. The insets show the calculated dispersion parameters $t_{1/2}$ to $t_{\max}$ versus delay time and the concentration of the extracted charge carriers, respectively.

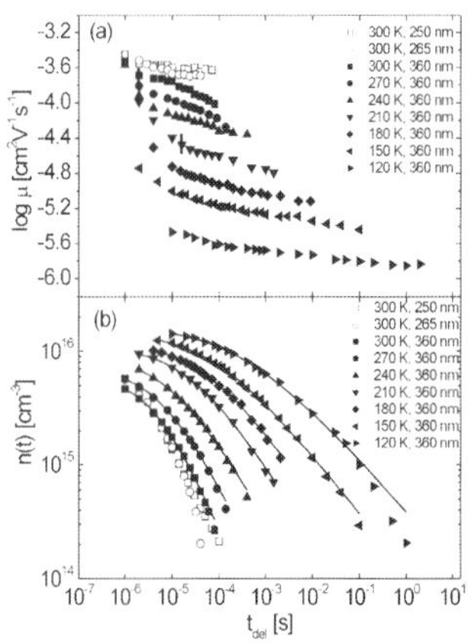

Fig. 20. Mobility (a) and the density (b) of extracted charge carriers versus the delay time for samples with different active layer thickness. Charge carrier mobility and density measured for the 360 nm device. Density relaxation of charge carriers fitted according to Eq. (25)

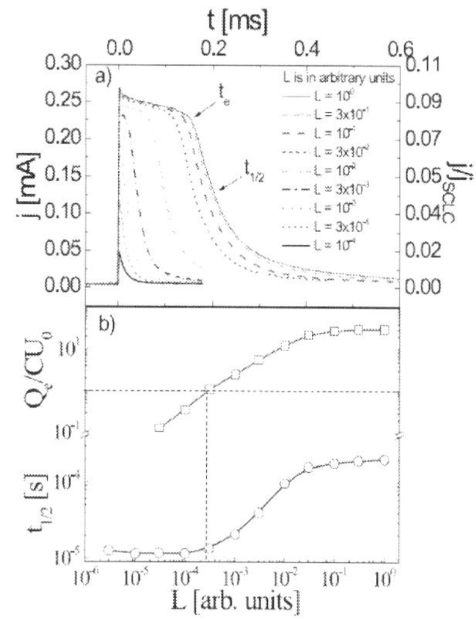

Fig. 21. Dependencies of integral TOF photocurrent transients (a), and of Q_e/CU_0 and $t_{1/2}$ on intensity of light pulse in RRP3HT:PCBM bulk heterojunction. Horizontal dotted line (b) corresponds $Q_e = CU_0$

demonstrate strong dependence on electric field. However, as is obvious from experimental results, this is not a case. The increasing of random potential also did not cause reduction of B/B_L ratio. In blend of segmented electron and hole transporting materials the meeting of charge carriers of opposite sign may be limited by charge carriers of lower mobility. However, this is insufficient to explain such big reduction of bimolecular recombination in RRP3HT:PCBM blend. This can be explained by, that the interface between polymer and acceptor materials, decreases Coulomb interaction, which suppress gemini and bimolecular recombination as it was proposed in (Arkhipov Heremans & Bässler, 2003).

Furthermore, double injection current transients (DoI) and photo-CELIV measurements revealed, that the reduced B depends on the charge carrier density as in the case of Auger recombination (Juška et al, 2008). The same conclusion followed from transient photo voltage and transient photo absorption spectroscopy experiments (Shuttle et al, 2008). In the recent transient absorption spectroscopy experiments (Nelson, 2003) it was suggested that this type of relaxation is caused by a stochastic transport attributed to an exponential tail of localized states. However photo-CELIV and TOF experiments showed that the photocurrent relaxation is caused by the charge carrier's recombination (Pivrikas et al, 2005).

In this work, we are demonstrating that the reduction of B and its dependence on charge carriers density is caused by the two dimensional Langevin recombination (Juška et al, 2009).

Sirringhaus (Sirringhaus et al, 1999) showed that the mobility across and along the lamellar structure differs more than 100 times, which led to the fact that the recombination of charge carriers is mainly taking place in the two-dimensional lamellar structure. When spacing between lamellas $l << r_m$, r_m is determined by $\pi r_m^2 l = 1/n$. Then the recombination probability

$$f_{2D} = \frac{1}{t_m} = \frac{3\sqrt{\pi}}{4} \cdot \frac{e(\mu_n + \mu_p)}{\varepsilon\varepsilon_0} (l \cdot n)^{3/2} = \gamma_{2D} n^{3/2} . \quad (27)$$

where γ_{2D} is 2D recombination parameter. Hence, in two dimensional case, the bimolecular recombination coefficient will be reduced in comparison with one of the three dimensional case as

$$\frac{B_{2D}}{B_{3D}} = \frac{3\sqrt{\pi}}{4} l^{3/2} n^{1/2} .$$

For RRP3HT $l \cong 1.6$ nm (Sirringhaus et al, 1999) and, for example, when $n = 10^{16}\text{cm}^{-3}$, $\beta_{2D}/\beta_{3D} = 6\times10^{-3}$, and that is close to experimental results of Ref. (Pivrikas et al, 2007).
In the 2D recombination case, using Eq. (27), the equation governing the decay of the charge carriers is:

$$\frac{dn}{dt} = G - \gamma_{2D} n^{5/2} . \quad (28)$$

where G is the rate of the photogeneration or double injection. Similar dependencies were observed experimentally: $dn/dt \propto n^{2/6}$ (Shuttle et al, 2008). According to Eq. (28), after excitation by short pulse of light, the decay of the density of charge carriers is described as:

$$n(t) = \left(n_0^{-3/2} + \frac{3}{2}\gamma_{2D} t \right)^{-2/3} \propto t^{-2/3} \Big|_{n_0 \to \infty} , \quad (29)$$

here n_0 is initial density of photogenerated charge carriers. The similar dependence is observed using photo-CELIV technique (Fig.22).
In the case of 3D Langevin recombination $n(t) \propto t^{-1}$. It is worth to notice, that the slower than $n(t) \propto t^{-1}$ dependence can be observed due to the mobility dependence on time (stochastic transport), as it was shown in regiorandom poly(3-hexylthiophene) (Pivrikas et al, 2007). However, it is established by photo-CELIV that the mobility does not depend on the delay time after excitation (Fig. 22). That is why this explanation is not valid for the RRP3HT:PCBM blends. Another technique, which allows investigation of recombination process, is DoI current transient technique (Juška et al, 2007).

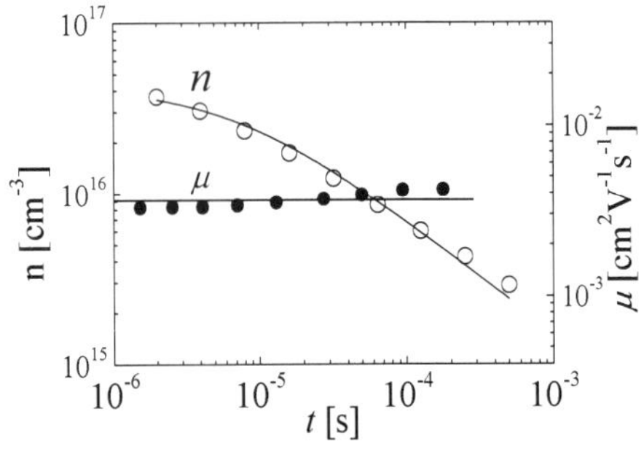

Fig. 22. Measured using photo-CELIV dependencies of mobility and density of electrons on time in annealed RR-P3HT:PCBM bulk heterojunction.

In the case of 3D Langevin recombination volt-ampere characteristics and DoI current transients corresponds to the sum of the space charge limited currents of electrons and holes, because the injected charge carriers will recombine completely within the interelectrode distance. In Fig. 23a numerically modelled DoI current transients are shown for the both 3D and 2D Langevin recombination cases for different distance l and ratios between fast and slow charge carriers mobilities μ_f / μ_{sl}.

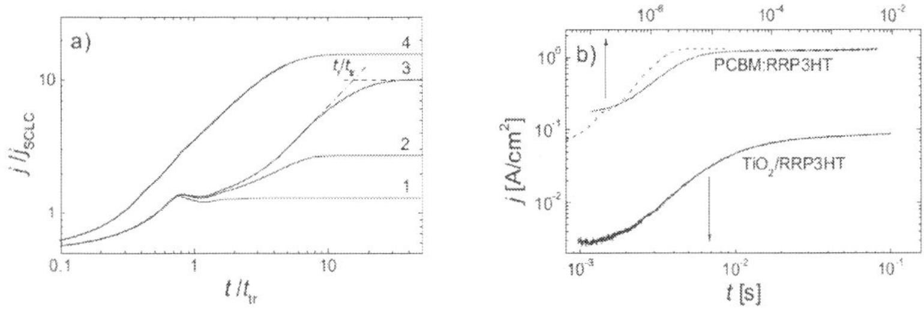

Fig. 23. Numerically modelled DoI current transients: a) in the case of 3D Langevin recombination (1) and 2D Langevin recombination for [(2) l = 10 nm, (3) l = 1 nm and μ_f / μ_{sl} = 10, (4) l = 1 nm and μ_f / μ_{sl} = 1]; both the time scale and current are normalized to transit time t_{tr} and SCLC of faster charge carriers, respectively. (b) DoI current transients: solid lines - measurements of RRP3HT:PCBM bulk heterojunction (d = 1.4 μm, U = 9 V) and TiO_2/RRP3HT (d = 0.6 μm, U = 4 V) structure; dashed line - numerically modelled DoI current transient for RRP3HT:PCBM structure ($\mu_n = 10^{-2}$ cm²/Vs, $\mu_p = 2.5\times10^{-3}$cm²/Vs, l = 1.6 nm)

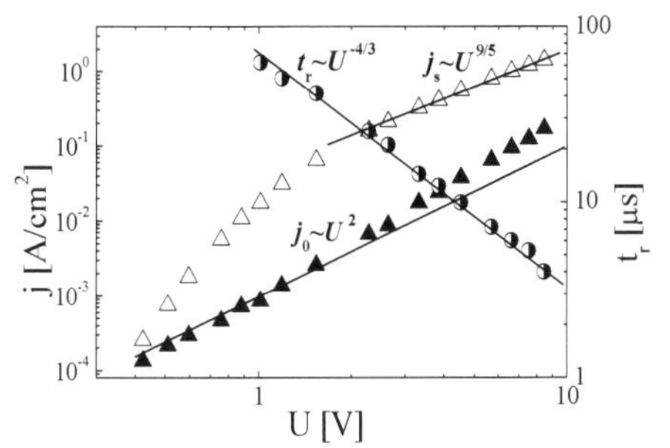

Fig. 24. The dependencies of initial current j_0, saturation current j_s and recombination time t_r on voltage in the case of double injection.

In the case of 2D Langevin recombination the current-voltage characteristics can be obtained from Eq. (16) in the same way as in Ref. (Juška et al, 2006): $j \propto U^{9/5} / d^{13/5}$, while the saturation time of the DoI current transient is $t_r \propto U^{-6/5}$. The observed experimental results (Juška et al, 2007) are very close to these dependencies.

The comparison of DoI current transients obtained experimentally and numerically modelled using $\mu_n = 10^{-2}$ cm²/Vs, $\mu_p = 2.5\times10^{-3}$ cm²/Vs and l = 1.6 nm (obtained from X-rays studies (Sirringhause et al, 1999) of RRP3HT:PCBM blend is presented in Fig. 23b. The ratios between initial and stationary currents are in good agreement with experiment giving the same *l* value, and the discrepancy in current rise times could be explained by not taken into account dispersion of the charge carrier transport.

So, the observed increase of DoI current in RRP3HT:PCBM bulk heterojunction and TiO_2/RRP3HT structures and good fit of numerical modelling with experimental data in RRP3HT:PCBM, proves that recombination takes place in RRP3HT. The slower rise of current in TiO_2/RRP3HT structure is caused by the lower electron mobility and deep trapping (Juška et al, 2008).

By the integral mode TOF method, where the *RC* time constant of the measurement setup is much larger than the transit time of the charge carriers ($RC >> t_{tr}$) we can determine the 2D recombination parameter γ_{2D} in lamellas and its temperature dependencies in more convenient (Pivrikas et al, 2005) and straightforward way because it is independent on material's parameters. In the case of 3D Langevin recombination, the current transient saturates as a function of light-intensity and the amount of extracted charge slightly exceeds *CU* (when $\alpha d >> 1$; $Q_{ex} = CU$, therefore $t_{ex} = 0$). In the case of 2D Langevin recombination, the charge carrier extraction time t_{ex}, when collected charge saturates with light intensity, is estimated in the similar way as in the case of reduced bimolecular recombination (Juška et al, 1995):

$$t_{ex} = \left(\frac{2}{3\gamma_{2D}}\right)^{2/5} \left(\frac{ed}{j_{ex}}\right)^{3/5} \propto j_{ex}^{-3/5}, \tag{30}$$

where j_{ex} is extraction current, which could be varied by changing loading resistor or applied voltage. In the case of the reduced bimolecular recombination $t_{ex} \propto j_{ex}^{-1/2}$.

In Fig. 25b the extraction time as a function of the density of extraction current in different structures containing RRP3HT is shown. Since t_{ex} shows the same dependence on the extraction current density j_{ex}, it can be concluded that the recombination is taking place in RRP3HT and it is governed by the 2D Langevin recombination.

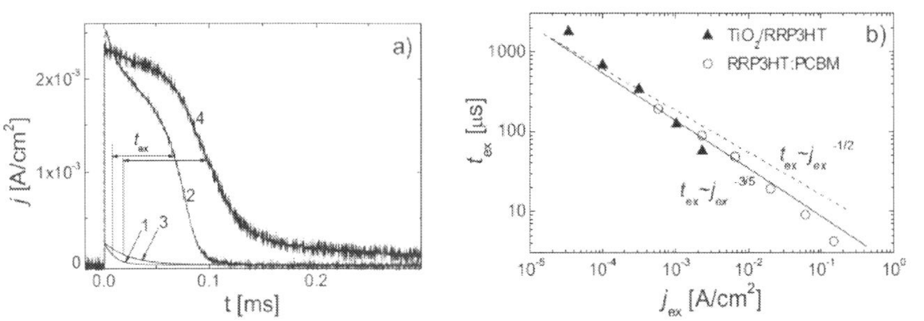

Fig. 25. Current transients of charge carrier extraction (a) observed by integral TOF: small charge drift current (1, 3) and transient of saturated on light intensity photocurrent (2, 4) in TiO_2/RRP3HT structures and RRP3HT:PCBM bulk heterojunction, respectively. Measurement of extraction time is indicated. Dependencies of extraction time (b) on the extraction current density in TiO_2/RRP3HT structures and RRP3HT/PCBM bulk heterojunction

4. Conclusion

In this work there are demonstrated the methods of investigation of charge carrier recombination in organic solar cells, where stochastic transport of charge carriers complicates interpretation of experimental results: charge carriers extraction by linearly increasing voltage (CELIV), which allows independently measure relaxation of density and mobility of photoexcited charge carriers; double injection current transient, which is additional method of investigation of charge carrier recombination and, which allows to measure dependence of recombination coefficient on electric field; integral time of flight (SCLC), using of which allows easily estimate the temperature dependence of recombination coefficient.

Experimentally it is shown, that the decay of the density of photogenerated charge carriers in the blend of MDMO-PPV:PCBM is of 3D Langevin-type, which is typical for organic materials, and in annealed samples of RRP3HT and bulk heterojunction solar cells of RRP3HT:PCBM it is of 2D Langevin-type recombination in the lamellar structure.

5. Referencies

Adriaenssens, G. J. & Arhipov, V. I. Non-Langevin recombination in disordered materials with random potential distributions. *Solid State Communications*, Vol. 103, Issue 9 (September 1997) 541- 543. ISSN 0038-1098.

Arkhipov, V. I.; Heremans, P. & Bässler, H., Why is exciton dissociation so efficient at the interface between a conjugated polymer and an electron acceptor? *Applied Physics Letters*, Vol. 82, Issue 25 (June 2003) 4605 1 - 3, ISSN 0003 6951.

Juška, G. ; Viliūnas, M.; Arlauskas, K.; Kočka, J. Space-charge-limited photocurrent transients: The influence of bimolecular recombination. *Physical Review B*, Vol. 51, No. 23 (June 1995) 16668 - 16676, ISSN 1098-0121.

Juška, G.; Viliūnas, M.; Arlauskas, K.; Stuchlik, J. & J. Kočka. Ultrafast Charge Carrier Recombination in a-Si:H and μc-Si:H. *Physica status solidi (a)*, Vol. 171, No. 2 (February 1999) 539 - 547, ISSN 1682-6300.

Juška, G.; Arlauskas, ; Viliūnas, M. & Kočka, J.. Extraction Current Transients: new method of study of charge transport in microcrystalline silicon. *Physical Review Letters*, Vol. 84, No. 21, (May 2000) 4946-4949, ISSN 0031-9007, a.

Juška, G.; Arlauskas, K.; Viliūnas, M.; Genevičius, K.; Österbacka, R. & Stubb, H. Charge transport in π-conjugated polymers from extraction current transients. *Physical Review B*, Vol. 62, No. 24 (December 2000) 16235-16238, ISSN 1098-0121, b.

Juška, G.; Genevičius, K.; Arlauskas, K.; Österbacka, R. & Stubb, H.. Features of charge carrier concentration and mobility in π-conjugated polymers. *Macromolecular Symposia*, Vol. 212, No. 1 (May 2004) 209-217, ISSN 1022-1360.

Juška, G.; Arlauskas, K.; Sliaužys, G.; Pivrikas, A.; Mozer, A. J.; Sariciftci, N. S.; Scharber, M. & Österbacka, R. Double injection as a technique to study charge carrier transport and recombination in bulk-heterojunction solar cells. *Applied Physics Letters*, Vol. 87, No. 22 (November 2005) 222110 1 – 3, ISSN 0003-6951.

Juška, G.; Sliaužys, G.; Genevičius, K.; Arlauskas, K.; Pivrikas, A.; Scharber, M.; Dennler, G.; Sariciftci, N. S. & Österbacka R. Charge-carrier transport and recombination in thin insulating films studied via extraction of injected plasma. *Physical Review* B, Vol. 74, No. 11 (September 2006) 115314 1-5, ISSN 1098-0121.

Juška, G.; Genevičius, K.; Sliaužys, G.; Pivrikas, A.; Scharber, M. & Österbacka, R. Double-injection current transients as a way of measuring transport in insulating organic films. *Journal of Applied Physics,* Vol. 101, No. 11 (June 2007) 114505 1-5, ISSN 0021-8979.

Juška, G.; Genevičius, K.; Sliaužys, G.; Nekrašas, N. & Österbacka, R.. Double injection in organic bulk-heterojunction. *Journal of Non-Crystalline Solids,* Vol. 354, Issues 19-25 (May 2008) 2858-2861, ISSN 0022-3093.

Juška, G.; Genevičius, K.; Nekrašas, N.; Sliaužys, G. & Österbacka, R.. Two dimensional Langevin recombination in regioregular poly(3-hexylthiophene), *Applied Physics Letters,* Vol. 95, No. 1 (July 2009), 013303 1-3, ISSN 0003-6951.

Mozer, A. J.; Dennler, G.; Sariciftci, N. S.; Westerling, M.; Pivrikas, A.; Österbacka, R. & Juska, G. Time-dependent mobility and recombination of the photoinduced charge carriers in conjugated polymer/fullerene bulk heterojunction solar cells, *Physical Review B,* Vol. 72, No. 3 (July 2005), 035217 1-10, ISSN 1098-0121.

Nelson, J. Diffusion-limited recombination in polymer-fullerene blends and its influence on photocurrent collection, *Physical Review B,* Vol. 67, No. 15 (April 2003) 155209 1-10, ISSN 1098-0121.

Österbacka, R.; Pivrikas, A.; Juška, G.; Genevičius, K.; Arlauskas, K. & Stubb, H. Mobility and density relaxation of photogenerated charge carriers in organic materials. *Current Applied Physics,* Vol. 4, No. 5 (August 2004) 534-538, ISSN 1567-1739.

Pivrikas, A.; Juška, G.; Österbacka, R.; Westerling, M.; Viliūnas, M.; Arlauskas, K. & Stubb, H. Langevin recombination and space-charge-perturbed current transients in regiorandom poly(3-hexylthiophene). *Physical Review B,* Vol. 71, No. 12, (March 2005) 125205 1-5, ISSN 1098-0121.

Pivrikas, A.; Juška, ; Mozer, A. J.; Scharber, M.; Arlauskas, K.; Sariciftci, N. S.; Stubb, H. & Österbacka, R. Bimolecular recombination coefficient as a sensitive testing parameter for low-mobility solar-cell materials. *Physical Review Letters,* Vol. 94, No. 17, (May 2005) 176806 1 - 4, ISSN 0031-9007.

Pivrikas, A.; Sariciftci, N. S.; Juška, G. & Österbacka, R. A review of charge transport and recombination in polymer/fullerene organic solar cells. *Progress in Photovoltaics:. Research and Applications* , Vol. 15 (July 2007) 677-696, ISSN 1062-7995.

Sirringhaus, H.; Brown, P. J.; Friend, R. H.; Nielsen, M. M.; Bechgaard, K.; Langeveld-Voss, B. M. W.; Spiering, A. J. H.; Janssen, R. A. J.; Meljer, E. W.; Herwig, P. & de Leeuw, D. M. Two dimensional charge transport in self-organized, high mobility conjugated polymers. *Nature,* Vol. 401 (October 1999), 685 - 688, ISSN 0028-0836.

Shuttle, G.; O'Regan, B.; Ballantyne, A. M.; Nelson, J.; Bradley, D. D. C.; de Mello, J. & Durrant, J. R. Experimental determination of the rate law for charge carrier decay in a polythiophene: Fullerene solar cell. *Applied Physics Letters,* Vol. 92, No. 9, (March 2008), 093311 1 - 3, ISSN 0003-6951.

14

Numerical Simulation of Solar Cells and Solar Cell Characterization Methods: the Open-Source on Demand Program AFORS-HET

Rolf Stangl, Caspar Leendertz and Jan Haschke
Helmholtz-Zentrum Berlin für Materialien und Energie, Institut für Silizium Photovoltaik, Kekule-Str.5, D-12489 Berlin Germany

1. Introduction

Within this chapter, the principles of numerical solar cell simulation are described, using AFORS-HET (**a**utomat **f**or **s**imulation of **het**erostructures). AFORS-HET is a one dimensional numerical computer program for modelling multi layer homo- or heterojunction solar cells as well as some common solar cell characterization methods.

Solar cell simulation subdivides into two parts: optical and electrical simulation. By optical simulation the local generation rate $G(\mathbf{x},t)$ within the solar cell is calculated, that is the number of excess carriers (electrons and holes) that are created per second and per unit volume at the time t at the position $\mathbf{x}$ within the solar cell due to light absorption. Depending on the optical model chosen for the simulation, effects like external or internal reflections, coherent superposition of the propagating light or light scattering at internal surfaces can be considered. By electrical simulation the local electron and hole particle densities $n(\mathbf{x},t)$, $p(\mathbf{x},t)$ and the local electric potential $\varphi(\mathbf{x},t)$ within the solar cell are calculated, while the solar cell is operated under a specified condition (for example operated under open-circuit conditions or at a specified external cell voltage). From that, all other internal cell quantities, such like band diagrams, local recombination rates, local cell currents and local phase shifts can be calculated. In order to perform an electrical simulation, (1) the local generation rate $G(\mathbf{x},t)$ has to be specified, that is, an optical simulation has to be done, (2) the local recombination rate $R(\mathbf{x},t)$ has to be explicitly stated in terms of the unknown variables n, p, φ , $R(\mathbf{x},t) = f(n, p, \varphi)$. This is a recombination model has to be chosen. Depending on the recombination model chosen for the simulation, effects like direct band to band recombination (radiative recombination), indirect band to band recombination (Auger recombination) or recombination via defects (Shockley-Read-Hall recombination, dangling-bond recombination) can be considered.

In order to simulate a real measurement, the optical and electrical simulations are repeatedly calculated while changing a boundary condition of the problem, which is specific to the measurement. For example, the simulation of a i-V characteristic of a solar cell is done by calculating the internal electron and hole current (the sum of which is the total current) as a function of the externally applied voltage.

Most solar cells, which are on the market today, can be described as a one dimensional sequence of different semiconductor layers. If they are uniformly illuminated, a one dimensional solar cell modelling is sufficient (the internal electron/hole current can flow only in one direction). This is the case for most wafer based silicon solar cells as well as for most thin film solar cells on glass as long as the integrated series connection shall not be explicitly modelled, see Fig.1 (left).

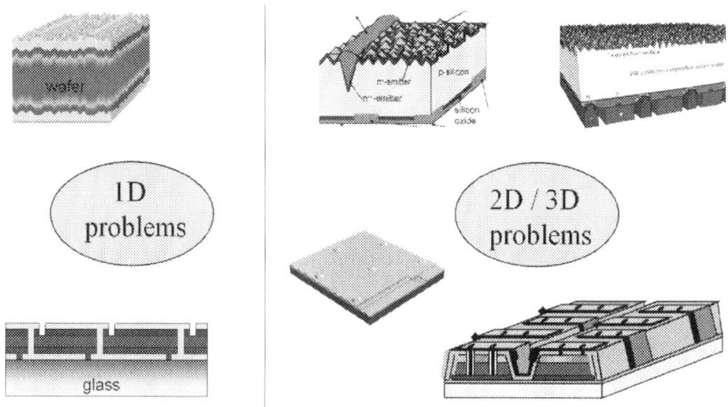

Fig. 1. solar cell structures which can be treated as a one dimensional problem (left), or which have to be treated as a two or even three dimensional problem (right).

However, in order to minimize contact recombination, stripe- or point-like metallic contacts which are embedded within an insulating passivation layer (i.e. silicon nitride, silicon oxide) are sometimes introduced. These contacts can either be placed on both sides of the solar cell or favourably only at the rear side of the solar cell, thereby avoiding shadowing due to the contacts. In these cases, the resulting solar cells have to be modelled as two or even three dimensional problems (the internal electron/hole current can flow in 2 or even 3 directions), see Fig.1 (right). In the current version 2.4 of AFORS-HET only 1D simulations are possible; however, there is a 2D mode under development.

Another possibility to reduce contact recombination is the use of heterojunctions, that is different semiconductors are used to form the solar cell absorber (photon collecting area), the electron extracting area and the hole extracting area of the solar cell. Ideally, the excess carriers of the solar cell absorber (electrons and holes) should be selectively attracted/repelled towards the contacts, see Fig. 2. These selective contacts can be either conventionally realized by doping/counter doping of the solar cell absorber, leading to a formation of an internal electric field by which the selective excess carrier separation is achieved. In this case, homojunctions will form, i.e. there are no band offsets, as the absorber and the electron/hole extracting areas of the solar cell consist of the same semiconductor. In principle, if different semiconductors with appropriately matched work functions are used to form the electron/hole extracting areas, heterojunctions can be formed having the same internal electric field as the homojunction, but with additional band offsets that enhance the repelling character of the contacts, see Fig. 2 (right).

A heterojunction solar cell will thus have a higher open circuit voltage compared to a homojunction solar cell. Less excess carriers of the repelled type are transported into the

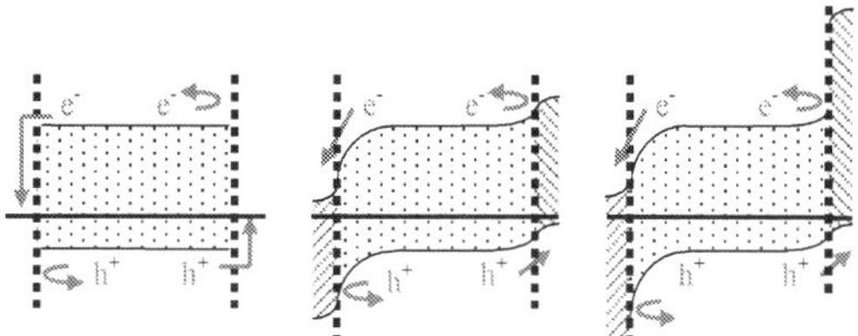

Fig. 2. schematic sketch of selective absorber contacts (band diagrams of a p-type semiconductor used as an absorber material). Ideal contacts (left), homojunction contacts (middle) and ideally aligned heterojunction contacts (right). NOTE: The dimensions of the x axis are schematic and not in scale!

electron/hole collecting regions, and thus the contact recombination at the metallic contacts is reduced. However, an essential pre-requisite is not to create too many interface defects during the formation of the heterojunction at the interface between the absorber and the electron/hole collecting area, which will otherwise act as additional recombination centres.

A realistic computer program for solar cell modelling should therefore be able to handle homojunctions as well as heterojunctions, and it should be able to consider interface defects and the corresponding interface recombination $R^{it}(t)$. Depending on the physical assumption how to describe an electron/hole transport across a heterojunction interface, a distinct interface model has to be chosen. For example, within the current version of AFORS-HET 2.4 a drift-diffusion and a thermionic emission interface model can be chosen, allowing the placement of interface defects but neglecting tunnelling. Tunneling interface models are under development.

To assure a numerical simulation with reliable results, a good model calibration, i.e. a comparison of simulation results to a variety of different characterisation methods, is necessary. The solar cell under different operation conditions should be compared to the simulations. Also different characterisation methods for the solar cell components, i.e. for the individual semiconductor layers and for any sub stacks should be tested against simulation. Only then the adequate physical models as well as the corresponding model input parameters can be satisfactory chosen. Thus a good solar cell simulation program should be able to simulate the common characterisation methods for solar cells and its components.

In this chapter, we describe AFORS-HET (**a**utomat **for s**imulation of **het**erostructures), a one dimensional numerical computer program to simulate solar cells as well as typical solar cell characterisation methods. Thus a variety of different measurements on solar cell components or on the whole solar cell can be compared to the corresponding simulated measurements in order to calibrate the parameters used in the simulations.

All optical and electrical models, which can be used in AFORS-HET, are discussed and their mathematical and physical background is stated. Furthermore, many solar cell characterisation methods, which can be simulated by AFORS-HET, are sketched. The difference in modelling thick film (wafer based) or thin film solar cells on glass will be investigated in order to choose the appropriate model. The basic input parameters of the corresponding models are described. Some selected results in modelling wafer based amorphous/crystalline silicon solar cells illustrate the concepts of numerical solar cell simulation within practical applications.

2. Brief description of AFORS-HET

The current version 2.4 of AFORS-HET solves the one dimensional semiconductor equations (Poisson´s equation and the transport and continuity equation for electrons and holes) with the help of finite differences under different conditions, i.e.: (a) equilibrium mode (b) steady state mode, (c) steady state mode with small additional sinusoidal perturbations, (d) simple transient mode, that is switching external quantities instantaneously on/off, (e) general transient mode, that is allowing for an arbitrary change of external quantities. A multitude of different physical models has been implemented. The generation of electron/hole pairs (optical models of AFORS-HET) can be described either by Lambert-Beer absorption including rough surfaces and using measured reflection and transmission files, or by calculating the plain surface incoherent/coherent multiple internal reflections, using the complex indices of reflection for the individual layers. Different recombination models can be considered within AFORS-HET: radiative recombination, Auger recombination, Shockley-Read-Hall and/or dangling-bond recombination with arbitrarily distributed defect states within the bandgap. Super-bandgap as well as sub-bandgap generation/recombination can be treated. The following interface models for treating heterojunctions are implemented: Interface currents can be modelled to be either driven by drift diffusion or by thermionic emission. A band to trap tunnelling contribution across a hetero-interface can be considered. The following boundary models can be chosen: The metallic contacts can be modelled as flatband or Schottky like metal/semiconductor contacts, or as metal/insulator/semiconductor contacts. Furthermore, insulating boundary contacts can also be chosen.

Thus, all internal cell quantities, such as band diagrams, quasi Fermi energies, local generation/recombination rates, carrier densities, cell currents and phase shifts can be calculated. Furthermore, a variety of solar cell characterisation methods can be simulated, i.e.: current voltage, quantum efficiency, transient or quasi-steady-state photo conductance, transient or quasi-steady-state surface photovoltage, spectral resolved steady-state or transient photo- and electro-luminescence, impedance/admittance, capacitance-voltage, capacitance-temperature and capacitance-frequency spectroscopy and electrical detected magnetic resonance. The program allows for arbitrary parameter variations and multidimensional parameter fitting in order to match simulated measurements to real measurements.

AFORS-HET, version 2.4, is an open source on demand program. If you want to contribute send an e-mail to AFORS-HET@helmholtz-berlin.de, specifying in detail what you would like to implement. It is distributed free of charge and it can be downloaded via internet:

http://www.helmholtz-berlin.de/forschung/enma/si-pv/projekte/asicsi/afors-het/index_en.html

3. Basic input parameter of AFORS-HET and associated physical models

3.1 Optical parameter (super bandgap generation optical models)

The incoming spectral photon flux $\Phi_0(\lambda,t)$, that is the number of incident photons of wavelength λ at the time t, has to be stated. In order to calculate the local super-bandgap generation rate $G(x,t)$ within the semiconductor stack, that is the number electrons and holes that are created per second and per unit volume at the time t at the position x due to super-bandgap light absorption, there are two optical models available: (1) Lambert-Beer absorption and (2) coherent/incoherent internal multiple reflections. For both models, the thicknesses L_i and the dielectric properties of the semiconductor layers have to be specified, i.e. the complex refractive indices, $\tilde{n}_i(\lambda) = n_i(\lambda) - i\,k_i(\lambda)$ with refractive index $n(\lambda)$ and extinction coefficient $k(\lambda)$. If the model Lambert-Beer absorption is chosen, a measured reflectivity $R(\lambda)$ of the semiconductor stack can be specified, and the resulting absorption $A(\lambda, x, t)$ within the semiconductor stack will be calculated, assuming Lambert Beer absorption by using the specified values for $k_i(\lambda)$ only and performing a ray tracing in order to account for textured surfaces and multiple bouncing of the radiation within the stack. If the model coherent/incoherent internal multiple reflections is chosen, the reflectivity $R(\lambda)$, the transmisivity $T(\lambda)$ and the absorption $A(\lambda, x, t)$ of the semiconductor stack is calculated from the specified values $n_i(\lambda)$, $k_i(\lambda)$, assuming plain surfaces within the stack but taking coherent internal multiple reflections into account, if desired. For both models, $G(x,t)$ is calculated from $A(\lambda, x, t)$ by integration over all wavelengths of the incident spectrum. In order to model optical sub-bandgap generation, optical electron/hole capture cross sections $\sigma_{n,opt} \neq 0$, $\sigma_{p,opt} \neq 0$ for the Shockley-Read-Hall defects have to be specified.

3.2 Layer parameter (semiconductor bulk models)

For each semiconductor layer, the thickness L, the electron/hole mobilities μ_n, μ_p, the effective valence/conduction band densities N_V, N_C, the electron/hole thermal velocities v_n, v_p, the electron affinity χ, the relative dielectric constant ε, the doping profile $N_D(x)$, $N_A(x)$ and the bandgap E_g of the semiconductor has to be specified. In order to describe recombination within the semiconductor, up to four different recombination models can be chosen, (1) radiative recombination, (2) Auger recombination, (3) Shockley-Read-Hall recombination, (4) dangling bond recombination. For radiative recombination, the radiative band to band rate constant r^{bb} has to be specified (Sze & Kwok, 2007). For Auger recombination, the electron/hole Auger rate constants r_n^{Aug}, r_p^{Aug} have to be specified (Sze & Kwok, 2007). For Shockley-Read-Hall recombination, the defect density distribution within the bandgap of the semiconductor $N_{trap}(E)$ and two capture cross sections σ_n, σ_p and if needed also two optical capture cross sections σ_n^{opt}, σ_p^{opt} for the electron/hole capture have to be specified (Sze & Kwok, 2007). For dangling bond recombination, the defect distribution within the bandgap of the semiconductor $N_{trap}(E)$, four capture cross sections σ_n^+, σ_p^0, σ_n^0, σ_p^- and the correlation energy U have to be specified (Sah & Shockley, 1958). Optical capture is not yet implemented in case of dangling bond recombination.

3.3 Interface parameter (semiconductor/semiconductor interface models)

The electron/hole current transport across a semiconductor/semiconductor interface can be described by three different interface models, (1) no interface, (2) drift diffusion interface, (3) thermionic emission interface. If no interface is chosen, no additional interface defects can be specified. Otherwise, an interface defect distribution $N_{trap}^{it}(E)$ can be specified. If the drift diffusion interface is chosen, an interface thickness L_{it} and interface capture cross sections σ_n^{it}, σ_p^{it} have to be specified. For both models (1) and (2), transport across the semiconductor/semiconductor interface is treated according to the drift-diffusion approximation like in the bulk of the semiconductor layers (Sze & Kwok, 2007). If the thermionic emission interface is chosen, the interface is regarded to be infinitively thin and four capture cross sections $\sigma_{n,I}^{it}$, $\sigma_{n,II}^{it}$, $\sigma_{p,I}^{it}$, $\sigma_{p,II}^{it}$ and if needed also four optical capture cross sections $\sigma_{n,I}^{it,opt}$, $\sigma_{n,II}^{it,opt}$, $\sigma_{p,I}^{it,opt}$, $\sigma_{p,II}^{it,opt}$ for electron/hole capture from both sides of the interface have to be specified. Transport across the interface is then treated according to the theory of thermionic emission (Sze & Kwok, 2007).

3.4 Boundary parameter (back/front contact to semiconductor boundary models)

The boundaries of the semiconductor stack may either be metallic (usually constituting the contacts of the solar cell) or they may be insulating in order to simulate some specific measurements requiring insulator contacts. Four different boundary models can be chosen: (1) flatband metal/semiconductor contact, (2) Schottky metal/semiconductor contact, (3) insulator contact, (4) metal/insulator/semiconductor contact. If choosing the flatband metal/semiconductor contact, there will be no band banding induced within the semiconductor due to the contact (flatband contact). The electron/hole surface recombination velocities $S_n^{front/back}$, $S_p^{front/back}$ of the metallic contact have to be specified (Sze & Kwok, 2007). If choosing the Schottky metal/semiconductor contact, an additional work function $\phi^{front/back}$ of the metal contact has to be specified. A depletion or accumulation layer within the semiconductor due to the contact will then form according to Schottky theory (Sze & Kwok, 2007). If choosing the insulator/semiconductor or the metal/insulator/semiconductor contact, interface states between the insulator and the semiconductor can be stated, that is an interface defect distribution $N_{trap}^{it}(E)$ and interface capture cross sections σ_n^{it}, σ_p^{it} have to be specified (Kronik & Shapira, 1999). In case of the metal/insulator/semiconductor contact an additional interface capacity $C^{front/back}$ has to be specified (Kronik & Shapira, 1999). Due to the interface defects a band bending within the semiconductor can form.

3.5 Circuit elements

A series resistance R_s, a parallel resistance R_p, a parallel capacitance C_p and in case of an metal/insulator/semiconductor contact also a series capacitance C_s can be specified. If circuit elements are specified, the internal cell voltage V_{int} and the internal cell current I_{int} of the semiconductor stack will differ from the external cell voltage V_{ext} and external cell current I_{ext} of the modeled device.

3.6 External parameters

External parameters are defined to be parameters which are externally applied to the device under consideration and which can also be easily varied in a real experiment. These are the temperature T of the device, a spectral and a monochromatic illumination source leading to the spectral photon flux $\Phi_0(\lambda,t)$ required for the optical simulations, and the external cell voltage $V_{ext}(t)$ or the external cell current $I_{ext}(t)$ which is applied to the device. The remaining quantity, i.e. the external cell current $I_{ext}(t)$ or the external cell voltage $V_{ext}(t)$ respectively, will be calculated.

4. Mathematical description of the DGL system solved by AFORS-HET

In the following, the differential equations and corresponding boundary conditions, which are solved by AFORS-HET under the various conditions, are stated.

An arbitrary stack of semiconductor layers can be modeled. Within each semiconductor layer the Poisson equation and the transport and continuity equations for electrons and holes have to be solved. At each semiconductor/semiconductor interface and at the front and back side boundary of the stack the current transport through these interfaces/boundaries can be described by different physical models. It results a highly non-linear coupled system of three differential equations with respect to time and space derivatives. The electron density $n(x,t)$, the hole density $p(x,t)$, and the electric potential $\varphi(x,t)$ are the independent variables, for which this system of differential equations is solved. It is solved according to the numerical discretisation scheme as outlined by Selberherr (Selberherr, 1984) in order to linearize the problem and using the linear SparLin solver which is available in the internet (Kundert et. al., 1988).

It can be solved for different calculation modes: (1) EQ calculation mode, describing thermodynamic equilibrium at a given temperature, (2) DC calculation mode, describing steady-state conditions under an external applied voltage or current and/or illumination, (3) AC calculation mode, describing small additional sinusoidal modulations of the external applied voltage/illumination, and (4) TR calculation mode, describing transient changes of the system, due to general time dependent changes of the external applied voltage or current and/or illumination.

In case of using the EQ or the DC calculation mode, all time derivatives vanish, resulting in a simplified system of differential equations. The system of differential equations is then solved for the time independent, but position dependent functions, $n^{EQ/DC}(x)$, $p^{EQ/DC}(x)$, $\varphi^{EQ/DC}(x)$.

$$n(x,t) = n^{EQ}(x), \qquad n(x,t) = n^{DC}(x)$$

$$p(x,t) = p^{EQ}(x) \qquad p(x,t) = p^{DC}(x)$$

$$\varphi(x,t) = \varphi^{EQ}(x) \qquad \varphi(x,t) = \varphi^{DC}(x)$$

In case of using the AC calculation mode, it is assumed that all time dependencies can be described by small additional sinusoidal modulations of the steady-state solutions. All time dependent quantities are then modelled with complex numbers (marked by a dash ~), which allows to determine the amplitudes and the phase shifts between them. I.e., for the independent variables of the system of differential equations, one gets:

$$n(x,t) = n^{DC}(x) + \tilde{n}^{AC}(x)\, e^{i\omega t}$$

$$p(x,t) = p^{DC}(x) + \tilde{p}^{AC}(x)\, e^{i\omega t}$$

$$\varphi(x,t) = \varphi^{DC}(x) + \tilde{\varphi}^{AC}(x)\, e^{i\omega t}$$

In case of using the TR calculation mode, the description of the system starts with a steady-state (DC-mode) simulation, specifying an external applied voltage or current and/or illumination. An arbitrary evolution in time of the external applied voltage or current and/or illumination can then be specified by loading an appropriate file. Then, the time evolution of the system, i.e. the functions $n(x,t)$, $p(x,t)$, $\varphi(x,t)$ during and after the externally applied changes are calculated.

4.1 Optical calculation: super bandgap generation models

In order to describe the generation rate $G_n(x,t)$, $G_p(x,t)$ of electrons and holes due to photon absorption within the bulk of the semiconductor layers, a distinction between super-bandgap generation (for photons with an energy $E_{photon} = hc/\lambda \geq E_g$) and sub-bandgap generation (for photons with an energy $E_{photon} = hc/\lambda \leq E_g$) is made ($\lambda$: photon wavelength h: Planck´s constant, c: velocity of light, E_g: bandgap of the semiconductor layer in which the photon absorption takes place). Only the super-bandgap generation rate is calculated by optical modelling as it is independent of the local particle densities $n(x,t)$, $p(x,t)$. Sub-bandgap generation depends on the local particle densities and must therefore be calculated within the electrical modeling part.

The optical super-bandgap generation rate is equal for electrons and holes $G(x,t) = G_n(x,t) = G_p(x,t)$. It can either be imported by loading an appropriate file (using external programs for its calculation) or it can be calculated within AFORS-HET.

So far, two optical models are implemented in AFORS-HET, i.e. the optical model Lambert-Beer absorption and the optical model coherent/incoherent internal multiple reflections. The first one takes textured surfaces and multiple internal boundary reflections into account (due to simple geometrical optics) but neglects coherence effects. It is especially suited to treat wafer based crystalline silicon solar cells. The second takes coherence effects into account, but this is done only for plain surfaces. If coherence effects in thin film solar cells are observable it may be used.

4.1.1 Optical model: Lambert-Beer absorption

Using this model, the absorption within the semiconductor stack will be calculated assuming simple Lambert-Beer absorption, allowing for multiple for and backward traveling of the incoming light, however disregarding coherent interference. A (measured) reflectance and absorptance file of the illuminated contact $R(\lambda)$, $A(\lambda)$ can be loaded or constant values can be used. The incoming spectral photon flux $\Phi_0(\lambda,t)$ is weighted with the contact reflection and absorption, i.e. the photon flux impinging on the first semiconductor layer is given by $\Phi_0(\lambda,t)R(\lambda)A(\lambda)$. To simulate the extended path length caused by a textured surface, the angle of incidence δ of the incoming light can be adjusted. On a textured Si wafer with <111> pyramids, this angle is δ=54.74°, whereas δ=0° equals normal incidence. The angle γ in which the light travels through the layer stack depends on the wavelength of the incoming light and is calculated according to Snellius' law:

$$\gamma(\lambda) = \delta - \arcsin\left\{\sin(\delta)\cdot\frac{1}{n(\lambda)}\right\},$$

whereas $n(\lambda)$ is the wavelength dependent refraction index of the first semiconductor layer at the illuminated side. Note, that within this model, the change in $\gamma(\lambda)$ is neglected, when the light passes a semiconductor/semiconductor layer interface with two different refraction indices. Thus it is assumed that all photons with a specified wavelength cross the layer stack under a distinct angle γ.

Photon absorption is then calculated from the spectral absorption coefficient $\alpha_x(\lambda) = 4\pi\, k(\lambda)\,/\,\lambda$ of the semiconductor layer corresponding to the position x within the stack, which is calculated from the provided extinction coefficient $k(\lambda)$ of the layer. The super bandgap electron/hole generation rate for one single run trough the layer stack (no multiple passes) is then given by:

$$G(x,t)=\int_{\lambda_{\min}}^{\lambda_{\max}} d\lambda\ \Phi_0(\lambda,t)\,R(\lambda)\,A(\lambda)\ \alpha_x(\lambda)\ e^{\frac{-\alpha_x(\lambda)\,x}{\cos(\gamma)}}\ .$$

The minimum and maximum wavelengths $\lambda_{\min}$, $\lambda_{\max}$ for the integration are generally provided by the loaded spectral range of the incoming spectral photon flux, $\Phi_0(\lambda,t)$. However, if necessary, $\lambda_{\max}$ is modified in order to ensure that only super-bandgap generation is considered: $\lambda_{\max} \le hc/E_g$.

To simulate the influence of light trapping mechanisms, internal reflections at both contacts can be additionally specified. They can either be set as a constant value or wavelength dependant (a measured or calculated file can be loaded). The light then passes through the layer stack several times as defined by the user, thereby enhancing the absorptivity of the layer stack (the local generation rate). The residual flux after the defined number of passes is added to the transmitted flux at the contact, at which the calculation ended (illuminated or not-illuminated contact), disregarding the internal reflection definitions at this contact.

This model was designed to estimate the influence of light trapping of crystalline silicon solar cells and to adapt the simulation to real measurements. However, it neglects the internal multiple reflections and refractions within the layer stack.

4.1.2 Optical model: coherent/incoherent internal multiple reflections

Using this model, the absorption within the semiconductor stack will be calculated by modelling coherent or incoherent internal multiple reflections within the semiconductor stack. Additional non-conducting optical layers in front of the front contact/behind the back contact of the solar cell can be assumed, for example in order to model the effect of anti-reflection coatings. Normal incidence of the incoming illumination is assumed.

The reflectance, transmittance and absorptance of all layers (optical layers and the semiconductor layers) is calculated, using the concepts of complex Fresnel amplitudes. Each layer can be specified to be optically coherent or optically incoherent for a particular light beam (incident illumination). A layer is considered to be coherent if its thickness is smaller than the coherence length of the light beam that is incident on the system.

In order to be able to consider coherent effects, the specified incoming illumination $\Phi_0(\lambda,t)$ is modeled by an incoming electromagnetic wave, with a complex electric field component $\tilde{E}_0^+(\lambda,t)$ (front side illumination, electromagnetic wave traveling in positive direction towards the back contact, with $\Phi_0(\lambda,t)=Const\left|\tilde{E}_0^+(\lambda,t)\right|^2$), or $\tilde{E}_{N+1}^-(\lambda,t)$ respectively (back side illumination, electromagnetic wave traveling in negative direction towards the front contact, with $\Phi_0(\lambda,t)=Const\left|\tilde{E}_{N+1}^-(\lambda,t)\right|^2$). The complex electric field components of the travelling wave are raytraced according to the Fresnel formulas, and thus the resulting electromagnetic wave $\tilde{E}(x,\lambda,t)$ at any position x within the layer stack is calculated. An incoherent layer is modeled by a coherent calculation of several electromagnetic waves within that layer (specified by the integer $N_{incoherentIterations}$), assuming some phase shift between them, and averaging over the resulting electric field components.

4.2 Electrical calculation - bulk layers: semiconductor bulk models

Within the bulk of each semiconductor layer, Poisson's equation and the transport equations for electrons and holes are to be solved in one dimension. So far, there are two semiconductor bulk models available, i.e. the bulk model "standard semiconductor" and the bulk model "crystalline silicon". If using the standard semiconductor model, all bulk layer input parameters as specified in Chapter 3.2 can be individually adjusted. If using the crystalline silicon bulk model, most input parameters for crystalline silicon are calculated from few remaining input parameters, i.e. from the doping and defect densities $N_D(x)$, $N_A(x)$, N_{trap} of crystalline silicon. Thus effects like band gap narrowing or the doping dependence of the mobility or of the Auger recombination of crystalline silicon are explicitly modeled.

Within each layer, a functional dependence in space can be specified for the doping densities $N_D(x)$, $N_A(x)$. These input parameters can be chosen to be (1) constant, (2) linear, (3) exponential, (4) Gaussian like, (5) error function like decreasing or increasing as a function of the space coordinate x.

4.2.1 Bulk model: standard semiconductor

The doping densities $N_D(x)$, $N_A(x)$ of fixed donator/acceptor states at apposition x within the cell are assumed to be always completely ionized. Contrary, defects $N_{trap}(E)$ located at a specific energy E within the bandgap of the semiconductor can be locally charged/uncharged within the system. Defects can be chosen to be either (1) acceptor-like Shockley-Read-Hall defects, (2) donor-like Shockley-Read-Hall defects or (3) dangling bond defects. Depending on the defect-type chosen, these defects can either be empty, singly occupied with electrons or even doubly occupied with electrons (in case of the dangling bond defect). Acceptor-like Shockley-Read-Hall defects are negatively charged, if occupied and neutral, if empty. Donor-like Shockley-Read-Hall defects are positively charged, if empty, and neutral, if occupied. Dangling bond defects are positively charged, if empty, neutral, if singly occupied and negatively charged, if doubly occupied.

Poisson´s equation, which is to be solved within each layer, reads:

$$\frac{\varepsilon_0 \varepsilon_r}{q} \frac{\partial^2 \varphi(x,t)}{\partial x^2} = p(x,t) - n(x,t) + N_D(x) - N_A(x) + \sum_{trap} \rho_{trap}(x,t)$$

q being the electron charge and ε_0, ε_r being the absolute/relative dielectric constant. The defect density of charged defects $\rho_{trap}(x,t)$ will depend on the defect-type of the defect under consideration and on the local particle densities $n(x,t)$, $p(x,t)$ within in the system. It is described by a trap density distribution function $N_{trap}(E)$ of the defect, specifying the amount of traps at an energy position E within the bandgap and by some corresponding defect occupation functions $f_{0,trap}^{SRH}(E,x,t)$, $f_{1,trap}^{SRH}(E,x,t)$, $f_{+,trap}^{DB}(E,x,t)$, $f_{0,trap}^{DB}(E,x,t)$, $f_{-,trap}^{DB}(E,x,t)$, specifying the probability that traps with an energy position E within the bandgap are empty or singly or doubly occupied with electrons. Thus $\rho_{trap}(x,t)$ equates to

$\rho_{trap}(x,t) = -\int dE\ f_{1,trap}^{SRH}(E,x,t)\ N_{trap}(E)$ in case of acceptor-like Shockley-Read-Hall defects,

$\rho_{trap}(x,t) = +\int dE\ f_{0,trap}^{SRH}(E,x,t)\ N_{trap}(E)$ in case of donator-like Shockley-Read-Hall defects,

$\rho_{trap}(x,t) = +\int dE\ \left(f_{+,trap}^{DB}(E,x,t) - f_{-,trap}^{DB}(E,x,t)\right)\ N_{trap}(E)$ in case of dangling bond defects.

The explicit formulas for the defect occupation functions $f_{0,trap}^{SRH}(E,x,t)$, $f_{1,trap}^{SRH}(E,x,t)$, $f_{+,trap}^{DB}(E,x,t)$, $f_{0,trap}^{DB}(E,x,t)$, $f_{-,trap}^{DB}(E,x,t)$ are described later within this text.

The one dimensional equations of continuity and transport for electrons and holes, which have to be solved within each layer, read:

$$-\frac{1}{q}\frac{\partial j_n(x,t)}{\partial x} = G_n(x,t) - R_n(x,t) \quad -\frac{\partial}{\partial t}n(x,t)$$

$$+\frac{1}{q}\frac{\partial j_p(x,t)}{\partial x} = G_p(x,t) - R_p(x,t) \quad -\frac{\partial}{\partial t}p(x,t)$$

The electron/hole super-bandgap generation rates $G_n(x,t)$, $G_p(x,t)$ have to be determined by optical modeling, the corresponding recombination rates $R_n(x,t)$, $R_p(x,t)$ are described later in this text. The electron/hole currents $j_n(x,t)$, $j_p(x,t)$ are driven by the gradient of the corresponding quasi Fermi energy $E_{Fn}(x,t)$, $E_{Fp}(x,t)$. Using a Maxwell Boltzmann approximation for the Fermi-Dirac distribution function, the position dependent Fermi energies and the corresponding local electron/hole currents are explicitly:

$$E_{Fn}(x,t) = E_C(x) + kT\ln\frac{n(x,t)}{N_C(x)} = -q\chi(x) + q\varphi(x,t) + kT\ln\frac{n(x,t)}{N_C(x)}$$

$$E_{Fp}(x,t) = E_V(x) - kT\ln\frac{p(x,t)}{N_V(x)} = -q\chi(x) + q\varphi(x,t) - E_g(x) - kT\ln\frac{p(x,t)}{N_V(x)}$$

$$j_n(x,t) = q\ \mu_n\ n(x,t)\ \frac{\partial E_{Fn}(x,t)}{\partial x}$$

$$j_p(x,t) = q\ \mu_p\ p(x,t)\ \frac{\partial E_{Fp}(x,t)}{\partial x}$$

with the corresponding electron/hole mobilities μ_n, μ_p, the electron affinity χ, the bandgap E_g, the conduction/valence band energy E_C, E_V and the effective conduction/valence band density of states N_C, N_V of the semiconductor.

Recombination

Recombination from the conduction band into the valence band may occur directly, i.e. via radiative band to band recombination, $R_{n,p}^{BB}(x,t)$, or via Auger recombination, $R_{n,p}^{A}(x,t)$. It may also occur via defect states located within the bandgap of the semiconductor, i.e. via Shockley-Read-Hall recombination $R_{n,p}^{SHR}(x,t)$ or via dangling bond recombination, $R_{n,p}^{DB}(x,t)$:

$$R_{n,p}(x,t) = R_{n,p}^{BB}(x,t) + R_{n,p}^{A}(x,t) + R_{n,p}^{SHR}(x,t) + R_{n,p}^{DB}(x,t)$$

Optical sub-bandgap generation

Optical sub-bandgap generation (for $hc/\lambda < E_g$) is calculated using Shockley-Read-Hall recombination statistics. A negative electron/hole SHR recombination rate $R_n^{SRH}(x,t)$,

$R_p^{SRH}(x,t)$ means sub-bandgap generation of an electron/hole from a defect state (trap) into the conduction/valence band. Sub-bandgap generation can either be voltage driven and/or be driven by an optical excitation.

The SRH optical emission coefficients $e_{n,optical}^{trap}(E,x,t)$, $e_{p,optical}^{trap}(E,x,t)$ can be calculated from the optical electron/hole capture cross sections $\sigma_{n,optical}^{trap}$, $\sigma_{p,optical}^{trap}$:

$$e_{n,optical}^{trap}(E,x,t) = \int_{\lambda_{\min}}^{\lambda_{\max}} d\lambda \ \sigma_{n,optical}^{trap} \ N_C \ \Phi(\lambda,x,t) \ \vartheta(E_C - E - hc/\lambda)$$

$$e_{p,optical}^{trap}(E,x,t) = \int_{\lambda_{\min}}^{\lambda_{\max}} d\lambda \ \sigma_{p,optical}^{trap} \ N_V \ \Phi(\lambda,x,t) \ \vartheta(E - E_V - hc/\lambda)$$

with $\Phi(\lambda,x,t)$: spectral photon flux inside the semiconductor layers, of wavelength λ at the position x and at time t, N_C, N_V: effective conduction/valence band density, E_C, E_V: energy position of the conduction/valence band, and $\vartheta(E)$: step function, $\vartheta(E) = 1 \ for \ E \leq 0$, $\vartheta(E) = 0 \ for \ E > 0$.

Again, the minimum and maximum wavelengths $\lambda_{\min}$, $\lambda_{\max}$ for the integration are generally provided by the loaded spectral range of the incoming spectral photon flux, $\Phi_0(\lambda,t)$. However, if necessary, $\lambda_{\min}$ is modified in order to ensure that only sub-bandgap generation is considered: $\lambda_{\min} \geq hc/E_g$.

Radiative recombination

The radiative band to band rate constant r^{BB} has to be specified in order to equate the radiative band to band recombination rates $R_{n,p}^{BB}(x,t)$. The resulting electron and hole recombination rates are always equal:

$$R_{n,p}^{BB}(x,t) = r^{BB} \left\{ n(x,t)\, p(x,t) - N_C \ N_V \ e^{-E_g/kT} \right\}$$

In case of using the DC or AC calculation mode and neglecting second order terms in case of the AC calculation mode, this simplifies to

$$R_{n,p}^{BB}(x) = r^{BB} \left\{ n^{DC}(x)\, p^{DC}(x) - N_C \ N_V \ e^{-E_g/kT} \right\}$$

$$R_{n,p}^{BB}(x,t) = R_{n,p}^{BB}(x) + \widetilde{R}_{n,p}^{BB}(x) \ e^{i\omega t}$$

$$\widetilde{R}_{n,p}^{BB}(x) = r^{BB} \, n^{DC}(x) \ \widetilde{p}^{AC}(x) + r^{BB} p^{DC}(x) \ \widetilde{n}^{AC}(x)$$

Auger recombination

The electron/hole Auger rate constants r_n^A, r_p^A have to be specified in order to calculate the Auger recombination rates $R_{n,p}^A(x,t)$. Again, the resulting electron and hole recombination rates are always equal:

$$R_{n,p}^A(x,t) = \left[r_n^A \ n(x,t) + r_p^A \ p(x,t) \right] \left\{ n(x,t)\, p(x,t) - N_C \ N_V \ e^{-E_g/kT} \right\}$$

In case of using the DC or AC calculation mode, neglecting second order terms within the AC calculation mode, this simplifies to

$$R_{n,p}^{A}(x)=\left[r_n^A\, n^{DC}(x)+r_p^A\, p^{DC}(x)\right]\,\left\{n^{DC}(x)\,p^{DC}(x)-N_C\,N_V\,e^{-E_g/kT}\right\}$$

$$R_{n,p}^{A}(x,t)=R_{n,p}^{A}(x)+\widetilde{R}_{n,p}^{A}(x)\,e^{i\omega t}$$

$$\widetilde{R}_{n,p}^{A}(x)=\left[r_n^A\, n^{DC}(x)^2+2\,r_p^A\, n^{DC}(x)\,p^{DC}(x)\right]\widetilde{p}^{AC}(x)+\left[r_p^A\, p^{DC}(x)^2+2\,r_n^A\, n^{DC}(x)\,p^{DC}(x)\right]\widetilde{n}^{AC}(x)$$

Shockley Read Hall recombination

Shockley-Read-Hall recombination (Shockley & Read, 1952) requires specifying the character (acceptor-like or donor-like), the capture cross sections σ_n^{trap}, σ_p^{trap}, $\sigma_{n,optic}^{trap}$, $\sigma_{p,optic}^{trap}$ and the energetic distribution $N_{trap}(E)$ of the defect density within the bandgap of the semiconductor, of each defect. An arbitrary number of defects with either one of the following energetic distributions $N_{trap}(E)$ can be chosen:

1. point like distributed at a single energy E_{trap} within the bandgap:

$$N_{trap}(E)=N_{trap}^{point}\,\delta(E-E_{trap})$$

with N_{trap}^{point}: defect density of the point like defect, $\delta(E)$: delta function

2. constantly distributed within a specific region within the bandgap:

$$N_{trap}(E)=\left(E_{trap}^{end}-E_{trap}^{start}\right)\,N_{trap}^{const}\;\vartheta\left(E-E_{trap}^{end}\right)\,\vartheta\left(E_{trap}^{start}-E\right)$$

with E_{trap}^{start}, E_{trap}^{end}: start and end energy of the energy interval within the bandgap, where a constant defect density is assumed, N_{trap}^{const}: constant defect density per energy, $\vartheta(E)$: step function

3. exponentially decaying from the conduction/valence band into the bandgap:

$$N_{trap}(E)=N_{trap}^{C,tail}\,e^{-(E_C-E)/E_{trap}^{C,tail}},\quad N_{trap}(E)=N_{trap}^{V,tail}\,e^{-(E-E_V)/E_{trap}^{V,tail}}$$

i.e. conduction/valence band tail states, with $N_{trap}^{C,tail}$, $N_{trap}^{V,tail}$: tail state density per energy at the conduction/valence band, $E_{trap}^{C,tail}$, $E_{trap}^{V,tail}$: characteristic decay energy (Urbach energy) of the conduction/valence band tail state,

4. Gaussian distributed within the bandgap:

$$N_{trap}(E)=\frac{N_{trap}^{db}}{\sigma_{trap}^{db}\sqrt{2\pi}}\,e^{-\frac{\left(E-E_{trap}^{db}\right)^2}{2\sigma_{trap}^{db\,2}}}$$

i.e. dangling bond states, with N_{trap}^{db}: total dangling bond state density, E_{trap}^{db}: specific energy of the Gaussian dangling bond peak, σ_{trap}^{db}: standard deviation of the Gaussian dangling bond distribution.

For each defect, electron/hole capture coefficients $c_{n,p}^{trap}$ are equated

$$c_{n,p}^{trap} = v_{n,p}\,\sigma_{n,p}$$

with $v_{n,p}$: electron/hole thermal velocity, $\sigma_{n,p}$: electron/hole capture cross section of the defect. The corresponding electron/hole emission coefficients $e_{n,p}^{trap}(E,x,t)$ are then given by:

$$e_n^{trap}(E,x,t) = c_n^{trap}\,N_C\,e^{-(E_C-E)/kT} + e_{n,optic}^{trap}(E,x,t)$$

$$e_p^{trap}(E,x,t) = c_p^{trap}\,N_V\,e^{-(E-E_V)/kT} + e_{p,optic}^{trap}(E,x,t)$$

In case of using the DC or AC calculation mode, this simplifies to

$$e_n^{trap}(E,x) = c_n^{trap}\,N_C\,e^{-(E_C-E)/kT} + \int d\lambda\;\sigma_{n,optic}^{trap}\,N_C\,\Phi(\lambda,x)\;\vartheta(E_C - E - hc/\lambda) \quad \text{(DC mode)}$$

$$e_p^{trap}(E,x) = c_p^{trap}\,N_V\,e^{-(E-E_V)/kT} + \int d\lambda\;\sigma_{p,optic}^{trap}\,N_V\,\Phi(\lambda,x)\;\vartheta(E - E_V - hc/\lambda)$$

$$e_{n,p}^{trap}(E,x,t) = e_{n,p}^{trap}(E,x) + \tilde{e}_{n,p}^{trap}(E,x)\,e^{i\omega t} \quad \text{(AC mode)}$$

$$\tilde{e}_n^{trap}(E,x) = \int d\lambda\;\sigma_{n,optic}^{trap}\,N_C\,\tilde{\Phi}(\lambda,x)\;\vartheta(E_C - E - hc/\lambda)$$

$$\tilde{e}_p^{trap}(E,x) = \int d\lambda\;\sigma_{p,optic}^{trap}\,N_V\,\tilde{\Phi}(\lambda,x)\;\vartheta(E - E_V - hc/\lambda)$$

Finally, the Shockley-Read-Hall recombination rate due to the defects is

$$R_n^{SRH}(x,t) = \sum_{trap} \int dE\,\left\{c_n^{trap}\,n(x,t)\,N_{trap}(E)\,f_{0,trap}^{SRH}(E,x,t) - e_n^{trap}(E,x,t)\,N_{trap}(E)\,f_{1,trap}^{SRH}(E,x,t)\right\}$$

$$R_p^{SRH}(x,t) = \sum_{trap} \int dE\,\left\{c_p^{trap}\,p(x,t)\,N_{trap}(E)\,f_{1,trap}^{SRH}(E,x,t) - e_p^{trap}(E,x,t)\,N_{trap}(E)\,f_{0,trap}^{SRH}(E,x,t)\right\}$$

In case of using the DC or AC calculation mode, neglecting second order terms and assuming zero optical emission coefficients within the AC calculation mode (actual stage of the AFORS-HET development at the moment) this simplifies to

$$R_n^{SRH}(x) = \sum_{trap} \int dE\,\left\{c_n^{trap}\,n^{DC}(x)\,N_{trap}(E)\,f_{0,trap}^{SRH}(E,x) - e_n^{trap}(E,x)\,N_{trap}(E)\,f_{1,trap}^{SRH}(E,x)\right\}$$

$$R_p^{SRH}(x) = \sum_{trap} \int dE\,\left\{c_p^{trap}\,p^{DC}(x)\,N_{trap}(E)\,f_{1,trap}^{SRH}(E,x) - e_p^{trap}(E,x)\,N_{trap}(E)\,f_{0,trap}^{SRH}(E,x)\right\}$$

$$R_{n,p}^{SRH}(x,t) = R_{n,p}^{SRH}(x) + \tilde{R}_{n,p}^{SRH}(x)\,e^{i\omega t}$$

$$\tilde{R}_n^{SRH}(x) = \sum_{trap} \int dE\,\left\{c_n^{trap}\,N_{trap}(E)\,f_{0,trap}^{SRH}(E,x)\,\tilde{n}^{AC}(x) - \left(c_n^{trap} + e_n^{trap}(E,x)\right)N_{trap}(E)\,\tilde{f}_{1,trap}^{SRH}(E,x)\right\}$$

$$\tilde{R}_p^{SRH}(x) = \sum_{trap} \int dE\,\left\{c_p^{trap}\,N_{trap}(E)\,f_{1,trap}^{SRH}(E,x)\,\tilde{p}^{AC}(x) + \left(c_p^{trap} + e_p^{trap}(E,x)\right)N_{trap}(E)\,\tilde{f}_{1,trap}^{SRH}(E,x)\right\}$$

A positive electron/hole SHR recombination rate means recombination of an electron/hole from the conduction/valence band into the defect state (trap), a negative electron/hole SHR recombination rate means sub-bandgap generation of an electron/hole from the defect state (trap) into the conduction/valence band.

Dangling bond recombination

To calculate charge state and recombination of dangling bond defects in amorphous silicon the most exact description developed by Sah and Shockley (Sah & Shockley, 1958) is used. Three different occupation functions $f^{DB}_{+,trap}(E,x,t)$, $f^{DB}_{0,trap}(E,x,t)$, $f^{DB}_{-,trap}(E,x,t)$ for the positively, neutral and negatively charge states have to be derived, corresponding to the empty, single or double occupied electronic state. Four capture/emission processes with the capture cross sections σ_n^+, σ_p^0, σ_n^0, σ_p^- have to be defined as can be seen in Fig. 3. The two transition energies $E_{0/-}$, $E_{+/0}$ are separated by the correlation energy U, which accounts for the fact that the capture-emission process is influenced by the charge state of the dangling or by rearrangement of the lattice in the surrounding.

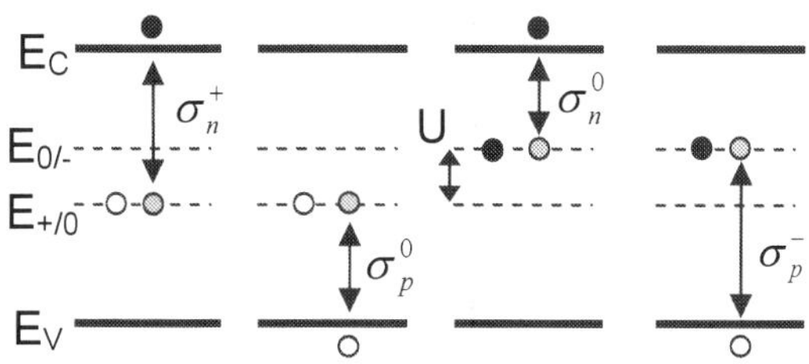

Fig. 3. dangling bond recombination according to Sah and Shockley

For each defect, electron/hole capture coefficients $c^{trap}_{n/p,+/0/-}$ are calculated with the specified electron/hole thermal velocity $v_{n/p}$:

$$c^{trap}_{n/p,+/0/-} = v_{n/p}\,\sigma^{+/0/-}_{n/p}$$

The emission coefficients for the DC calculation mode, neglecting optical emission are given by:

$$e^{trap}_{n,0}(E,x) = \frac{1}{2} c^{trap}_{n,+} N_C\, e^{-(E_C-E)/kT}$$

$$e^{trap}_{p,+}(E,x) = 2 c^{trap}_{p,0} N_V\, e^{-(E-E_V)/kT}$$

$$e^{trap}_{n,-}(E,x) = 2 c^{trap}_{n,0} N_C\, e^{-(E_C-(E+U))/kT}$$

$$e^{trap}_{p,0}(E,x) = \frac{1}{2} c^{trap}_{p,-} N_V\, e^{-(E+U-E_V)/kT}$$

Finally the dangling bond recombination coefficients are given by:

$$R_n^{DB}(x) = \sum_{trap} \int dE \left\{ c^{trap}_{n,+}\, n^{DC}(x) N_{trap}(E)\, f^{DB}_{+,trap}(E,x) - e^{trap}_{n,0}(E,x) N_{trap}(E) f^{DB}_{0,trap}(E,x) + c^{trap}_{n,0}\, n^{DC}(x) N_{trap}(E)\, f^{DB}_{0,trap}(E,x) - e^{trap}_{n,-}(E,x) N_{trap}(E) f^{DB}_{-,trap}(E,x) \right\}$$

$$R_p^{DB}(x) = \sum_{trap} \int dE \left\{ c_{p,-}^{trap}\, p^{DC}(x) N_{trap}(E)\, f_{-,trap}^{DB}(E,x) - e_{p,0}^{trap}(E,x) N_{trap}(E) f_{0,trap}^{DB}(E,x) + c_{p,0}^{trap}\, p^{DC}(x) N_{trap}(E)\, f_{0,trap}^{DB}(E,x) - e_{p,+}^{trap}(E,x) N_{trap}(E) f_{+,trap}^{DB}(E,x) \right\}$$

Dangling bond recombination is still under development. Especially time dependent recombination and optical defect to band emissions are not implemented at the current state of AFORS-HET development.

Defect occupation functions

The defect occupation functions $f_{0,trap}^{SRH}(E,x,t)$, $f_{1,trap}^{SRH}(E,x,t)$, $f_{+,trap}^{DB}(E,x,t)$, $f_{0,trap}^{DB}(E,x,t)$, $f_{-,trap}^{DB}(E,x,t)$ specify the probability for a specific defect (either Shockley-Read-Hall or dangling bond) that traps with an energy position E within the bandgap of the semiconductor are empty or singly or even doubly occupied with electrons.
In case of using the DC or AC calculation mode, they can be explicitly expressed in terms of the local particle densities $n^{DC}(x)$, $p^{DC}(x)$, $\tilde{n}^{AC}(x)$, $\tilde{p}^{AC}(x)$. In case of using the TR calculation mode, the defect occupation functions are generally determined by additional differential equations. Transient DB defect occupation functions have not been implemented in AFORS-HET yet (actual stage of AFORS-HET development).

Shockley Read Hall defect occupation functions

A Shockley-Read-Hall defect can be either empty or occupied by an electron, thus

$$f_{0,trap}^{SRH}(E,x,t) + f_{1,trap}^{SRH}(E,x,t) = 1$$

The Shockley-Read-Hall defect occupation function $f_{1,trap}^{SRH}(E,x,t)$ for electrons will be explicitly stated in case of using the EQ, DC, AC or the TR calculation mode. The Shockley-Read-Hall defect occupation function $f_{0,trap}^{SRH}(E,x,t)$ can then directly be equated.
Generally, a local change of the trapped charge stored in SRH defects must be determined by the difference between the local electron and hole SRH recombination rates:

$$\frac{d}{dt}\rho_{\text{trap}}(x,t) = R_p^{SHR}(x,t) - R_n^{SHR}(x,t)$$

This defines for each defect an additional differential equation for its SHR defect occupation function $f_{1,trap}^{SRH}(E,x,t)$ with respect to its time derivative:

$$\frac{d}{dt} f_{1,trap}^{SRH}(E,x,t) = \left(c_n^{trap}\, n(x,t) + e_p^{trap}(E,x,t)\right)\left(1 - f_{1,trap}^{SRH}(E,x,t)\right) - \left(c_p^{trap}\, p(x,t) + e_n^{trap}(E,x,t)\right) f_{1,trap}^{SRH}(E,x,t) \quad (\#)$$

In case of using the EQ or the DC calculation mode, the time derivative vanishes, and an explicit expression for the SHR defect occupation function, $f_{1,trap}^{SRH,DC}(E,x)$, which is no longer time dependant, can be derived:

$$f_{1,trap}^{SRH,DC}(E,x) = \frac{c_n^{trap}\, n^{DC}(x) + e_p^{trap}(E,x)}{c_n^{trap}\, n^{DC}(x) + e_n^{trap}(E,x) + c_p^{trap}\, p^{DC}(x) + e_p^{trap}(E,x)}$$

Of course, in case of using the EQ calculation mode, the SHR defect occupation function could be also equivalently be described by the Fermi-Dirac distribution function, $f_{1,\,trap}^{SRH,DC}(E,x) = f_{1,\,trap}^{SRH,EQ}(E,x)$, which implicitly determines the position independent Fermi energy E_F.

$$f_{1,trap}^{SRH,EQ}(E,x) = \frac{1}{1+e^{\frac{E-E_F}{kT}}}$$

In case of using the AC calculation mode, the differential equation (#) can be explicitly solved, assuming time independent optical emission coefficients within the AC calculation mode (actual stage of the AFORS-HET development at the moment) and assuming the time dependencies $n(x,t) = n^{DC}(x) + \tilde{n}^{AC}(x)\, e^{i\omega t}$, $p(x,t) = p^{DC}(x) + \tilde{p}^{AC}(x)\, e^{i\omega t}$. Neglecting second order terms, one gets for the SHR defect occupation function in the AC calculation mode, $f_{1,trap}^{SRH,AC}(E,x,t)$:

$$f_{1,trap}^{SRH,AC}(E,x,t) = f_{1,trap}^{SRH,DC}(E,x) + \tilde{f}_{1,trap}^{SRH,AC}(E,x)\, e^{i\omega t}$$

$$\tilde{f}_{1,trap}^{SRH,AC}(E,x) = \frac{c_n^{trap}\, f_{0,trap}^{SRH,DC}(E,x)\, \tilde{n}^{AC}(x) \; - \; c_p^{trap}\, f_{1,trap}^{SRH,DC}(E,x)\, \tilde{p}^{AC}(x)}{c_n^{trap}\, n^{DC}(x) + e_n^{trap}(E,x) + c_p^{trap}\, p^{DC}(x) + e_p^{trap}(E,x) + i\omega}$$

In case of using the TR calculation mode, the transient SRH defect occupation function $f_{1,trap}^{SRH,TR}(E,x,t_{i+1})$ at the time step t_{i+1} for an evolution of the system from the time point t_i towards the time point t_{i+1} can be stated by solving the differential equation (#) using a full implicit time discretisation scheme with respect to the particle densities and the emission rates:

$$\frac{d}{dt} f_{1,trap}^{SRH,TR}(E,x,t)$$
$$= \left(c_n^{trap}\, n(x,t_{i+1}) + e_p^{trap}(E,x,t_{i+1})\right)\left(1 - f_{1,trap}^{SRH,TR}(E,x,t)\right) - \left(c_p^{trap}\, p(x,t_{i+1}) + e_n^{trap}(E,x,t_{i+1})\right) f_{1,trap}^{SRH,TR}(E,x,t)$$

An analytical solution of this differential equation leads to:

$$f_{1,trap}^{SRH,TR}(E,x,t_{i+1}) = f_{1,trap}^{SRH,DCtr}(E,x,t_{i+1}) - \frac{f_{1,trap}^{SRH,DCtr}(E,x,t_{i+1}) - f_{1,trap}^{SRH,TR}(E,x,t_i)}{e^{dt\left(c_n^{trap}\, n(x,t_{i+1}) + e_n^{trap}(E,x,t_{i-1}) + c_p^{trap}\, p(x,t_{i-1}) + e_p(E,x,t_{i-1})\right)}}$$

with

$$f_{SRH}^{DCtr}(E,x,t_{i+1}) = \frac{c_n^{trap}\, n(x,t_{i+1}) + e_p^{trap}(E,x,t_{i+1})}{c_n^{trap}\, n(x,t_{i+1}) + e_n^{trap}(E,x,t_{i+1}) + c_p^{trap}\, p(x,t_{i+1}) + e_p^{trap}(E,x,t_{i+1})}$$

In the steady-state limit, i.e. for Limes $dt \to \infty$, $dt = t_{i+1} - t_i$, this formula converts to the well known steady state SRH defect occupation function $f_{1,trap}^{SRH,DC}(E,x) = f_{1,trap}^{SRH,DCtr}(E,x,t_{i+1})$.

Dangling bond defect occupation functions

A dangling bond defect can be either empty or singly or doubly occupied by an electron, hereby being in its positive, neutral or negative charged state, thus

$$f_{+,trap}^{DB}(E,x,t)+f_{0,trap}^{DB}(E,x,t)+f_{-,trap}^{DB}(E,x,t)=1$$

For the EQ and DC calculation mode the occupation functions are given by:

$$f_{+}^{DB}(E,x)=\frac{P^0P^-}{N^+P^- + P^0P^- + N^+N^0}$$

$$f_{0}^{DB}(E,x)=\frac{P^-N^+}{N^+P^- + P^0P^- + N^+N^0}$$

$$f_{-}^{DB}(E,x)=\frac{N^0N^+}{N^+P^- + P^0P^- + N^+N^0}$$

where:

$$N^+ = c_n^+ n^{DC} + e_p^+$$

$$P^0 = c_p^0 p^{DC} + e_n^0$$

$$N^0 = c_n^0 n^{DC} + e_p^0$$

$$P^- = c_p^- p^{DC} + e_n^-$$

Dangling bond defects are still under development and especially time dependent dangling bond occupation functions (to be used in the AC or TR calculation mode) are not implemented at the current state of AFORS-HET development.

4.2.2 Bulk model: crystalline silicon

If the bulk model "crystalline silicon" is used for a semiconductor layer, most layer input parameter as described in chapter 3.2 are calculated from the doping densities and the defect densities of crystalline silicon.

Thus a doping and temperature dependent material parameterization for crystalline silicon can be undertaken, i.e. it is possible to specify (1) the temperature dependence of the intrinsic carrier concentration of crystalline silicon (intrinsic carrier density model), (2) the doping dependence of the electron/hole mobilities in crystalline silicon (mobility model), (3) the doping dependence of the Auger recombination in crystalline silicon or even its dependence on local excess carriers (Auger recombination models), (4) the doping and the temperature dependence of the Shockley-Read-Hall recombination in crystalline silicon (SRH lifetime model) and (5) doping dependence of the bandgap in crystalline silicon (bandgap narrowing model). All these models are similar to the numerical computer simulation program for crystalline silicon solar cells PC1D (Basore & Clugston, 1997).

Only one single neutral SRH defect at a certain energy position E within the bandgap is assumed. The doping densities N_D, N_A and the amount of traps N_{trap} are specified by the user, all other layer input parameters are calculated according to the above mentioned models.

4.3 Electrical calculation - interfaces: semiconductor/semiconductor interface models

Each interface between two adjacent semiconductor layers can be described by three different interface models: (1) interface model: "no interface", (2) interface model: "drift-diffusion interface" and (3) interface model: "thermionic emission interface". If "no interface" is chosen, the transport across the interface is treated in complete analogy to the "drift diffusion" interface model, however, no interface defects can be specified. The "drift diffusion" interface model models the transport across the heterojunction interface in the same way as in the bulk layers, thereby assuming a certain interface thickness. The "thermionic emission" interface model treats a real interface which interacts with both adjacent semiconductor layers.

4.3.1 Interface model: no interface

Per default, the electron and hole currents across the semiconductor/semiconductor interface are assumed to be driven by drift diffusion, with no interface defects present at the interface.

The drift diffusion model assumes an interface layer of a certain thickness (which is given by the specified grid point to boundary distance within the numerical settings of AFORS-HET). Within this interface layer, the material properties change linearly from semiconductor I to semiconductor II. The elelctron/hole currents across the heterojunction interface can then be treated like in the bulk of a semiconductor layer (drift diffusion driven).

Denoting x_{it}^{I}, x_{it}^{II} the positions directly adjacent to the semiconductor I/II located left/right to the interface, $\mu_{n,p}^{I}$, $\mu_{n,p}^{II}$ the mobilities of the two adjacent semiconductors, and $\mu_{n,p}^{it} = \frac{\mu_{n,p}^{I} + \mu_{n,p}^{II}}{2}$, $n^{it} = \frac{n(x_{it}^{I}) + n(x_{it}^{II})}{2}$, $p^{it} = \frac{p(x_{it}^{I}) + p(x_{it}^{II})}{2}$, the corresponding electron/hole currents across the interface are (like in the bulk):

$$j_n^{it}(t) = q\ \mu_n^{it}\ n^{it}(t)\ \frac{\partial E_{Fn}(x,t)}{\partial x}$$

$$j_p^{it}(t) = q\ \mu_p^{it}\ p^{it}(t)\ \frac{\partial E_{Fp}(x,t)}{\partial x}$$

4.3.2 Interface model: drift-diffusion interface

The electron and hole currents across the semiconductor/semiconductor interface are assumed to be driven by drift diffusion, with additional interface defects present at the interface. Half of the interface states can be occupied by electrons or holes from semiconductor I, the other half from semiconductor II.

The interface defects (given in cm^{-2}) are distributed homogeneously within the interface layer (per cm^{-3}). I.e. the specified interface defect density $N_{it}(E)$ is converted into a homogeneous layer defect density of the interface layer, $N_t(E)$, selectively according to one of the two following formulas, either $N_t(E) = \frac{N_{it}(E)}{d_{it}}$ or $N_t(E) = (N_{it}(E))^{\frac{3}{2}}$. Thus half of the defects specified are recombination active within semiconductor I, the other half within semiconductor II. The calculation of the electron/hole currents across the heterojunction interface is then performed in the same way as when using the interface model "no interface" described above.

4.3.3 Interface model: thermionic emission interface

The electron and hole currents across the interface are assumed to be driven by thermionic emission (Anderson, 1962; Sze & Kwok, 2007). Additional interface defects can be present at the interface. These states can be filled with electrons or holes from both sides of the interface.

Lets denote x_{it}^{I}, x_{it}^{II} the positions directly adjacent to the semiconductor I/II located left/right to the interface, $\chi^{I,II}$, $E_g^{I,II}$, $v_{n,p}^{I,II}$ the electron affinity, the bandgap and the thermal velocities of the two semiconductors. According to Anderson theory (Anderson, 1962) the conduction/valence band offsets ΔE_C, ΔE_V, which determine the energetic barrier of the heterojunction interface to be overcome by thermionic emission, are:

$$\Delta E_C = \chi^{II} - \chi^{I}$$

$$\Delta E_V = E_g^{II} - E_g^{I} + \chi^{II} - \chi^{I}$$

The sign convention is such, that ΔE_C, ΔE_V is negative if E_C, E_V drops from the left side to the right side of the interface. The thermionic emission currents across the heterojunction interface $j_n^{TE,I\to II}(t)$, $j_p^{TE,I\to II}(t)$, $j_n^{TE,II\to I}(t)$, $j_p^{TE,II\to I}(t)$ are then explicitly written using the $\vartheta(E)$ step function, $\vartheta(E)=1\ for\ E\le 0$, $\vartheta(E)=0\ for\ E>0$:

$$j_n^{TE,I\to II}(t) = v_n^I\, n\left(x_{it}^I,t\right) e^{-\frac{|\Delta E_C|}{kT}\vartheta(-\Delta E_C)}, \qquad j_n^{TE,II\to I}(t) = v_n^{II}\, n\left(x_{it}^{II},t\right) e^{-\frac{|\Delta E_C|}{kT}\vartheta(\Delta E_C)}$$

$$j_p^{TE,I\to II}(t) = v_p^I\, p\left(x_{it}^I,t\right) e^{-\frac{|\Delta E_V|}{kT}\vartheta(\Delta E_V)}, \qquad j_p^{TE,II\to I}(t) = v_p^{II}\, p\left(x_{it}^{II},t\right) e^{-\frac{|\Delta E_V|}{kT}\vartheta(-\Delta E_V)}$$

The netto electron/hole current across the interface due to thermionic emission j_n^{it}, j_p^{it}, is:

$$j_n^{it} = j_n^{TE,I\to II} - j_n^{TE,II\to I}, \qquad j_p^{it} = j_p^{TE,I\to II} - j_p^{TE,II\to I}$$

Additional to the thermionic emission process across the heterojunction interface, there is recombination due to the interface defects. The interface electron/hole recombination rates from both sides of the interface $R_{n,I}^{it}(t)$, $R_{n,II}^{it}(t)$, $R_{p,I}^{it}(t)$, $R_{p,II}^{it}(t)$, are described with SRH recombination:

$$R_{n,I}^{it}(t)=\sum_{trap}\int dE\left\{c_{n,I}^{it}\, n\left(x_{it}^{I},t\right)N_{trap}^{it}(E)\left(1-f_{trap}^{it}(E,t)\right)-e_{n,I}^{it}(E,t)\,N_{trap}^{it}(E)\,f_{trap}^{it}(E,t)\right\}$$

$$R_{n,II}^{it}(t)=\sum_{trap}\int dE\left\{c_{n,II}^{it}\, n\left(x_{it}^{II},t\right)N_{trap}^{it}(E)\left(1-f_{trap}^{it}(E,t)\right)-e_{n,II}^{it}(E,t)\,N_{trap}^{it}(E)\,f_{trap}^{it}(E,t)\right\}$$

$$R_{p,I}^{it}(t)=\sum_{trap}\int dE\left\{c_{p,I}^{it}\, p\left(x_{it}^{I},t\right)N_{trap}^{it}(E)\,f_{trap}^{it}(E,t)-e_{p,I}^{it}(E,t)\,N_{trap}^{it}(E)\left(1-f_{trap}^{it}(E,x,t)\right)\right\}$$

$$R_{p,II}^{it}(t)=\sum_{trap}\int dE\left\{c_{p,II}^{it}\,p\left(x_{it}^{II},t\right)N_{trap}^{it}(E)\,f_{trap}^{it}(E,t)-e_{p,II}^{it}(E,t)\,N_{trap}^{it}(E)\left(1-f_{trap}^{it}(E,x,t)\right)\right\}$$

Thus, recombination at the interface is treated equivalently to bulk SRH recombination, with two exceptions: The interface defect density $N_{trap}^{it}(E)$ is now given in defects per $cm^{-2}\,eV^{-1}$ instead of $cm^{-3}\,eV^{-1}$, consequently, interface recombination is now a recombination current, given in $cm^{-2}\,s^{-1}$, instead of $cm^{-3}\,s^{-1}$. Furthermore, the interface defect distribution function $f_{trap}^{it}(E,t)$ changes compared to the bulk defect distribution function $f_{trap}(E,x,t)$, as the interface states can interact with both adjacent semiconductors. For the EQ or DC calculation mode, one gets explicitly :

$$f_{trap}^{it,DC}(E)=f_{trap}^{it,EQ}(E)=$$
$$\frac{c_{n,I}^{it}\,n\left(x_{it}^{I}\right)+c_{n,II}^{it}\,n\left(x_{it}^{II}\right)+e_{p,I}^{it}(E)+e_{p,II}^{it}(E)}{c_{n,I}^{it}\,n\left(x_{it}^{I}\right)+c_{n,II}^{it}\,n\left(x_{it}^{II}\right)+c_{p,I}^{it}\,p\left(x_{it}^{I}\right)+c_{p,II}^{it}\,p\left(x_{it}^{II}\right)+e_{n,I}^{it}(E)+e_{n,II}^{it}(E)+e_{p,I}^{it}(E)+e_{p,II}^{it}(E)}$$

Using the AC calculation mode, one gets:

$$f_{trap}^{it,AC}(E,t)=f_{trap}^{it,DC}(E)+\tilde{f}_{trap}^{it,AC}(E)\,e^{i\omega t}$$

$$\tilde{f}_{trap}^{it,AC}(E)=$$
$$\frac{\left\{1-f_{trap}^{it,DC}(E)\right\}\left\{c_{n,I}^{it}\,\tilde{n}^{AC}\left(x_{it}^{I}\right)+c_{n,II}^{it}\,\tilde{n}^{AC}\left(x_{it}^{II}\right)\right\}-f_{trap}^{it,DC}(E)\left\{c_{p,I}^{it}\,\tilde{p}^{AC}\left(x_{it}^{I}\right)+c_{p,II}^{it}\,\tilde{p}^{AC}\left(x_{it}^{II}\right)\right\}}{c_{n,I}^{it}\,n\left(x_{it}^{I}\right)+c_{n,II}^{it}\,n\left(x_{it}^{II}\right)+c_{p,I}^{it}\,p\left(x_{it}^{I}\right)+c_{p,II}^{it}\,p\left(x_{it}^{II}\right)+e_{n,I}^{it}(E)+e_{n,II}^{it}(E)+e_{p,I}^{it}(E)+e_{p,II}^{it}(E)+i\omega}$$

Within the actual stage of AFORS-HET development, interface states described by thermionic emission are only implemented within the EQ, DC and AC calculation mode, i.e. the transient defect distribution function of such states has not been implemented yet.
The heterojunction interface itself is treated as a boundary condition for the differential equations describing the semiconductor layers. Thus, six boundary conditions for the potential and the electron/hole currents at each side of the interface have to be stated, i.e.:
The potential is assumed to be equal on both sides of the interface (thereby neglecting interface dipoles):

1. $\varphi\left(x_{it}^{I}\right)=\varphi\left(x_{it}^{II}\right)$

The total charge stored in the interface states is equal to the difference in the dielectric displacements (a consequence of the Gauss law applied to the Poisson equation)

2. $\varepsilon_0\varepsilon_r^{I}\left.\dfrac{\partial\varphi(x)}{\partial x}\right|_{x_{it}^{I}}-\varepsilon_0\varepsilon_r^{II}\left.\dfrac{\partial\varphi(x)}{\partial x}\right|_{x_{it}^{II}}=q\sum_{defects}\rho_{it}$

The total current across the heterojunction interface $j_{ges}^{it}(t)=j_n^{it}(t)+j_p^{it}(t)$ under steady-state conditions is equal to the constant (that is position independent) total current left (or right) to the interface

3. $j_n^{it}(t)+j_p^{it}(t)=j_n\left(x_{it}^{I},t\right)+j_p\left(x_{it}^{I},t\right)$

The total electron/hole recombination rate from both sides of the interface is equal for electrons and holes (valid only for EQ, DC and AC calculation mode)

4. $R^{it}_{n,I}(t)+R^{it}_{n,II}(t)=R^{it}_{p,I}(t)+R^{it}_{p,II}(t)$

The electron/hole current left to the interface is equal to the netto electron/hole current across the heterojunction interface plus the interface recombination current

5. $j_n\left(x^I_{it},t\right)=j^{it}_n(t)+R^{it}_{n,I}(t)$

6. $j_p\left(x^I_{it},t\right)=j^{it}_p(t)-R^{it}_{p,I}(t)$

4.4 Electrical calculation - boundaries: front/back contact to semiconductor models

The electrical front/back contacts of the semiconductor stack are usually assumed to be metallic, in order to be able to withdraw a current. However, they may also be insulating in order to be able to simulate some specific measurement methods like for example quasi steady state photoconductance (QSSPC) or surface photovoltage (SPV). So far, four different boundary models for the interface between the contact and the semiconductor adjacent to the contact can be chosen: (1) "flatband metal/semiconductor contact" (2) "Schottky metal/semiconductor contact", (3) "insulator/semiconductor contact", (4) "metal/insulator/semiconductor contact". The boundaries serve as a boundary condition for the system of differential equations describing the semiconductor stack, thus three boundary conditions for the potential and the electron/hole currents at the front and at the back side of the stack have to be stated.

4.4.1 Boundary model: flatband metal/semiconductor contact

Per default, an idealized flatband metal/semiconductor contact is assumed at the boundaries. That is, only the effective electron/hole surface recombination velocities $S^{front/back}_{n/p}$ have to be specified. The metal work function of the front/back contact, $\phi^{front/back}$ is calculated in a way, that flatband conditions are reached according to Schottky theory (Sze & Kwok, 2007). Normally, flatband conditions are calculated within the thermal equilibrium EQ calculation mode, however, in case of using the DC, AC or TR calculation mode with an external illumination (optical super bandgap generation) enabled, they are recalculated in order to ensure flatband conditions independent from the applied illumination.

The interface between the metallic front/back contact and the semiconductor is treated as a boundary condition for the differential equations describing the semiconductor layers. Thus, for each contact, three boundary conditions involving the potential and electron/hole densities adjacent to the contact have to be stated. Denoting x^{front}_{it}, x^{back}_{it} the position within the semiconductor directly adjacent to the metallic contact, these are:

The electric potential is fixed to zero at one contact (for example the back contact).

1.a $\varphi\left(x^{back}_{it},t\right)=0$

At the other contact (for example the front contact) the external applied cell voltage $V_{ext}(t)$ or the external applied current density $j_{ext}(t)$ through the cell is specified (voltage controlled or current controlled calculation). The external solar cell resistances, i.e. the series resistance R^S_{ext} and the parallel resistance R^P_{ext}, which can optionally be specified, will affect the internal cell voltage $V_{\mathrm{int}}(t)$ at the boundary of the semiconductor stack and also the position independent internal current density $j_{\mathrm{int}}(t)$ through the semiconductor stack.

In case of a voltage controlled calculation, the internal cell voltage can be expressed by the specified external cell voltage $V_{ext}(t)$ and the position independent internal cell current $j_{\text{int}}(t) = Const(t) = j_n(x,t) + j_p(x,t)$:

$$V_{\text{int}}(t) = V_{ext}(t) + \frac{j_{\text{int}}(t)\,R_{ext}^S}{1+\frac{R_{ext}^S}{R_{ext}^P}} = V_{ext}(t) + \frac{\left\{j_n\left(x_{it}^{front},t\right) - j_p\left(x_{it}^{front},t\right)\right\}\,R_{ext}^S}{1+\frac{R_{ext}^S}{R_{ext}^P}}$$

Thus the potential at other contact can be specified:

1.b $\varphi\left(x_{it}^{front},t\right) = \phi^{front} - \phi^{back} + V_{\text{int}}(t)$

In case of a current controlled calculation, the internal cell current density can be expressed by the specified external cell current density $j_{ext}(t)$ and the internal cell voltage $V_{\text{int}}(t) = \varphi\left(x_{it}^{front},t\right) - \varphi\left(x_{it}^{back},t\right) - \phi^{front} + \phi^{back}$:

$$j_{\text{int}}(t) = j_{ext}(t) + \frac{V_{\text{int}}(t)}{R_{ext}^P} = j_{ext}(t) + \frac{\varphi\left(x_{it}^{front},t\right) - \varphi\left(x_{it}^{back},t\right) - \phi^{front} + \phi^{back}}{R_{ext}^P}$$

Thus the position independent total internal cell current $j_n(x,t) + j_p(x,t) = Const(t)$ can be specified:

1.b $j_n\left(x_{it}^{front},t\right) + j_p\left(x_{it}^{front},t\right) = j_{\text{int}}(t)$

Furthermore, the electron and hole particle densities at the interface, or the electron/hole currents into the metal contacts can be specified for both contacts.

In the EQ calculation mode, the majority carrier density at the interface under equilibrium $n^{EQ}\left(x_{it}^{front/back}\right)$ or $p^{EQ}\left(x_{it}^{front/back}\right)$ is given by the majority barrier height $\phi_{Bn}^{Schottky} = q\left\{\phi^{front/back} - \chi^{front/back}\right\}$, $\phi_{Bp}^{Schottky} = q\left\{E_g - \phi^{front/back} + \chi^{front/back}\right\}$ of the metal/semiconductor contact (with $\chi^{front/back}$ being the electron affinity of the semiconductor adjacent to the front/back contact):

2.a, 2b $n^{EQ}\left(x_{it}^{front/back}\right) = N_C\, e^{-\frac{\phi_{Bn}^{Schottky}}{kT}}$ or $p^{EQ}\left(x_{it}^{front/back}\right) = N_V\, e^{-\frac{\phi_{Bp}^{Schottky}}{kT}}$

The corresponding minority carrier density under equilibrium $p^{EQ}\left(x_{it}^{front/back}\right)$ or $n^{EQ}\left(x_{it}^{front/back}\right)$ is then given by the mass action law:

3.a, 3.b $p^{EQ}\left(x_{it}^{front/back}\right) = \frac{N_C\, N_V\, e^{-\frac{E_g}{kT}}}{n^{EQ}\left(x_{it}^{front/back}\right)}$ or $n^{EQ}\left(x_{it}^{front/back}\right) = \frac{N_C\, N_V\, e^{-\frac{E_g}{kT}}}{p^{EQ}\left(x_{it}^{front/back}\right)}$

As flatband conditions are chosen, the metal work function is calculated to give a zero build in voltage due to the metal/semiconductor contact:

$$\phi^{front/back} = E_C - E_F\left(x_{it}^{front/back}\right) + \chi^{front/back}$$

In all other calculation modes (DC, AC, TR), the electron/hole currents into the metal contact, $j_n^{it,\,front/back}(t)$, $j_p^{it,\,front/back}(t)$ are specified:

2.a $j_n^{it,front}(t) = \quad q\, S_n^{front} \left\{ n\left(x_{it}^{front}, t\right) - n^{EQ}\left(x_{it}^{front}\right) \right\}$

2.b $j_n^{it,back}(t) = -q\, S_n^{back} \left\{ n\left(x_{it}^{back}, t\right) - n^{EQ}\left(x_{it}^{back}\right) \right\}$

3.a $j_p^{it,front}(t) = -q\, S_p^{front} \left\{ p\left(x_{it}^{front}, t\right) - p^{EQ}\left(x_{it}^{front}\right) \right\}$

3.b $j_p^{it,back}(t) = \quad q\, S_p^{back} \left\{ p\left(x_{it}^{back}, t\right) - p^{EQ}\left(x_{it}^{back}\right) \right\}$

Furthermore, if using the DC, AC or TR calculation mode with an external illumination (optical super bandgap generation) enabled, an illumination dependent metal work function is calculated, in order to ensure illumination independent flatband conditions: Assuming a zero internal current density (no netto current through the semiconductor stack), the metal work function is now iteratively calculated from the majority quasi Fermi energy $E_{Fn}\left(x_{it}^{front}, t\right)$ or $E_{Fp}\left(x_{it}^{front}, t\right)$ instead from the Fermi energy, in order to ensure a zero build in voltage due to the metal/semiconductor contact

$$\phi^{front/back}(t) = E_C - E_{Fn}\left(x_{it}^{front/back}, t\right) + \chi^{front/back} \qquad \text{or}$$

$$\phi^{front/back}(t) = E_C - E_{Fp}\left(x_{it}^{front/back}, t\right) + \chi^{front/back}$$

4.4.2 Boundary model: Schottky metal/semiconductor contact

This boundary model can describe metal/semiconductor contacts, which drive the semiconductor into depletion or into accumulation (Sze & Kwok, 2007). Explicit values of the metal work function $\phi^{front/back}$ can be specified in order to fix the majority barrier height of the metal/semiconductor contact $\phi_{Bn}^{Schottky} = q\left\{\phi^{front/back} - \chi^{front/back}\right\}$, $\phi_{Bp}^{Schottky} = q\left\{E_g - \phi^{front/back} + \chi^{front/back}\right\}$. Otherwise, this boundary model is totally equivalent to the flatband metal/semiconductor boundary model described above.

4.4.3 Boundary model: insulator contact

If the boundary of the semiconductor stack is considered to be insulating, additional interface states can be defined, as according to (Kronik & Shapira, 1999). They are treated equivalent to the bulk, but with densities given in cm^{-2} instead of cm^{-3}. For an insulator/semiconductor contact at the front the three boundary conditions are:

1. $\quad 0 = -\varepsilon_0 \varepsilon_r \left.\dfrac{\partial \varphi(x,t)}{\partial x}\right|_{x_{front}} - q \sum_{trap} \rho^{trap}$

2. $\quad 0 = j_n\left(x_{front}\right) - R_{it,n}^{front}$

3. $\quad 0 = -j_p\left(x_{front}\right) - R_{it,p}^{front}$

4.4.4 Boundary model: metal/insulator/semiconductor contact

In case of using a metal/insulator/semiconductor MIS contact, the insulator capacity C has to be additionally specified. At the insulator/semiconductor interface additional interface

defects can be defined, which are treated equivalent to the bulk but with densities given in cm^{-2} instead of cm^{-3}, as according to (Kronik & Shapira, 1999). Depending on whether the MIS contact is defined on only one or on both boundaries two different cases have to be discussed. If both boundaries have an MIS contact, the capacities C^{front}, C^{back} of the front and back boundaries can be defined separately. Furthermore, one has to define the voltage fraction f that drops at the front MIS contact compared to the fraction that drops at the back MIS contact. For a given external voltage this defines how the different metal layers are charged. Time dependent boundary conditions (AC or TR calculation mode) for the MIS contact on are not implemented at the current state of AFORS-HET development.
Both semiconductor potentials $\varphi(x_{front})$ and $\varphi(x_{back})$ at the front/back boundary of the semiconductor stack and the metal work functions $\phi^{front/back}$ of the front/back contact enter the boundary condition for the electric potential. Also enters the net charge ρ_{it} of the interface, which has to be calculated by summing over all interface defects. The electron/hole currents into the interface defects $j_n(x_{front})$, $j_p(x_{front})$, are given by the recombination rates $R_{it,n}^{front}$, $R_{it,p}^{front}$ of the interface defects. The three boundary conditions for a MIS contact read:

1. $$0 = C\left\{V_{ext} - \left[\varphi(x_{back}) - \varphi(x_{front})\right] + \left[\phi^{back} - \phi^{front}\right]\right\} - \varepsilon_0\varepsilon_r \left.\frac{\partial\varphi(x)}{\partial x}\right|_{x_{front}} - q\sum_{trap}\rho_{it}^{trap}$$

2. $$0 = j_n(x_{front}) - R_{it,n}^{front}$$

3. $$0 = -j_p(x_{front}) - R_{it,p}^{front}$$

in the case that only one MIS contact at the front boundary is chosen, and

1.a $$0 = f\,C^{\text{front}}\left\{V_{ext} - \left[\varphi(x_{back}) - \varphi(x_{front})\right] + \left[\phi^{back} - \phi^{front}\right]\right\} - \varepsilon_0\varepsilon_r \left.\frac{\partial\varphi(x)}{\partial x}\right|_{x_{front}} - q\sum_{trap}\rho_{it}^{trap}$$

2.a $$0 = j_n(x_{front}) - R_{it,n}^{front}$$

3.a $$0 = -j_p(x_{front}) - R_{it,p}^{front}$$

1.b $$0 = (1-f)\,C^{\text{back}}\left\{V_{ext} - \left[\varphi(x_{back}) - \varphi(x_{front})\right] + \left[\phi^{back} - \phi^{front}\right]\right\} - \varepsilon_0\varepsilon_r \left.\frac{\partial\varphi(x)}{\partial x}\right|_{x_{back}} - q\sum_{trap}\rho_{it}^{trap}$$

2.b $$0 = j_n(x_{back}, t) - R_{it,n}^{back}$$

3.b $$0 = -j_p(x_{back}, t) - R_{it,p}^{back}$$

in case that two MIS contacts at both boundaries are chosen.

5. Characterization methods simulated by AFORS-HET

In the following it is described how the most common solar cell characterization methods are simulated within AFORS-HET, i.e. current-voltage (IV), quantum efficiency (QE), quasi-steady-state photoconductance (QSSPC), impedance (IMP, ADM, C-V, C-T, C-f), surface photovoltage (ID-SPV, VD-SPV, WD-SPV) and photo-electro-luminescence (PEL).

5.1 Measurement model: current-voltage characteristic (IV)

This measurement varies the external voltage at the boundaries and plots the resulting external current through the semiconductor stack in order to obtain the current-voltage characteristic of the simulated structure. For each voltage value the total current through the structure (the sum of the electron and hole current at a boundary gridpoint) is calculated. This can be done in the dark or under an illumination. The measurement model can iterate the specific data points maximum-power point (mpp), open-circuit voltage (Voc), short-circuit current (Isc) and thus calculate the fill-factor FF and the efficiency Eff of the solar cell, whereas the illumination power density $P_{\text{illumination}}$ in W/cm^2 is calculated from the incident photon spectrum:

$$FF = \frac{V_{mpp}\, I_{mpp}}{V_{oc}\, I_{sc}}, \qquad Eff = \frac{V_{mpp}\, I_{mpp}}{P_{\text{illumination}}} = \frac{FF\; V_{oc}\; I_{sc}}{P_{\text{illumination}}}$$

5.2 Measurement model: quantum efficiency (QE)

In order to simulate quantum efficiencies, the semiconductor stack is additionally illuminated with a monochromatic irradiation at a certain wavelength λ, and the difference ΔI_{SC}^{irrad} of the resulting short circuit current with and without the additional irradiation is computed. A quantum efficiency $QE(\lambda)$ can then defined as

$$QE(\lambda) = \frac{number.of.electrons.in.the.external.circuit}{number.of.photons} = \frac{\Delta I_{SC}^{irrad} / q}{number.of.photons}$$

Different quantum efficiencies are calculated, depending on the number of photons which are considered: (1) external quantum efficiency (EQE): all photons of the additional irradiation, which are incident on the semiconductor stack, whether they are reflected, absorbed or transmitted, are counted. (2) internal quantum efficiency (IQE): only the absorbed photons of the additional irradiation are counted. Note, that like in a real measurement, photons which are absorbed in the contacts are also counted, despite the fact that they do not contribute to the current. (3) corrected internal quantum efficiency (IQE1): only the photons of the additional irradiation which are absorbed in the semiconductor stack are counted.

5.3 Measurement model: quasi steady state photoconductance (QSSPC)

The excess carrier density dependant lifetimes $\tau_{n\_all}(\Delta n)$, $\tau_{p\_all}(\Delta p)$ for a semiconductor stack of the thickness L under a given external illumination are calculated according to the following equations:

$$\tau_{n\_all}(\Delta n) = \frac{\Delta n}{\Delta G} \qquad \tau_{p\_all}(\Delta p) = \frac{\Delta p}{\Delta G}$$

$$\Delta n = \left(\int dx \{n_{\text{illuminated}}(x)\} - \int dx \{n_{\text{dark}}(x)\}\right) / L$$

$$\Delta p = \left(\int dx \{p_{\text{illuminated}}(x)\} - \int dx \{p_{\text{dark}}(x)\}\right) / L$$

$$\Delta G = \left(\int dx \{G_{\text{illuminated}}(x)\} - \int dx \{G_{\text{dark}}(x)\}\right) / L$$

The average dark and illuminated carrier densities and the average generation rate are calculated by integrating over the whole structure. Thus the excess carrier densities Δn, Δp and the corresponding change in generation rate ΔG can be calculated. Within the measurement model, the external illumination intensity is varied and the resulting excess carrier dependant lifetimes $\tau_{n\_all}(\Delta n)$ and $\tau_{p\_all}(\Delta p)$ are plotted.
For typical structures that have a c-Si layer with low mobility passivation layers at the front and back additionally c-Si carrier lifetimes τ_{n_c-Si} and τ_{p_c-Si} are calculated by only integrating over the c-Si layer. To model the typical QSSPC measurements of passivated c-Si wafers done with the commercially available setup by Sinton Consulting (Sinton & Cuevas, 1996), an effectively measured carrier lifetime τ_{qss} is calculated by the following equation:

$$\tau_{qss} = \frac{\Delta n_{c-Si}\mu_{n\_c-Si} + \Delta p_{c-Si}\mu_{p\_c-Si}}{\mu_{n\_c-Si} + \mu_{p\_c-Si}} / \Delta G_{c-Si}$$

5.4 Measurement model: impedance, capacitance (IMP, ADM, C-V, C-T)

Both boundaries must be described by a voltage controlled metal-semiconductor contact. Additional to the time independent external DC voltage V_{ext}^{DC} an alternating sinusoidal AC voltage is superimposed, $\tilde{V}_{ext}(x,t) = V_{ext}^{DC}(x) + V_{ext}^{AC}(x)\, e^{i\omega t}$, with a small amplitude V_{ext}^{AC} and a given frequency f, $\omega = 2\pi f$. The resulting external current through the semiconductor stack in the limes of a sufficiency small amplitude is calculated, $\tilde{I}_{ext}(x,t) = I_{ext}^{DC}(x) + I_{ext}^{AC}(x)\, e^{i\omega(t+\delta)} = I_{ext}^{DC}(x) + \tilde{I}_{ext}^{AC}(x)\, e^{i\omega t}$. It is also sinusoidal and of the same frequency f, with an AC-amplitude I_{ext}^{AC} and a phase shift δ, or with a complex amplitude $\tilde{I}_{ext}^{AC}$ respectively.
The impedance is defined to be the complex resistance of the semiconductor stack, i.e. the quotient of ac-voltage to ac-current. The admittance is defined to be the complex conductivity of the semiconductor stack, i.e. the quotient of ac-current to ac-voltage. It can be equivalently represented by a parallel circuit of a conductance G and a capacitance C.

$$\widetilde{IMP} = \frac{V_{ext}^{AC}}{\tilde{I}_{ext}^{AC}}, \qquad \widetilde{ADM} = \frac{\tilde{I}_{ext}^{AC}}{V_{ext}^{AC}} = G + i\,2\pi f C$$

Depending on the measurement chosen, the frequency is varied and the amplitude and phase shift of the impedance is plotted (measurement IMP), or the capacitance, conductance and conductance divided by frequency is plotted (measurement ADM). Furthermore, for a fixed frequency f, the capacitance can be plotted as a function of the external DC-voltage (measurement C-V) or as a function of the temperature (measurement C-T).

5.5 Measurement model: surface photovoltage (ID-SPV, VD-SPV, WD-SPV)

In order to simulate a steady-state surface photovoltage (SPV) signal (Kronik & Shapira, 1999), the front side boundary should usually be a metal-insulator-semiconductor contact. The semiconductor stack is additionally illuminated with a monochromatic irradiation at a certain wavelength and intensity. The potential difference Δf with and without monochromatic illumination at the front (first grid point) and at the back (last grid point) of the stack is computed and as output the SPV signal $V_{SPV\_front/back}$ is calculated.

$$V^{SPV}_{front/back} = \Delta\varphi_{\text{front/back}} = \left(\varphi^{\text{illuminated}}_{\text{front/back}} - \varphi^{\text{dark}}_{\text{front/back}}\right)$$

Note that only one quantity φ_{front} or φ_{back} will change upon illumination, as the potential is fixed to $\varphi = 0$ either at the front side or at the back side. Depending on the measurement, either the intensity of the monochromatic illumination is varied (ID-SPV, intensity dependant surface photovoltage), or the external voltage is varied (VD-SPV, voltage dependant surface photovoltage), or the wavelength of the monochromatic illumination is varied (WD-SPV, wavelength dependant surface photovoltage).

5.6 Measurement model: photo electro luminescence (PEL)

When an external illumination and/or an external voltage are applied the emitted radiation can be calculated according to the generalized Plank equation (Würfel, 1982).

$$I(\lambda) = 2c\int dx \left\{ \frac{\alpha(\lambda, x)}{\lambda^5} \cdot \frac{1}{\exp\left(\left(\frac{hc}{\lambda} - \left(E_{Fn}(x) - E_{Fp}(x)\right)\right)/kT\right) - 1} \right\}$$

By integration over the whole structure the wavelength dependant emitted intensity to the front and back is calculated taking photon re-absorption into account. For a given absorption coefficient α and a given wavelength λ the spectra $I(\lambda)$ of the emitted photons is determined by the splitting of the quasi-Fermi levels of electrons and holes E_{Fn}, E_{Fp}. The external working conditions like external illumination and/or applied voltage that cause the quasi-Fermi level splitting have to be specified. Furthermore the wavelength region for which the emitted intensity is calculated can be selected.

6. Selected examples on AFORS-HET simulations

To illustrate the concepts of numerical solar cell simulation, some selected examples simulating a simple amorphous/crystalline silicon solar cell are shown. The absorber of the solar cell (designed for photon absorption) is constituted by a 300 μm thick p-doped textured silicon wafer, c-Si, whereas the emitter of the solar cell (designed for minority carrier extraction, that is electron extraction) consists of an ultra thin 10 nm layer of n-doped, hydrogenated amorphous Silicon, a-Si:H, see Fig. 4. In order to support the lateral electron transport, a transparent conductive oxide layer, TCO, is used as a front side contact. For the sake of simplicity, majority carrier extraction that is hole extraction, is realized as a simple metallic flatband contact to the p-type absorber. Please note, that this solar cell structure is not a high efficiency structure, as a back surface field region, BSF, for hole extraction in order to avoid contact recombination, has not been used. However, this structure has been chosen, as it clearly reveals the properties of an amorphous/crystalline heterojunction interface.

This interface is crucial for the performance of an amorphous/crystalline heterojunction solar cell: By an adequate wet-chemical pre-treatment of the wafer prior to the deposition of a-Si:H onto the surface of the silicon wafer, one has to ensure that an a-Si:H/c-Si heterocontact with a low a-Si:H/c-Si interface state density, D_{it}, will form. The influence of D_{it} on the solar cell performance as well as on various solar cell characterisation methods will be shown. Thus a sensitivity analysis of different measurement methods in order to measure an unknown D_{it} is performed my means of numerical simulation.

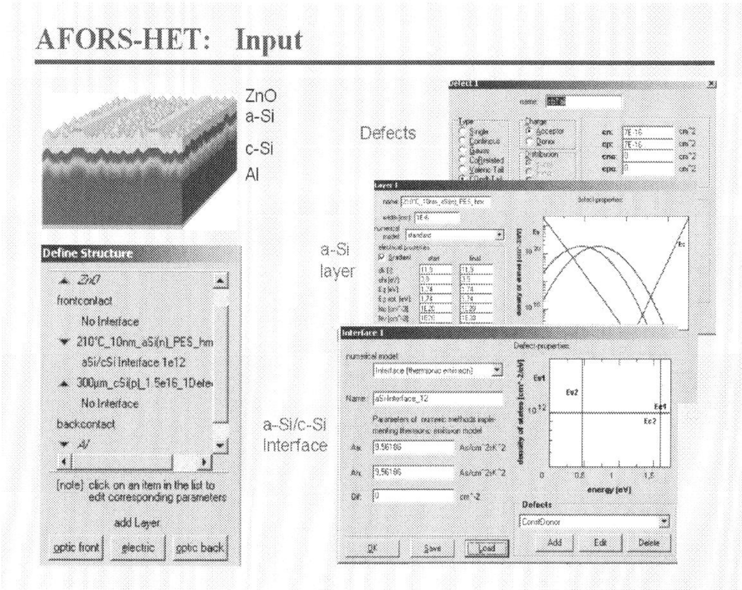

Fig. 4. Screenshots of typical AFORS-HET input: Simulation of TCO/a-Si:H(n)/c-Si(p)/Al heterojunction solar cells. (left) layer sequence, (right) defect distributions $N_{trap}(E)$ of the a-Si:H(n) layer and of the a-Si:H(n)/c-Si(p) interface.

Fig. 4 shows typical screenshots of an AFORS-HET input while modelling the above mentioned TCO/a-Si:H(n)/c-Si(p)/Al heterojunction solar cell. In order to model the c-Si absorber, the bulk model "crystalline silicon" is chosen, specifying the appropriate doping (i.e. $N_A = 1.5\ 10^{16}\ cm^{-3}$) and the appropriate lifetime of the wafer (i.e. specifying a defect density of a single midgap defect $N_t = 1.\ 10^{10}\ cm^{-3}$, which corresponds to a mean lifetime of $1\,ms$ as indicated in the input window). In order to model the a-Si:H emitter, the bulk model "standard semiconductor" is chosen, specifying the measured density of state distributions within the bandgap of a-Si:H (Korte & Schmidt, 2008), see Fig. 4. I.e. the measured Urbach tail states and the measured dangling bond states of a-Si:H have to be stated and the doping density N_D has to be adjusted to a value which leads to the measured Fermi level to valence band distance $E_F - E_V = 250\,meV$ of a-Si:H (Korte & Schmidt, 2008). Furthermore, the electron affinity of a-Si:H has to be adjusted to a value in order to represent the measured valence band offset $\Delta E_V = E_V^{cSi} - E_V^{aSi} = 450\,meV$ (Korte & Schmidt, 2008). For modeling the a-Si/c-Si interface, the interface model "drift diffusion interface" is chosen, assuming a simple constant distribution of interface defects within the bandgap, exhibiting a donor like character below midgap and an acceptor like character above midgap, see Fig.4. The TCO layer at the front is modelled as an optical layer, thus at the front contact the measured TCO absorption (Schmidt et. al., 2007) as well as the measured solar cell reflection due to the surface texturing (Schmidt et. al., 2007) is specified. Therefore, for the optical calculation the optical model "Lambert-Beer absorption" has to be specified.

6.1 Optical calculation

Fig. 5 shows the resulting spectral absorptions of the incoming AM 1.5 illumination within the different layers of the solar cell: More than half of the low wavelength radiation ($\lambda \leq 350\,nm$) is absorbed within the 80 nm thick TCO layer and is therefore lost for solar energy conversion. Also the defect-rich, ultra-thin a-Si:H emitter is significantly absorbing photons up to $\lambda \leq 600\,nm$. All photons with $\lambda \leq 600\,nm$, which are not absorbed, are reflected. Most photons with $\lambda \geq 800\,nm$, which are not absorbed by the solar cell absorber, are transmitted, some of them are reflected, a few of them are absorbed in the TCO layer due to free carrier absorption. After exceeding the bandgap of the c-Si absorber (for $\lambda \geq 1120\,nm$) there is no more photon absorption in the absorber.

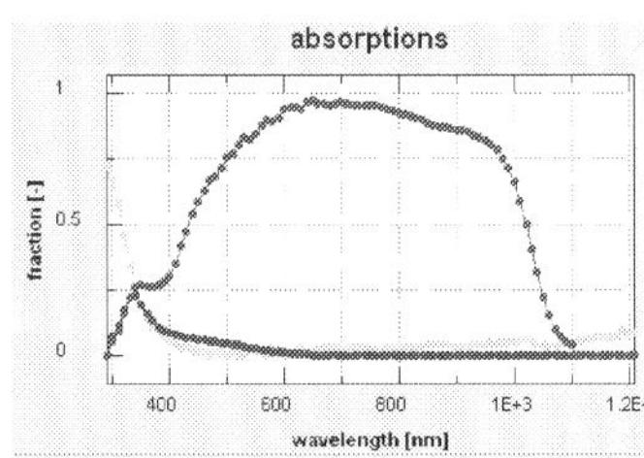

Fig. 5. Screenshot of the spectral absorption within the different solar cell layers (yellow: TCO layer, blue: a-Si:H emitter layer, red: c-Si absorber layer).

6.2 Equilibrium band diagrams

Fig. 6 shows the resulting equilibrium band diagrams (conduction band energy, valence band energy and Fermi energy as a function of the position within the solar cell) assuming different interface state densities D_{it}, after an electrical calculation has been performed.

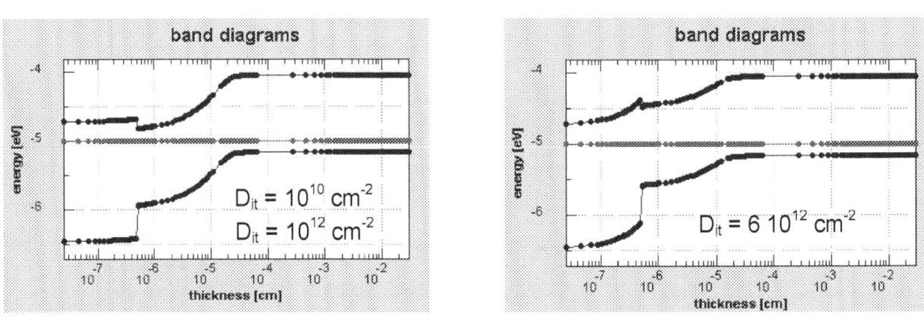

Fig. 6. Screenshots of equilibrium band diagrams (red: Fermi energy, black: valence and conduction band energy) for three different a-Si:H/c-Si interface state densities D_{it}.

Note that the equilibrium band diagram does not change until $D_{it} \geq 2\ 10^{12}\ cm^{-2}$.

6.3 Current-voltage characteristics

However, if one looks at the solar cell performance, i.e. if one calculates the corresponding current-voltage characteristics, D_{it} will reduce the open-circuit voltage of the solar cell for $D_{it} \geq 1\ 10^{10}\ cm^{-2}$, see Fig. 7. Even if interface states in a comparatively low concentration are formed, i.e. $1\ 10^{10}\ cm^{-2} \leq D_{it} \leq 5\ 10^{10}\ cm^{-2}$, this will significantly reduce the solar cell efficiency.

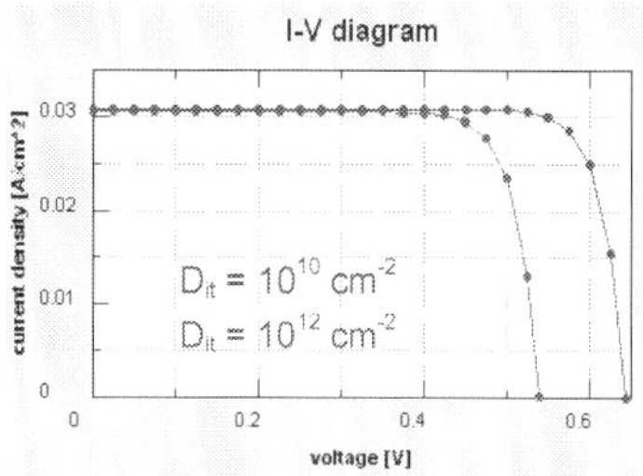

Fig. 7. Screenshot of a current-voltage simulation under AM 1.5 illumination for two different a-Si:H/c-Si interface state densities D_{it}.

6.4 Quantum efficiency

The influence of D_{it} is not noticeable in a quantum efficiency measurement, as the short-circuit current density is not affected due to a D_{it} variation, and quantum efficiency is a measure for the excess carrier collection efficiency under short circuit conditions. In Fig. 8 internal as well as external quantum efficiency is shown (IQE, EQE), whereas the difference of the two results from the measured reflection losses.

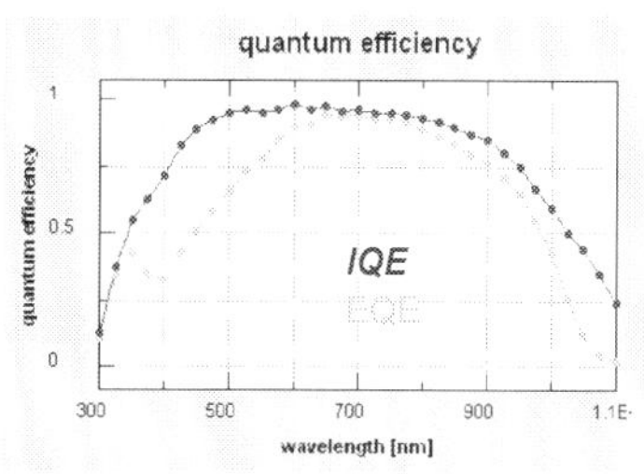

Fig. 8. Screenshot of a quantum efficiency simulation (there is no difference for different a-Si:H/c-Si interface state densities D_{it}).

6.5 Impedance, capacitance

If one monitors temperature dependent impedance in the dark (i.e. if one calculates the resulting conductance and capacitance as a function of temperature), the onset of the change of the equilibrium band bending due to an increasing D_{it} can be detected. As soon as the increasing D_{it} starts to change the equilibrium band bending, an additional peak in the conductance spectra evolves (Gudovskikh et. al., 2006), see Fig. 9. Thus, dark capacitance-temperature (C-T) measurements are sensitive to interface states only for $D_{it} \geq 2.10^{12} cm^{-2}$.

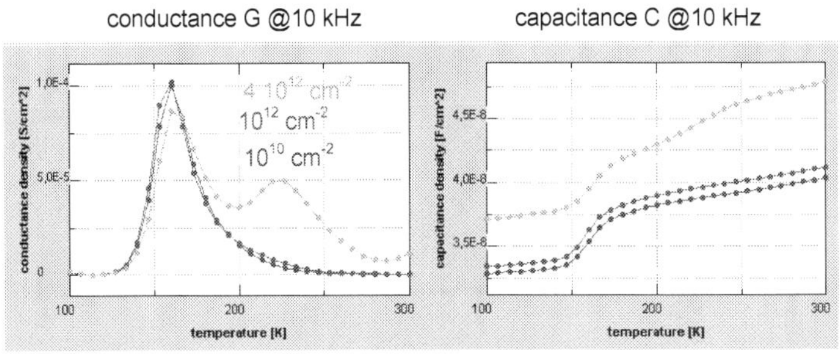

Fig. 9. Screenshot of a capacitance-temperature simulation at an AC frequency of 10 kHz for three different a-Si:H/c-Si interface state densities D_{it}.

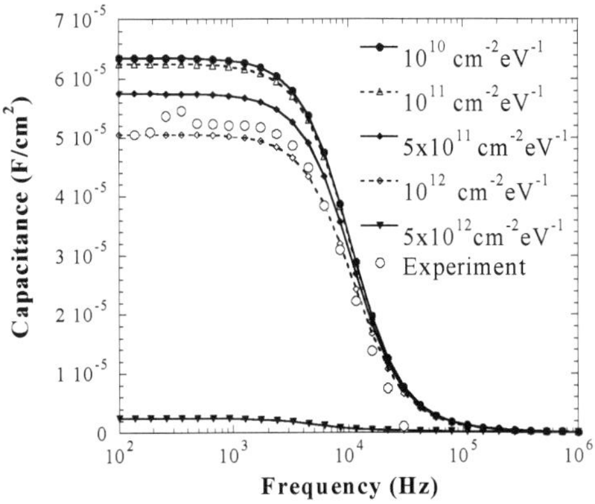

Fig. 10. Comparison of simulated and measured capacitance-frequency measurements under AM1.5 illumination for different a-Si:H/c-Si interface state densities D_{it}. Data from (Gudovskikh et. al., 2006).

In order to enhance the sensitivity towards D_{it}, measurements under illumination have to be performed. Fig. 10 shows an example of an illuminated capacitance-frequency (C-f) measurement, where the corresponding simulations are compared to a real experiment (Gudovskikh et. al., 2006). According to the simulation, the D_{it} of the solar cell under investigation was in the range $D_{it} \approx 8.10^{11} cm^{-2}$. A sensitivity analysis of this measurement technique indicates a sensitivity towards D_{it} for $D_{it} \geq 1.10^{11} cm^{-2}$. However, this is still not sufficient in order to characterize well passivated solar cells with a low D_{it} in the range $1\ 10^{10}\ cm^{-2} \leq D_{it} < 1\ 10^{11}\ cm^{-2}$.

6.6 Photoluminescence

Photoluminescence proofs to be quite sensitive to D_{it}. This is because this measurement performs without current extraction. As an example, Fig. 11 shows the simulated steady-state photoluminescence spectra as well as the transient photoluminescence decay (after an integration of the spectra) due to a pulse-like excitation for two different values of $D_{it} = 1.10^{10} cm^{-2}$ and $D_{it} = 1.10^{12} cm^{-2}$. If one integrates the spectra, the simulated measurement signals differ for more than one order of magnitude.

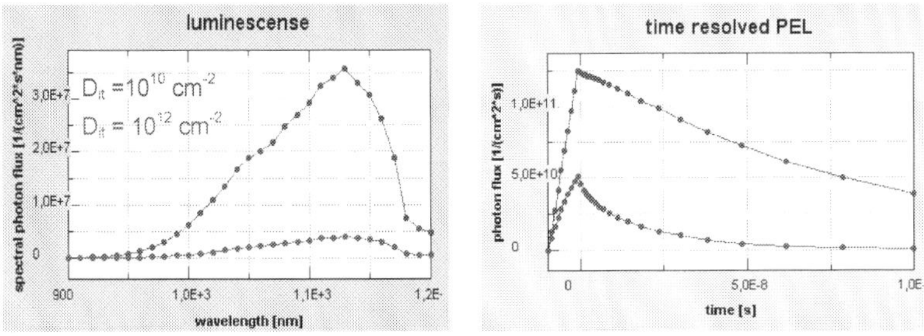

Fig. 11. Screenshots of photoluminescence simulations for two different a-Si:H/c-Si interface state densities D_{it}. (left) steady-state photoluminescence spectra, (right) transient photoluminescence decay after a pulse-like excitation.

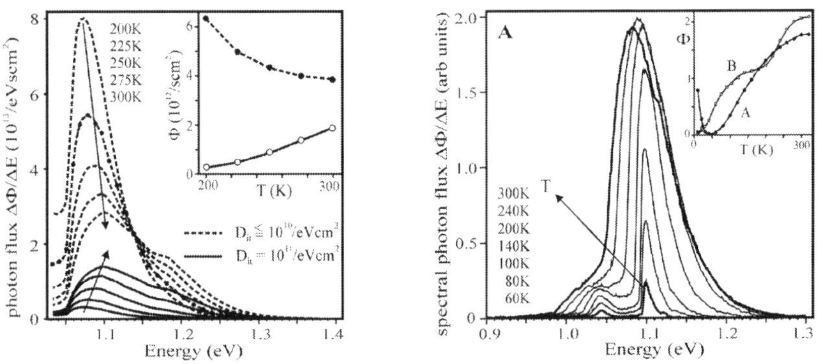

Fig. 12. (left) Simulated temperature dependant photoluminescence measurements for different a-Si:H/c-Si interface state densities D_{it}. (right) Measured temperature dependant photoluminescence. Data from (Fuhs et. al, 2006).

The sensitivity towards D_{it} can even be more enhanced, if one performs temperature dependant photoluminescence measurements, see Fig. 12. Here the character of the measurement even changes if D_{it} is in the range $1\,10^{10}\,cm^{-2} \leq D_{it} < 1\,10^{11}\,cm^{-2}$. For $D_{it} \leq 1.10^{10} cm^{-2}$ the spectral emission decreases with increasing temperature, see Fig. 12, thus indicating a non noticeable amount of interface defects, whereas for example for $D_{it} = 1.10^{11} cm^{-2}$ an increasing spectral emission with increasing temperature is observed (Fuhs et. al, 2006).

7. Conclusion

A mathematical description of AFORS-HET, version 2.4, a one dimensional computer program for the simulation of solar cells and solar cell characterization methods has been stated. Some selected examples, simulating amorphous/crystalline silicon heterojunction solar cells and investigating the sensitivity of various measurement methods towards the interface state density D_{it}, were presented.

8. References

Anderson, R. L. (1962). Experiments on Ge-GaAs Heterojunctions, *Solid State Electron.*, 5 (1962), 341-51

Basore, P.; Clugston, D. A. (1997). PC1D, Version 5.9, Copyright 2003, University of New South Wales, latest publication describing the program: PC1D Version 5: 32-bit Solar Cell Simulation on Personal Computers, *26th IEEE Photovoltaic Specialists Conf. (Sept 1997)*

Fuhs, W.; Laades, L.; v.Maydell, K.; Stangl, R.; Gusev, O.B.; Terukov, E.I.; Kazitsyna-Baranovski, S.; Weiser, G. (2006). Band-edge electroluminescence from amorphous/crystalline silicon heterostructure solar cells, *Journal of Non-Crystalline Solids*, 352 (2006) 1884–1887

Gudovskikh, A.S.; Kleider, J.P.; Stangl, R. (2006). New approach to capacitance spectroscopy for interface characterization of a-Si:H/c-Si heterojunctions, *Journal of Non-Crystalline Solids*, 352 (2006) 1213-1216

Korte, L.; Schmidt, M. (2008). Investigation of gap states in phosphorous-doped ultra-thin a-Si:H by near-UV photoelectron spectroscopy, in: *J. Non. Cryst. Sol.* 354 (2008) 2138-2143

Kronik, L.; Shapira, Y. (1999). Surface Photovoltage Phenomena: Theory, Experiment, and Applications, *Surface Science Reports*, 37 (1999), 5-206

Kundert et. al. (1988). A sparse linear equation solver, department of electrical engineering and computer science, Berkeley, CA, USA, 1988, available from: http://www-rab.larc.nasa.gov/nmp/nmpCode.htm

Sah, C.; Shockley, W. (1958). Electron-Hole Recombination Statistics in Semiconductors through Flaws with Many Charge Conditions, *Physical Review*, 109 (1958), 1103

Schmidt, M.; Korte, L.; Laades, A.; Stangl, R.; Schubert, Ch.; Angermann, H.; Conrad, E.; v.Maydell, K. (2007). Physical aspects of a-Si:H/c-Si hetero-junction solar cells, *Thin Solid Films* 515 (2007), p. 7475-7480

Selberherr, S. (1984). Analysis and simulation of semiconductor devices, *Springer Verlag*, 1984

Shockley, W.; Read, W. T. (1952). Statistics of the Recombinations of Holes and Electrons, *Physical Review*, 87 (1952), 835

Sinton, R.A.; Cuevas, A. (1996). Contactless determination of current-voltage characteristics and minority-carrier lifetimes in semiconductors from quasi-steady-state photoconductance data, *Applied Physics Letters*, 69 (1996), 2510-2512

Sze, S. M.; Kwok, K. N. (2007). Physics of Semiconductor Devices, *John Wiley & Sons, Inc., Hoboken, New Jersey*, 2007

Würfel, P. (1982). The chemical potential of radiation, *Journal of Physics C*, 15 (1982), 3967-3985

15

Amorphous Silicon Carbide Photoelectrode for Hydrogen Production from Water using Sunlight

Feng Zhu[1], Jian Hu[1], Ilvydas Matulionis[1], Todd Deutsch[2], Nicolas Gaillard[3], Eric Miller[3], and Arun Madan[1]
[1]MVSystems, Inc., 500 Corporate Circle, Suite L, Golden, CO, 80401
[2]Hawaii Natural Energy Institute (HNEI), University of Hawaii at Manoa, Honolulu, HI 96822,
[3]National Renewable Energy Laboratory (NREL), Golden, CO 80401, USA

1. Introduction

Hydrogen is emerging as an alternative energy carrier to fossil fuels. There are many advantages of hydrogen as a universal energy medium. For example, it is non-toxic and its combustion with oxygen results in the formation of water to release energy. In this chapter, we discuss the solar to hydrogen production directly from water using a photoelectrochemical (PEC) cell; in particular we use amorphous silicon carbide (a-SiC:H) as a photoelectrode integrated with a-Si tandem photovoltaic (PV) cell. High quality a-SiC:H thin film with bandgap ≥2.0eV was fabricated by plasma enhanced chemical vapor deposition (PECVD) technique using SiH_4, H_2 and CH_4 gas mixture. Incorporation of carbon in the a-SiH film not only increased the bandgap, but also led to improved corrosion resistance to an aqueous electrolyte. Adding H_2 during the fabrication of a-SiC:H material could lead to a decrease of the density of states (DOS) in the film. Immersing the a-SiC:H(p)/a-SiC:H(i) structure in an aqueous electrolyte showed excellent durability up to 100 hours (so far tested); in addition, the photocurrent increased and its onset shifted anodically after 100-hour durability test. It was also found that a SiO_x layer formed on the surface of a-SiC:H, when exposed to air led to a decrease in the photocurrent and its onset shifted cathodically; by removing the SiO_x layer, the photocurrent increased and its onset was driven anodically. Integrating with a-Si:H tandem cell, the flat-band potential of the PV/a-SiC:H structure shifts significantly below the H_2O/O_2 half-reaction potential and is in an appropriate position to facilitate water splitting and has exhibited encouraging results. The PV/a-SiC:H structure produced hydrogen bubbles from water splitting and exhibited good durability in an aqueous electrolyte for up to 150 hours (so far tested). In a two-electrode setup (with ruthenium oxide as counter electrode), which is analogous to a real PEC cell configuration, the PV/a-SiC:H produces photocurrent of about 1.3 mA/cm^2 at zero bias, which implies a solar-to-hydrogen (STH) conversion efficiency of over 1.6%. Finally, we present simulation results which indicate that a-SiC:H as a photoelectrode in the PV/a-SiC:H structure could lead to STH conversion efficiency of >10%.

2. Principles and status of using semiconductor in PEC

In general, hydrogen can be obtained electrolytically, photo-electrochemically, thermochemically, and biochemically by direct decomposition from the most abundant material on earth: water. Though a hydrogen-oxygen fuel cell operates without generating harmful emissions, most hydrogen production techniques such as direct electrolysis, steam-methane reformation and thermo-chemical decomposition of water can give rise to significant greenhouse gases and other harmful by-products. We will briefly review the solid-state semiconductor electrodes for PEC water splitting using sunlight. Photochemical hydrogen production is similar to a thermo-chemical system, in that it also employs a system of chemical reactants, which leads to water splitting. However, the driving force is not thermal energy but sunlight. In this sense, this system is similar to the photosynthetic system present in green plants. In its simplest form, a photoelectrochemical (PEC) hydrogen cell consists of a semiconductor as a reaction electrode (RE) and a metal counter electrode (CE) immersed in an aqueous electrolyte, and PEC water splitting at the semiconductor-electrolyte interface drove by sunlight, which is of considerable interest as it offers an environmentally "green" and renewable approach to hydrogen production (Memming, 2000).

2.1 Principles of PEC

The basic principles of semiconductor electrochemistry have been described in several papers and books (Fujishima & Honda, 1972; Gerscher & Mindt, 1968; Narayanan & Viswanathan, 1998; Memming, 2000; Gratzel, 2001).

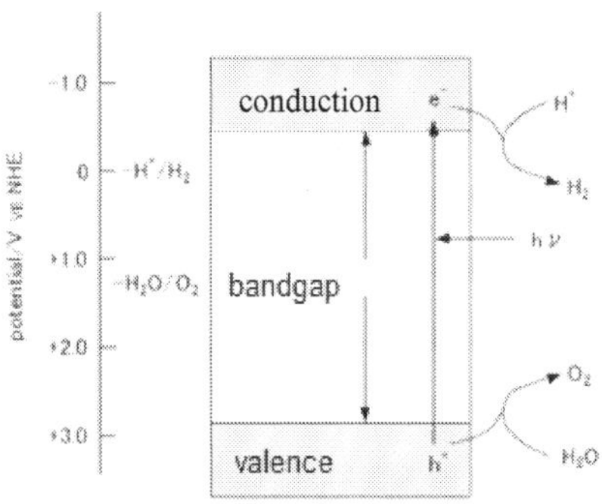

Fig. 1. The band diagram of the PEC system. The conduction band edge needs to be located negative (on an electrochemical scale) high above the reduction potential of water, the valence band edge positive enough below the oxidation potential of water to enable the charge transfer. NHE stands for "normal hydrogen electrode".

The only difference between a photoelectrochemical and a photovoltaic device is that in the PEC case, a semiconductor-electrolyte junction is used as the active layer instead of the solid-state junctions in a photovoltaic structure. In both cases, a space charge region is formed where contact formation compensates the electrochemical potential differences of electrons on both sides of the contact. The position of the band edges of the semiconductor at the interface can be assumed in a first approximation to be dependent only on the pH of the solution and independent of the potential (Fermi level) of the electrode or the electrolyte (Memming, 2000; Kuznetsov & Ulstrup, 2000). For direct photoelectrochemical decomposition of water, several primary requirements of the semiconductor must be met: the semiconductor system must generate sufficient voltage (separation of the quasi Fermi levels under illumination) to drive the electrolysis, the energetic of the semiconductor must overlap that of the hydrogen and oxygen redox reactions (saying the band positions at the semiconductor-electrolyte interface have to be located at an energetically suitable position as shown in Fig.1), the semiconductor system must be stable in aqueous electrolytes, and finally the charge transfer from the surface of the semiconductor must be fast enough not only to prevent corrosion but also reduce energy losses due to overvoltage (Gerscher & Mindt, 1968; Narayanan & Viswanathan, 1998; Memming, 2000).

Neglecting losses, the energy required to split water is 237.18 kJ/mol, which converts into 1.23 eV, i.e. the PV device must be able to generate more than 1.23 Volts. The STH conversion efficiency in PEC cells can be generally expressed as

$$\text{Efficiency}=\frac{\text{chemical energy in hydrogen produced in a PEC cell}}{\text{energy in the sunlight over the collection area}}=\frac{J_{ph}V_{WS}}{E_S} \qquad (1)$$

where J_{ph} is the photocurrent density (in mA/cm^2) generated in a PEC cell, V_{WS} = 1.23 V is the potential corresponding to the Gibbs free energy change per photon required to split water, and Es is the solar irradiance (in mW/cm^2). Under AM1.5 G illumination, a simple approximation for the STH efficiency is J_{ph} times 1.23(in %) (Memming, 2000; Miller & Rocheleau, 2002).

2.2 Status of using semiconductor in PEC

Although as early as in 1839 E. Becquerel (Memming, 2000) had discovered the photovoltaic effect by illuminating a platinum electrode covered with a silver halide in an electrochemical cell, the foundation of modern photoelectrochemistry has been laid down much later by the work of Brattain and Garret and subsequently Gerischer (Bak, et al., 2002; Mary & Arthru, 2008), who undertook the first detailed electrochemical and photoelectrochemical studies of the semiconductor–electrolyte interface. From then on, various methods of water splitting have been explored to improve the hydrogen production efficiency. So far, many materials that could be used in the PEC cell structure have been identified as shown in Fig.2. However, only a few of the common semiconductors can fulfil the requirements presented above even if it is assumed that the necessary overvoltage is zero. It should be noted that most materials have poor corrosion resistance in an aqueous electrolyte and posses high bandgap, which prevents them from producing enough photocurrent (Fig.5).

Photoelectrolysis of water, first reported in the early 1970's (Fujishima, 1972), has recently received renewed interest since it offers a renewable, non-polluting approach to hydrogen production. So far water splitting using sunlight has two main approaches. The first is a two-step process, which means sunlight first transform into electricity which is then used to split water for hydrogen production (Tamaura, et al., 1995; Hassan & Hartmut, 1998). Though only about 2V is needed to split water, hydrogen production efficiency depends on large current via wires, resulting in loss due to its resistance; the two-step process for hydrogen production is complex and leads to a high cost.

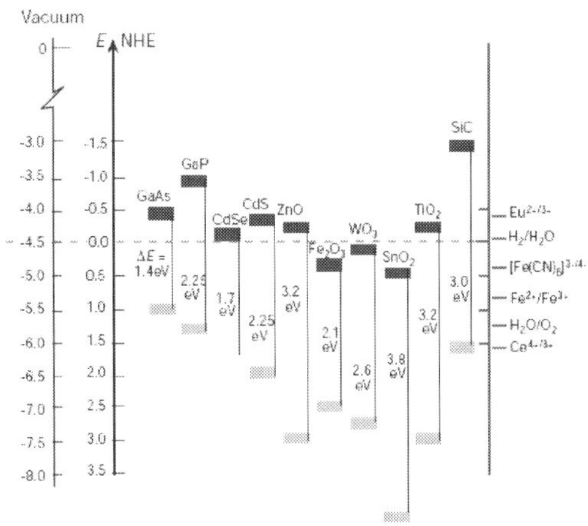

Fig. 2. Band positions of some semiconductors in contact with aqueous electrolyte at pH1. The lower edge of the conduction band (red colour) and the upper edge of the valence band (green colour) are presented along with the bandgap in electron volts. For comparison, the vacuum energy scale as used in solid state physics and the electrochemical energy scales, with respect to a normal hydrogen electrode (NHE) as reference points, are shown as well as the standard potentials of several redox couples are presented against the standard hydrogen electrode potential on the right side (Gratzel, 2001).

Another approach is a one-step process, in which there are no conductive wires and all the parts are integrated for water splitting, as shown in Fig.3. In this structure as there are no wires, hence no loss. Another advantage is that the maintenance is low compared to the two-step process discussed above.

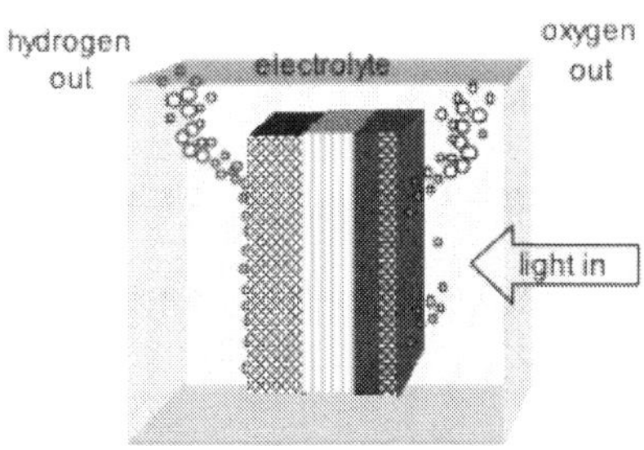

Fig. 3. Generic Planar Photoelectrode Structure with Hydrogen and Oxygen Evolved at Opposite Surfaces (Miller & Rocheleau, 2002)

In 1972, Fujishima and Honda used n type TiO_2 as the anode and Pt as the cathode to form the PEC structure and achieved 0.1% of STH efficiency (Fujishima & Honda, 1972). In this system TiO_2 absorbed the sunlight to produce the current while its bandgap (~3.2eV) provided the needed voltage for water splitting. Although TiO_2 is corrosion resistant in an aqueous electrolyte, but because of its high band gap leads to absorption of sunlight in the short wavelength range only, resulting in a small current and hence a low STH efficiency. In order to increase the current, some researchers are attempting to narrow its bandgap to enhance its absorption, and with limited success (Masayoshi, et al., 2005; Nelson & Thomas, 2005; Srivastava, et al. 2000).

In 1975, Nozik first reported using $SrTiO_3$(n) and GaP(p) photoelectrodes as the anode and cathode respectively (Nozik, 1975) and obtained a STH efficiency of 0.67%. In 1976, Morisaki's group introduced utilizing a solar cell to assist the PEC process for hydrogen production (Morisaki, et al., 1976). Silicon solar cell was integrated with TiO_2 in series to form a PEC system, which exhibited higher photo current by absorbing more sunlight and higher voltage. Later, Khaselev and Turner in 1998 reported 12.4% of STH efficiency using p-$GaInP_2$/n-GaAs/p-GaAs/Pt structure (Khaselev & Turner, 1998); in this, surface oxygen was produced at the p-$GaInP_2$ side and hydrogen from the Pt side. Although this structure exhibited high STH efficiency, the corrosion resistance of p-$GaInP_2$ in an aqueous electrolytes was very poor, and was almost all etched away within a couple of hours. (Deutsch et al., 2008).

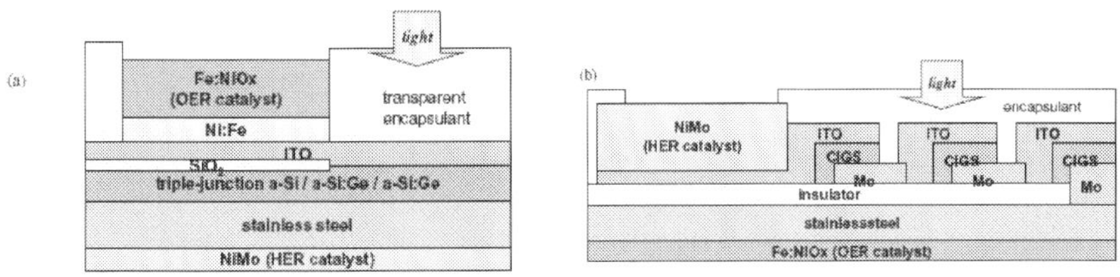

Fig. 4. (a) A-Si triple PV junctions and (b) CIGS PV cell integrated into a PEC system (Miller, et al., 2003)

Richard et al., reported 7.8% of STH efficiency by using NiMo or CoMo as cathode, Ni-Fe-O metal as anode and integrating with a-Si/a-Si:Ge/a-Si:Ge triple junctions solar cell as shown in Fig.4 (a) (Richard, et al., 1998). They also used copper indium gallium selenide (CIGS) module to replace a-Si triple junctions to produce even higher photo current as shown in Fig.4 (b).

Yamada, et al., also used a similar structure (Co–Mo and the Fe–Ni–O as the electrodes) and achieved 2.5% STH efficiency (Yamada, et al., 2003). More notably, a STH efficiency of 8% was reported by Lichta, et al., using AlGaAs/Si RuO_2/Pt black structure (Lichta, et al., 2001). In this structure, solar cell was separated from the aqueous electrolyte to avoid being corroded; it should be noted that the fabrication process for the device was very complicated. The non-transparent electrode had to cover the active area of the solar cell in order to enlarge electrode-electrolyte contact to as large area as possible.

In 2006, a "hybrid" PEC device consisting of substrate/amorphous silicon (nipnip)/ ZnO/WO_3, which would lead to ~3% solar-to-hydrogen (STH) conversion efficiency, was reported (Stavrides, et al., 2006). In this configuration, transparent WO_3 prepared by sputtering technique acted as the photoelectrode, whereas the amorphous silicon tandem solar cell was used as a photovoltaic device to provide additional voltage for water splitting at the interface of photoelectrode-electrolyte. In this structure, primarily the UV photons are absorbed by WO_3 while the green to red portion of the AM1.5 Global spectra was absorbed in the a-Si tandem photovoltaic device. Due to a high bandgap (Eg) (2.6-2.8eV) of the WO_3 photoelectrode, the photocurrent density of this hybrid PEC device is limited to no more than 5 mA/cm^2 (as shown in Fig.5), resulting in low STH efficiency.

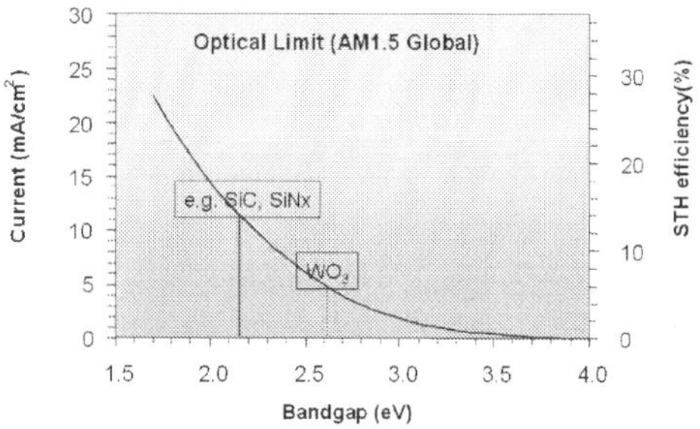

Fig. 5. Maximum current available as a function of the bandgap (E_g) of various materials under Global AM1.5 illumination (assumptions are that all the photons are absorbed for energy in excess of the band gap and the resulting current is all collected)

The US Department of Energy has set a goal to achieve STH conversion efficiency of 10% by 2018 (Miller & Rocheleau, 2001). To reach this goal, a photocurrent > 8.1mA/cm^2 is needed in PEC devices as deduced from equation (1). As shown in Fig.5, materials with narrower bandgap could produce higher photo current, such as a-SiC:H and a-SiN_x:H which can be routinely grown using plasma enhanced chemical vapor deposition (PECVD) technique and their bandgaps can be tailored into the ideal range by the control of stoichiometry, i.e., ≤ 2.3eV. In addition to generating enough photocurrent, necessary for STH conversion efficiency higher than 10%, a-SiC:H when in contact with the electrolyte, could also produce a significant photovoltage as other semiconductors (Nelson&Thomas, 2008), which would then reduce the voltage that is needed from the photovoltaic junction(s) for water splitting. Further, incorporation of carbon should lead to a more stable photoelectrode compared to pure amorphous silicon, which has poor resistance to corrosion when in contact with the electrolyte (Mathews, et al., 2004; Sebastian, et al., 2001).

3. a-SiC:H materials and its application as absorber layer in solar cells

A-SiC:H films were fabricated in a PECVD cluster tool system specifically designed for the thin film semiconductor market and manufactured by MVSystems, Inc. The intrinsic a-SiC:H films were deposited using CH_4, SiH_4, and H_2 gas mixtures at 200°C substrate temperature. The detail deposition parameters were presented in the reference [Zhu, et al., 2009].

3.1 a-SiC:H materials prepared by RF-PECVD

Fig.6 shows the bandgap (E_g), photoconductivity (σ_{ph}) and gamma factor (γ) as function of $CH_4/(SiH_4+CH_4)$ gas ratio used during a-SiC:H growth. As $CH_4/(SiH_4+CH_4)$ gas ratio increases, Eg increases from ~1.8eV to over 2.0eV (Fig.6 (a)) while the dark conductivity (σ_d) decreases to <1.0 x 10^{-10} S/cm (not shown here), which is the limit of the sensitivity of our measurement technique. We also note that σ_{ph} decreases from about 1.0 x 10^{-5} to 1.0 x 10^{-8} S/cm when $CH_4/(SiH_4+CH_4)$ gas ratio increases.

Here, the γ, is defined from $\sigma_{ph} \propto F^{\gamma}$, where σ_{ph} is the photoconductivity and F is the illumination intensity; we infer the density of defect states (DOS) of the amorphous semiconductor from this measurement (Madan & Shaw, 1988). High-quality a-Si materials

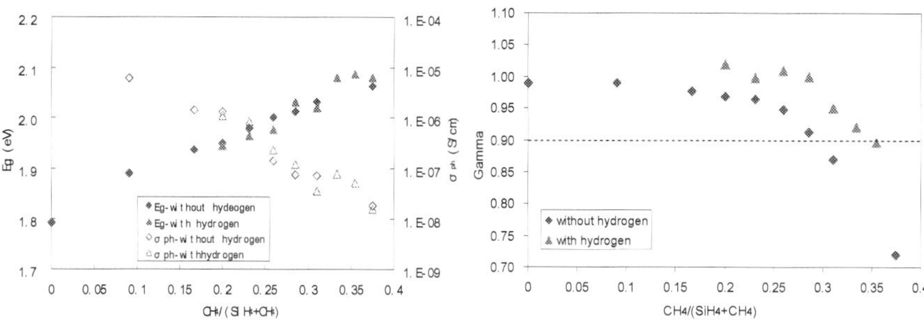

Fig. 6. (a) Bandgap (E_g), photoconducitivty (σ_{ph}) and (b) γ are plotted as function of the $CH_4/(SiH_4+CH_4)$ gas ratio used during the fabrication of a-SiC:H materials with/without H_2.

generally exhibits γ >0.9. As shown in Fig.6 (b), when $CH_4/(SiH_4+CH_4)$ gas ratio is > 0.35, γ decreases to a low value of ~0.7, indicative of a material with high defect states. For $CH_4/(SiH_4+CH_4)$ gas ratio < 0.3, γ > 0.9, which indicates the DOS in materials is low. Fig.6 also shows the effect of hydrogen on a-SiC:H films. E_g and σ_{ph} of a-SiC:H films prepared with 100sccm H_2 flow during deposition process have similar value as that prepared without H_2 as shown in Fig.6 (a). It should be noted that use of H_2 during fabrication led to an increase of γ, as shown in Fig.6 (b), indicates that the DOS in the film is decreased due to removal of weak bonds due to etching and passivation (Yoon, et al., 2003; Hu, et al., 2004;).

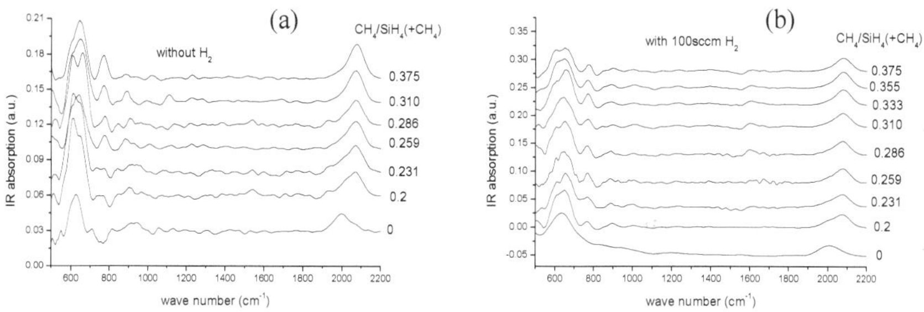

Fig. 7. IR spectra of a-SiC:H as function of $CH_4/(SiH_4+CH_4)$ gas ratio without (a) and with H_2 dilution (b).

Evidence of carbon incorporation in the films can be discerned from infrared (IR) spectroscopy. As shown in Fig.7, regardless of H_2 addition during deposition process, the peak at 2000 cm^{-1} (related to Si-H stretching vibration mode) always shifts towards 2080 cm^{-1}, once $CH_4/(SiH_4+CH_4)$ gas ratio is greater than 0.2. This shift of Si-H stretching vibration mode is mainly caused by incorporation of C atoms, and probably due to the back-bonding of the Si atoms to carbon (Hollingsworth, et al., 1987). In addition, it is found that in the a-SiC:H films using 100sccm H_2 dilution, the ratio of the absorption peak at 2080 cm^{-1} to 2000 cm^{-1} is smaller than that without the use of H_2 dilution, implying less defective films (Hu, et al., 2004). It is also seen from Fig.7 (a) that in the a-SiC:H without H_2 dilution, the IR peak at 780 cm^{-1}, which is related to Si-C stretching mode, increases as the $CH_4/(SiH_4+CH_4)$ gas ratio increases; whereas films produced with 100sccm H_2 dilution, this peak keeps almost constant (as shown Fig.7(b)). This is likely due to the removal of SiH_2 and a decrease of carbon clusters in the films (Desalvo, et al., 1997). It was also found that at a fixed $CH_4/(SiH_4+CH_4)$ =0.2, σ_{ph} is enhanced from 4.0 x 10^{-7} S/cm to 3.2 x 10^{-6} S/cm as the H_2 flow increased from 0 to 150sccm.

The decrease in σ_{ph} with increasing $CH_4/(SiH_4+CH_4)$ gas ratio as shown in Fig. 6(a) is unlikely be due to an increase of the recombination centers related to defects since the γ factor is >0.9.The decrease of σ_{ph} results from a reduction in the absorption coefficient as Eg increases. In order to further evaluate this, the nominal photocurrent, Ip, at certain wavelength, under uniform bulk absorption (here we select wavelength 600nm) can be measured and the photocurrent be expressed as,

$$I_p = e.N_{ph\,(\lambda)}\,(1-R_\lambda)\,[1 - \exp(-\alpha_\lambda d)]\eta\tilde{\imath}/t_t \quad (2)$$

Where, $N_{ph(\lambda)}$ is the photon flux, R_λ is the reflection coefficient, α_λ the absorption coefficient, d the film thickness, η is the quantum efficiency of photo generation, $\tilde{\imath}$ is the recombination lifetime and t_t is the transit time. Assuming that η, $\tilde{\imath}$, t_t and $(1-R_\lambda)$ are constant for different films (i.e. different E_g), then to the first order approximation, the normalized photocurrent, $I_p/[1-\exp(-\alpha_\lambda d)]$, can account for the changes in the absorption coefficient as E_g varies (Madan & Shaw, 1988). It was indeed shown that the normalized photocurrent does not change significantly as E_g increases (Hu, et al., 2008). This is in contrast to the decrease in σ_{ph} with E_g as shown in Fig.6 (a), suggesting a low DOS, consistent with high γ (>0.9) throughout the range.

3.2 Photothermal deflection spectroscopy (PDS) spectrum of a-SiC:H films

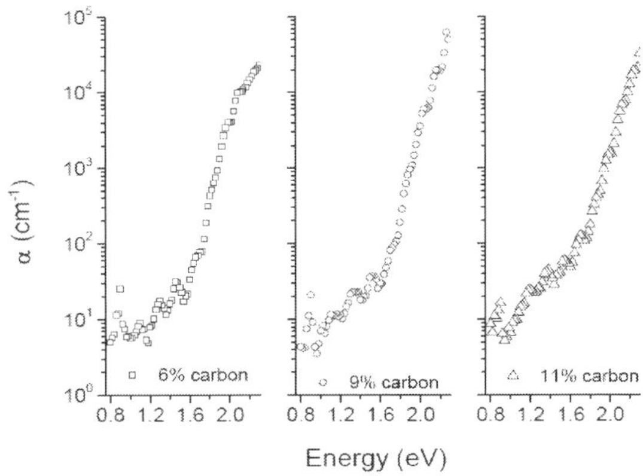

Fig. 8. Absorption coefficient curves of three a-SiC:H films, with differing carbon concentrations, measured using photothermal deflection spectroscopy (PDS)

Fig.8 shows the absorption coefficient of the three chosen films with differing carbon concentrations, prepared with H_2 dilution, measured by the photothermal deflection spectroscopy (PDS). Using energy dispersive x-ray spectroscopy on a JEOL JSM-7000F field emission scanning electron microscope with an EDAX Genesis energy dispersive x-ray spectrometer their carbon concentrations are 6, 9, and 11% (in atomic), corresponding to methane gas ratio, used in the fabrication, of 0.20, 0.29 and 0.33 respectively. The signal seen here is a convolution of optical absorption from every possible electronic region including extended, localized and deep defect states. In the linear region between about 1.7-2.1eV, the absorption coefficient primarily results from localized to extended state transitions and is known as the Urbach tail. This region can be described by $\alpha=\alpha_0 \exp(E/E_0)$ where E is the excitation energy and E_0 is the Urbach energy which is the inverse slope of the data when plotted versus $\ln(\alpha)$. Since the absorption coefficient here directly depends on the density of localized states, E_0 is considered to be a measure of the amount of disorder (Cody, et al., 1981). Their bandgap values are presented in Fig.6 (a) as previously discussed and E_0 is 78, 85, and 98 meV for carbon concentrations of 6, 9, and 11%, respectively. For comparison, a typical value for device grade a-Si:H is ~50 meV(Madan & Shaw, 1988). As the carbon concentration increases, so too does the value of E_0. This is expected as the density of localized states is increasing with more disorder created by introducing more carbon. Also, there is an increase in the bandgap from E_{04} = 2.06-2.18eV with carbon concentration (E_{04} is defined as the energy value where the absorption coefficient $\alpha = 10^4\ cm^{-1}$). This is known to be a result of at least some of the carbon being incorporated in the form of sp^3 carbon which is essentially an insulator (Solomon, 2001). The feature at 0.88eV in Fig. 8 is an overtone of an O-H vibrational stretch mode from the quartz substrate.

As the bandgap increases with carbon incorporation, as evidenced from the PDS data, the Urbach energies are 50% to 100% higher than is typically seen in device grade a-Si:H. This is typically interpreted as an increase in localized states within the bandgap region just above the valence band and below the conduction band resulting from structural disorder. It is believed that the carbon is incorporated into our films as a mixture of sp^2 and sp^3 carbon from ESR test (Solomon, 2001; Simonds (a), et al., 2009).

3.3 a-SiC single junction devices

The previous results suggest that high quality a-SiC:H can be fabricated with Eg ≥2.0eV. To test the viability of a-SiC:H material in device application, we have incorporated it into a p-i-n solar cell in the configuration, glass/Asahi U-Type SnO_2:F/p-a-SiC:B:H/i-a-SiC:H/n-a-Si/Ag as shown in Fig.9. The Ag top contact defines the device area as $0.25cm^2$. The thickness of i-layer is ~300nm.

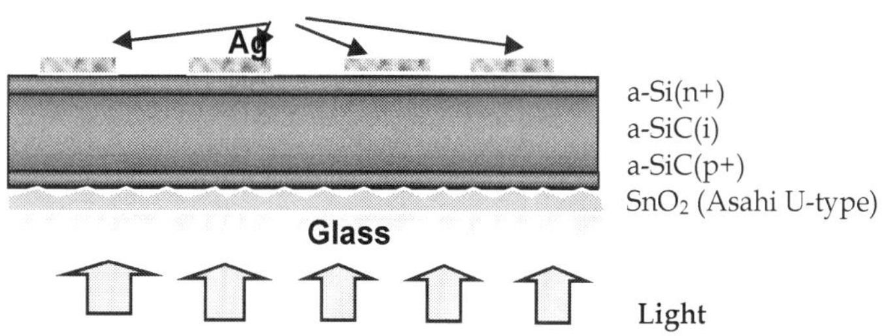

Fig. 9. Configuration of p-i-n single junction solar cell

Three a-SiC:H i-layers with different carbon concentration were used in single junction solar cells. Fig.10 (a) and (b) show their J-V and quantum efficiency (QE) curves, respectively. As mentioned above, the three films with carbon concentration of 6%, 9%, and 11%, correspond to bandgaps of approx.2.0eV, 2.1eV and 2.2eV respectively. Though the bandgap increases with carbon concentration, the performance of a-SiC devices deteriorates quickly, especially the fill factor, implying an increase of the defects from carbon inclusion. As the carbon concentration increases, the QE peak shifts toward the short wavelength region and becomes smaller (Fig. 10 (b)), resulting from higher defects density with bandgap (Madan & Shaw, 1988). The influence of defects resulting from increased carbon can also be seen in the dark J-V curves. Here carrier transport is only affected by the built-in field and the defects in the films. As the carbon concentration increases, the diode quality factor deduced from the dark J-V curves also increases, which also implies loss due to increased defect densities (Simonds (b), et al., 2009). The device performances variation is consistent with the PDS data discussed above, where E_0 is 78, 85, and 98 meV for carbon concentrations of 6, 9, and 11%, respectively; increasing E_0 is indicative of increased defect state density (Madan & Shaw, 1988).

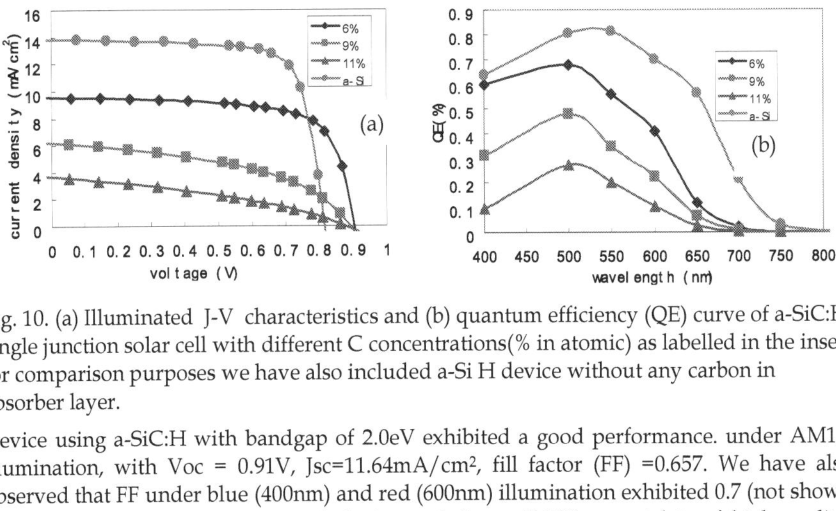

Fig. 10. (a) Illuminated J-V characteristics and (b) quantum efficiency (QE) curve of a-SiC:H single junction solar cell with different C concentrations(% in atomic) as labelled in the inset: for comparison purposes we have also included a-Si H device without any carbon in absorber layer.

Device using a-SiC:H with bandgap of 2.0eV exhibited a good performance. under AM1.5 illumination, with Voc = 0.91V, Jsc=11.64mA/cm^2, fill factor (FF) =0.657. We have also observed that FF under blue (400nm) and red (600nm) illumination exhibited 0.7 (not shown here), which indicates it is a good device and that a-SiC:H material is of high-quality. Compared with the normal a-Si:H devices (Eg~1.75eV), the QE response peak shifts towards a shorter wavelength; asis to be expected at long wavelength the QE response is reduced due to the increase in its Eg. Jsc of ~8.45mA/cm^2 has been obtained with reduced a-SiC:H intrinsic layer thickness (~100nm). This implies that it is possible to use a-SiC:H as a photoelectrode in PEC devices for STH efficiency >10%. Here a-SiC:H with bandgap of 2.0eV is selected to be used as the photoelectrode in PEC.

4. a-SiC:H used as a photoelectrode in PEC devices

An intrinsic a-SiC:H (~200nm) and a thin p-type a-SiC:H:B layer (~20nm) was used as the photoelectrode (Fig.11) to form a PV(a-Si tandem cell)/a-SiC:H device. In general, the a-

SiC:H behaves as a photocathode where the photo generated electrons inject into the electrolyte at the a-SiC:H/electrolyte interface to reduce the H^+ for hydrogen evolution. This way, anodic reaction and thus corrosion on a-SiC:H layer can be mitigated.

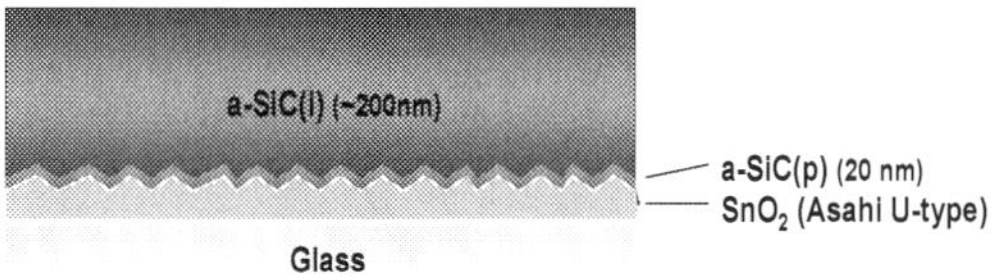

Fig. 11. Configuration of a-SiC:H photoelectrodes.

The current-potential characteristics of a-SiC:H photoelectrodes were measured with a three-electrode setup, with either saturated calomel electrode (SCE) or Ag/AgCl as the reference electrode (RE) and Pt as the counter electrode (CE). The samples were illuminated through the intrinsic a-SiC:H side under chopped light, using either a xenon or tungsten lamp (both calibrated to Global AM1.5 intensity calibrated with reference cell). However, due to difference in spectrum, the photocurrent of photoelectrodes varied depending on the light source used (Murphy, et al., 2006). Typical current-potential characteristics of a-SiC:H photoelectrodes are shown in Fig.12. The photocurrent density (J_{ph}) is defined as the difference between the current density without illumination (J_{dark}) and the current density under illumination. Initial experiments, using aqueous 0.0–0.5pH sulfuric acid electrolyte led to significant degradation during 10-minute test. The analysis of the initial results pointed towards using less acidic solutions (higher pH), which was described in elsewhere (Matulionis, et al., 2008). It was found that a better photoelectrode performance (diminished corrosion and higher J_{ph}) was achieved in a pH2 electrolyte.
Durability tests were carried out at NREL which involved a constant current density of -3 mA/cm^2 applied to the a-SiC:H photoelectrode. A tungsten lamp was used as the light source (calibrated to Global AM1.5 intensity with a 1.8eV reference cell). Electrolyte used was pH2 sulphamic acid/potassium biphthalate solution and a Triton X-100 surfactant.

4.1 Durability of a-SiC:H photoelectrodes

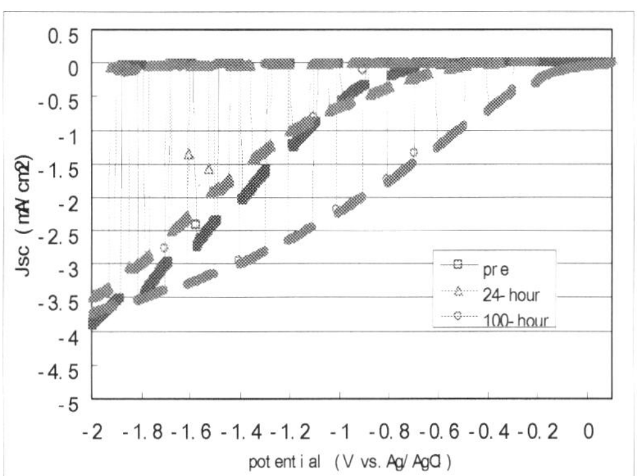

Fig. 12. J-V curves before and after 24- and 100-hour durability test on textured SnO_2 substrate

Fig.12 shows J-V curves of a-SiC:H photoelectrode on textured SnO_2 substrate before (blue curve) and after 24- (orange curve) and 100-hour (pink curve) durability tests. It is seen that after the 100-hour durability test, J_{dark} remains very low, suggesting the a-SiC:H photoelectrode is stable in electrolyte up to 100-hour (so far tested). Photo images of the surface morphology of tested a-SiC:H photoelectrodes also verify that they remain largely intact. Interestingly, both J_{ph} and its onset change noticeably after the durability test, particularly after the 100-hour durability test. For instance, the photocurrent onset shifts anodically (towards a lower absolute potential) by ~0.6V. This means the extra voltage needed to overcome various overpotential and the non-ideal energy band edge alignment to the H_2O/O_2 redox potential is lower after the durability test. We noted that increase in the test-duration time resulted in larger onset shift and increase in J_{ph} . The exact reason for such a behavior is not known at the moment. Possible causes could be, (i) the native-grown SiO_x on the surface of a-SiC:H film is probably eliminated (as described in more details later), and (ii) modification of the surface of a-SiC:H photoelectrode.

Fig.13 shows the surface morphology of the a-SiC:H photolelectrode on Asahi U type SnO_2 before and after the 100-hour durability test. One can see that after the test, the surface

Fig. 13. SEM images of surface morphology of a-SiC:H photoelectrode (a) before and (b)after 100-hour durability test .

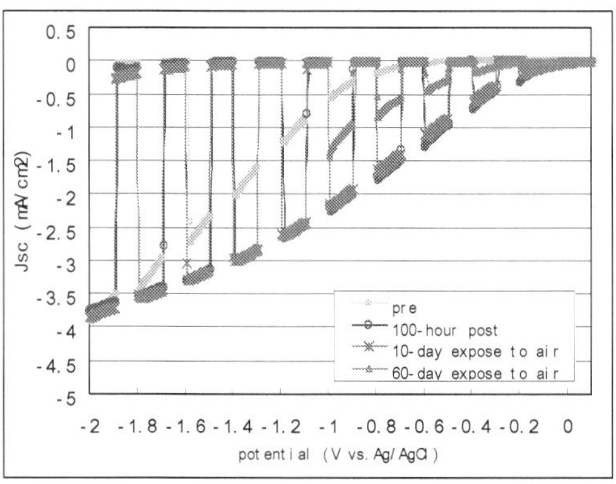

Fig. 14. J-V curves of a-SiC:H photoelectrode before and after 100-hour durability test, and then exposed to air for 10- and 60-day and tested.

morphology of a-SiC:H looks similar. The only difference is that after the test there are many tiny motes on the surface of a-SiC:H photoelectrode and it appears as if something deposited on it as shown the insert magnified image in Fig.13(b), while before the test the surface is smooth (insert magnified image in Fig.13.(a)). More work is needed to understand this.

We have also noted changes, in the J-V characteristic after the 100-hour test, a-SiC:H photoelectrode when exposed to air as shown in Fig.14. We note that after 10-day exposure to air, J-V characteristic remains unchanged. After exposing to air for 60 days J_{ph} decreases and photocurrent onset shifts cathodically. The reason probably is that extended exposure to air could leads to SiO_x formation on the surface of a-SiC:H photoelectrodes and its thickness could be time dependent.

4.2 Effect of SiO_x on the surface of a-SiC:H

As seen in Fig.12 and 14, the J_{ph} and its onset of a-SiC:H photoelectrodes could vary under different conditions. As discussed above, there is a possibility of SiOx layer formation on the surface of a-SiC:H photoelectrodes. Using X-ray photoelectron spectroscopy (XPS), we have investigated the surface of a-SiC:H films. Fig.15 shows the XPS spectra for an a-SiC:H film conditions of "as-is" and after etching with hydrofluoric acid (HF concentration of 48%) for different etching times, 10 to 60 seconds. It is clearly seen that there is a very thin SiO_x (a few nm thick) which exists on the a-SiC:H surface, as evident by the peak around 104eV which is associated with the Si-O bonds in SiO_x. The SiO_x layer becomes thinner as the HF etching time become longer, and disappears eventually after an HF dip for 30 seconds. The peak around 101eV related to Si peak, remains the same.

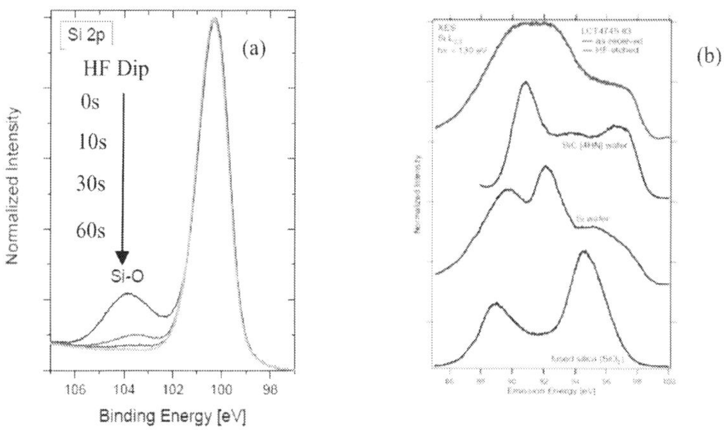

Fig. 15. Changes in XPS (a) and XES (b) curves of a-SiC:H films with HF etching of surface.

The XES curves of before and after HF dip are completely superimposed, which suggests that HF dip does not change the composition of a-SiC:H, as shown in Fig. 15(b). XES curves of crystal SiC wafer, Si wafer and SiO_2 are also shown in Fig. 15 (b) as well as that of a-SiC:H. Crystal SiC wafer has characteristic peaks at 91eV and 98eV while crystal Si wafer exhibits peaks at 90eV and 92eV. One can see that the data for a-SiC:H curve includes crystal Si and SiC characteristic peaks but not SiO peaks which would be at 87eV and 95eV a-SiC:H.

This data suggests that SiO_x grows on the surface of a-SiC:H after deposition and when exposed to air and not during the fabrication process.
In order to investigate the effect of thin SiO_x on the J-V characteristic of the a-SiC:H photoelectrodes, the thickness of SiO_x layer was systematically reduced using HF dip. Before the test, a-SiC:H photoelectrodes were measured, then dipped in HF for 10 to 30 seconds, and measured again. Fig.16 shows a comparison of J-V characteristics before and after the HF dip. It is seen that after HF dip, the J_{ph} increases noticeably (absolute value), e.g., from -2.92mA/cm^2 to -5.54mA/cm^2 for HF-dip for 10-second, and -6.3mA/cm^2 for HF-dip for 30-second at -1.4V vs.Ag/AgCl. Meanwhile, the photocurrent onset shifts anodically by about 0.23V for HF-dip for 30 seconds. Further increasing the dip time beyond 30 seconds, corrosion of a-SiC:H film was clearly evident (as seen clearly by naked eyes). Interestingly, after the a-SiC:H photoelectrode was removed from the electrolyte and exposed to air for 1.5 hours, the J-V characteristics was the same as after the HF dip (not shown here); however, after exposing to air for 67 hours, J_{ph} degraded and completely returned to its initial value as shown in Fig.16 (red one). These results confirm without any doubt that the thin SiO_x layer on the surface of a-SiC:H photoelectrodes indeed affects both J_{ph} and its onset.

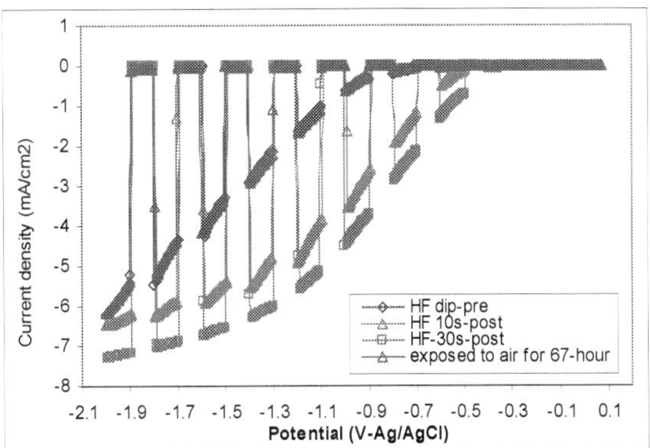

Fig. 16. J-V characteristics of a-SiC:H photoelectrode before and after HF-dip

Same HF dip experiments were repeated by the group in Hawaii (HNEI), in which SCE was used as RE and Xenon lamp calibrated to AM1.5 as light source. The results are the same as shown above in Fig.16, except that higher J_{ph}, (> 8mA/cm^2) was observed and possibly due to the use of a different light source.
Comparing Fig.16 with Fig.14, though the effect of SiO_x on J-V characteristics of a-SiC:H photoelectrodes is verified to some extent, J_{ph} and its onset variations are significantly different. J-V characteristic of a-SiC:H photoelectrode after 100-hour durability test and exposed to air for 60 days returns, but not comes back its original value, while after HF-dip and exposure to air only for 67 hours J-V characteristic reverts to its initial value. Apparently, the durability test has modified the surface of a-SiC:H photoelectrodes, resulting in a favorable interface which facilitates photocurrent generation. More work is underway to understand the surface variation during durability test and its effect on the J_{ph} and its onset.

4.3 Integration of the a-SiC:H photoelectrode with a-Si tandem device

The above results show that the a-SiC:H photoelectrode exhibits high photocurrent (i.e., up to ~8 mA/cm^2), and good durability in a pH2 electrolyte for up to ~100 hours. Its main drawback, however, is the non-ideal surface band structure. Our theoretical analysis showed that the hydrogen evolution reaction is thermodynamically allowed at the surface of the a-SiC:H photoelectrode, since the photogenerated electrons are of energy which is higher than the redox potential of H^+/H_2 half-reaction. On the other hand, to promote oxygen evolution at the counter electrode, a minimum external bias of ~ -1.4 V is needed to bring the quasi-Fermi energy level of photogenerated holes below H_2O/O_2 redox potential (Hu, et al., 2008).

In order to solve this non-ideal valence band edge alignment problem, an a-Si:H tandem solar cell was integrated into the PEC cell to form a hybrid PV/a-SiC:H configuration, as shown in Figure 17(a). The substrate used for the hybrid PEC device was typically Asahi U-type fluorine-doped SnO_2 coated glass. Other types of substrates such as stainless steel and ZnO coated glass were also used for comparison purposes. The a-Si:H tandem solar cell which was used in the hybrid PEC device when fabricated into a solid state device exhibited Voc = 1.71 V, Jsc=7.02 mA/cm^2, FF=0.74, and efficiency of ~9%, as shown in Figure 17(b).

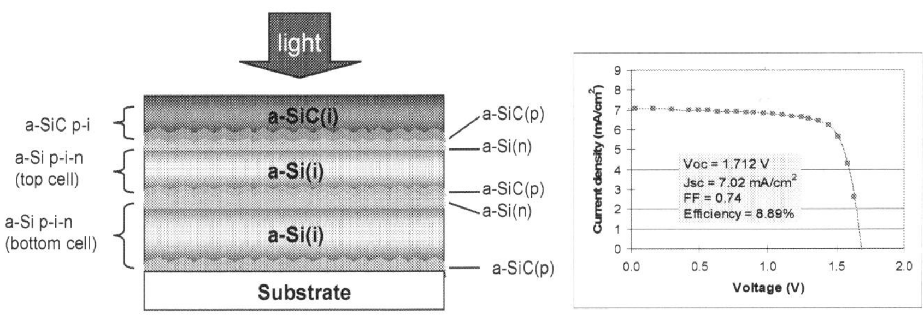

Fig. 17. (a) Configuration of the hybrid PEC device and (b) J-V curve of a-Si tandem device

4.4. Flat-band potential

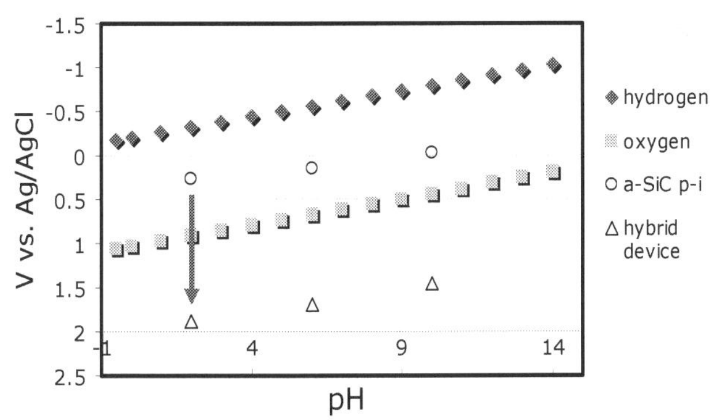

Fig. 18. V_{fb} vs. pH for the a-SiC:H photoelectrode and the hybrid PEC device

Fig.18 shows the flat-band potential (V_{fb}), determined by the illuminated open-circuit potential (OCP) method, as a function of pH of electrolyte for both the a-SiC:H photoelectrode (open circles) and the hybrid PEC device (open triangles). The change of the V_{fb} with pH nearly exhibits a slope of ~60 mV/pH, as predicted by the Nernst equation (Memming, 2000). We note in Fig. 18, that at pH2, V_{fb} =+0.26 V (vs. Ag/AgCl) for the a-SiC:H photoelectrode, whereas in the case of the hybrid device, the V_{fb} significantly by ~ +1.6 V (as indicated by the red arrow) and is below the H_2O/O_2 half-reaction potential (by +0.97 V) and is in an appropriate position to facilitate water splitting. Figs.19 (a) and (b) show the current density vs. potential characteristics for hybrid PEC device fabricated on different substrates, SnO_2 and ZnO coated glass and SS, and measured in the pH2 buffered electrolyte (sulphamic acid solution with added potassium biphthalate) using the 3-electrode and 2-electrode setup respectively. In the 2-electrode setup, there was no reference electrode and contained only the working electrode (hybrid PEC device) and the counter electrode which was ruthenium oxide (RuO_2) rather than conventional platinum (Pt).

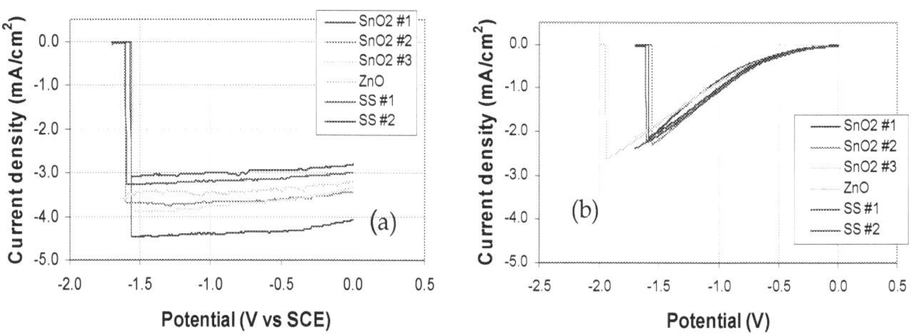

Fig. 19. Current density vs. potential characteristics measured in (a) 3-electrode and (b) 2-electrode setup.

From Fig.19 (a), we see the saturated photocurrent of the hybrid cell using different substrates is in the range of 3-5 mA/cm^2. The larger photocurrent using SnO_2 (>4 mA/cm^2) coated glass substrate is due to the inherent texture of the SnO_2 which enhances internal photon absorption. More significantly, we see that the photocurrent density of ~0.3 mA/cm^2 occurs at a zero potential using the 2-electrode setup (Fig.19 (b)). Hydrogen production was observed in a short-circuit condition (Hu, et al., 2009).

It should be noted that compared with the 3-electrode case the photocurrent measured in the 2-electrode setup (even using RuO_2 counter electrode) is much lower, suggesting limiting factors. We have noted that the over-potential loss in the 2-electrode setup can be due to, (1) type of electrolyte used, (2) type of counter electrode used and (3) formation of thin SiO_x layer on the a-SiC:H surface. Initial results have shown that, after dipping the hybrid device into 5% hydrofluoric (HF) acid for 30 seconds and using RuO_2 as the counter electrode, the photocurrent is enhanced from 0.33 to 1.33 mA/cm^2 at zero bias in two-electrode setup.

4.5 Durability of the hybrid PEC device

The test was performed in the pH2 buffered electrolyte, with Pt as the counter electrode. During test, a constant current density of 1.6 mA/cm^2 was applied to the device, while the voltage (potential) across the sample was recorded over a 148-hour period. The current density vs. potential characteristics of the device prior to and after 22, 48 and 148-hour tests were measured in both the 3- and 2-electrode setups. Throughout these durability tests, H_2 production from the hybrid device occurred. Fig.20 (a) shows the current vs. potential curves measured prior to and after 148-hour test. These results show that the dark current shows almost no change, and hence no corrosion occurs in the hybrid device, after the 148-hour test, as is evident also in Fig.20 (b) and (c).

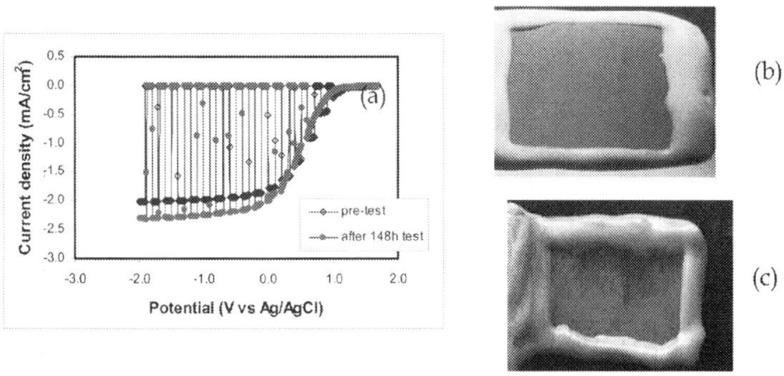

Fig. 20. (a) Current density vs. potential characteristics measured prior to and 148-hour test. (b) Photo images of the hybrid PEC device prior to and (c) after a 148-hour test in pH2 electrolyte.

5. Pathway to 10% STH efficiency in PV/a-SiC:H device

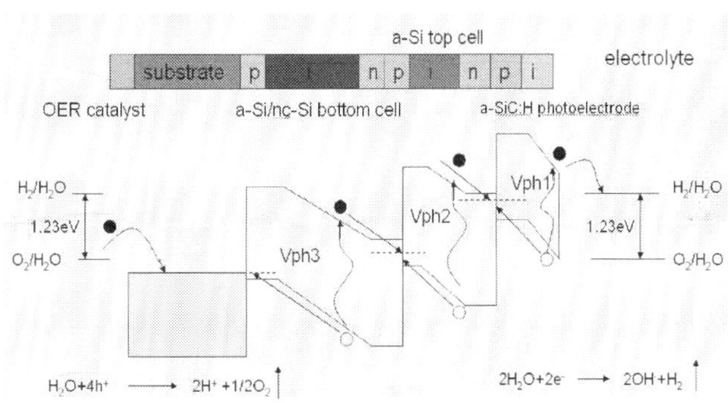

Fig. 21. The schematic energy band diagram of a hybrid PV/PEC cell containing a-SiC:H photoelectrode and a-Si :H/a-Si :H (or nc-Si :H) tandem solar cells

In the previous sections, we have shown that a-SiC:H photoelectrodes are durable in electrolyte for up to 100 hours (so far tested). The J_{ph}, measured in a three-electrode setup, could be larger than $6mA/cm^2$ (at -1.4V vs. Ag/AgCl) after removing SiO_x layer on the surface. The extra bias is needed because, (1) the non-ideal energy band edge alignment to the H_2O/O_2 redox potential, as confirmed by the flatband voltage measurement and (2) various over-potential losses, e.g., due to the surface SiO_x barrier layer and various interfaces. Incorporating a-Si tandem photovoltaic device could eliminate the external bias as suggested by previous data. The schematic energy band diagram for a PV/a-SiC:H cell is shown in Fig.21. Since a-Si:H tandem solar cell could provide a photovoltage (V_{ph2}+V_{ph3}) of >1.8V (Shah, et al., 2004), while the a-SiC:H photoelectrode could provide a photovoltage (V_{ph1}) >0.5V (as shown by the initial tests), hence the total photovoltage in such a configuration (sum of V_{ph1} + V_{ph2} + V_{ph3}) is expected to be >2.3 V. Thus, such a PV/a-SiC:H structure can provide needed voltage for water splitting, as confirmed by the results described in the previous sections.

Further enhancement in the STH efficiency can be achieved by employing a nano-crystalline silicon single junction solar cell (nc-Si:H) in the tandem solar cell, and integrated with the a-SiC:H photoelectrode as described in Fig.21. Although the Voc of nc-Si single junction is ~0.55V (Mai, et al., 2005), lower than that in a-Si single junction, but the sum of V_{ph1} + V_{ph2} + V_{ph3} is expected to be higher than 1.5V, still large enough for water splitting. It is possible that with such a configuration, a J_{ph}>8mA/cm^2 could be generated, leading to STH efficiency up to 10% as shown in table 1.

PEC (a-SiC:H)		PV cell		STH (%)
Eg	J_{ph} (mA/cm2)	configuration	J_{ph} after filtered by a-SiC:H (mA/cm^2)	
2	8.85	a-Si/a-Si	6.7	8.24
2.3	8.55	nc-Si/a-Si	8.85	10.52

Table 1. The PV/a-SiC:H structure PEC simulation results

6. Conclusion

State-of-the-art a-SiC:H films have been prepared using RF-PECVD deposition technique. Incorporation of carbon in amorphous silicon network increases the bandgap to >2.0eV and adding H_2 during fabrication has led to a material with low defects. A-SiC:H with Eg=2.0eV used as the active layer in single junction solar cell led to an efficiency of ~7%, which also indicated that a-SiC:H is high-quality and that it has potential to be used as photoelectrode. Immersing in pH2 sulphamic acid electrolyte a-SiC:H photoelectrodes exhibit durability for up to 100 hours (so far tested). J_{ph} increases as well as photocurrent onset shifts towards anodically after durability test. This behavior could be due to a change in the surface structure of the a-SiC:H photoelectrode, or partially due to elimination of the surface SiO_x layer. HF etch experiment confirmed that the SiO_x layer on the surface of a-SiC:H indeed affects both J_{ph} and its onset. After removing SiO_x layer, a-SiC:H photoelectrode exhibited a J_{ph} over 6 mA/cm^2 at potential -1.4V (vs. Ag/AgCl), compare to that less than 4mA/cm^2 (vs. Ag/AgCl). More analysis needs to be done to understand the mechanisms and improve the interface between a-SiC:H and electrolyte and hence to increase J_{ph}. Our initial PV/a-SiC:H used as photoelectrode has exhibited ~1.33mA/cm^2 current density under zero volt external bias, and the photocurrent onset shifts enormously, from ~ -0.6V to ~ +1.2V, or by a net ~1.6v. Hydrogen production has been demonstrated in this type of hybrid PEC cell. It exhibits good durability in aqueous electrolytes for up to ~150 hours. Work on further increasing the photocurrent in such a PV/a-SiC:H device is underway. We have also shown by simulation that it is possible to achieve STH efficiency >10% using such PV/a-SiC:H devices.

7. Acknowledgments

The work is supported by US Department of Energy under contract number DE-FC36-07GO17105. The authors would like to thank Ed Valentich for his assistance in sample fabrication.

8. Reference

Bak, T.; Nowotny, J.; Rekas, M.; Sorrell, C. C. (2002). Photo-electrochemical hydrogen generation from water using solar energy. Materials-related aspects, *international Journal of hydrogen energy*, 27 (October 2002) 991-1022, ISSN: 0360-3199

Cody, G.D.; Tiedje, T.; Abeles, B.; Moustakas, T.D.; Brooks, B.; and Goldstein, Y.; (1981); Disorder and the optical absorption edge of hydrogenated amorphous silicon, *Journal De Physique* colloque C4, (October 1981) C4-301-304, ISSN:1155-4304

Desalvo, A.; Giorgis, F.; Pirri, C.F.; Tresso, E.; Rava, P.; Galloni, R.; Rizzoli, R.; and Summonte, C.; (1997), Optoelectronic properties, structure and composition of a-SiC:H films grown in undiluted and H_2 diluted silane-methane plasma, *Journal Applied Physics* 81 (June 1997) 7973-7980, ISSN: 0021-8979

Deutsch, T.G. ; Head, J. L.; Turner, J. A.; (2008). Photoelectrochemical Characterization and Durability Analysis of GaInPN Epilayers, *Journal of Electrochemcal Society*, Vol. 155, (July 2008), B903-B907, ISSN: 0013-4651Fujishima, Akira; & Honda, Kenichi; (1972), Electrochemical Photolysis of Water at a Semiconductor Electrode, *Nature*, 238 (July 1972) 37-38, ISSN: 0028-0836

Gerischer, H. & Mindt, W. (1968).The Mechanisms of the Decomposition of Semiconductors by Electrochemical Oxidation and Reduction, Electochimica Acta, 13 (June 1968) 1329-1341, ISSN: 0013-4686

Gratzel, M. (2001). Photoelectrochemical cells, *Nautre* Vol.414 (November 2001) 338-344, ISSN: 0028-0836

Hassan, A. & Hartmut, S. (1998). The German-Saudi HYSOLAR program, *International Journal of hydrogen energy* 23 (June 1998) 445-449, ISSN: 0360-3199

Helmut, T. (2008). Photovoltaic hydrogen generation, *International Journal of Hydrogen energy*, 33 (November 2008) 5911-5930, ISSN: 0360-3199

Hu, Z. H.; Liao,X. B.; Diao, H. W.; Kong, G.L.; Zeng, X. B.; and Xu, Y.Y.; (2004), phous silicon carbide films prepared by H2 diluted silane-methane plasma, *Journal of Crystal Growth*, Vol. 264 (March 2004) 7-12, ISSN: 0022-0248

Hu, J.; Zhu, F.; Matulionis, I.; Kunrath, A.; Deutsch, T.; Miller, L. E.; Madan, A.; (2008), Solar-to-Hydrogen Photovoltaic/Photoelectrochemical Devices Using Amorphous Silicon Carbide as the Photoelectrode, *Proc. 23rd European Photovoltaic Solar Energy Conference, Feria Valencia, 1-5 September 2008, #1AO.5.6*

Hu, J.; Zhu, F.; Matulionis, I.; Deutsch, T.; Gaillard, N.; Miller, L. E.; Madan, A.; (2009), Development of a hybrid photoelectrochemical (PEC) device with amorphous silicon carbide as the photoelectrode for water splitting, *Symposium S: Materials in Photocatalysis and Photoelectrochemistry for Environmental Applications and H2 Generation , MRS Spring Meeting, San Francisco, April 13-17, 2009, # 1171-S03-05*

Hollingsworth, R.; Bhat, P.; and Madan, P.; (1987); Amorphous silicon carbide solar cells, *IEEE Photovoltaic Specialists Conference, 19th*, New Orleans, LA, May 4-8, 1987, Proceedings (A88-34226 13-44) 684-688

Khaselev, O. & Turner, J. A.; (1998), A Monolithic Photovoltaic-Photoelectrochemical Device for Hydrogen Production via Water Splitting, *Science* Vol.280, (April 1998), 425-427, ISSN: 0036-8075

Kuznetsov, A.M. & Ulstrup, J.; (2000). Theory of electron transfer at electrified interfaces, *Electrochimica Acat* 45 (May 2000) 2339-2361, ISSN: 0013-4686

Lichta, O. S.; Wang, B. ; Mukerji, S. ; Soga, T.; (2001). Over 18% solar energy conversion to generation of hydrogen fuel; theory and experiment for efficient solar water splitting, *International Journal of Hydrogen Energy* 26(July 2001), 653–659, ISSN: 0360-3199

Madan, A. & Shaw, M. P., (1988), The Physics and Applications of Amorphous Semiconductors, *Academic* Press, ISBN-13: 9780124649606

Mai, Y.; Klein, S.; Carius, R.; Stiebig, H.; Geng, X.; (2005); Open circuit voltage improvement of high-deposition-rate microcrystalline silicon solar cells by hot wire interface layers, *Appllied Physics Letters 87* (2005) 073503, ISSN:0003-6951

Mary, D.A. & Arthur, J.N. (2008). *Nanostrustrued and photoelectrochemical Systems for Solar Photo Conversion*, Imperial College Press, ISBN-13: 978-1-86094-255-6

Masayoshi, U.; Atsuko, K.; Mihoko, O.; Kentarou, H.; and Shinsuke, Y., (2005). Photoelectrochemical study of lanthanide titanium oxides, $Ln_2Ti_2O_7$ (Ln = La, Sm, and Gd), *Journal of Alloys and Compounds*, 400, (September 2005) 270-275, ISSN: 0925-8388

Mathews, N.R.; Miller, Eric, L.; Sebastian, P.J.; Hernandez, M.M.; Mathew, X.; and Gamboa, S.A.; (2004), Electrochemical characterization of a-SiC in different electrolytes, *International Journal of Hydrogen Energy*, 29 (August 2004) 941-944, ISSN: 0360-3199

Matulionis, I.; Zhu, F.; Hu, J.; Deutsch, T.; Kunrath, A.; Miller, L. E.; Marsen, B.; Madan,A; (2008).Development of a corrosion-resistant amorphous silicon carbide photoelectrode for solar-to-hydrogen photovoltaic/photoelectrochemical devices, *proceedings of the SPIE Conference on Solar Hydrogen and Nanotechnology*, Vol. 7044, 7044-11, ISBN: 9780819472649, San Diego, CA, USA, August 11-15, 2008, San Diego

Memming, Rudiger. (2000). *Semiconductor Electrochemistry*, Wiley-VCH, ISBN-13: 978-3527301478

Miller, E. L. & Rocheleau, E. R. (2001), Photoelectrochmical hydrogen production, *Proceedings of the 2001 DOE Hydrogen Program Review NREL/CP-570-3053*

Miller, E. L.& Rocheleau, E. R. (2002). Photoelectrochemical production of hydrogen, Proc. 2002 US DOE Hydrogen Program Review, NREL/CP-610-32405

Miller, E. L. & Rocheleau, E. R.; Deng, X.; (2003). Design considerations for a hybrid amorphous silicon/photoelectrochemical multijunction cell for hydrogen production, International Journal of Hydrogen Energy, 28 (June 2003) 615-623, ISSN: 0360-3199

Morisaki, H.; Watanabe, T.; Iwase, M.; and Yazawa,; (1976), Photoelectrolysis of water with TiO_2-covered solar-cell electrodes, *Applied Physics Letter*, Vol, 29, (September1976) 338-340, ISSN: 0003-6951

Murphy, A.B.; Barnes, P.R.F.; Randeniya, L.K.; Plumb I.C.; Grey I.E.; Horne M.D.; and Glasscock, J.A.; (2006). Efficiency of solar water splitting using semiconductor electrodes, *International Journal of Hydrogen energy* 31 (2006) 1999-02017, ISSN: 0360-3199

Narayanan, R. & Viswanathan, B. (1998). *Chemical and Electrochemical Energy Systems*, University Press (India) Limited, ISBN: 8173710694

Nelson, A. K. & Thomas, L. G., (2005). Design and characterization of a robust photoelectrochemical device to generate hydrogen using solar water splitting, *International Journal of hydrogen energy* 31 (September 2006) 1658-1673, ISSN: 0360-3199

Nelson, A. K.& Thomas L.G.; (2008). Design and characterization of a robust photoelectrochemical device to generate hydrogen using solar water splitting, *International Journal of Hydrogen Energy 31* (November 2008) 1658-1673, ISSN: 0360-3199

Nozik, A. J.; (1975), Photoelectrolysis of water using semiconducting TiO2 crystals, *Nature*, 257 (October 1975) 383-385, ISSN: 0028-0836

Richard, R.E.; Eric, L. M.; Anupam, M.; (1998). High-effciency photoelectrochemical hydrogen production using multijunction amorphous silicon photoelectrodes, *Energy & Fuels*, 12 (January 1998); 12:3–10, ISSN: 0887-0624

Shah, A.V; Schade, H.; Vanecek, M.; (2004); Thin-film silicon solar cell technology, *Progress in Photovoltaics: Research and Applications*, 12 (March 2004) 113-142, ISSN: 1062-7995

Simonds, B. J. (a); Zhu, F. Zhu; Gallon, J.; Hu, J.; Madan, A.; and Taylor, C.; (2009); Defects in Hydrogenated Amorphous Silicon Carbide Alloys using Electron Spin Resonance and Photothermal Deflection Spectroscopy, *Mater. Res. Soc. Sym. Proc. Vol.1153, April 13-17, 2009, San Francisco, CA*, paper # 1153-A18-05

Simonds, B. J. (b); Zhu, F. Zhu; Hu, J.; Madan, A.; and Taylor, C.; (2009); Defects in amorphous silicon carbide and their relation to solar cell device performance, *proceedings of the SPIE Conference on Solar Hydrogen and Nanotechnology*, Vol. 7409, 7408-3, ISBN: 9780819476999, San Diego, CA, USA, August 2009, San Diego

Sebastian, P. J.; Mathews, N. R.; Mathew, X.; Pattabi, M.; and Turner, J.; (2001), Photoelectrochemical characterization of SiC, *International Jounal of Hydrogen Energy* 26 (February 2001) 123-125, ISSN: 0360-3199

Solomon, I.;(2001); Amorphous silicon–carbon alloys: a promising but complex and very diversified series of materials, *Applied Surface Science* 184 (December 2001) 3-7, ISSN: 0169-4332

Srivastava, O.N.; Karn, R.K.; and Misra, M.; (2000), Semiconductor-septum photoelectrochemical solar cell for hydrogen production, *International Journal of Hydrogen Energy*, 25 (June 2000) 495-503, ISSN: 0360-3199

Stavrides, A.; Kunrath, A; Hu, J.; Matulionis, I.; Marsen, B.; Cole, b.; Miller, E.; and Madan, A.; (2006), Use of amorphous silicon tandem junction solar cells for hydrogen production in a photoelectrochemical cell, *proceedings of the SPIE Conference on Solar Hydrogen and Nanotechnology*, Vol. 6340, 63400K, ISBN: 9780819464194, San Diego, CA, USA, August 2006, SPIE, San Diego

Tamaura, Y.; Steinfeld, A.; Kuhn, P.; and Ehrensberger, P. (1995). Production of Solar Hydrogen by a Novel, 2-step, Water-splitting Thermochemical cycle, *Energy* Vol. 20, No.4 (1995) 325-330, ISSN: 0360-5442

Tokio O. (1979). *Solar-Hydrogen Energy Systems*, Pergamon Press (Oxford, New York), ISBN: 0080227139

Yamada, Y.; Matsuki, N.; Ohmori, T.; Mametsuka, H.; (2003). One chip photovoltaic water electrolysis device, *International Journal of Hydrogen Energy* 28 (November, 2003) 1167- 1169, ISSN: 0360-3199

Yoon, D. H.; Suh, S. J.; Kim, Y. T.; Hong, B.; Jang, G. E.; (2003); Influence of Hydrogen on a-SiC:H Films Deposisted by RF PECVD and Annealing Effect, *Journal of Korean Physic Society* Vol. 42 p 943-946, ISSN 0374-4884

Zhu, F.; Hu, J.; Matulionis, I.; Deutsh,T.; Gailard, N.; Kunrath, A.; Miller, L.E.; Madan, A.; (2009), Amorphous silicon carbide photoelectrode for hydrogen production directly from water using sunlight, *Philosophical Magazine* 89:28 (October 2009) 2723-2739, ISBN: 1478-6443

16

Contact Definition in Industrial Silicon Solar Cells

Dr. Luis Jaime Caballero
Isofoton S.A.
Spain

1. Introduction

The incredible development that industrial silicon based photovoltaic devices have followed for the last decades has been related to the consecution of a simple, easy and economically feasible way to define the electrical contacts of the devices. Due to the size of silicon photovoltaic device in relation to its substrate wafer size (one device per wafer), traditional microelectronic means to define contact (using metal evaporation in vacuum and photolithographic processes) are not economically appropriate for the mass production devices. Screen-printing technique has represented the perfect means to allow the production cost reduction, and the strong introduction of photovoltaic devices for terrestrial applications in the global market.

Although similar primitive printing techniques have been known by the mankind thousand of years ago, the industrial introduction of the screen-printing technique had to wait till the end of the XIX century, beginning of the XX century. Its application in the textile industry started the beginning of its intensive use. It was in the second half of the XX century when electronic industry started to employ this technique in the field of the hybrid circuits for the deposition of dielectric and conductive layers; but its first applications in the photovoltaic industry dates from 1975-6 (Ralph, 1975); (Haigh, 1976). Since then, improvements of the screen-printing techniques and metal pastes for the creation of contacts have been crucial for the development and improvement that industrial produced silicon solar cells have followed, becoming the heart of its fabrication processes.

But the technical characteristics of the screen-printed contacts are far from the ones obtained when a metal is deposited on the silicon surface with a microelectronic process in vacuum conditions, introducing limitations to the final energy conversion efficiency that devices can reach. The strong interest in increasing the conversion efficiency of industrial solar cells is encouraging the appearance of new research focused on overcoming this limitations. For this reason, advances and new developments in other techniques seem to be the future way to get improved contacts in mass production device fabrication. The application of these could mean changes in the nowadays standard processing technology that would be introduced by the industry in coming years.

This chapter aims to present the curent contact definition technique, reviewing the screen-printing technology, and the ways followed to optimize its results, getting higher conversion efficiencies in devices; some design topics for the front grid patterns will be discussed, finishing with a quick introductory view on the future of the industrial silicon solar cell with special attention to its contact definition.

2. Screen printing of metallic contacts basic fundamentals

The screen printing process consists in the transference of an ink or paste (with a specific viscosity) through a screen that allows its pass in a defined pattern, thanks to the pressure applied by a squeegee fixed in a moving part. Basic parts of a system for a solar cell screen-printing definition process are:

- The screen, comprising a frame that holds a stretched fabric with a photo stencil attached to the mesh with the required design of the grid pattern.
- A squeegee, comprising a holder with a fixed, flexible, resilient blade.
- A metallic conductive paste that is transferred to the device surface,
- And a silicon substrate located in a chuck aligned with the pattern to be transferred.

After a flooding of the screen with paste, the moving of the squeegee produces the deposition of the paste in the wafer surface as can be seen in the Fig. 1 where all the steps in a printing cycle are shown.

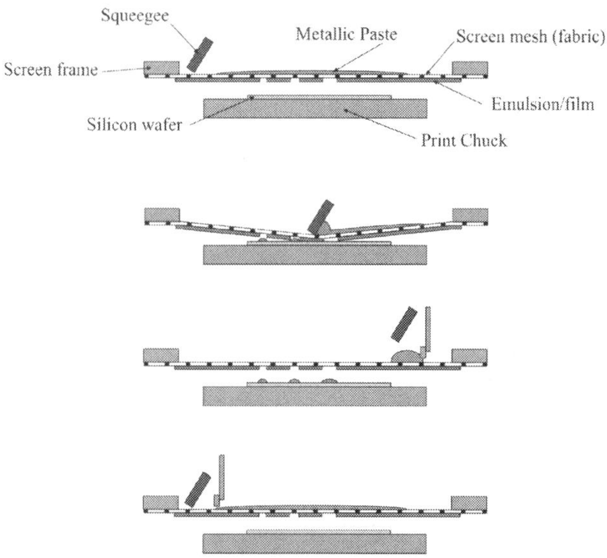

Fig. 1. Printing cycle steps, main parts of screen-printing machine are detailed along with the process sequence followed to define the contact

The result of this printing process that affects the final solar cell efficiency is highly dependent on a big number of different parameters, not only related to the raw materials involved in this production step, such as the properties of metallic pastes (viscosity and rheology) or wafers characteristics such as homogeneity in thickness or flatness, but also related to the used screens, squeegees, and the set of processing parameters related to the printing process. All of these parameters are related and require a practical optimization in order to get an improved result of the printing definition.

It can be said that the more homogeneous the device structure (in the whole surface area), the more efficient the final solar cell. Therefore a homogeneous definition of the front contact is important in order to get an optimal industrial solar cell.
Among the processing parameter are the pressure applied by the squeegee, the speed of its printing movement and the related position of the wafer surface with respect to the screen position. While among the parameter related with the screens that define the pattern, it can be found the material of the fabric (typically stain-steel due to its good response for mass production), its constituent wires' diameter and mesh density (wires per centimetre), the thickness of the fabric, its attached emulsion or film, and the stretching of the mesh of the screen.
All this set of parameters and its effects (studied early in the literature of the thick film technology field (Holmes & Loasby, 1976)) makes the printing a complex process to be optimised, which must be analysed on the industrial environment with the help of statistical tools based in the final device performance, and the optical and electrical characterization of the final contact definition of the produced solar cells; always inside specific designs of experiments where parameter are changed in defined working ranges (using a DOE analysis tool).

3. Composition of metallic pastes

3.1 Front paste composition

Metallic contact nowadays employed by the industry are based on the use of silver as a conductive metal due to its good contact properties with n type silicon, its good conductivity and its excellent solderability (needed for the later interconnection of cells). To make the deposition of this metallic contact with a screenprinting technique possible, with the goal of reaching a final correct mechanical adhesion to the surfaces and good electrical properties, special metallic pastes have been developed for the solar cell industry.
Silver powder represents the 70-85% in weight of the commercial pastes with a mixture of different shaped particles, of different sizes (as spherical powder grains or flakes), that are responsible for the paste conductivity and final cohesion of the contact.
The rest of the paste components are extremely sensitive to the device surface that want to be contacted (with different dielectrics layers of different thickness that pastes must go through), and to the specific processing of contact creations such as temperature profiles and processing times. These are:

- Powder of glass frits.
  Glass frits are metal oxides that play the most important role in the formation of the contact because its function is to melt the dielectric layers (by forming eutectic alloys of lower melting point), that are deposited or grown on the silicon emitter, allowing the metal particles to reach the silicon surface. Additionally its content determines the adhesion of the paste to the silicon substrate.
  An example of the different glass frits contain of two pastes is shown in Table 1 (Firor & Hogan, 1981)
- Organic compounds used as a vehicle to transport the suspended silver and glass particles, allowing its disposal with the screen printing techniques. Among these organic compounds are:
  Organic solvent to allow the mixture to be used as a paint, and

Organic binders to maintain the particles joined once the solvents have been evaporated (cellulosic resins) after transferring of pattern.

- Other additives to modify the rheological properties of the mixture and its interaction with the substrate surfaces (wetting agents).

Composition of the glass frits contain (% weight)		
Glass	Paste 1 Phosphate glass based	Paste 2 Borosilicate glass based **(the most important type)**
Al_2O_3	11.6	14.6
B_2O_3	--	2.1
BaO	0.2	--
CdO	--	0.6
CaO	8.6	0.2
CuO	7.7	0.6
Fe_2O_3	1.4	--
Na_2O	4.1	--
P_2O_5	65.8	4.4
PbO	--	51.8
SiO_2	0.61	25
ZnO	--	0.8

Table 1. Different glass frit contain, two examples of different pastes are evaluated

Although the exact formulations of metallization pastes are kept as industrial secrets it is possible to summarise the general components of a typical front contact paste as it is shown in table 2

Components	**(Wt.%)**
Silver	70 - 85
Glass Frits	0 - 5
Cellulosic Resin	3 - 15
Solvent (Pine Oil or Glycol Ethers)	3 - 15
Additives (Rheological Modifiers and Surfactants)	0 - 2

Table 2. Components of industrial screen-printing pastes for the definition of the front contact

All the components are dispersed and intimately mixed by using agitators, three roll mills or any similar equipment.

Front Pastes are typically designed to find a compromise between a situation that obtains a good final diode quality after processing (reached with higher metal particle sizes and frits with a higher melting temperature points), and a situation that makes easy the electrical contact with the emitter (reached with lower metal particle sizes and frits with a lower melting temperature points).

3.2 Rear paste composition

As for the front contact, the rear contact is also deposited by screen-printing techniques in nowadays industrial solar cells. In this case, as the type of silicon that must be contacted is p, the metal that constitutes the paste for the rear contact is Aluminium instead of Silver. But due to the problems of solderability that this material presents, with the formation of its surface oxide, industrial definition of the rear contact is carried out with a double step printing that defines:

- First the area that will be used, later on, to bond the connection ribbons (tabs), using a paste based on silver or silver and aluminium mixtures;
- And a second printing process that must be aligned with the first, with a paste based on aluminium particles to create the rear structure of the industrial solar cells knows as *bsf* (back surface field), due to the electrical field that aluminium doping generates in the silicon bulk near the surface.

Apart from the different metal particles of this pastes, the rest of the components that constitutes the final mixture are more or less the same because frits to improve adhesion and to remove residual dielectric layers, that can appear due to processing, are also needed, although in a lesser extend.

3.3 Supplied pastes properties

Among the parameters typically reported by the paste suppliers it is possible to find:

- The 'Viscosity', whose characterization should report the measured value along with equipment used to measure viscosity (spindle or cone/plate) and temperature.
- Possible 'Thinners' that are listed in the event of needed viscosity adjustments.
- The 'Solids Content' that is reported for combined inorganic content (in absence of organic) and 'Metal Content'.
- The 'Fineness of Grind' (FOG) that is reported for the first scratch (≤).
- Some after processing parameters such as: the 'Dried Thickness' and 'Fired Thickness' that are given as a range reported in μm, and the 'Resistivity' (for a fired thickness given in mΩ/square).
- Some processing parameters such as the 'Drying Profile' giving applicable ranges and necessary durations, the 'Peak temperature' that gives the optimal firing range, the 'Peak Duration' reporting the optimal value or a range and the 'Firing Atmosphere' that is typically air.
- And other supplied parameters such as the 'Shelf Life' of the paste that is listed for a properly sealed container.

4. Contact formation, the co-firing process

As both contacts of solar cells are nowadays deposited using the same technique, pastes are designed to follow a common thermal final treatment that creates the contact in both faces of the cell. This process is called 'co-firing' and industrially is carried out with in-line belt furnaces in a (typically) air atmosphere during processing. The sequence to produce the final cell structure is:

- Printing of the front (silver) contact
- Drying of the front contact
- Printing of the back contact (aluminium contact with two printing processes as was introduced in previous section, with a drying process before the second printing process)

- Drying of the Back contact
- Co-firing process, to create the contact, and a
- Laser isolation process (if wafers have not been previously isolated chemically)

It is important to keep the defined order in these steps to avoid any possible aluminium contamination of the front side of devices that would destroy the performance of the final cells (by creating a serious shunting problem of the p-n junction).

During the co-firing process wafers follow a fast process with a defined temperature profile similar to the one that is shown in Fig. 2, resulting in chemical and structural changes inside the printed pastes and substrate surfaces.

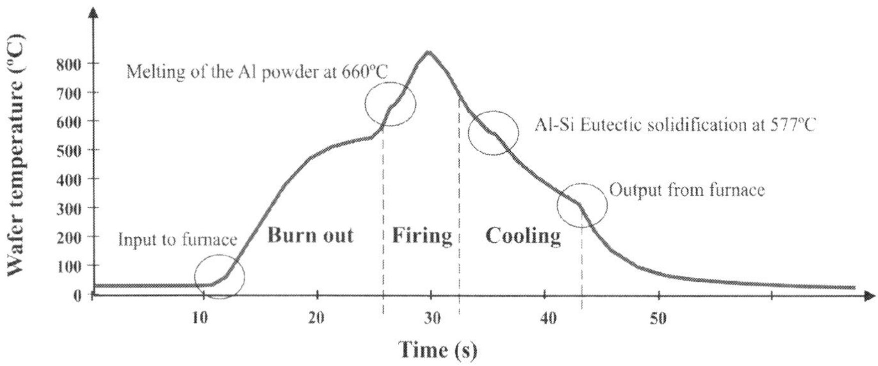

Fig. 2. Temperature profile of the furnace for the co-firing process, and time of processing

The organic components that keep the dried metallic pastes attached to the surfaces are burn out at the beginning of the process at temperatures below 600° C, step that is also used to produce the diffusion of passivating hydrogen from the antireflective layer to the silicon bulk. After this step the formation of the contact takes place at higher temperatures during the firing peak. At this point glass frits (PbO-BO_3-SiO_2) melt the silicon nitride while silver particles are transported through this melted mixture and suffer a sintering process that creates a conductive film.

Liquid Pb generated during the chemical reaction etches the silicon surface in places where, during cooling, silver uses to recrystallize creating silver crystallites that are responsible for the ohmic contact with the emitter.

After these steps the silver and glass frits mixture solidify creating a conductive contact that combines a direct contact between the sintered silver and the silver crystallites, a tunnelling contact through frits, and regions without electrical contact.

On the other hand, at the rear surface aluminium paste follow the same burn out of organic components as a first step. When temperature are above 660°C the aluminium melt and can go through the oxide that covers the particles (its oxide). The melted aluminium reaches the silicon surface, and silicon is solved following the phase diagram of the Si-Al alloy. During cooling down, silicon is rejected from the melt and recrystallize as an epitaxial Al-doped layer creating a *bsf*.

When temperature are below 577°C, the melted alloy solidify with the eutectic composition creating a layer whose contact with the silicon produces the typical bow of this aluminium *bsf* device structure.

Controlling the bowing produced at the creation of the contact is one of the most important issues for the industry nowadays. The most extended technique to keep controlled the bowing is to control the weight of aluminium deposited through the weighting of the deposited paste (between 6 and 10 mg/cm^2 of dried paste).

5. Practical industrial optimization of the metallization process

Industrial optimization of the firing process are carried out by means of a simplified DOE (Design of Experiment) that tries to find the optimal firing peak temperature for a specific thickness of substrates with a defined weight of rear aluminium paste deposited, once the wafers are right printed, and pastes are dried (in-line furnaces must be previously adjusted to ensure a perfect drying that eliminates many in other way existing problems).
Fig. 3 shows the typical shape of the optimal metallization process working window as a function of the processing speed (speed of the furnace belt) and the furnace firing zone temperature; the effect of the substrate thickness on the right working window and its orientation inside the furnace during processing is also shown.

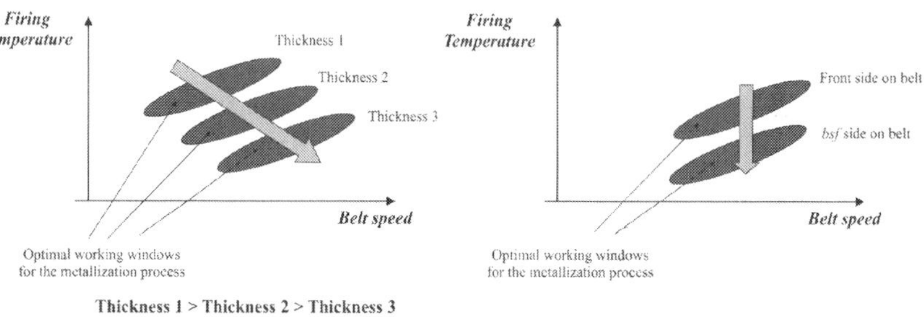

Fig. 3. Optimal Firing working window variation, picture on the left shows the dependency of the thickness of substrates, while picture on the right shows that when the front side is put on the belt, higher firing peak temperatures are needed

A set of wafers with the same characteristics are split in groups that will be processed at a different firing temperatures, always with the same chain speed that is previously fixed. Taking the electrical results of the processed cells (Maximum Power and Fill Factor) it can be established the position of the optimal processing temperature in each case. Fig. 4 shows a practical example of contact optimisation.
Other characterisation techniques such as Corescan (Van der Heide et al., A 2002); (Van der Heide et al., B 2002) can also be used with optimization purposes, its measurements allow to determining local problems related to the existence of cool spots during firing, contamination problems, and processing defects in other steps of the production that can generate problems during contact creation.

6. Series resistance analytical modelling and optimal design for the front grid of industrial solar cells

The continuous rises in current that industrial photovoltaic devices have been undergoing due to the continuous improvement of materials, processing, device design and over all, substrate wafer size increase, make its associated device power losses also increase. Thus,

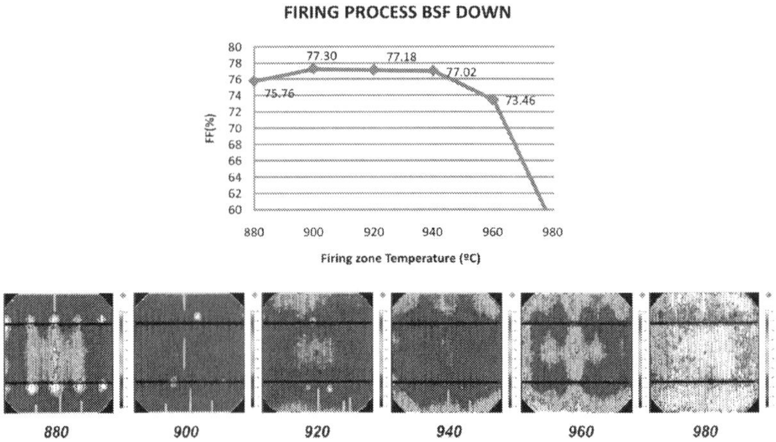

Fig. 4. Firing process optimization, example of a firing optimization attending to the final Fill Factor of resulting cells processed with its *bsf* rear contact on the furnace belt during firing. Corescan analysis for each temperature is also shown

series resistance of cells that introduces this power loss constitutes a key factor that must be taken into account in order to reach higher energy conversion efficiencies. Series resistance components characterization and its associated power losses have a great importance in the cell and module fabrication field (Luque et al., 1986); (Roberts et al., 2000). It has a potential application in grid design, metallization paste research and development, solder or conductive adhesives contact resistance requirements, and number of solder points (Caballero et al., 2006) and interconnection ribbon thickness selection.

Impact of the grid pattern design in the final series resistance of an industrial screen-printed silicon solar cell is known to be one of the strongest, so the correct design of the typical H pattern industrial front grids (a set of fingers orthogonal to two main collecting bus-bars) can improve the final device performance.

This section will deal with this task, showing how it is possible to optimise the design of front grid through the use of an analytical modelling derived from the theoretical expressions of the power loss for each part of the device. The presented modelling will take into account only the contributions that compose the series resistance of a solar cell with total aluminium rear *bsf*, leaving out of this text the cell interconnection inside a module and the analysis of all the additional contributions to the module series resistance that can appear, just in order to get a simple reference tool for further future analysis.

6.1 Cell series resistance components

When a cross-section of a simple cell is analysed it can be easily distinguished all the layers responsible for the different components of the series resistance, each one produces a power loss associated to the pass of the current through its volume. The total series resistance is the addition of all these components as it is plotted in the equivalent circuit of Fig. 5.

The proposed analytical model calculate the series resistance of a square cell as the result of the parallel association of multiple unitary basic cells (neglecting the interconnection resistance between them) in which all the components of the series resistance are calculated.

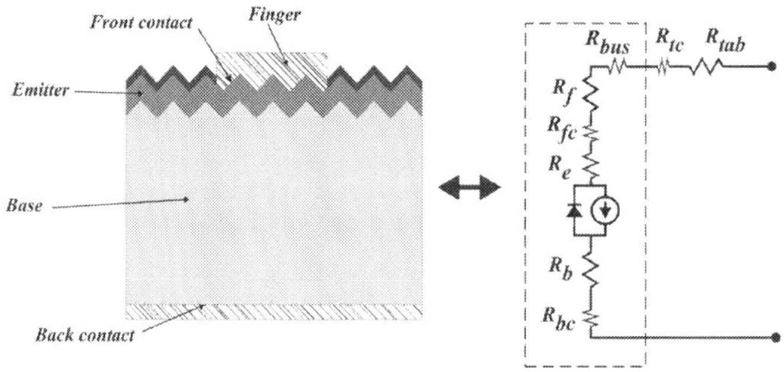

Fig. 5. Components of the cell's series resistance, equivalent circuit for an Aluminium *bsf* solar cell are shown inside the dashed box

Considering that current is extracted from the soldered points that join the bus bar with the interconnection conductive ribbons (tabs), and these points are homogeneously distributed along the bus bar, it is possible to define a basic unitary cell and the number of them that are needed to get the total series resistance (being each soldered point in contact with four unitary cells), as it is shown in Fig. 6.

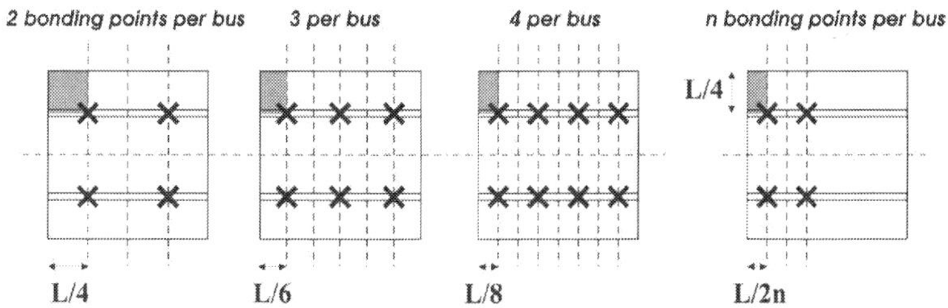

Fig. 6. Basic calculus cell for two bus-bar, it is shown the basic unitary cell size for the case of having two collecting bus-bars with a variable number of bonding points per bus-bar in the solar cell

The higher the number of soldered points per bus-bar the more accurate the result of this simple modelling, because in that way, the real current flow in an interconnected solar cell produces a natural distribution of the device in regions that are more similar to the basic cell proposed by this modelling.

Components of the unitary basic cell's series resistance are detailed in Table 3; expressions such as the emitter, finger and bus components are deduced from matching the power loss integrated along the current path (increasing the current along this path) with the power loss associated to an equivalent resistance crossed by the total current ($I_{Total}^2 R_{equivalent}$).

It must be noted that series resistance associated to the BSF layer has been neglected due to its low value.

Equations that are summarized in Table 3 include characteristic parameters from the different materials, such as ρ_{base} (bulk resistivity of the wafer), ρ_{metal}, $R_{FrontPaste}$ and $R_{BackPaste}$ (resistivity of the metallic grid in Ωcm and semiconductor-metal contact specific resistivities in Ωcm²); characteristic parameters from the process and wafer such as R_e (Emitter layer resistance in Ω/square), L(wafer side) and w_{base} (base width); and grid design parameters such as

Component	Expression
Emitter	$R_{Emitter} = \frac{n \cdot s^2}{3L} \frac{R_e}{(L/2 - w_{bus})}$
Base	$R_{base} = \rho_{base} \frac{8n \cdot w_{base}}{L^2}$
Metallic finger	$R_{finger} = \frac{n \cdot s}{3L} \cdot \frac{\rho_{metal}}{w_f h_f}\left(\frac{L}{2} - w_{bus}\right)$
Bus bar	$R_{bus} = \frac{\rho_{metal}}{3n} \cdot \frac{L}{w_{bus} h_{bus}}$
Rear contact resistance	$R_{fc} = \frac{8n \cdot s \cdot R_{FrontPaste}}{L \cdot (w_f \cdot L + 2w_{bus} \cdot (s - w_f))}$
Front contact resistance	$R_{bc} = \frac{8n \cdot R_{BackPaste}}{L^2}$

Table 3. Analytical expressions of the series resistance components for a two bus-bar cell

$w_{bus,\,f}$ (width of buses or fingers), $h_{bus,\,f}$ (high of buses or fingers) and s (separation between fingers). All the expressions include the number of bonding points per busbar referred as n. As the total series resistance for the solar cell is the parallel of all the unitary basic cells, it can be expressed as:

$$R_{Series\,Total} = \frac{\sum R_{Components}}{8n}$$

With a set of values typical from the industrial environment as could be:

L	156 mm
w_{base}	200-240 μm
ρ_{base}	1 Ωcm
w_{bus}	1.8-2 mm
w_f	100-150 μm
h_{bus}	25 μm
h_f	12 μm
s	1.8-2.5 mm
R_e	45 Ω/ sqr
$R_{FrontPaste}$ (Recart,2001)	~10 mΩcm^2
$R_{BackPaste}$ (Recart,2001)	~10 mΩcm^2
ρ_{metal} (Recart,2001)	2-3 ·10^{-6} Ωcm

Table 4. Typical technological values for the industrial solar cell

The modelling of the series resistance is complete and can be used to include its effect in the simulation of a device performance.

With these parameters it is possible to evaluate the effect of the number of solder spots (contact points with the interconnection tab) in the series resistance as it is shown in Fig. 7. This analysis shows that a number of contact points higher than eight points per bus in industrial solar cells produces an increment in series resistance of devices that can be neglected.

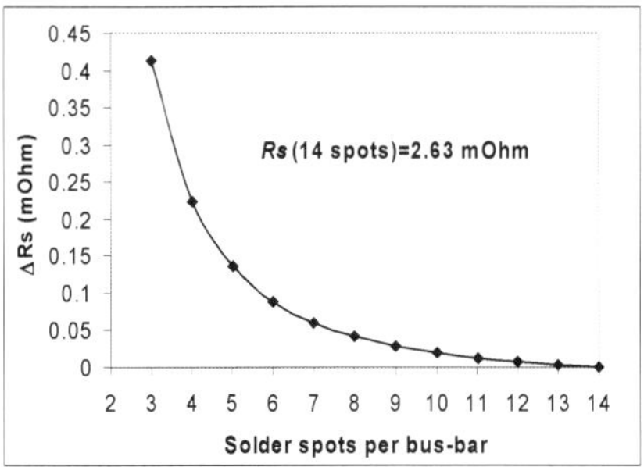

Fig. 7. Reduction of the series resistance with the increment of bonding points per bus-bar, result of the modelling are presented as an increment with respect to the situation of having 14 bonding point.

Joining this series resistance parameters and the response of an ideal diode, inside a modelling of solar cell, allows to optimise the design of front grids for each specific case, knowing for example the optimal finger separation that must be selected to get the highest conversion efficiency. So using the mathematical expressions proposed elsewhere (Luque et al., 1986), it is possible to have a modelling whose input parameters are just the photo-generated current density (J_{SC}) without front grid, the open circuit voltage (V_{OC}) reached by the device and the parameters needed to model the series resistance. An example of the results of such a modelling is shown in Fig. 8 where the effect of finger separation on efficiency is studied for the case of having a J_{SC}=36.5 mA/cm2, V_{OC}= 616 mV and the set of defining parameters presented previously in this section.

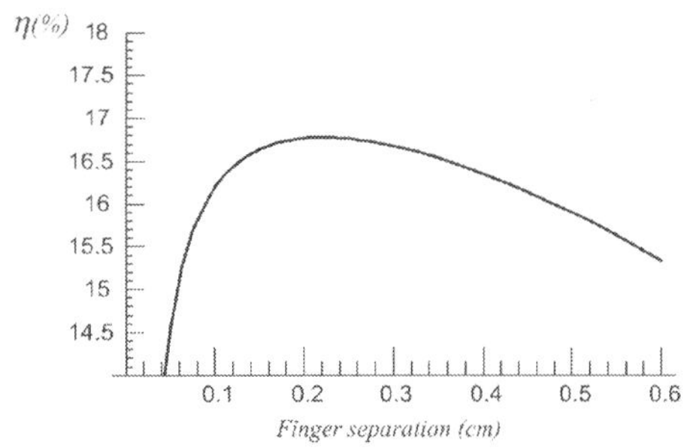

Fig. 8. Efficiency as a function of finger separation, the resulting efficiency of the mathematical modelling is plotted as a function of the finger separation when the set of parameters of Table 4 is used (with a 100 μm fingers width)

6.2 Three buses vs. two buses design for an industrial solar cell

The modelling of the previous subsection can be used as a easy tool to evaluate changes in the industrial development of solar cells, not only the improvement in material properties, but also changes of the design of the metallic contact.

The increase in size solar cells have followed last years has motivated the appearance of new designs with three bus-bars to collect the generated current instead of two, but, what is the effect of this change of the front grid design in the performance of the solar cells?. This section will discuss on this topic. A comparison is carried out based on the analytical modelling results of the two alternative designs that nowadays it is possible to find in the industry (two and three bus-bars). The proposed model needs to be adapted to introduce the calculus of the resistance of the three bus-bars cell. Table 5 summarises the expressions of the components of the resistance, while Fig. 9 presents the efficiencies of both cases as a function of the shadowing factor for the optimal finger separation in each point. In both cases fingers of the grid have 100 μm width, so to get a specific shadowing factor, just the bus-bar width can be modify (as it is plotted in Fig. 9). The rest of design parameters are the same presented in Table 4 with a w_{base} of 200 μm and a ρ_{metal} of 3 ·10-6 Ωcm.

Component	Expression
Emitter	$R_{Emitter} = \frac{n \cdot s^2}{3L} \frac{R_e}{\left(\frac{L}{3} - w_{bus}\right)}$
Base	$R_{base} = \rho_{base} \frac{12 \cdot n \cdot w_{base}}{L^2}$
Metallic finger	$R_{finger} = \frac{n \cdot s}{3L} \cdot \frac{\rho_{metal}}{w_f h_f} \left(\frac{L}{3} - w_{bus}\right)$
Bus bar	$R_{bus} = \frac{\rho_{metal}}{3n} \cdot \frac{L}{w_{bus} h_{bus}}$
Rear contact resistance	$R_{fc} = \frac{12 \cdot n \cdot s \cdot R_{FrontPaste}}{L \cdot \left(w_f \cdot L + 3 \cdot w_{bus} \cdot (s - w_f)\right)}$
Front contact resistance	$R_{bc} = \frac{12 \cdot n \cdot R_{BackPaste}}{L^2}$

Table 5. Analytical expressions of the series resistance components for a three bus-bar cell

With a total series resistance that is:

$$R_{Series\,Total} = \frac{\sum R_{Components}}{12n}$$

From Fig. 9 it is clear that three bus-bars design presents an improved performance that produce an increase of 0.1 points in efficiency. This efficiency increase comes from the reduction of the finger resistivity as can be seen in the resistance distribution carried out for a fixed shadowing factor of 7.34% that is shown in Fig. 10, this is why it is important to notice that an improvement in material parameters, such as conductivity or contact resistance of pastes, can reduce the difference between the results of the two patterns (for example a reduction of paste resistivity to 1·10-6 Ωcm and a contact resistance of metal-semiconductor of 1 mΩcm² would result in a difference of just 0.05 points in efficiency).

In addition to the efficiency increase the three bus-bars design has another advantage that is the reduction of interconnection power loss in the final module, although the width of the bus-bar is reduced in the three buses design to have the same metal covering factor (when finger width is kept as a constant), as can be seen in Fig. 9, it must be taken into account that the total interconnection width (addition of the width of all the bus-bars) is higher, reducing the resistance associated to the tabbing interconnection (that have the same width than the bus-bar) and thus the final module power loss. For example, for a shadowing factor of 7.34% in the presented case of Fig. 9, the relative reduction of power loss in the tab interconnection can be estimated in 12.4% for a series association of cells independently of the tab thickness used (when the same tab thickness is used in both two and three buses case).

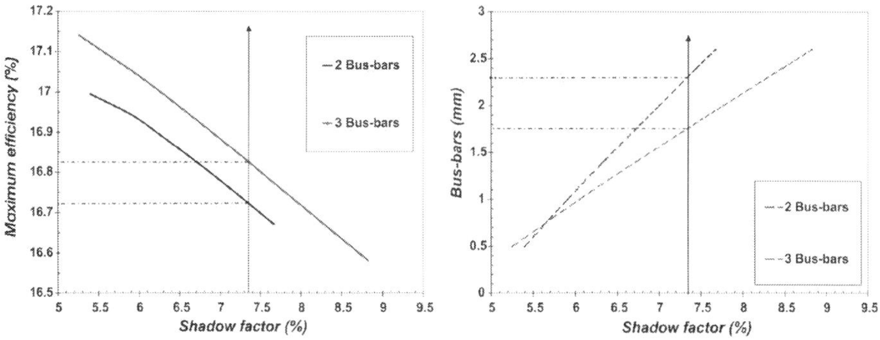

Fig. 9. Comparison between two and three bus-bar cell behaviour for the case of having fingers of 100μm width (fixed), picture on the left shows the efficiency as a function of shadowing factor of the front grid, while picture on the right shows the bus-bar width needed to reach an specific shadowing factor

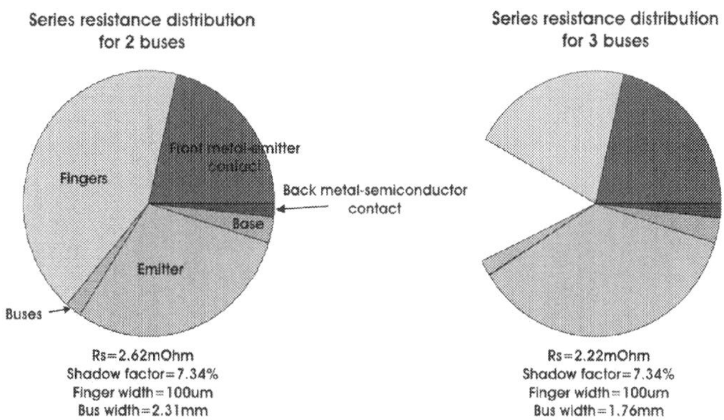

Fig. 10. Series resistance component distribution for cells with two and three bus-bars, cells with the same shadowing factor of the grid and finger width (different bus-bar width) are studied for the optimal finger separation of each case. The reduction of total resistance is mainly due to a reduction of the component associated to fingers

6.3 Multi-bus bar solar cell concept

Taking the results shown in the previous section, it is clear that solar cell performance is improved with the addition of one bus more to its front contact grid, but what happens when a bigger number of buses are added to the front grid design? Can we expect further improvement?. To answer these questions we need to modify the mathematical expressions of the analytical model to evaluate the new situation that is shown in Fig. 11.

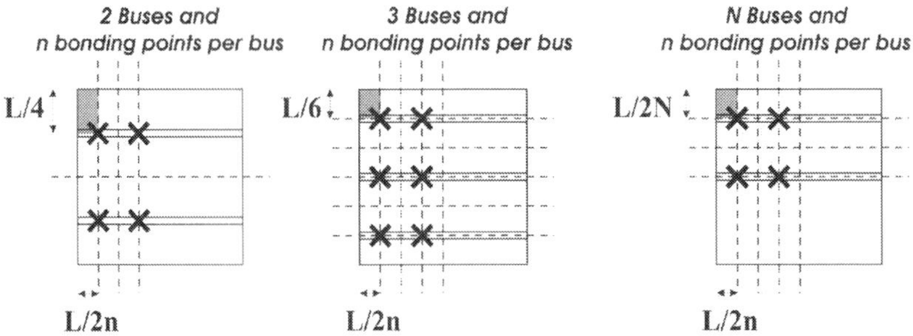

Fig. 11. Basic calculus cell for a variable number of bus-bar, it is shown the basic unitary cell size for the case of having N collecting bus-bars with n bonding points per bus-bar in the solar cell

With this configuration, expressions of the different components of the series resistance will change as it is shown in Table 6, where N is the number of buses and the rest of symbols represent the same parameter as in Table 3.

Component	Expression
Emitter	$R_{Emitter} = \frac{n \cdot s^2}{3L} \frac{R_e}{\left(\frac{L}{N} - w_{bus}\right)}$
Base	$R_{base} = \rho_{base} \frac{4N \cdot n \cdot w_{base}}{L^2}$
Metallic finger	$R_{finger} = \frac{n \cdot s}{3L} \cdot \frac{\rho_{metal}}{w_f h_f} \left(\frac{L}{N} - w_{bus}\right)$
Bus bar	$R_{bus} = \frac{\rho_{metal}}{3n} \cdot \frac{L}{w_{bus} h_{bus}}$
Rear contact resistance	$R_{fc} = \frac{4N \cdot n \cdot s \cdot R_{FrontPaste}}{L \cdot \left(w_f \cdot L + N \cdot w_{bus} \cdot \left(s - w_f\right)\right)}$
Front contact resistance	$R_{bc} = \frac{4N \cdot n \cdot R_{BackPaste}}{L^2}$

Table 6. Analytical expressions of the series resistance components for a multi bus-bar cell (with a number of bus-bars equal to N)

Total solar cell series resistance corresponds in this case to the expression:

$$R_{Series\,Total} = \frac{\sum R_{Components}}{4n \cdot N}$$

With these new analytical modelling and the set of values for each technological parameter presented in the previous section, it is possible to extend the study of the performance of cells with two and three buses to a bigger number of buses, as it is plotted in Fig. 12, where the maximum possible efficiencies (reached with the optimal design of grid) are shown as a function of the covering factor for a grid with 100 microns finger width.

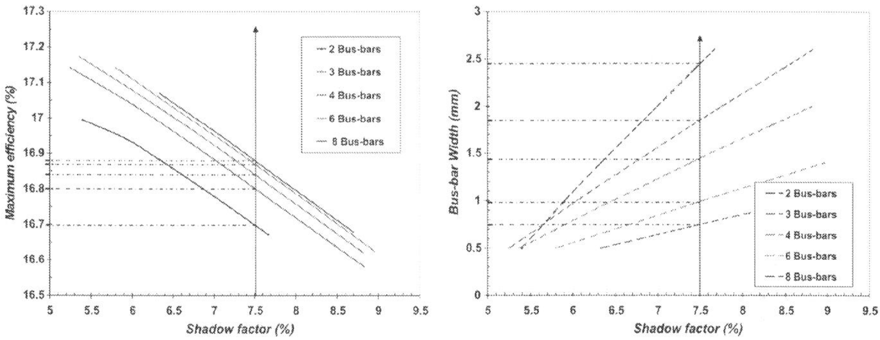

Fig. 12. Comparison between behaviour of cells with a different number of bus-bars for the case of having fingers of 100μm width (fixed), picture on the left shows the efficiency as a function of shadowing factor of the front grid, while picture on the right shows the bus-bar needed to reach an specific shadowing factor

When the number of bus-bars increases an improvement in efficiency is reached, but this improvement, due to the addition of one new bus-bar, makes lower when the total number of buses increase as can be seen in Fig. 12.

As in the case of using three bus-bars, when the number of bus-bars is increased a further reduction of the interconnection module power losses is obtained. It can be calculated from the example referred in Fig. 12 (for a shadowing factor of 7.5%) that a relative power loss reduction of 11.7% (with respect to the two bus-bars configuration) would be reach with three bus- bars, while it would be 15.5%, 18.3%, and 18.3% with four, six and eight bus-bars respectively.

Nowadays technological parameters make possible the improvement of solar cells with the addition of more buses mainly thanks to the reduction of the finger resistance. The improvement is not so high for more than three or four buses, and it must be taken into account that a future improvement in conductivity of metal pastes or its resulting finger cross section's aspect ratio would produce a lower improvement, due to the addition of more buses.

6.4 Modelling the effects of future improvements in metallization of solar cells

The proposed modelling of the series resistance constitutes a powerful tool to know how industrial solar cells can evolve thanks to the improvement of metallic pastes, or new passivation processing for bulk and surface. With this purpose the modelling will be used to evaluate the efficiency of a solar cell (with an optimal H-pattern grid design) when it happens a change in the technological parameters related to the metallic contact (resistivity of metal and contact resistance) or in the *Voc* (parameter that is related to the passivation of volume and surfaces). All the rest of parameters of the modelling have been kept constant for this analysis with the values referred in Table 7.

L	156 mm
Jsc(without grid shadow)	36.5 mA/cm^2
Number of bus-bars	2 and 3
Number of contacts per bus-bar	8
w_{base}	200 μm
ρ_{base}	1 Ωcm
w_{bus}	2.45 mm (2 bus-bars) 1.85 mm (3 bus-bars)
w_f	100 μm
h_{bus}	25 μm
h_f	12 μm
R_e	45 Ω/sqr

Table 7. Model parameters used in the analysis of the evolution of the cell behaviour when the technological parameters associated to the screen-printing technology are modified

Results of the modelling are plotted in Fig. 13 for two different situations, with a *Voc* of 616 mV and for a *Voc* of 630 mV.

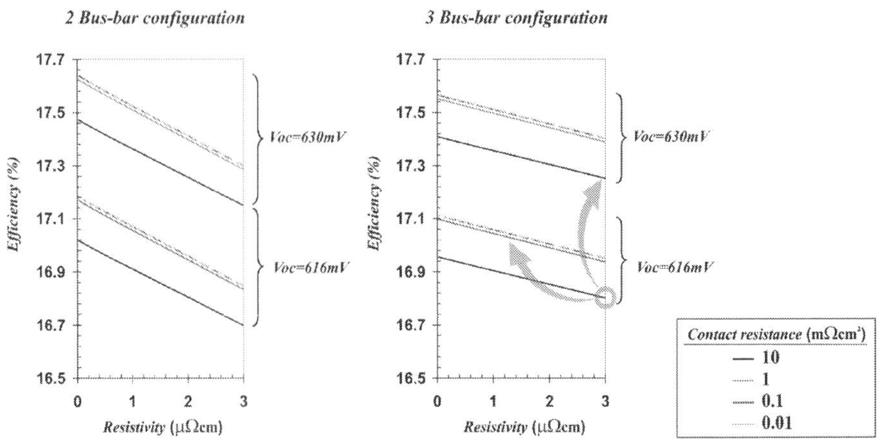

Fig. 13. Modelling results for the efficiency as a function of the finger resistivity and contact resistance for cells of two and three bus-bars with the optimal design of grid in each case, when generated current are kept constant and two different *Voc* are considered. Rest of parameters for the modelling are taken from Table 7

As it is shown in the figure a jump in *Voc* of 14mV has an stronger impact in cell efficiencies than any possible improvement in the material of the front metallic grid; even when it is use the resistivity of the pure silver (1.63 μΩcm (Lide, 1974)) for the optimal grid.
When an increase in generated current is studied, as it is shown in the Fig. 14, a similar result is obtained, showing that although it is possible to improve the solar cell performance thanks to the pattern optimization and enhancement of the materials properties for the front grid contact, the improvement of the device design and materials under the front contact have a bigger importance in the road to the industrial solar cells improvement.

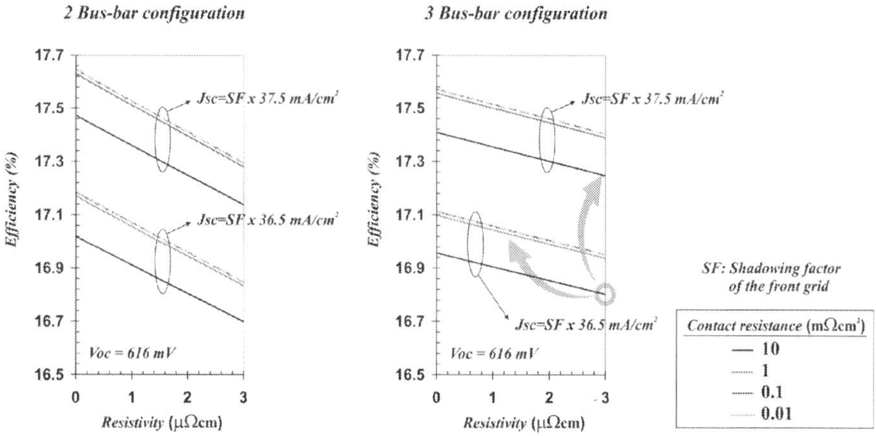

Fig. 14. Modelling results for the efficiency as a function of the finger resistivity and contact resistance for cells of two and three bus-bars with the optimal design of grid in each case, when *Voc* are kept constant and two different generated current are considered (36.5 and 37.5 mA/cm^2 without grid shadowing). Rest of parameters for the modelling are taken from Table 7

7. Coming future for the industrial front face cell definition

7.1 Changes of the device structure

As can be extracted from the previous section, the coming future for the solar cells front face definition, to get an effective improvement on the energy conversion efficiency, is not only related to the use of an optimal metal grid definition, but is also related to the introduction of different device's structures that will improve the performance of cells.
The Different working lines that can be followed by industry are related to the introduction in production lines of high efficiency concepts that are being successfully tested by research institutes and universities, among these lines it can be found several approach such as:

- To change the silicon surface topology introducing new more efficient texturization processing (Zhao et al., 1999), (Kumaravelu et al., 2002).
- And improving the optical performance of these with and optimised design of the anti-reflective coatings and rear reflector in order to get higher generated current (Nilsen et al., 2005); (Glunz et al., 2007).

- Improving the front, bulk and rear surface electrical passivation in order to enhance the final open circuit voltage of the devices, by means of better aluminium rear pastes, using gettering steps or introducing more radical changes in the cell design as a rear local contact with an improved electrical passivation. (Glunz et al., 2007).

Among all the high efficiency concepts, the idea of a 'selective emitter' in the devices is prone to be one of the first concepts introduced by the industry in its production lines without the need of radical changes in the production processes. This is why a special attention must be devoted to this concept.

The concept of 'selective emitter' consist in rising the surface doping level in the emitter area where metal grid will be deposited to improve later on its electrical contact (as it is plotted in Fig. 15), and keeping a low surface doping level in the rest of the front side of the cell (area that in that way will be later electrically better passivated) (Green, 1995). This concept not only improves the generated current due to the improvement of the low wavelength response of the resulting cell emitters, but also makes the open circuit voltage to rise due to the improvement in the front surface passivation.

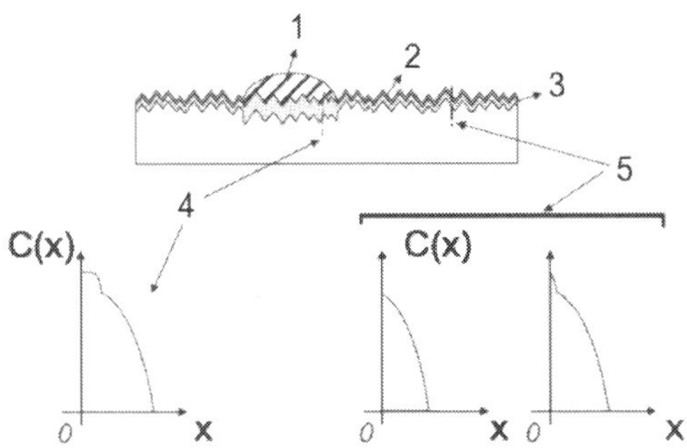

Fig. 15. Selective emitter concept, picture shows a cross section of the upper part of a solar cell where it can be observe the metal fingers (1); antireflective coating (2) and diffused layer (3). A different Phosphorus dopant concentration as a function of depth must be appreciated under the metallic grid (4) and in the non-contacted area (5)

But technical difficulty in obtaining this structure has made the selective emitter concept typical from fabrication processes of research labs, or implemented in solar cells with more complex structures and complicated fabrication processes due to the alignment requirements (as the LGBC or Laser Grove Buried Contact solar cell (Wenham & Green, 1993)), leaving this concept out of simple processing production due to its high implementation cost.

Apart from the typical microelectronic way to get a selective emitter with several diffusion steps and using diffusion barriers deposited on the (later on) non contacted areas (with an expensive photolithographic process due to the restrictive alignment between patterns and the high number of processing steps), new innovative techniques to get a selective emitter with screen-printing contacts are appearing recently due to the great industrial interest on improving photovoltaic device efficiencies (Raabe et al., 2007). All these new techniques can be classified in two big groups according to the way the selective emitter structure is obtained:

A. Selective emitter developed without masking processes, carrying out local phosphorus diffusion on the wafer surface.
   Several alternatives for this kind of processes has been studied, among these, it could be emphasized:
   1. The local deposition of phosphorus sources on the wafer surface by using a screen-printing process previous to the high temperature diffusion step, generating most highly doped areas in a self-aligned process that only requires the right location of the contact grid during the metallization step. Different examples of these processes can be found in (Horzel et al., A 1997); (Horzel et al., B 1997); (Salami et al., 2004).
   2. The use of lasers to create, after a soft phosphorus diffusion step in all the cell area, a higher doped contact area (Besi-Vetrella et al., 1997).
   3. Or the use of special metallization pastes that include doping material in its composition, thereby these pastes are used as a source of dopants for the contact area during the firing step of the metallization, creating a selective emitter structure when they are deposited on wafers with a low doped emitter in a self-aligned fabrication process. Examples of these processes can be found in (Rohatgi et al., 2001); (Porter et al., 2002); (Hilali et al., 2002).
B. Selective emitter developed using masking processes to protect, with barriers, part of the front area from the diffusion step (creating zones with a softer diffusion due to these barriers) (Bultman et al., 2000); (Bultman et al., 2001). Or masking to protect the surface from a selective etching. As it is refer in (Ruby et al., 1997); (Zerga, A. et al., 2006); (Haverkamp et al., 2008).

The alternatives exposed in the first group produce a correct selective emitter structure but have a drawback, after the phosphorus diffusion and the gettering step it introduces, when it is carried out in super-saturation conditions, the impurities concentrations in the silicon bulk keep constant, because impurities remain mainly trapped in the 'dead layer' that appears near the surface, without been effectively removed from the device, reducing the potential impact of the improvement this step could have.

Among the different alternatives of the second group, however, it exists processes that can carry out an effective reduction of impurities when these include a surface etching, what result in a better device performance; but also present some drawbacks related to the needed mask treatments and processing (such as deposition, curing and removing steps), giving slightly more complex fabrication process.

All the developed alternatives in both groups (with exception of the use of self-doped metallization pastes) present a fundamental problem for the selective emitter structure, that is the need of an alignment with the next processing steps for the contact definition, complicating the fabrication routes. This problem gets worse when it is taken into account the random deformation screen-printing technique presents for the transferred patterns with the increase in the number of prints (deformation that is associated to the relaxation of the fabrics, that compose the screens, and gets a maximum value in the mass production environment). But appearance of new alignment relaxed device structures concepts (Caballero, 2009) can help to develop new and easier industrial fabrication processes that could finally result in the implantation of the selective emitter as a common part of the typical industrial solar cell device structures. The introduction of the selective emitter structure would force the re-design of the front contact grid that could present a different optimal finger separation (with closer fingers) due to the increase in the emitter resistance of the new cells, but no other additional important changes.

7.2 The future of the metallic contact definition

Apart from the improvement of the general parameters of the device (*Voc* and *Jsc*), as previous sections has shown, improving the front contact of solar cells is possible and it can produce an increase in the final efficiency of the industrial solar cells. In this section it will be reviewed the strategies that research centres and industry are following for a future improvement of the front grid.

The approach for improving the front grid is based on increasing the aspect ratio of the cross-section of fingers as it is shown in Fig. 16, reducing the seepages that increase the shadowing factor of the grid without reducing the grid resistance, increasing the finger height, and reducing its width in order to produce a lower grid shadowing.

Fig. 16. Future improvement of the metal finger cross-section

Several strategies can be followed or will be followed by the industry with this purpose; from the introduction of slight modifications or changes of the nowadays production technology, such as:

1. Optimising the paste composition with different combinations of the silver particles with different shapes and sizes, in order to maximise the finger heights after printing.
2. Optimising the printing process, modifying the fabrics and emulsions characteristics in the screens, and the processing parameters to reduce the paste seepages during printing.
3. Introducing a heated chuck inside the standard printer units in order to produce an increase in resolution by heating the wafer substrates during printing process. Reaching finger width as lower as 50 microns (Erath et al., 2009).
4. Substituting the standard pastes by the Hotmelt technology pastes (Williams et al., 2002) that produce an improved aspect ratio of the final metal fingers. This technology changes the traditional solvent of the paste by a long Chain alcohol with a melting point in the range of 40 to 90°C, so final paste needs a heated screen to be disposed.

To the introduction of additional processing steps, that needs the addition of new machines in the production lines, such as:

5. Growing pure silver over the screen-printed contact in an electrolytic bath, reducing the resistance of the fingers and improving the contact resistance in its edges, due to the silver filling of the empty space of the fingers volume.
   Several approaches can be found in the market, based on the classical electrolytic growing, that needs a current contact with the front grid, or based on the LIP technique (Pysch et al., 2008); (Glunz et al., 2008) that doesn't need any contact. In both cases grid fingers width increases after processing.

Or the complete change of the technology for the contact definition using new techniques nowadays under development, in research projects of several companies and institutions, such as:

6. Growing the complete finger in an electrolytic processing, but defining previously a grid which is used as a seed for the electrolytic grown with narrow lines to avoid an excessive increase of the final finger width.
   This seed for growing could be produced with an electroless nickel plating over the bare emitter silicon (Glunz et al., 2008) (processing that would introduce previously, a masking of the front silicon nitride layer using an inkjet printing system to deposit the mask with the needed definition of lines; and an etching of the masked structure to open the nitride layer), or defined with new promising techniques such as inkjet printing (Mills & Branning, 2009), aerosol jet printing (technique able to reach finger definition of 40 microns) (Hörteis et al., 2008), or laser direct-printing (able to reach a finger definition below 20 microns) (Shin et al., 2008); (Arnold et al., 2004) over the silicon nitride layer (with its later firing through previous to the electrolytic growing) or also over the bare silicon emitter (processing that would need also a previously masked etchings of the nitride).

The introduction of the plating of new materials to grow the contact grid can reduce additionally the final cost of the metallization step. Thus, copper with a good conductive properties (97.61% of the silver conductivity (Brady et al., 2002)), is stirring up the industrial interest on the creation of new multiple metal layer contact such as:

Seed of Nickel/ Copper/ Tin
Seed of Silver/ Nickel/ Cooper/ Tin or
Directly Copper without seed/ Tin

On the other hand, situation for the back aluminium contact of cells is different because it is not related with the improvement of resolution, but with the need of an improvement in *Voc* to increase the final cell efficiencies. In this case the possible paths for industry to follow are:

- Improving the characteristics of the Aluminium pastes for the total *bsf* device structure, designing new formulations of the used pastes and removing the limitation that nowadays the aluminium contact has, related with the bowing of devices during the firing process, by means of, for example, adding special thermal treatments processing (Huster, 2005).
- Changing the total rear contact by a local rear contact where non contacted areas must be passivated with a new deposited layer with improved properties of passivation for p type silicon. Laser techniques for the creation of the local contacts (Schneiderlöchner et al., 2002); (Tucci et al., 2008) are taking a good position to be industrialised for the creation of this kind of local contacted structures.

8. Conclusions

This chapter has presented the technique used by the mass production industry to define the contacts of the silicon solar cell, its basic principles and factors that have an influence in its result with the main aim of giving an introductory view of a technique responsible for the development and expansion of the nowadays photovoltaic terrestrial market.

It has been shown how through a simple analytical modelling the performance of different designs for the front contact in the commercial solar cells can be optimised and compared. And a quick review of the coming changes in the device structure design, and the future techniques that are under research to substitute the screen-printing technology have been done in order to give an idea of how the industry can evolve in the coming years.

9. References

Arnold, C. B.; Sutto, T. E.; Kim, H.; Piqué, A., 'Direct-write laser processing creates tiny electrochemical systems', *Laser Focus World*, May 2004.

Besi-Vetrella, U. et al., 'Large area, screen printed Silicon solar cells with selective emitter made by laser overdoping and RTA spin-on glasses', U. Besi-Vetrella et al., 26th IEEE PVSC, Anaheim, CA, 1997.

Brady, G. S.; Clauser, H. R.; Vaccari, J. A., *Materials handbook 15th Edition*, Ed. Mc Graw Hill, pg 1106, 2002

Bultman, J. H. et al., 'Single step selective emitter using diffusion barriers', *Proceedings of the 16th EPVSEC*, Glasgow, 2000.

Bultman, J. H. et al., 'Ideal single diffusion step selective emitters: a comparison between theory and practice', *Proceedings of the 17th EPVSEC*, Munich, 2001.

Caballero, F. J., *Patent pending concept*, 'Structure of selective emitter in solar cell of easy alignment between patterns for its industrial mass production, with possible local phosphorus gettering process integration and able for in-line processing', 2009.

Caballero, L.J.; Sánchez-Friera, P.; Lalaguna, B.; Alonso, J.; Vázquez, M.A., 'Series Resistance Modelling Of Industrial Screen-Printed Monocrystalline Silicon Solar Cells And Modules Including The Effect Of Spot Soldering', *Proceedings of the 4th WCPVSC*, Hawaii, 2006.

Erath, D.; Filipovic, A.; Retzlaff, M.; Goetz, A. K.; Clement, F.; Biro, D.; Preu, R., 'Advanced screen printing technique for high definition front side metallization of crystalline silicon solar cells', *Solar Energy Materials & Solar Cells*,V5,Issue 18, 2009.

Firor, K.; Hogan, S., 'Effects of processing parameters on thick film inks used for solar cell front metallization', *Solar Cells*, vol. 5, Dec. 1981, p. 87-100.

Glunz, S. W. et al., 'High-Efficiency Crystalline Silicon Solar Cells', Advances in OptoElectronics, Volume 2007, Article ID 97370, Hindawi Publishing Corporation, 2007

Glunz, S. W. et al., 'Progress in advanced metallization technology at Fraunhofer ISE', *Proceedings of the 33rd IEEE PVSC*, San Diego, 2008.

Green , M. A., *'Silicon solar cells, Advanced principles & Practice'*, Centre for photovoltaic devices and systems UNSW, Sydney, 1995.

Haigh, A. D. (Ferranti Ltd.), 'Developments in Polycrystalline Silicon Solar Cells And A Novel Form Of Printed Contact'. *Proceedings of the International Conference on Solar Electricity*. France, 1976.

Haverkamp, H. et al., 'Minimizing the electrical losses on the front side: development of a selective emitter process from a single diffusion', *Proceedings of the 33rd IEEE PVSC*, San Diego, CA, 2008.

Hilali, M. et al., 'Optimization of self-doping Ag Paste Firing to achieve high Fill Factors on screen-printed Silicon solar cells with 100Ω/sq. Emitter', *Proceedings of the 29th IEEE PVSC*, New Orleans, 2002.

Holmes, P. J.; Loasby, R. G., *Handbook of Thick Film Technology*, Electrochemical Publications Limited, 1976.

Horzel, J. et al., A, 'Novel method to form selective emitters in one diffusion step without etching or masking', *Proceedings of the 14th EPVSEC*, Barcelona, 1997.

Horzel, J. et al., B, 'A simple processing sequence for selective emitters', *Proceedings of the 26th IEEE PVSC*, Anaheim, CA, 1997.

Hörteis, M.; Richter, P. L.; Glunz, S. W., 'Improved front side metallization by aerosol jet printing of Hotmelt inks', *Proceedings of the 23rd EPVSEC*, Valencia, 2008.

Huster,F., Aluminium-Back Surface Field: Bow investigation and Elimination, *Proceedings of the 20th EPVSEC*, Barcelona, pp. 635-638, 2005.

Kumaravelu, G.; Alkaisi, M. M.; Bittar, A., 'Surface texturing for silicon solar cells using reactive ion etching technique', *Proceedings of the 29th IEEE Photovoltaic specialists conference*, New Orleans, Louisiana, USA, 2002.

Lide, D. R., *Handbook of Chemistry and Physics*, CRC Press 1974.

Luque A. et al. *Solar Cells and Optics for Photovoltaic Concentration*, chapter 4, Adam Hilger series on optics and optoelectronics, 1989.

Mills, R. N.; Branning, P., 'Inkjet systems for use in photovoltaic production', *Photovoltaic World*, September 2009.

Nilsen , D.; Stensrud, E. and Holt , A., Double layer anti-reflective coating for silicon solar cells, Proceedings of the 31st IEEE PVSC, Orlando, Florida, USA, pp.1237-1240, 2005.

Porter L. M.. et al., 'Phosphorus-Doped, Silver-Based Pastes for Self-Doping Ohmic Contacts for Crystalline Silicon Solar Cells', *Sol. Energ. Mat. and Sol. Cells* 73 (2), 209-219, 2002.

Pysch, D.; Mette, A.; Filipovic, A.; Glunz, S. W., 'Comprehensive analysis of advanced solar cell contacts consisting of printed fine-line seed layers thickened by silver plating', *Progress in Photovoltaics*, V17, Issue 2, pp. 101-114, 2008.

Raabe, B. et al. 'Monocrystalline Silicion-Future Cell Concepts', *Proceedings of the 22nd EPVSEC*, Milan, 2007.

Ralph, E. L. (Spectrolab Inc.), 'Recent advancements in low-cost solar cell processing', *Proceedings of the 11th IEEE PVSC*, 1975.

Recart, F.,*Evaluación de la serigrafía como técnica de metalización para células solares eficientes´*,.PhD Thesis, Universidad del Pais Vasco, 2001.

Roberts, S.; Heasman, K. C.; Bruton T. M., 'The reduction of module power losses by optimisation of the tabbing ribbon', *Proceedings of the 16th EUPVSEC*, Glasgow, 2000.

Rohatgi, A. et al., 'Self-aligned self-doping selective emitter for screen-printed silicon solar cells', *Proceedings of the 17th EPVSEC*, Munich, Germany, 2001.

Ruby, D. S. et al., 'Recent progress on the self-aligned, selective-emitter Silicon solar cell', *Proceedings of the 26th IEEE PVSC*, Anaheim, CA, 1997.

Salami, J. et al., 'Characterization of screen printed phosphorus diffusion paste for Silicon solar cells', *Proceedings of the PVSEC-14*, Bangkok, Thailand, 2004.

Shin, H.; Lee, H.; Sung, J. and Lee, M., 'Parallel laser printing of nanoparticulate silver thin film patterns for electronics', *Appl. Phys. Lett.* 92, 233107 (2008).

Schneiderlöchner, E.; Preu, R.; Lüdemann, R.; Glunz, S. W., Laser-fired rear contacts for crystalline silicon solar cells, *Progress in Photovoltaics*, V. 10 Issue 1, Pages 29 - 34, 2002

Tucci M.; Talgorn, E.; Serenelli L.; Salza, E.; Izzi, M. and Mangiapane, P., 'Laser fired back contact for silicon solar cells', Thin Solid Films, Volume 516, Issue 20, pp. 6767-6770, 2008.

Van der Heide A. S. H., Bultman J. H., Hoornstra J., et al. 'Locating losses due to contact resistance, shunts and recombination by potential mapping with the Corescan'.

Proceedings of the 12th NREL Workshop on Crystalline Silicon Solar Cells, Materials and Processes, Breckenrige (CO), USA, A, 2002

Van der Heide A. S. H., Bultman J. H., Hoornstra J., et al., 'Optimizing the front side metallization process using corescan', *Proceedings of the 29th IEEE PVSC*, New Orleans, USA, B, 2002.

Williams, T.; McVicker, K.; Shaikh, A.; Koval, T.; Shea, S.; Kinsey, B. and Hetzer, D., 'Hot Melt Ink Technology for Crystalline Silicon Solar Cells', 29th IEEE PVSC, New Orleans, 2002.

Wenham, S. R. and Green, M.A. Australian Patent 570309, 1993.

Zerga, A. et al., 'Selective emitter formation for large-scale industrially MC-Si solar cells by hydrogen plasma and wet etching', *Proceedings of the 21st EPVSEC*, Dresden, 2006.

Zhao, J.; Wang, A.; Campbell, P.; and Green, M. A., 'A 19.8% Efficient Honeycomb Multicrystalline Silicon Solar Cell with Improved Light Trapping', *IEEE Transactions On Electron Devices*, Vol. 46, No. 10, pp 1978-1983, 1999.

17

Aerostat for Solar Power Generation

G. S. Aglietti, S. Redi, A. R. Tatnall, T. Markvart and S.J.I. Walker
University of Southampton
United Kingdom

1. Introduction

One of the major issues in the use of ground based photovoltaic (PV) panels for the large scale collection of solar energy is the relatively low energy density. As a result a large area is required on the ground to achieve a significant production. This issue is compounded by the fact that the power output of the devices is strongly dependent on the latitude and weather conditions. At high latitudes the sun is relatively low on the horizon and a large part of the solar energy is absorbed by the atmosphere. Countries situated at high latitudes, with climates such as the UK, are therefore challenged in their exploitation of solar energy as the average number of Peak Solar Hours (PSH - numerically equal to the daily solar irradiation in kWh/m^2) is relatively low. In Europe, typical annual average PSH values for horizontal surfaces range from about 2.5 h in northern England to 4.85 h in southern Spain (Markvart & Castañer, 2003). As, roughly speaking, the cost of the energy produced is inversely proportional to the average PSH, northern European countries are at a considerable economical disadvantage in the exploitation of solar energy with respect to other regions. On the other hand, areas with high ground solar irradiations (e.g. African deserts, see Kurokawa, 2004) are remote from most users and the losses over thousands of miles of cables and the political issues entailed in such a large project, severely reduce the economic advantages.

A different approach to address most of the shortcomings of ground based solar energy production was proposed by Glaser et al., 1974 and his idea has captured the imagination of scientists up to this day. The basic concept was to collect solar energy using a large satellite orbiting the Earth. This satellite would be capable of capturing the full strength of the solar radiation continuously and transmit it to the ground using microwave radiation. The receiving station would then convert the microwave radiation into electric energy for widespread use.

The original concept was revisited in the late 90's (Mankins, 1997) in view of the considerable technological advances made since the 70's and research work on this concept is still ongoing. However a mixture of technical issues (such as the losses in the energy conversions and transmission), safety concerns (regarding the microwave beam linking the satellite with the ground station) and cost have denied the practical implementation of this concept. The latter is a substantial hurdle as the development of Satellite Solar Power (SSP) cannot be carried out incrementally, in order to recover part of the initial cost during the development and use it to fund the following steps, but it requires substantial funding upfront (tens of billions of dollars according to Mankins, 1997) before there is any economical return.

As a compromise between Glaser's (SSP) and ground based PV devices it is possible to collect the solar energy using a high altitude aerostatic platform (Aglietti et al., 2008a, b). This approach allows most of the weather related issues, except for very extreme weather conditions, to be overcome as the platform will be above the cloud layer. As the platform is also above the densest part of the troposphere, the direct beam component from the sun will travel through considerably less air mass than if it was on the ground (in particular for early morning and evening) and this will further improve the energy output. Therefore this method enables considerably more solar power to be collected when compared to an equivalent ground based system. In addition, the mooring line of the platform can be used to transmit the electric energy to the ground in relative safety and with low electrical losses. Although this approach would capture between 1/3 and 1/2 of the energy that could be harvested using a SSP, the cost of the infrastructure is orders of magnitude lower, and this approach allows an incremental development with a cost to first power that is a few orders of magnitudes smaller than that necessary for SSP.

Most researchers up till now have proposed harvesting energy at high altitude by exploiting the strong winds existing in the high atmosphere such as the jet streams (Roberts et al., 2007). This would be achieved using Flying Electrical Generators, that are essentially wind turbines collecting wind power at altitudes from few hundred meters (www.magenn.com) to over 10 km.

The extraction of this energy using the type of machines proposed by Roberts et al. 2007, although feasible and most probably economically viable, is relatively complex in mechanical terms. One of the issues is that in low wind the machine (that is heavier than air) needs to reverse its energy flow and take energy from the ground to produce enough lift to support itself and the tether. Alternative designs like the MAGENN (www.magenn.com) overcome this problem using a lighter-than-air approach so that the buoyancy keeps it in flight all the time. However the mechanical complications are still considerable.

The exploitation of solar energy at high altitude may therefore be simpler in engineering/mechanical terms, and provide a very predictable/reliable source. One of the crucial steps to demonstrate the viability of the concept is a reliable calculation of the solar energy available as a function of the altitude. After a brief introduction on aerostatic platforms, the energy available at different altitudes is investigated. The concept of the Aerostat for Solar Power Generation (ASPG) is then described together with the equations that link its main engineering parameters/variables, and a preliminary sizing of an ASPG, based on realistic values of the input engineering parameters is presented.

2. Aerostatic platforms

Lighter-than-air craft (aerostats) have been progressively neglected by the main stream research in Aerospace Engineering during the second half of the past century after having made remarkable technological progress that culminated in the 1930's with the construction of over 200m long airships (Dick & Robinson, 1992, Robinson, 1973). There have been some developments of historical interest (Kirschner, 1986) but little of significance.

However, in the last few years, aerostats have attracted a renewed interest. Their typical market niches (scientific ballooning, surveillance/reconnaissance (Colozza & Dolce, 2005)) are expanding and more researchers have proposed several different applications, ranging from high altitude aerostats as astronomical platforms (Bely & Ashford, 1995) to infrastructures for communication systems (Badesha, 2002).

Amongst the most recent achievements in scientific ballooning are the Ultra-High Altitude Balloon (UHAB) developed for NASA (launched in 2002 with a volume of nearly 1.7 million cubic metres and reached an altitude of 49 km) and the ultra-thin film high altitude balloon constructed by the Institute of Space and Astronautical Science (ISAS) of Japan, which successfully carried a 10 kg payload to a world-record altitude of 53 km.

Tethered aerostats are limited to lower altitudes due to the weight of the tether, which increase linearly with height. Commercial aerostats fly up to 8km but various studies have been conducted to prove that considerably greater altitudes can be reached. For example the Johns Hopkins University Applied Physics Laboratory (JHU/APL) has conducted a successful feasibility study (although not experimentally demonstrated) on a high altitude (20 km) tethered balloon-based space-to-ground optical communication system (Badesha, 2002). The US Airforce has made extensive use of aerostats as a surveillance system, and there are aerostats available on the market like the Puma Tethered Aerostat (www.rosaerosystems.pbo.ru) or the TCOM's 71M (www.tcomlp.com) that can fly up to approximately 5 km tethered with payloads of 2250 kg and 1600 kg respectively. These aerostats have a mooring cable (i.e. their tether) that supplies the aerostat onboard systems and the payload with electric power, and they are designed to be able to withstand lightning strikes and strong winds.

Concerning the size, today's airships are considerably smaller than those constructed in the 1930's and that is mainly due to the economics of their typical functions. However, from a technical point of view, the state of the art in the relevant technologies would allow the construction of aerostats much larger than those currently in operation.

The possibility of using solar power as source of energy for the airship propulsion and/or to supply energy to on board systems has been investigated by Khoury and Gillett, 2004. The "sunship" that he proposed was a very simple and conventional envelop design, filled with helium, with thin film solar arrays covering appropriate areas of the external surface. The electrical power produced by the cells was then used for the propulsion and on board electrical system, with part of the energy stored in suitable units with high storage energy to weigh ratio. Notwithstanding the quality of the case made by the author, the "sunship" was never built.

However changes in the economy, driven by politics and/or technical factors (limitation of resources or scientific advances) transform the markets and the viability of certain technologies may change as a result. A typical case is that of wind turbines, whose technology has been available for decades, but only in the last few years have become a viable method to produce large quantities of electric energy.

3. High altitude solar radiation

The first step in the development of the ASPG concept is to evaluate how much solar energy is available as a function of altitude. This allows a direct assessment of the potential of the ASPG when compared to an equivalent ground based system.

This section presents a set of calculations to enable this comparison and considers the possible influence of cloud layers at different heights above the ground (up to 12 km). The results obtained are based on existing models for ideal clear sky conditions and are integrated with experimental data acquired by scientific instrumentation at a specific site in the south of the United Kingdom. Although this makes the analysis very location specific, the general conclusions about the potential of high altitude solar collectors can be extended to other countries at similar latitudes and with similar climatic conditions. Moreover it must be noted that the assumptions made in the calculation process are quite conservative in order to avoid a possible overestimate of the potential of the ASPG concept.

The attenuation that a solar beam experiences as it travels through the atmosphere is called extinction and it is mainly due to two different kinds of processes, absorption and scattering, the former being the conversion of photon energy into thermal energy and the latter involving the deflection of the photons after the interaction with atmospheric molecules or larger particles suspended in the air. These two processes cause the radiation falling on a surface to be divided into two components: direct (or beam) and diffuse. The contribution of the diffuse radiation becomes proportionally more important when the collector is located at lower altitudes, in particular under cloudy sky conditions.

Since several models that describe the characteristics of the clear atmosphere at various altitudes have been developed in the past, existing publications can be considered when dealing with these conditions. Some of these models have also been embedded in specific software (such as SMARTS (Donovan & Van Lammeren, 2001) hosted by the National Renewable Energy Laboratory), which are widely used now as design tools in the PV industry. For the present purposes the sky is considered clear above an altitude of 12 km, since it is rare to find clouds at higher layers of the atmosphere and the impact on the final results is assumed to be minimal. Therefore, the clear sky radiation falling on a sun tracking surface at 12 km can be directly determined with the use of SMARTS (Gueymard, 1995) using the site location and time of year as inputs. For the initial calculations only the beam contribution of the solar radiation is included in the analysis.

The influence of clouds is incorporated into the calculation process using the extinction parameter, determined from experimental data. The data used was acquired by radar/lidar systems (www.cloud-net.org) at a station located at Chilbolton Observatory (51.1445 N, 1.4370 W) in the South of the UK. These experimental measurements have been elaborated (see Redi 2009) to provide the extinction parameter profile in actual sky conditions, which relates the attenuation of the solar radiation to its path through the atmosphere, considering the possible presence of cloud layers. The observations were performed almost everyday of the year (from April 2003 to September 2004), 24 hours a day, in the height range between 0 and 12 km. Averaging the data for the different layers of the atmosphere and in different months of the year, it is possible to obtain the extinction parameter profile at various altitudes above the ground. As an example the values obtained for the month of March is presented in Fig. 1.

Having determined the extinction parameter the Lambert-Beers' attenuation law (Liou, 2002) can be used to calculate the loss of intensity of a solar beam as it travels through the atmosphere. By dividing the atmospheric path along the vertical in segment of length Δh_i and defining the extinction parameter for each segment α_i (m^{-1}), the variation of intensity I (the irradiance in W/m^2) can be expressed as:

$$I(\vartheta_Z) = I_{12km} \exp\left(-AM_{REL}(\vartheta_Z) \cdot \sum(\alpha_i \cdot \Delta h_i) \right) \tag{1}$$

where I_{12km} is defined as the solar irradiance at 12 km estimated with SMARTS and $AM_{REL}(\vartheta_Z)$ is the relative air mass, which describes the path length relative to that at the Zenith and it is therefore a function of the solar Zenith angle ϑ_Z.

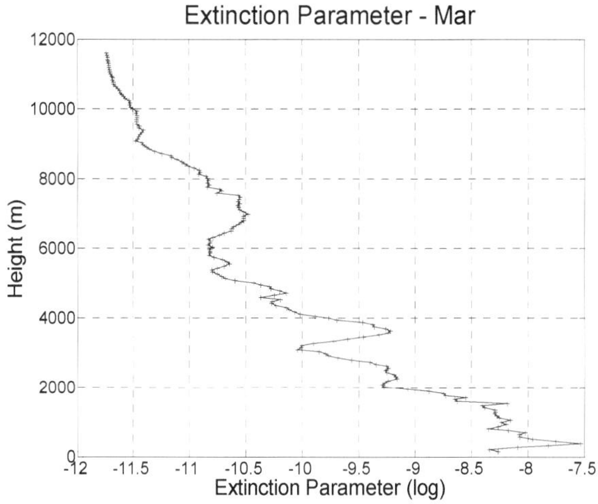

Fig. 1. Daily Mean Extinction (log) for the month of March - Chilbolton Observatory (51.1445 N, 1.4370 W)

Models like MODTRAN (Berk et al., 1989) or LibRadtran (Mayer & Kylling, 2005) that are able to integrate information about the cloud structure with the clear sky data, could be considered as a more accurate alternative to the one proposed here. However these models are quite sophisticated and they are generally more oriented towards atmospheric physics studies rather than engineering ones.

Starting from the irradiance values obtained with SMARTS at 12 km, Eq. 1 is applied to the extinction parameter values in actual sky for different months, in order to get an estimate of the irradiance below an altitude of 12 km in atmospheric conditions including possible clouds. The results obtained for the month of March at 12 km, 6 km and on the ground are presented in Fig. 2.

As a final step, it is necessary to integrate the irradiance values during the day (from sunrise to sunset) to calculate the total beam energy (beam irradiation E_B) falling on the high altitude solar collector:

$$E_B(h) = \int_{SR}^{SS} I_B(h)dt \tag{2}$$

where *SS* and *SR* are the time of Sunset and Sunrise.

The total beam energy can now be used to evaluate the potential of the ASPG system when compared to an equivalent ground based PV array.

Considering an altitude of 6 km, and integrating graphs like the ones shown in Fig. 3 gives value of about 3600 kWh/m^2 for the energy reaching the sun tracking platform in one year. It must be noticed that due to various conservative assumptions in the calculations this shoud be a rather conservative estimate. A general overview of the beam energy that can be collected by a platform located at an altitude up to 12 km is provided in Table 1.

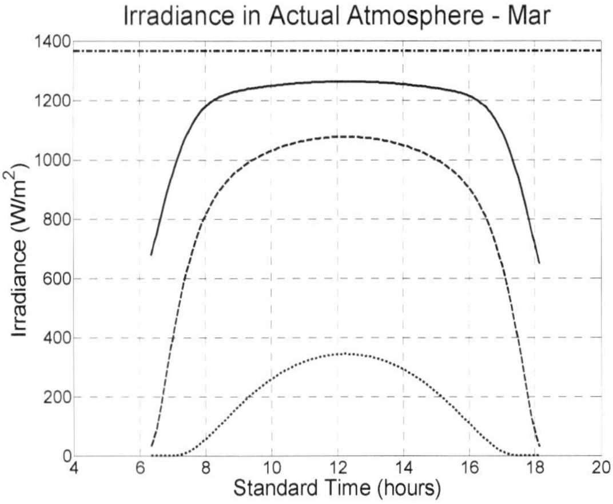

Fig. 2. Irradiance at 12 km (solid line),and at 6 km (dashed line) and on the ground (dotted line) including possible clouds, solar constant (dash-dotted line) - Chilbolton Observatory (51.1445 N, 1.4370 W)

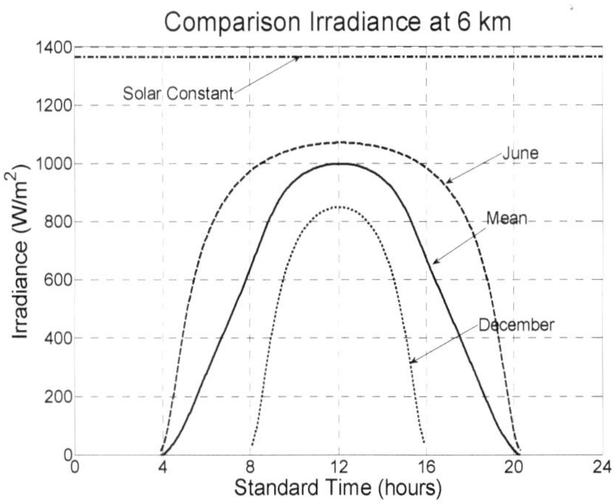

Fig. 3. Comparison between irradiance at 6km for different monthly means - Chilbolton Observatory (51.1445 N, 1.4370 W)

Altitude [km]	Total Year Beam Irradiation (including possible clouds/) [kWh/m^2]	Total Year Beam Irradiation (clear sky) [kWh/m^2]
6	3600	4530
9	4710	4800
12	5310	5310

Table 1. Year Beam Irradiation at different altitudes - Chilbolton Observatory (51.1445 N, 1.4370 W)

The study has been focused so far on the evaluation of the beam (direct) component of the solar radiation that can reach the surface. However, the contribution of the diffuse part of the radiation can be not negligible. Having considered this, the contribution of the diffuse radiation is estimated at different altitudes, with the use of SMARTS in ideal clear sky conditions. This assumption is expected to be extremely conservative since clouds can increase significantly the amount of this component which can become therefore higher particularly if actual sky conditions are considered.
The ratio between diffuse and global irradiation is estimated to range from about 5% (12 km) to 6% (6 km). This contribution is then summed to the values presented in Table 1, leading to the determination of the total radiation reaching a sun pointing surface located at an altitude up to 12 km which is presented in Table 2. Here, the comparison between the results obtained and the typical irradiation value expected for a PV array based on the ground (facing south and tilted at a fixed angle close to the latitude of its location) is given. The value of the energy collected by the solar power satellite (Glaser et al., 1974) is also included as a limit solution.
In addition to the contribution of the diffuse component of the solar radiation, some considerations about the albedo flux should be made. Considering that the albedo radiation factor can reach a value of 10 % for a satellite in Low Earth Orbit (Jackson, 1996), this component can significantly contribute to the total radiation estimate. In our specific case it is difficult during this preliminary analysis to give an estimate of this component with a sufficient degree of confidence. It must be noticed though, as a caveat, that the values presented in Table 2 are conservative and they are expected to increase when the albedo is included, especially in presence of cloud layers below the platform when the sun is at high zenith angles.

Altitude [km]	Year Global Irradiation including clouds (conservative estimate) [kWh/m^2]	Year Global Irradiation (clear sky) [kWh/m^2]
Ground Based (Chilbolton)	1150	
6	3830	4819
9	4985	5080
12	5590	5590
Solar Power Satellite	12000	

Table 2. Comparison between the year global irradiation (including diffuse) values at different altitudes for Chilbolton Observatory (51.1445 N, 1.4370 W)

The results obtained can give a preliminary idea of the gain that a high altitude solar generator could bring in terms of energy collected, if compared to the same generator located on the ground.
Due to the very conservative assumptions the values in the first column can be considered as "minimum" values, and the one in the last column as maximum, although the inclusion of the albedo should give results even higher.
Moreover a collector placed at an altitude of 12km could collect around 45% of what could be collected by the same PV system in a geostationary position (i.e. Solar Power Satellite).
The study presented is preliminary and it involves several assumptions that have been made to simplify the analysis and provide useful results to support the following phases of

the project. In particular the extinction parameter data used are acquired in a defined and limited time period and they are relative to a precise location in the south of the UK, which is relatively well placed to collect solar energy on the ground. For this reason and the other discussed previously, the figures presented are conservative and the advantage with respect to ground installations is expected to increase as installations at northern latitude are considered and the values concerning the estimate of the diffuse and albedo contributions are revised.

4. High altitude winds

The knowledge of the mean wind speed at a certain altitude (and its statistical properties) is essential to calculate the aerodynamic forces acting on the aerostat and in particular to determine the forces along the mooring cable.

The wind speed data described in this section were provided by the Natural Environment Research Council (NERC), from the Mesosphere-Stratosphere-Troposphere (MST) Radar station located at Capel Dewi (52.42°N, 4.01°W), near Aberystwyth in west Wales (UK). This facility can provide vertical and horizontal wind speed data, covering an altitude range from 2 to 20 km, with 300 m resolution. However for this study only the data up to 10km were processed. The particular set of data described here covers the period January-December 2007 and measurements were acquired every day continuously. The radar is located near the coast, where the wind speed is expected to be slightly higher than inland and therefore the estimate should be conservative.

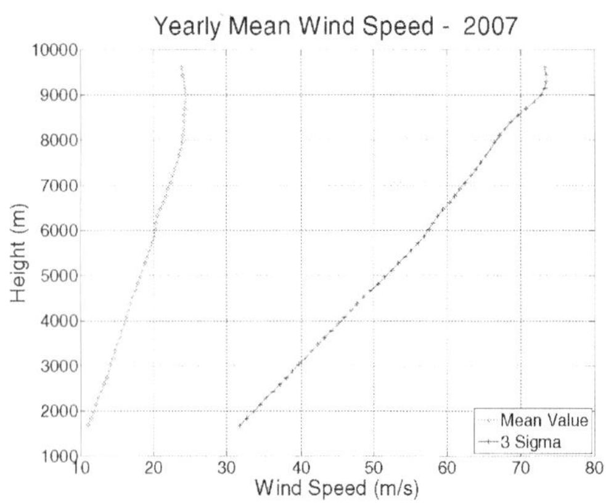

Fig. 4. Wind speed variation with altitude, year mean and 3 sigma values shown

Fig. 4 shows the mean wind speed and the 3 sigma value, and it is possible to notice that at 6km altitude these values are 20m/s and 55m/s respectively.

Another factor to be considered is that, in the time domain the wind speed is quite variable with relatively rapid transients (i.e. gusts). As discussed in Aglietti, 2009 the tethered aerostat is a non-linear system, with considerable damping, and therefore rapid transients of the input in reality produce a response that is significantly lower than what is calculated simply using the maximum wind speed in a static analysis.

5. Concept design

5.1 Tethered aerostat

For an aerostat to support PV devices, this has to be able to produce enough lift via its buoyancy to overcome its weight, the weight of the solar cells plus any control system and that of the tether, still leaving enough margin to produce an appropriate tension in the tether to avoid excessive sag.

Neglecting any aerodynamic lift which could be generated by the shape of the Aerostat, the lifting force due to the aerostat buoyancy is:

$$B = (\rho_{air} - \rho_{gas})gVol \tag{3}$$

where *Vol* is the volume of the aerostat and ρ_{air} and ρ_{gas} the densities of air and gas (helium or hydrogen can be used) filling the aerostat envelop at the specific conditions of operations (e.g. pressure, altitude), and g is the gravity acceleration (9.81m/s^2). Here it is assumed that there is a negligible pressure differential between inside and outside the aerostat envelope, and for simplicity is also taken that the whole volume of the envelope is occupied by the gas (i.e. ballonets for altitude control completely empty).

Fig. 5. Schematic configuration of an Aerostat for Electrical Power Generation, as a gimballed tethered balloon - the grey area represents PV cells cladding.

Typically aerostats have streamlined bodies to reduce the aerodynamic drag, however when such shaped aerostats are moored they then tend to rotate to the oncoming flow direction like a weathervane. Here the aerostat is required to maintain its orientation towards the Sun, therefore a spherical shape pointed through a system of gimbals seems more appropriate (see Fig. 5). A spherical aerostat generates more aerodynamic drag and clearly would require a more substantial structure and tether, but these issues can be tackled by its structural design (Miller & Nahon, 2007). A tethered sphere also suffers substantial vortex induced vibrations (Williamson & Govardhan, 1997). However a previous study (Aglietti, 2009) has shown that due to the non-linearity of the structural problem (mainly the sag of the tether) and the very slow frequency response characterized by a high value of damping, the force oscillations in the tether line (produced by relatively rapid force transients, e.g. gusts) are relatively small. The resultant rotations of the aerostat are only a few degrees which in turn produces a drop in the energy production of less than 1%.

Given its spherical shape, from the volume it is possible to calculate the surface area, and from this, taking an appropriate material area density, it is possible to estimate the weight of the envelope. The area density of the material for the skin can then be increased by 33% as suggested in (Khoury & Gillett, 2004), to account for the weight of various reinforcements, support for the payload etc.

With a similar approach the weight of the PV cells can be estimated by the surface covered, and assuming that a fraction γ of the whole aerostat envelope is covered by the cells, knowing the area density of the cells (also here including wiring etc), it is possible to estimate the weight of the cells. Therefore the weight of aerostat and PV devices can be written as:

$$W_{Aero} = (1.33\delta_{aero} + \delta_{cells}\gamma) g\, 4\pi R^2 \tag{4}$$

where δ_{aero} and δ_{cells} are the area density of the envelope material and PV cells respectively, γ is the fraction of the envelope surface that is covered by the PV cells, g is gravity acceleration and R is the radius of the balloon.

To assess the weight of the tether it is necessary to estimate the weight of the electrical conductors (taken as aluminum for this high conductivity over mass ratio) plus that of the strengthening fibers (e.g. some type of Kevlar). The size of the required conductor can be estimated from the electrical current (that is the ratio between the power generated by the PV devices on the aerostat and the transmission voltage) and setting the electrical losses permitted in the cable to a specific value. Therefore the cross section of the conductor will be

$$A_{cond} = r_{Al} \frac{S}{\eta_{trans}} \frac{P_{gen}}{V^2} \tag{5}$$

where r_{Al} is the resistivity of the aluminum, S is the overall length of the conductor, η_{trans} is the ratio between the power lost in the cable and that generated by the PV devices (that is P_{gen}), and V is the voltage.

The power generated by the PV system can be estimated from the area covered by the cells (that is a fraction γ of the whole aerostat surface), their efficiency (η_{cells}), an efficiency parameter (η_{area}) that considers that the cells are on a curved surface and therefore the angle of incidence of the sun beam varies according to the position of the cells and finally the solar flux Φ at the aerostat operational altitude that is the irradiance discussed in the previous sections:

$$P_{gen} = 4\pi R^2 \gamma \eta_{cells} \eta_{area} \Phi \tag{6}$$

Finally the weight of the conductor will be its cross sectional area multiplied by length and by its specific weight (density times g), so substituting equation (6) in (5) the weight of the conductor can be written as:

$$W_{cond} = \delta_{Al} g r_{Al} \frac{S^2}{\eta_{trans}} \frac{4\pi R^2 \gamma \eta_{cells} \eta_{area} \Phi}{V^2} \tag{7}$$

The weight of the reinforcing fibres can be calculated from the strength necessary to keep the aerostat safely moored.
The maximum tensile force on the tether can be calculated as:

$$T = \sqrt{(B - W_{Aero})^2 + D^2} \tag{8}$$

where D is the aerostat drag force, equal to:

$$D = \frac{1}{2} \rho_{air} v^2 C_d \pi R^2 \tag{9}$$

In the above expression v is the maximum wind velocity and C_d is the drag coefficient. From the maximum expected tension in the tether, knowing the fibres strength (σ_u) it is possible to calculate the required cross section and from that the weight of the reinforcing fibres.

$$W_{fib} = \delta_{fib} S_T \frac{T}{\sigma_u} \tag{10}$$

where δ_{fib} is the density of the fibers and S_T the length of the tether.
So that the overall weight of the tether will be:

$$W_{Tether} = \delta_{cond} g r_{cond} S^2 \frac{\eta_{cells} \eta_{area} A_{cells} \Phi}{\eta_{trans} V_{gen}^2} + \delta_{fib} S_T \frac{T}{\sigma_u} \tag{11}$$

Aerodynamic forces will also act on the tether line, and they will produce further sagging (see Aglietti, 2009). However this effect does not modify significantly the maximum tension in the tether that will still be at the attachment between the balloon and the tether (equation 8).

5.2 Engineering parameters

In the previous sections, the equations that govern the preliminary sizing of the aerostatic platform have been derived and these equations can be combined, for example, to design a facility with a specified power output. Overall, as the lift and the weight are proportional to the aerostat volume and surface respectively, it will always be possible to design an aerostat large enough to "fly". Here the volume as been set at 179,000 m^3 (that is a 35 m radius sphere) which gives a suitable ratio between lift and drag (using helium as a gas filler would give a buoyancy of 1 MN).

In order to reduce the interference with the aviation industry and international air traffic the maximum altitude will be set to 6 km.
The values of the specific engineering parameters which appear in the equations (like for example the area density of the skin) have a crucial role in defining the size of the aerostat. In this section realistic and sometimes conservative values for these parameters will be discussed and utilized in the equations to size a viable platform.
Starting with the solar cells, there are various types available on the market. These range from light weight amorphous silicon triple junction cells (with an efficiency of up to 7%) that could be directly integrated on the skin (see for example Amrani et al., 2007), with a mass penalty that could be as low as 25 g/m^2, to heavier but more efficient cells (e.g. Triple-Junction with Monolithic Diode High Efficiency Cells (www.emcore.com) efficiency 28%), which require some rigid backing and could be used with a mass penalty that can be in the region of 850 g/m^2. These types of cells could be mounted on light weight carbon fibre reinforced plastic tiles that would be used to clad part of the aerostat envelope. Although amorphous silicon cells seem more appropriate, judging by the efficiency over area density ratio, there are issues concerning the ease of installations, repairs, amount of surface available and finally costs that have to be considered. In this study, an efficiency of 15% and an overall area density of the PV cells δ_{cells} (including connectors etc.) of 1 kg/m^2 will be considered.
Taking a maximum peak solar irradiation of 1.2 kW/m^2, from the equations in the previous section it is possible to calculate W_{Aero} (18.9x10^3 kg) and the peak power generated (~0.5 MW). However to size the conductor in the tether it is necessary to set a transmission voltage V, and this should be high enough in order to reduce the losses in the cable. One option is to connect the solar arrays to obtain a voltage in the region of 500V DC and use a converter to bring it up to a few kV. However the converter will introduce some electrical losses and its weight might be an issue as it has to be supported by the aerostat (although the weight of the converter might be compensated by a lighter cable). The other option is to "simply" connect identical groups of solar arrays in series, to maintain the same current and obtain a DC voltage in the region of 1.5-3kV. The solar panels would be provided with bypass and blocking diodes and other circuitry that might be necessary to protect the elements of the system. Setting the transmission voltage at 3kV and allowing for 5% electrical losses in the cable (i.e. η_{trans} = 0.05) enables the cross section of the aluminium conductor and its weight to be calculated as 388 mm^2 and 13.0x10^3 kg respectively.
Using the results in the previous section and taking a maximum wind speed of 55 m/s (3 sigma value) and using equation 9 it is possible to calculate the weight of the fibres as 2.7x10^3 kg, so that the overall weight of the tether will be 15.7x10^3 kg. It should be stressed that the 55 m/s value is quite conservative, in fact this corresponds to the peak wind speed during a gust and due to the highly non linear behaviour of the tether system (see Aglietti, 2009) the force in the tether will be considerably smaller. On the other hand this level of conservatism is more than justified by the catastrophic effect that the tether rupture would have.

6. Conclusion

This chapter has investigated the possibility of using a high altitude aerostatic platform to support PV modules to increase substantially their output by virtue of the significantly enhanced solar radiation at the operating altitude of the aerostat.

Although the figures presented for the analysis of the radiation have been obtained for a specific set of data relative to a well defined location in the UK (and the calculations presented involve some approximations, justified by the preliminary character of the analysis). The results obtained illustrate the advantages, in terms of irradiation, of collecting solar energy between 6km and 12 km altitude, rather than on the ground. The general conclusions can be extended, with a certain degree of approximation, to other countries at the same latitude and with similar climates.

Based on realistic values for the relevant engineering parameters that describe the technical properties of the materials and subsystems, a static analysis of the aerostat in its deployed configuration has been carried out. The results of the computations, although of a preliminary nature, demonstrate that the concept is technically feasible.

As the AEPG requires minimum ground support and could be relatively easily deployed, there are several applications where these facilities could be advantageous respect to other renewables.

It is acknowledged that the concept mathematical model and its concept design are of a preliminary nature. However they do indicate that there is the potential for a new facility to enter the renewable energy market, and further work should be carried out to investigate this possibility more in depth.

7. References

Aglietti, G.S. (2009). Dynamic Response of a High Altitude Tethered Balloon System. AIAA Journal Of Aircraft, Vol. 46, No. 6, page 2032-2040, doi: 10.2514/1.43332, November-December 2009.

Aglietti, G.S.; Markvart, T.; Tatnall, A.R. & Walker S.J.I. (2008). Aerostat for Electrical Power Generation Concept feasibility. *Proceedings of the IMechE Part G; Journal of Aerospace Engineering*, Vol 222, pp 29-39, ISBN0954-4100; DOI: 10.1243/09544100JAERO258 - Feb 2008.

Aglietti, G.S.; Markvart, T.; Tatnall, A.R. & Walker S.J.I. (2008) Solar Power Generation Using High Altitude Platforms Feasibility and Viability. *Progress in Photovoltaics: Research and Applications*, Vol. 16 pp 349-359, 2008.

Badesha, S.S. (2002). SPARCL: A high-altitude tethered balloon-based optical space-to-ground communication system, *Proceedings of the SPIE - The International Society for Optical Engineering*, Volume 4821, 2002, ISSN: 0277-786X.

Bely, P., & Ashford, R. L. (1995). High-altitude aerostats as astronomical platforms, *Proceedings of SPIE - The International Society for Optical Engineering*, Volume 2478, 1995, Pages 101-116.

Berk, A.; Bernstein L.S. & Robertson, D.C. (1989). MODTRAN: A moderate resolution model for LOWTRAN 7, Report GL-TR-89-0122, Geophysics Laboratory, Air Force Systems Command, United States Air Force, Hanscom, AFB, MA 01731

Colozza, A. & Dolce, J.L. (2005). High-Altitude, Long-Endurance Airships for Coastal Surveillance, *NASA Technical Report*, NASA/TM-2005-213427, 2005.

Dick, H.G. & Robinson, D.H. (1992). *The Golden Age of the Great Passenger Airships: Graf Zeppelin and Hindenburg*, Prentice Hall & IBD; ISBN-13: 978-1560982197, Nov 1992.

Donovan, D. & Van Lammeren A. (2001). Cloud effective particle size and water content profile retrievals using combined lidar and radar observations 1. Theory and examples. *Journal of Geophysical Research*. 106, 27425-27448.

El Amrani, A.; Mahrane, A.; Moussa, F. Y. & Boukennous, Y. (2007). Solar Module Fabrication. *International Journal of Photoenergy*, Vol. 2007, Article ID 27610

Glaser, P.E.; Maynard, O.E.; Mackovciak, J. & Ralph, E.L., Arthur D. Little, Inc., Feasibility study of a satellite solar power station, NASA CR-2357, NTIS N74-17784, Feb. 1974

Gueymard, C.A. (1995). SMARTS, A Simple Model of the Atmospheric Radiative Transfer of Sunshine: Algorithms and Performance Assessment. Technical Report No. FSEC-PF-270-95. Cocoa, FL: Florida Solar Energy Center.

Jackson, B. (1996). A software power model for a spin-stabilized LEO spacecraft utilizing V/T charge control. *Proceedings of the Aerospace Applications Conference IEEE*. 3. 219-227.

Khoury G.A. & Gillett J.D., (2004). *Airship Technology*, Cambridge Aerospace Series, Edited by Gabriel A. ISBN-10: 0521607531 2004.

Kirschner, E.J. (1986). *Aerospace Balloons: From Montgolfiere to Space*, pp. 21, TAB Books, Inc., Blue Ridge Summit, Pennsylvania, 1986.

Kurokawa, K. (2004). *Energy from the Desert: Feasibility of Very Large Scale Photovoltaic Power Generation (VLS-PV) Systems*, Earthscan Publications Ltd, ISBN-10: 1902916417, May 2004

Liou, K.N. (2002). *An Introduction to Atmospheric Radiation*, second ed. Elsevier Science

Mankins, J.C. (1997) A Fresh Look at Space Solar Power: New Architectures, Concepts and Technologies, *Proceedings of the 38th International Astronautical Congress*, IAF paper no IAF-97-R.2.03, 1997.

Markvart, T. & Castañer, L. (2003). *"Practical Handbook of Photovoltaics: Fundamentals and Applications"*, Elsevier, Oxford.

Mayer B. & Kylling A. (2005). Technical note: The libRadtran software package for radiative transfer calculations - description and examples of use. Atmos. Chem. Phys. 5, 1855-1877.

Miller J.I. & Nahon M. (2007). Analysis and Design of Robust Helium Aerostats. *AIAA Journal of Aircraft*, Vol. 44, No. 5, pp. 1447-1458, 2007

Redi S.; Aglietti, G.S.; Tatnall A.R.; Markvart T. (2009) An Evaluation of a High Altitude Solar Radiation Platform *ASME-Journal of Solar Energy Engineering* - in print.

Roberts, B.W.; Shepard, D.H.; Caldeira, K.; Cannon, M.E.; Eccles, D.G.; Grenier, A.J. & Freidin, J.F. (2007). Harnessing High-Altitude Wind Power, *IEEE Transactions On Energy Conversion*, Vol. 22, No. 1, March 2007.

Robinson, D.H. (1973). *Giants in the Sky: A History of the Rigid Airship*, University of Washington Press ISBN-13: 978-0295952499, 1973.

Williamson, C.H.K. & Govardhan, R.G. (1997). Dynamics And Forcing Of A Tethered Sphere In A Fluid Flow, *Journal of Fluids and Structures* (1997) 11 , 293 - 305

http://rosaerosystems.pbo.ru/english/products/puma.html, accessed 25/09/09

http://www.cloud-net.org/, accessed 25/09/09

http://www.emcore.com/assets/photovoltaics/btjm.solar.ds.pdf, accessed 25/09/09

http://www.magenn.com/news.php, accessed 25/09/09

http://www.tcomlp.com/aerostats.html, accessed 25/09/09

18

Photon Management in Dye Sensitized Solar Cells

Silvia Colodrero, Mauricio E. Calvo and Hernán Míguez
Instituto de Ciencia de Materiales de Sevilla
Consejo Superior de Investigaciones Científicas-Universidad de Sevilla
Spain

1. Introduction

Solar energy is nowadays one of the most promising future energy resources due to the depletion of fossil fuels, which supply the major part of all energy consumed worldwide. Among the different types of solar cell technologies, dye sensitization of mesoporous oxide based films has attracted a great deal of interest in the last few years because of the possibility it offers to achieve moderate efficiency devices at very low cost, being therefore an interesting alternative to conventional p-n junction solar cells. Dye sensitized solar cells (DSSC) consist of a nanocrystalline wide band gap semiconductor (usually TiO_2), which is deposited onto a transparent conductive substrate, and on whose surface a dye is adsorbed. The cell is completed with a counterelectrode, and both electrodes are put into electrical contact by infiltrating a liquid electrolyte in between them. Light is absorbed by the dye and charges are separated at the interface between the dye and the metal oxide it is anchored to. The optimization of the conversion efficiency, that is the fraction of light intensity that is converted into electrical power, is a key issue for this type of solar systems. In this way, different modifications of the originally proposed cell have been made in order to improve its performance, most of them based on the use of different semiconductors, dyes or ionic conductor. There is also an increasing interest in employing nanostructures to improve solar energy conversion devices (Kamat, 2007).

Another interesting route to enhance the cell efficiency is to modify its optical design in order to improve the light harvesting efficiency (LHE) or absorptance within the cell. The approach that has been explored the most has been the use of a diffuse scattering layer made of large TiO_2 colloids that are either deposited onto the nanocrystalline electrode or mixed with the nanocrystalline titania (nc-TiO_2) slurry. In both cases, increase of the optical light path within the absorbing layer rises the matter-radiation interaction time thus enhancing the probability of photon absorption. It has to be bear in mind that any structure introduced in the cell must permit the electrical contact between the electrolyte and the sensitized semiconductor slab, which forces it to have a porosity capable of sustaining the flow of charges. In recent times, different alternatives to light management in DSSC are being proposed and realized due to the development of novel porous periodic photonic nanostructures that can be easily integrated in these devices. The aim of this chapter is to give a brief review of the efforts performed to improve the light harvesting in DSSCs through the optimization of their optical design and provide a detailed description of the new emerging possibilities based on coupling dye-sensitized electrodes to different periodic photonic nanostructures. It will be shown that they have a great potential for the manipulation of light propagation. A detailed description of the integration processes used, of the different mechanisms of light harvesting enhancement that take place, and of actual examples showing the improvement of performance achieved will be presented in this chapter.

2. Brief description of DSSC

Photovoltaic devices have become a promising alternative energy source in the last decades. They are expected to increasingly and significantly contribute to overall energy production over the coming years. The photovoltaic field, dominated mainly by inorganic solid-state junction cells, is now being challenged by the emergence of new devices based on nanocrystalline and conducting polymer films, which offer a very low-cost fabrication and attractive features such as transparency, flexibility, etc. that might facilitate the market entry. Among all of them, dye sensitized solar cells (DSSC) are devices that have shown to reach moderate efficiencies, thus being feasible competitors to conventional cells.

DSSC combine the optical absorption and charge-separation processes by the association of a sensitizer as light-absorbing material with a wide band-gap semiconductor (usually titanium dioxide). As early as the 1970s, it was found that titanium dioxide (TiO_2) from photoelectrochemical cells could split water with a small bias voltage when exposed to light (Fujishima & Honda, 1972). However, due to the large band-gap for TiO_2, which makes it transparent for visible light, the conversion efficiency was low when using the sun as illumination source. Dye sensitization of semiconductor electrodes dates to the 1960s (Gerischer & Tributsch, 1968). This pioneering research involved an absorption range extension of the system into the visible region, as well as the verification of the operating mechanism by injection of electrons from photoexcited dye molecules into the conduction band of the n-type semiconductor. Since only a monolayer of adsorbed dye molecules was photoactive, light absorption was low and limited when flat surfaces of the semiconductor electrode were employed. This inconvenience was solved by introducing polycrystalline TiO_2 (anatase) films with a surface roughness factor of several hundreds (Desilvestro et al., 1985; Vlachopoulos et al., 1988). The amount of adsorbed dye was increased even further by using mesoporous electrodes, providing a huge active surface area thereby, and cells combining such electrodes and a redox electrolyte based on iodide/triiodide couple yielded 7% conversion efficiencies in 1991 (O´Regan & Gratzel, 1991). The current highest energy conversion efficiency is over 11% (Chiba et al., 2006), and further increase of the efficiency is possible by designing proper electrodes and sensitization dyes.

Figure 1 shows both a scheme and an energy level diagram of a liquid electrolyte dye sensitized solar cell. They usually consist of one electrode made of a layer of a few micrometers of titanium dioxide nanocrystals (average crystal size around 20 nm), that have been sintered together to allow electronic conduction to take place. A monolayer of a sensitizer dye, typically a ruthenium polypyridyl complex, is attached to the surface of the nanocrystalline electrode. This mesoporous film is deposited onto a conductive, transparent substrate, typically indium tin oxide (ITO) or fluorinated SnO_2 (FTO), and soaked with a redox electrolyte, typically containing I^-/I_3^- ion pairs. This electrolyte is also in contact with a colloidal platinum catalyst coated counter-electrode. Sunlight is harvested by the dye producing photo-excited electrons that are injected into the conduction band of the nanocrystalline semiconductor network, and then into the conducting substrate. At the same time the redox electrolyte reduces the oxidized dye and transports the electron acceptors species (I_3^-) to the counter-electrode, where the I_3^- is reduced back to I^- and the electrical circuit is completed via electron migration through the external load.

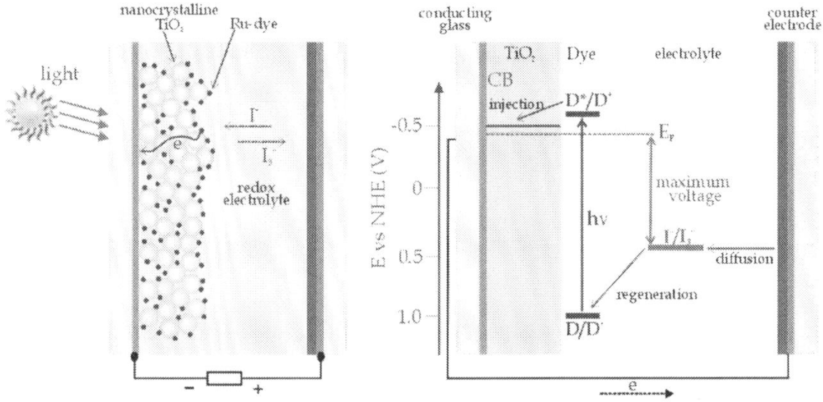

Fig. 1. Cross section view of the design of a dye sensitized solar cell under illumination conditions (left), and energy levels of the different components of the cell that represent the energetics of operation of such devices (right).

In contrast to silicon devices, charge separation is primarily driven by the oxidation/reduction potentials of the different species at the TiO_2/dye/electrolyte interface, being screened out any electric field gradient in the TiO_2 electrode due to the high concentration of mobile ions employed in the liquid electrolyte (Zaban et al., **1997)**. Photoinduced charge separation takes place at the TiO_2/electrolyte interface. Thus, electron injection requires the dye excited state to be more reducing than the TiO_2 conduction band. In the same way, regeneration of the dye ground state by the redox couple requires the dye cation to be more oxidizing than the I^-/I_3^- redox couple (Mori & Yanagida, 2006). The voltage output of the device is approximately given by the splitting between the TiO_2 Fermi level and the chemical potential of the redox electrolyte, being the former related with the density of injected electrons and the density of charge traps in the band gap of TiO_2. Under illumination conditions, the density of electrons injected into the semiconductor conduction band increases, raising the Fermi level towards the conduction-band edge and generating a photovoltage in the external circuit.

Charge transport processes within the cell are considered to be diffusive (Södergren et al., 1994), (Cao et al., 1996), (Schwarzburg & Willig, 1999) and are driven by concentration gradients generated in the device, thus making electrons to go towards the working electrode and triiodide ions towards the counter electrode. During the diffusion process, photogenerated electrons can recombine with acceptors species, such us dye cations and triiodide ions. Another loss pathway includes decay of the dye excited state to ground (Huang et al., 1997), (Nelson et al., 2001). Kinetic competitions between the different forward and loss pathways are therefore critical to determine the quantum efficiencies of charge separation and collection, and so the conversion efficiency. A diagram showing the kinetics of a DSSC is presented in Figure 2. It should be noticed that not only energetics but also kinetics must be taken into account, and they constitute the key issues to achieve high energy conversion devices.

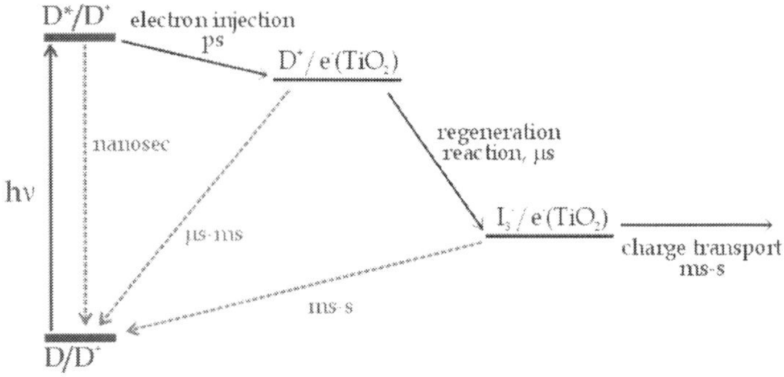

Fig. 2. Kinetics and competitive processes involved in the conversion of light to electric power in DSSC.

The overall conversion efficiency (η) of the dye-sensitized solar cell is determined by the photocurrent density measured at short circuit (J_{SC}), the open-circuit photovoltage (V_{oc}), the fill factor (ff), and the power of the incident light (P_{in}). These values can be extracted from the photocurrent density-voltage characteristics (IV curves) under AM 1.5 full sunlight (P_{in}=100 mWcm^{-2}). The relation between J and V is determined by varying the resistance of the outer circuit, being J_{SC} obtained when the resistance of the outer circuit is zero (thus voltage is zero) and V_{oc} when the resistance is maximum (thus photocurrent is zero). The output power of the device equals the product of J and V, and the fill factor expresses the efficiency of the device compared to that of and ideal cell. P_{max} is commonly reported as the output power of the commercial device and corresponds to the maximum value that can reach the output power. The performance of DSSC can be therefore estimated using the following equations:

$$\eta(\%) = \frac{J_{SC} \cdot V_{OC} \cdot ff}{P_{in}} \tag{1}$$

Where,

$$ff = \frac{P_{max}}{J_{SC} \cdot V_{OC}} \tag{2}$$

A typical IV curve corresponding to a 7 μm thick dyed-TiO_2 electrode measured under 1 sun illumination is displayed in Figure 3 (left). The solar radiation and the ruthenium dye absorptance spectra are shown in Figure 3 (right).
For a detailed description of DSSCs, we refer the reader to M. Graetzel (Graetzel, 2000) and M. Graetzel and J. Durrant (Graetzel & Durrant, 2008).
In the next section, we analyze two different approaches that contribute to the enhancement of DSSC efficiencies through the control of photon absorption into the cell. We put special emphasis to describe the integration of new materials known as porous one-dimensional photonic crystal due to their ease of integration and demonstrated promising performance.

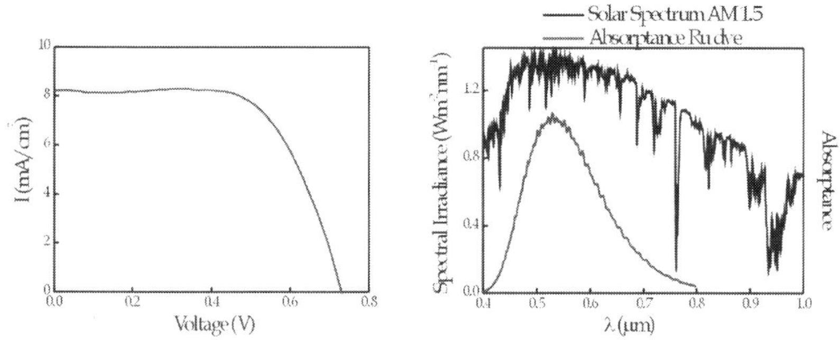

Fig. 3. IV curve for a 7 μm thick TiO_2 electrode, for which N719 as dye and a I^-/I_3^- redox couple based liquid electrolyte have been employed (left), and mismatch between the dye absorption spectrum and that for the AM 1.5 solar spectrum (right).

3. Approaches to light management in DSSC

In the last years, many attempts using different modifications of the originally proposed cell have been made in order to improve its performance, most of them based on the use of different semiconductors (Tennakone et al., 1999), dyes (Wang et al., 2005) or ionic conductors (Wang et al., 2004). However, the extremely delicate sensibility of the charge transport and recombination dynamics to any alteration of the nature of the interfaces present in the cell should be considered (Haque et al., 2005). For instance, some of the most important routes of research have focused on the molecular engineering of suitable dyes having broader absorption spectra that show a better matching to the solar spectrum and higher molar extinction coefficients (Wang et al., 2005), thus yielding higher short circuit currents. However, further improvements in terms of cell stability and durability should be done. On this respect, the performance of cells using solid state hole conductor based DSSC (Bach et al., 1998) to increase the long-term stability still remains far from that achievable when liquid organic electrolytes are employed.

The quantification of the electrical kinetic parameters of the cell has attracted the attention of many research groups and a great effort has been made in this direction. However, less interest has been paid to the study and development of optical elements that could be introduced in the cell for boosting the optical path of light, thus increasing the probability for the photons to be absorbed. Although it was well-known that by optical means the output power of the cell can be enhanced through a higher photogenerated current, since it depends on the number of photons collected by the dye (Tachibana et al., 2002), it was also clear that the introduction of optical elements that can enhance light harvesting in DSSC was not straightforward. First, they are typically made of dense materials, which would block the flow of charged species in solution. Second, the standard fabrication and integration processes usually employed to make optical materials did not seem to be compatible with the colloidal chemistry approaches normally taken to prepare a DSSC. The first and most successful approach to light management in DSSC was based on the introduction of a diffuse scattering layer, as described below, which largely enhances the photon path length through the working electrode, thus increasing the probability of optical absorption to take place. More recently, approaches based on periodic structures are also being investigated with promising results. Apart from the large enhancements of efficiency these latter structures gives rise to, they present the added advantage of allowing for the precise selection of the spectral range at which optical absorption is amplified, leading to both control over the aspect and the semi-transparency of the cell.

3.1 Effect of increase optical absorptance on the efficiency of the cell

Optical approaches to raise J_{SC}, and therefore efficiencies, are based on the increase of optical absorption caused by either an enlargement of the photon path length through the working electrode or light trapping effects occurring within the TiO_2 electrodes. J_{SC} can be attained by integrating the product of the ratio between the solar spectral irradiance and the photon energy density, $F(\lambda)$, and the photon-to-current conversion efficiency, IPCE, of the cell over the wavelength of the incident light:

$$J_{SC} = \int q\xi(\lambda)F(\lambda)IPCE(\lambda)d\lambda \qquad (3)$$

Here q is the electron charge and $\xi(\lambda)$ is a factor that accounts for the losses at the air-substrate interface. IPCE can be expressed as the product of light harvesting efficiency (LHE) and the electron-transfer yield $\Phi(\lambda)$, that is the product of the electron injection yield and the charge collection efficiency.

$$IPCE(\lambda) = LHE(\lambda)\Phi(\lambda) \qquad (4)$$

The LHE or optical absorptance, A, at a certain wavelength is defined as the fraction of incident photons that are absorbed by the dyed electrode: $A=I_A/I_0$, where I_0 is the incident intensity and I_A the intensity absorbed. Therefore the relationship between LHE and J_{SC} is given by the expression:

$$J_{SC} = \int q\xi(\lambda)F(\lambda)LHE(\lambda)\Phi(\lambda)d\lambda \qquad (5)$$

In a first approximation, absorptance for a standard dyed electrode is related with the extinction coefficient and the concentration of the dye. Since only a monolayer of dye molecules is attached to the surface of the TiO_2 nanoparticles, its total amount is directly related to the oxide layer thickness. For example, a 7 μm thick dye sensitized (N719) mesoscopic electrode can absorb nearly the 80% of incident photons at the maximum absorption wavelength. However, photons still in the visible range but of lower energy are weakly absorbed. Looking for a dye absorption enhancement, devices having thicker electrodes (around 10 μm) have been previously reported (Ito et al, 2006). Nevertheless, the thickness of TiO_2 layer cannot be increased at will without affecting its mechanical properties, reaching mass transport limitations in the electrolyte or/and reducing the photovoltage of the cell. In addition, electrons injected in the conduction band must travel a longer distance to reach the back contact, increasing the probability of recombination at grain boundaries and diminishing both current and voltage of the cell, as experimentally demonstrated (Ito et al., 2008). Finally, another disadvantage to scale-up the device would appear due to the high cost of sensitizer dyes (approx. 1000€/g). The thicker the electrodes, the higher the dye loads required, therefore raising the final cost of the DSSC. All these non-desirable features make it preferable to increase the absorption of the cell for a given dye and film thickness, modifying the optical path length within the film and improving the spectral response of the photoelectrode. Keeping in mind that a standard dye sensitized layer of around 7-8 μm thick will not absorb light strongly, these achievements can be obtained by reflecting light back to the dyed electrode. In what follows, we provide an overview of the different approaches taken towards the integration of optical passive components in order to increase the power conversion efficiency of DSSCs.

3.2 Diffuse scattering layer

TiO_2 working electrodes used in DSSC are composed of 20 nm size crystallites. These electrodes are essentially transparent since visible light is not scattered for titania particles of sizes on the order of the few tens of nanometers. In fact, the incident photons that are not absorbed by the dye sensitized electrode are either lost through the counter electrode or partially absorbed by the electrolyte solution. From a photo-chemical point of view, this implies that part of the reagent (light) is wasted. The first attempt to collect these escaping photons were based on the use of polydisperse packings of sub-micron size spheres as highly diffusive reflecting layers (see figure 4). The intensity of the scattering effect depends on the size and refractive index of the particles as well as on the refractive index of the medium surrounding them. Hence, particle sizes between 300 nm and 1000 nm made of transition metal oxides with high refractive index, such as TiO_2 (rutile n=2.8 or anatase, n=2.4) or ZrO_2 (n=2.1) can be used as efficient scatterers. The introduction of a reflecting layer in DSSC to scatter photons and re-inject them into the electrode was proposed in a theoretical work by Usami (Usami, 1997). In other attempt, Ferber and Luther simulated the scattering process for a mixture of small and large particles, concluding that an enhancement in photon absorption was produced (Ferber & Luther, 1998). Subsequent simulations using different approaches were made (Rothenberger et al., 1999), for which a 6% of increase of the photon flux was predicted.

The integration of scattering centre particles in DSSC can be experimentally done under different architectures. In one approach, they are jointly included with the TiO_2 nanocrystallites that form the electrode (Tachibana et al., 2002). In another, they are deposited as a second layer on top of the dye sensitized electrode in a well-known configuration referred to as "double-layer" (Hore et al., 2006). The latter configuration is normally the most employed. Other ways of integrating scattering layers made of submicron size disordered particles have also been reported (Wang et al., 2004), (Zhang et al., 2007). In general, the scattering layer is deposited using methods similar to those used for the TiO_2 electrode deposition, such as doctor blade, screen printing, etc. A porous network connection between both layers is needed to allow the dye load and a proper diffusion of the charge carriers.

Improvements of efficiency around 20% in average are reported using the arrangements before described for nc-TiO_2 layers with thicknesses between 5 and 7 microns. In fact, the efficiency record attained for a DSSC corresponds to a cell that incorporates a highly scattering layer (Chiba et al., 2006). The boost in efficiency is mainly a consequence of the increase of J_{SC} up to 20%. On the other hand, IPCE values can be incremented between 20% and 50% depending on the spectral region considered. The highest improvement is obtained in the red part of the spectrum, where the dye extinction coefficient is small. Additionally, this light-scattering layer has been shown to act not only as a photon-trapping system but also to be an equally active layer in photovoltaic generation (Zhang et al., 2007).

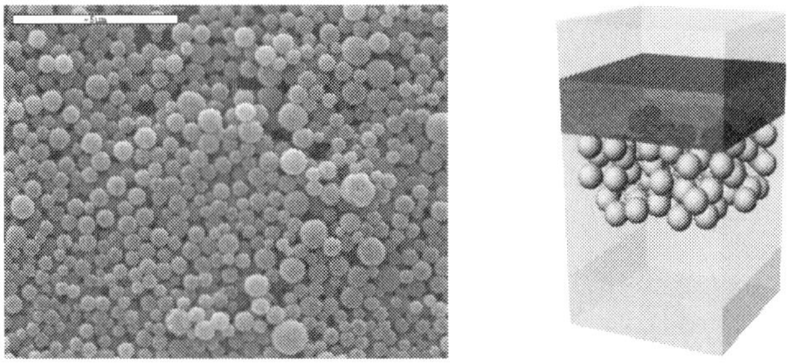

Fig. 4. Left: Scanning electron microcospy image of a slab made of polydisperse sub-micrometer TiO_2 particles that can be used as diffuse scattering layer in DSSC. Right: Scheme representing a DSSC with a "double layer" architecture.

The disadvantage of employing diffuse light scattering layers or mixed light scattering particles in DSSC is the loss of transparency of the cell. Unfortunately, the cells turn opaque, leaving them useless as window modules or any other application where transparency, one of the added values of these cells, was required. Also, the thickness of the electrode largely increases, particularly in the case of the double layer configuration, which might cause an increase of the resistance of the cell and a reduction of the voltage.

3.3 Periodic structures

An alternative to the use of disordered structures to enhance diffuse light scattering is the introduction of porous materials in which a periodic variation of the refractive index has been built up. As it will be shown next, periodic structures allows to achieve high reflectance within the cell at targeted and well-defined wavelength ranges, which may prevent the drawback of the loss of transparency. Also, in some cases, highly reflecting structures can be only a few hundreds of nanometres thick, which reduces the potential problems of increase of resistance and reduction of the photovoltage. Depending on the spatial dimensions where the modulation of refractive index is found, we will refer these structures as one, two, or three dimensional photonic crystals (1DPC, 2DPC, or 3DPC, respectively) (Joannopoulos et al. 1995). The interference effects associated with these periodic dielectrics give rise to the opening of a photonic band gap whose effect is detected as a maximum in the specular reflectance spectrum of the structure. In the following section, the effect of integrating both 3DPC and 1DPC in DSSC is described.

3.3.1 Three dimensional ordered structures

In 2003, Mallouk and co-workers proposed the use of a novel type of optical elements to improve the efficiency of DSSC. Their approach was based on the coupling of a particular type of 3DPC to dye-sensitized nc-TiO_2 films (Nishimura et al., 2003). This approach consisted of integrating a 3 μm layer of a TiO_2 ordered porous structure, known as inverse opal (see figure 5). By doing so, IPCE was shown to increase with respect to that of a standard cell used as reference (Figure 6a). Although the origin of this enhancement was first attributed to the reduction of the group velocity of photons near the edge of a stop band or photonic pseudogap, which implies the increase of the probability of absorption, it was later found that the absorption enhancement effect was partly due to the diffuse scattering caused by the imperfections present in the opal film (Halaoui et al., 2005) and, mainly, to the coupling between a standard electrode and the ordered structure, as explained next.,

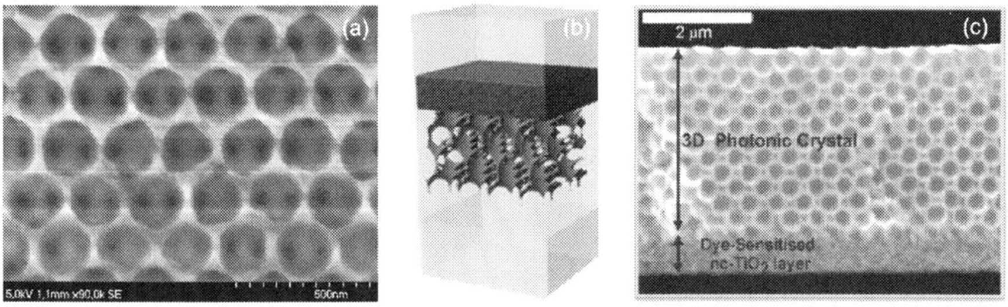

Fig. 5. (a) SEM cross section image of an inverse TiO_2 opal. 300 nm latex spheres were used as template, and then removed by thermal treatment. (b) Scheme and (c) SEM cross section image of a TiO_2 3D inverse opal deposited on top of a TiO_2 nanocrystalline electrode.

The theoretical analysis of the phenomena (Mihi & Míguez, 2005) demonstrated that the enhancement of the photocurrent in bilayer structures formed by a dye sensitized nanocrystalline TiO_2 film coupled to an inverse titania opal is mainly due to the surface resonant modes confined within the overlayer (figure 6b). This is a consequence of the mirror effect of the photonic crystal. The theoretical results identified an increase in the absorptance spectrum of the modelled bilayer and an estimated photocurrent enhancement factor very similar to that reported in Mallouk experiments (Figure 6c). Later experimental results confirmed convincingly the model proposed by Mihi & Míguez (Lee et al., 2008; Mihi et al., 2008) and as they predicted, the improvement was found for frequencies comprised within the photonic pseudogap of the inverse opal structure. In that way, the dependence of the increased IPCE with the position of the forbidden interval range can be employed to enhance absorption at desired spectral ranges. The selection of this region is made through the lattice parameter of the photonic crystal which is controlled as well by the size of the templating spheres and the infiltration degree. In these specific experiments, DSSC required an opal with a lattice parameter that gives red photocurrent enhancement.
Additionally it is possible to couple two or more 3DPC with different lattice parameters to enhance photocurrent in a wider spectral range (Mihi et al., 2006). Since this effect was confirmed theorically, we can mention that the disadvantage of these materials is the several steps involved to achieve these types of structures. This leads to defects within the 3D photonic crystal, and therefore a lower reflectivity is obtained.

3.3.2 DSSC coupled to 1DPC

Although the coupling of inverse opals to dye sensitized electrodes demonstrated an increased IPCE with respect to that of a reference cell (Nishimura et al., 2003), the main drawback of these 3D structures is the difficult assembly process to achieve reasonable reflecting periodic materials, which leads usually to thick structures (between 5-10 micron thick). This might have a deleterious effect on charge transport and recombination through the cell. Very recently, new types of one-dimensional photonic crystals (1DPC) have been prepared by alternate deposition of either mesoporous (Choi et al., 2006); (Fuertes et al., 2007) or nanoparticles (Wu et al., 2007); (Colodrero et al., 2008) based films. These structures

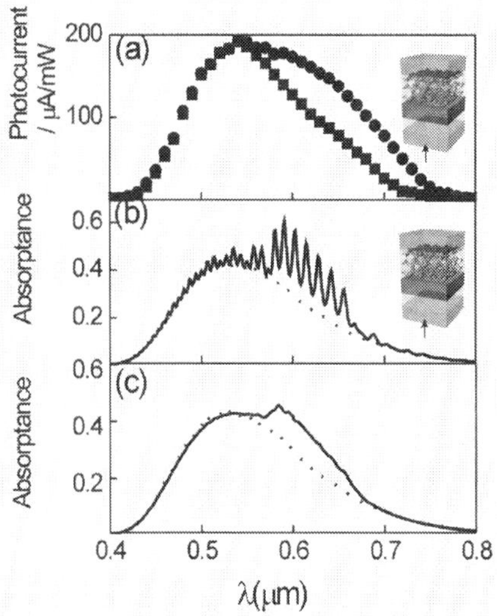

Fig. 6. (a) Circles: Photogenerated current observed for a bilayer DSSC like the one shown in the inset when illuminated from the rear side. Squares: Photocurrent corresponding to a non periodically structured, standard DSSC. These data have been extracted from Nishimura et al., 2004. The curve in (b) shows the calculated absorptance (or LHE) for the structure shown in the inset under rear illumination. (c) Averaged absorptance of bilayer DSSCs formed by nc-TiO_2 inverse opals of different width (from 3 to 17 sphere monolayers), each one of them having in turn different nc-TiO_2 layer thickness on top (from 6.5 μm to 7.5 μm). Dotted lines in (b) and (c) are the calculated absorptance spectra of standard DSSC having the same amount of absorbing material than in the bilayer system. The insets show schemes of the modelled structure. The corresponding illumination direction is indicated by an arrow. *(Extracted with permission from Mihi & Míguez, 2005)*

are usually easier to build and integrate than those abovementioned of higher dimensionality and present attractive features, such as very intense and wide Bragg reflections and reduced thickness (less than a micron versus the several micron thickness of opals). Furthermore, the advantage of such lattices lies on the wide range of materials available to be deposited as multilayers, which implies accurate control over the optical properties of the periodic ensemble, and on the high structural and optical quality attainable. These nanostructures could therefore be a potentially interesting alternative to other type of light scattering layers used within the solar cell field, having created high expectations due to the large improvement of the performance achieved for this type of devices.

In this section, we will focus on mesostructured Bragg reflectors in which the building blocks are nanoparticles of different sort (Colodrero et al., 2008) that can be easily coupled to DSSC to enhance the optical absorption. The novelty of these nanostructures is mainly the large and highly accessible interconnected mesoscopic porosity that they can present, which makes them suitable for this type of solar devices. In fact, some of the most successful approaches developed to improve the LHE in silicon photovoltaic systems are based on the implementation of coherent scattering devices such as highly reflecting distributed Bragg reflectors (Johnson et al., 2005), surface gratings (Llopis & Tobias, 2005), or a combination of both (Zeng et al., 2006). However, the implementation of such structures in DSSC had been no possible due to the need for porous back reflectors that allowed a proper flow of the electrolyte through the cell and, at the same time, due to the complicated deposition process of solid layers from colloidal suspensions.

The fabrication of DSSC containing nanoparticle based 1DPC involves two basic steps: first, the deposition of the nanocrystalline TiO_2 layer that acts as electrode onto a transparent conducting substrate, and second, the stacking of layers made of nanoparticles of different kind deposited alternately by spin-coating onto the sintered electrode. In this case, silica and titanium dioxide suspensions are employed because of the very high refractive index contrast they present, which allows achieving broad and intense Bragg reflections. The nanoparticle multilayer is periodic with a period of around a hundred nanometers, the thickness of these layers being controlled through either the concentration of the precursor suspensions or the rotation speed of the substrate during the spin coating process. Figure 7 shows a scheme of the described 1DPC based cell, as well as FESEM images corresponding to a cross section of both the nanocrystalline-TiO_2 electrode and the periodic structure deposited onto the former. The uniformity in the thickness of both types of layers composing the 1DPC, and even the different morphology of the particles employed, can be clearly distinguished in the picture below. The total thickness of the photonic crystal can vary between 0.5 to less than 2 microns, depending mainly on the lattice parameter of the structure and the number of layers that compose the periodic stack.

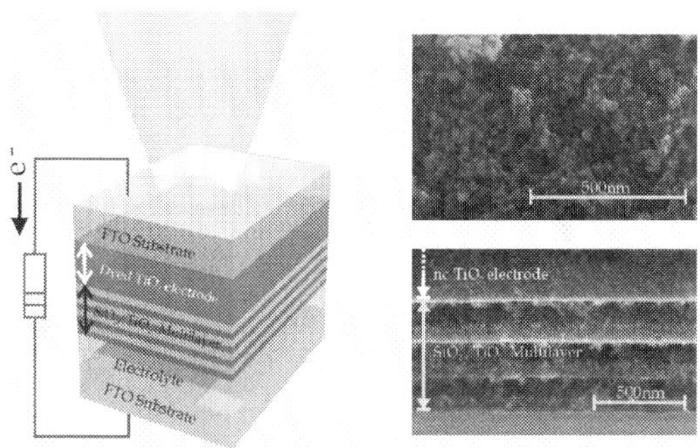

Fig. 7. Left: Design of a DSSC coupled to a nanoparticle based 1DPC. Right: FESEM images showing the TiO_2 nanocrystals forming the solar cell electrode (top), on which the porous periodic stack made of nanoparticles of different kind is deposited (bottom). In this case, a six layer photonic crystal has been implemented.

The procedure that follows to complete the solar cell is the same than the one usually employed for standard DSSC. It should be noticed that the nanoparticle multilayer integrated into the solar cell in this way behaves as a distributed Bragg reflector, providing the cell with a brilliant metallic reflection whose colour can be tuned by varying the thickness of the layers forming the periodic nanostructure. This can be readily seen in the photographs shown in Figure 8, in which the appearance of a reference cell and the same cell including two different 1DPC under perpendicular illumination are shown. Another remarkable issue from these systems is that the multilayer implemented like that does not alter significantly the cell semi-transparency, contrary to what happens when other scattering layers made of large titania nanoparticles are employed in DSSC to increase the photogenerated current. When these diffuse scattering layers are used, the solar cell becomes almost completely opaque as a consequence of the lack of spectral selectiveness of the incoherent scattering by slurries with a wide particle-size distribution. The comparison between the optical transmission spectra for the case of a standard reference cell (7.5 micron thick) and those corresponding to solar cells possessing the same TiO_2 electrode thickness but coupled to different 1DPC and to a diffuse scattering layer are also included in Figure 8.

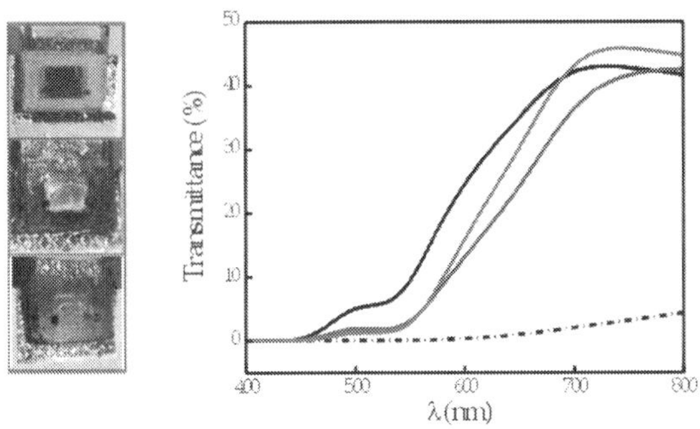

Fig. 8. Left: Images of a reference cell and different photonic crystal based DSSC. The brilliant colours displayed by the cell (bottom) arise from the periodic structures with different lattice parameter coupled to the dyed electrode (top image). Right: Transmittance spectra of a DSSC composed of a 7.5 micron thick electrode (black curve) and of the same electrode coupled to periodic structures with different lattice parameters (green and red curves). For comparison, the transmittance spectrum of a DSSC with the same electrode thickness but coupled to a 7.5 micron thick porous diffuse scattering layer is also plotted (black dashed line). *(Extracted with permission from Colodrero et al., 2009 [b])*

As explained in section 3.3.1, enhancement of optical absorption is primarily due to the partial localization of photons of certain narrow frequency ranges within the dyed TiO_2 electrode (that acts as absorbing layer) as a result of its coupling to the photonic crystal, which acts as a porous low-loss dielectric mirror (Mihi & Míguez, 2005). These optical modes could, in principle, be recognized as narrow dips in the reflectance spectra at frequencies located within the photonic band gap, the enhancement range being determined by the spectral width of the photonic band gap (Mihi et al., 2005). The first experimental demonstration of the mechanism of light harvesting enhancement that takes place in DSSC coupled to photonic crystals has been recently reported using the nanostructures under the scope of this section (Colodrero et al., 2009[a]). The effect of well defined optical absorption resonances was detected both in optical spectroscopy and photogenerated current experiments of very thin and uniform dye-sensitized TiO_2 electrodes coupled to high quality porous 1DPC, an unambiguous correspondence between them being established. This study demonstrated that light trapping within absorbing electrodes is responsible for the absorption enhancement that had previously been reported. Figure 9 shows the spectral response of the IPCE for three DSSC having increasing electrode thicknesses range from 350 nm thick to 1.5 micron thick but the same 1DPC implemented. In each case, an enhancement factor γ was calculated as the ratio between the IPCE of the 1DPC based cell and that of the reference one. The spectral behaviour of γ for each cell is compared to its corresponding optical reflectance measured under front-side illumination. It can be clearly seen that peaks of photocurrent correspond to the dips in reflectance are obtained, which are the fingerprint of optical resonant modes localized in a film coupled to a photonic crystal.

For these modes, matter-radiation interaction times are much longer; thus, the probability of optical absorption, and therefore the photogenerated current, is enhanced. As the thickness of the dye-sensitized electrode increases, the number of localized modes rises and so does the number of peaks in the γ curve. The presence of a photonic crystal not only enhances the photogenerated current but also allows one to vary the spectral photoelectric response of thin electrodes in a controlled manner. For instance, in the example shown in Figure 9a, the largest current is attained at λ = 470 nm instead of at λ = 515 nm, where the dye absorption curve reaches its maximum. Thus, the photonic crystal allows tailoring to measure the enhanced absorption window of the dye, and thus, its overlap with the solar spectrum.

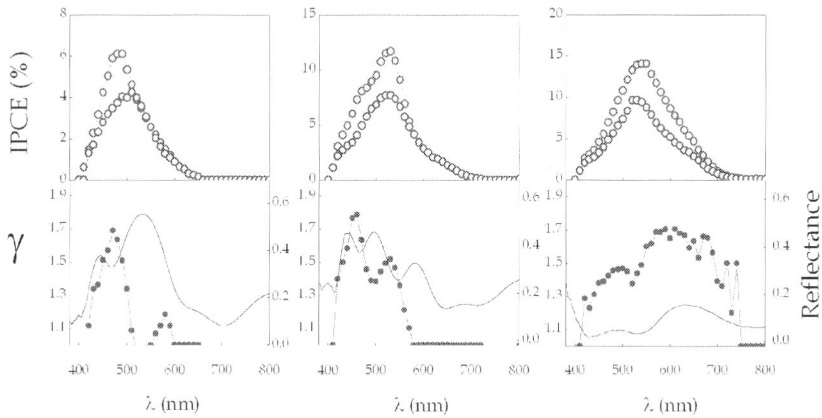

Fig. 9. Top: IPCE versus wavelength for cells containing the same 6 layer-1DPC coupled to dyed electrodes of increasing thicknesses in each case (from left to right). It is also plotted the IPCE for reference cells of the same electrode thickness without photonic crystal (blue circles). Bottom: Reflectance spectra of the 1DPC based solar cells (solid line) and the corresponding photocurrent enhancement factor (red circles). (*Extracted with permission from Colodrero et al., 2009 [a]*)

On the other hand, besides the experimental demonstration and confirmation of the light harvesting enhancement mechanism achieved using 1DPC based solar cells, great improvements in power conversion efficiency (η) have also been observed in this type of solar devices coupled to highly reflecting nanostructures (Colodrero et al., 2009 [a]), (Colodrero et al., 2009 [b]). After analyzing the photocurrent density-voltage (J-V) curves under 1 sun illumination of DSSC, on which photonic crystals reflecting different ranges of wavelengths were coupled, it was found that the photocurrent was largely improved while leaving the open-circuit voltage almost unaltered. The magnitude of this effect depends mainly on two factors: first, the spectral width and position of the photonic band gap relative to the absorption band of the ruthenium dye; second, the degree of optical coupling to the dye-sensitized electrode, which depends in turn on the thickness of that electrode. The magnitude of the photocurrent enhancement effect caused by the coupling to the 1DPC is therefore expected to be lower as the thickness of the electrode increases, since more photons are absorbed by the dyed nc-TiO_2 layer when they first pass through it. For this case, red reflecting 1DPCs might perform better, since the ruthenium dye captures less effectively solar radiation precisely for λ>600 nm. Results on the power conversion efficiency (η) for DSSC with a 7.5 micron thick electrode coupled to 1DPC reflecting different ranges of wavelengths (green and red) are shown in figure 10. An enhancement in the efficiency close to 20% with respect to that of the reference cell was obtained for the 1DPC based solar cell showing a better matching with the absorption spectrum of the dye.

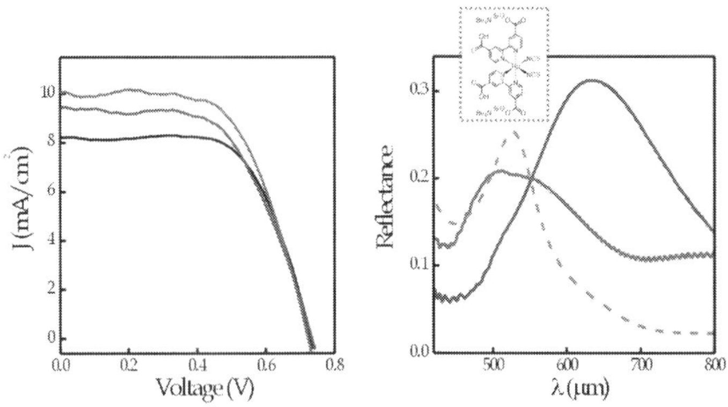

Fig. 10. Left: IV curves of a 7.5 micron thick electrode coupled to different 1DPC under 1 sun illumination. The corresponding IV curve for a reference cell is also plotted (black line). Right: Reflectance spectra measured under frontal illumination conditions of the PC-based solar cells together with the absorption spectrum of the Ru-dye (arbitrary units).

The photocurrent enhancement reported using PC based solar cells could be even larger at lower incident radiation intensities, reaching up to 30% of the reference value under 0.1 sun for the samples above described. This is mainly due to the decrease of density of carriers when so does the incident light intensity, which has a positive effect on electron transport and recombination through the cell. Besides, any resistance potentially introduced by the photonic crystal will have a minor effect at lower illumination conditions, since its effect increases with the number of carriers. In order to illustrate this effect, values of efficiency, photogenerated current and open-circuit voltage obtained for DSSC having 7.5 micron thick electrodes coupled to different 1DPC are presented in Figure 11. The variation of J_{sc} and V_{oc} with intensity of the incident radiation confirms that the presence of the PC enhances the photocurrent significantly, but has a minor effect on the photovoltage. The linear and logarithmic dependence observed for Jsc and Voc, respectively, versus incident radiation intensity are in good agreement with theoretical predictions (Nazeerudin et al., 1993); (Södergren et al., 1994).

To conclude this section and in order to prove the performance of these nanoparticle based structures as light harvesters, DSSC based on both 1DPC and diffuse scattering layers were evaluated and compared. For this purpose, a 7.5 micron diffuse scattering layer made of titania spheres 130 nm in diameter mixed with a paste similar to that employed to prepare the nanocrystalline titania layer was deposited onto a 7 micron thick reference electrode. A similar electrode was coupled to a 700 nm thick highly reflecting 1DPC. In order to perform a comparison of the effect on light harvesting these different architectures have, the 7 micron thick diffuse scattering layer was electrically isolated from the dye-sensitized electrode by introducing a thin layer of silica spheres 30 nm in diameter between them. By doing so, no contribution to the photocurrent from the different scattering layers employed is measured, since the 1DPC is also based on alternate layers of SiO_2 and TiO_2 nanoparticles, the first layer deposited onto the electrode being insulating. The effect of the PC on the short circuit photocurrent is observed to be similar and comparable to that of a diffuse scattering layer, provided that a suitable PC is chosen, as displayed in figure 12. Furthermore, the open circuit voltage is slightly higher in the case of the PC, which might be due to its much smaller width. It should be reminded that the enhancement in the case of PC is based on the partial confinement of light of a selected frequency range within the absorbing electrode, whereas in the case of diffuse scattering layers, the increase in efficiency is based on the random and non-selective scattering of visible light in all directions.

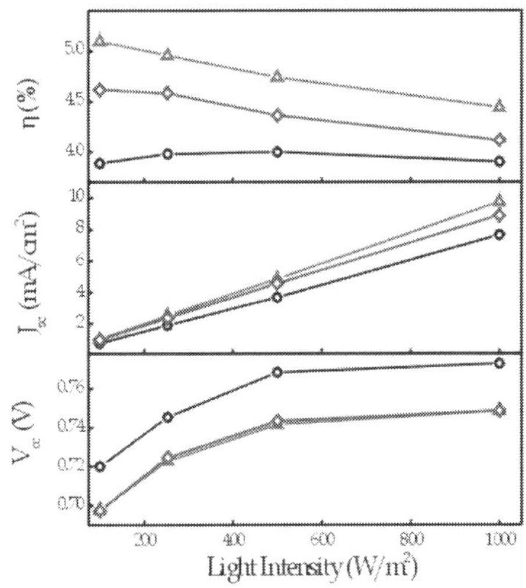

Fig. 11. Efficiency (η), short-circuit current density (J_{sc}) and open-circuit voltage (V_{oc}) for a reference cell (open black circles) and for those PC based cells having the same electrode thickness coupled to 1DPC with different lattice parameters (green and red symbols) under illumination at different light intensities. *(Extracted with permission from Colodrero et al., 2009 [b])*

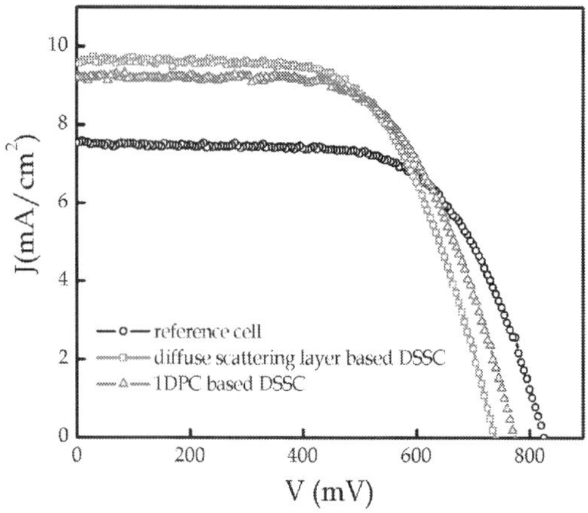

Fig. 12. Comparison between the efficiencies for a DSSC made of a 7 micron thick electrode and those corresponding to the same electrode thickness coupled to a diffuse scattering layer and a 1DPC. The thicknesses employed for the diffuse scattering layer and the periodic structure are 7.5 micron and 700 nm, respectively. (*Extracted with permission from Colodrero et al., 2009 [b]*)

4. Conclusions

Colloidal chemistry approaches are suitable for implementing optical devices of high quality in DSSC in order to improve their performance. This opens the door for the conscious optimization of the photonic design of DSSC, as is commonly done for their silicon counterparts. This may also open the way to amplifying the absorption of other dyes with low extinction coefficients that cover other regions of the visible and near-IR solar spectrum. In this respect, a thorough analysis in terms of the interplay between the effect of the electrode thickness, the dye absorption spectrum, and the characteristics of the Bragg reflection, such as intensity, spectral position and width, is needed for designing 1DPC based DSSC of optimized performance. On the other hand, the greater enhancement of efficiency attained for thin electrodes coupled to these photonic structures highlight the potential that they might have in other cells using very thin absorbing layers, in which the main source of loss of efficiency is frequently the low amount of light absorbed.

5. References

Bach, U.; Lupo, D.; Comte, P.; Moser, J.E.; Weissortel, F.; Salbeck, J.; Spreitzer, H. & Graetzel, M. (1998). Solid-state dye-sensitized mesoporous TiO2 solar cells with high photon-to-electron conversion efficiencies. *Nature*, 395, 6702 (October 1998) 583-585, 0028-0836

Baruch, P. (1985). A 2-level system as a model for a photovoltaic solar cells. *Journal of Applied Physics*, 57, 4, (February 1985) 1347-1355, 0021-8979

Cao, F.; Oskam, G.; Meyer, G.J. & Searson, P.C. (1996). Electron transport in porous nanocrystalline TiO2 photoelectrochemical cells. *Journal of Physical Chemistry* 100, 42, (October 1996) 17021-17027, 0022-3654

Colodrero, S.; Ocaña, M. & Míguez, H.(2008) Nanoparticle-based one-dimensional photonic crystals. *Langmuir* 24, 9, (May 2008) 4430-4434, 0743-7463

Colodrero, S.; Mihi, A.; Anta, J.A.; Ocaña, M. & Míguez, H. (2009) [a]. Experimental Demonstration of the Mechanism of Light Harvesting Enhancement in Photonic-Crystal-Based Dye-Sensitized Solar Cells. *Journal of Physical Chemistry C, 113, 4, (January 2009) 1150-1154,* 1932-7447

Colodrero, S.; Mihi, A.; Anta, J.A.; Ocaña, M. & Míguez, H. (2009) [b]. Porous One-Dimensional Photonic Crystals Improve the Power-Conversion Efficiency of Dye-Sensitized Solar Cells. *Advanced Materials,* 21,7, (February 2009) 764-768, 0935-9648

Chiba, Y.; Islam, A.; Komiya, R.; Koide, N. & Han, L. (2006). Conversion efficiency of 10.8% by a dye-sensitized solar cell using a TiO_2 electrode with high haze. *Applied Physics Letters,* 88, 22 (May 2006) 223505, 0003-6951

Choi, S. Y.; Mamak, M.; Freymann von, G.; Chopra, N. & Ozin, G. A. (2006). Mesoporous Bragg stack color tunable sensors. *Nano Letters* 6, 11 (November 2006) 2456-2461, 1530-6984

Desilvestro, J.; Gratzel, M.; Kavan, L.; Moser, J. & Augustynski, J. (1985). Highly Efficient Sensitization of Titanium Dioxide. *Journal of the American Chemical Society,* 107, 10, (May 1985) 2988-2990, 0002-7863

Ferber, J. & Luther, J. (1998). Computer simulations of light scattering and absorption in dye-sensitized solar cells. *Solar Energy Materials and Solar Cells,* 54, 1-4, (August 1998) 265-275, 0927-0248

Fuertes, M. C.; López-Alcaraz, F. J.; Marchi, M. C.; Troiani, H. E.; Míguez, H. & Soler Illia, G. J. A. A. (2007). Photonic crystals from ordered mesoporous thin-film functional building blocks. *Advanced Functional Materials, 17,* 8, (May 2007) 1247-1254, 1616-301X

Fujishima, A. & Honda, K. (1972). Electrochemical photolysis of water at a semiconductor electrode. *Nature,* 38, 5358, (1972) 37, 0028-0836

Gerisher H. & Tributsch, H. (1968). Electrochemistry of ZnO monocrystal spectral sensivity. *Berichte der Bunsen-gesellschaft fur physikalische chemie,* 72, 3, (March 1968) 437

Graetzel, M. (2000). Perspectives for dye-sensitized nanocrystalline solar cells. *Progress in Photovoltaics: Research and Applications,* 8, 1, (February 2000), 171-185, 1062-7995

Graetzel, M & Durrant, J. (2008) Dye Sensitized Mesoscopic Solar Cells, In: *Nanostructured and Photoelectrochemical Systems for Solar Photon Conversion,* Mary D Archer & Arthur J Nozik, (Ed.), (503-536), Imperial College Press., 978-1-86094-255-6, England

Halaoui, L.I.; Abrams, N.M. & Mallouk, T. (2005) Increasing the conversion efficiency of dye-sensitized TiO2 photoelectrochemical cells by coupling to photonic crystals. *Journal of Physical Chemistry B,* 109, 13 (April 2005), 6334-6342, 1520-6106

Haque, S.A.; Palomares, E.; Cho, B.M.; Green, A.N.M; Hirata, N.; Klug, D.R. & Durrant, J.R. (2005). Charge separation versus recombination in dye-sensitized nanocrystalline solar cells: the minimization of kinetic redundancy. *Journal of the American Chemical Society,* 127, 10, (March 2005) 3456-3462, 0002-7863

Hore, S.; Vetter, C.; Kern, R.; Smit, H. & Hinsch, A. (2006). Influence of scattering layers on efficiency of dye-sensitized solar cells. *Solar Energy Materials and Solar Cells,* 90, 9, (May 2006) 1176-1188, 0927-0248

Huang, S.Y.; Schlichthorl, G.; Nozik, A.J.; Graetzel, M. & Frank, A.J. (1997). Charge recombination in dye-sensitized nanocrystalline TiO2 solar cells. *Journal of Physical Chemistry B,* 101, 41, (April 1997) 2576-2582, 1089-5647

Ito, S.; Nazeeruddin, M.K.; Liska, P.; Comte, P.; Charvet, R.; Pechy, P.; Jirousek, M.; Kay, A.; Zakeeruddin, S.M. & Graetzel, M. (2006).Photovoltaic Characterization of Dye-sensitized Solar Cells: Effect of Device Masking on Conversion Efficiency. *Progress in photovoltaics,* 14, 7, (November 2006) 589-601, 1062-7995

Ito, S.; Murakami, T.N.; Comte, P.; Liska, P.; Graetzel, C.; Nazeerudin, M.K. & Graetzel, M. (2008). Fabrication of thin film dye sensitized solar cells with solar to electric power conversion efficiency over 10%. *Thin solid films,* 516, 14, (May 2008) 4613-4619, 0040-6090

Joannopoulos, J. D.; Meade, R. D. & Winn, J. N. (1995). *Photonic Crystals: Molding the Flow of Light*; Princeton University Press, 0691-03744-2, Princeton, NJ, USA.

Johnson, D. C.; Ballard, I.; Barnham, K. W. J.; Bishnell, D. B.; Connolly, J. P.; Lynch, M. C.; Tibbits, T. N. D.; Ekins-Daukes, N. J.; Mazzer, M.; Airey, R.; Hill, G. &. Roberts, J.S.(2005). Advances in Bragg stack quantum well solar cells. *Solar Energy Materials and Solar Cells,* 87, 1-4, (May 2005) 169-176, 0927-0248

Kamat, P. (2007). Meeting the clean energy demand: Nanostructure architectures for solar energy conversion. *Journal of Physical Chemistry C,* 111, 7, (February 2007) 2834-2860, 1932-7447

Lee, S.H.A.; Abrams, A.; Hoertz, P.G.; Barber, G.D.; Halaoui, L.H. & Mallouk, T.H. (2008). Coupling of Titania Inverse Opals to Nanocrystalline Titania Layers in Dye-Sensitized Solar Cells. *Journal of Physical Chemistry B,* 112, 46, (November 2008) 14415-14421, 1520-6106

Llopis, F. & Tobias, I. (2005). The role of rear surface in thin silicon solar cells. *Solar Energy Materials and Solar Cells,* 87, 1-4, (May 2005) 481-492, 0927-0248

Mihi, A. & Miguez, H. (2005). Origin of light-harvesting enhancement in colloidal-photonic-crystal-based dye-sensitized solar cells. *Journal of Physical Chemistry B,* 109, 33, (August 2005) 15968-15976, 1520-6106

Mihi, A.; Míguez, H.; Rodríguez, I.; Rubio, S.; Meseguer, F. (2005). Surface resonant modes in colloidal photonic crystals. *Physical Review B,* 71, 12 (March 2005) 125131, 1098-0121

Mihi, A; López-Alcaraz, F.J.; Míguez, H. (2006). Full spectrum enhancement of the light harvesting efficiency of dye sensitized solar cells by including colloidal photonic crystal multilayers. *Applied Physics Letters,* 88, 19 (May 2006), 193110, , 0003-6951

Mihi, A.; Calvo, M. E.; Anta, J. A. & Míguez, H. (2008). Spectral response of opal-based dye-sensitized solar cells. *Journal of Physical Chemistry C,* 112, 1, (January 2008) 13-17, 1932-7447

Mori, S. and Yanagida, S. (2006) Dye Sensitized Solar Cells, In: *Nanostructured Materials for Solar Energy Conversion,* Tatsuo Soga, (Ed.), (193-225), Springer B.V., 0-444-52844-X, England

Nazeerudin, M. K.; Kay, A.; Rodicio, I.; Humphry-Baker, R.; Müller, E.; Liska, P.; Vlachopoulos, N. & Graetzel, M. (1993). Conversion of light to electricity by cis-X2bis(2,2'-bipyridyl-4,4'-dicarboxylate)ruthenium(II) charge-transfer sensitizers (X = Cl-, Br-, I-, CN-, and SCN-) on nanocrystalline titanium dioxide electrodes *Journal.of the American Chemical Society* 115, 14, (July 2003) 6382-6390, 0002-7863

Nelson, J.; Haque, S.A.; Klug, D.R. & Durrant, J.R. (2001). Trap-limited recombination in dye-sensitized nanocrystalline metal oxide electrodes. *Physical Review B*, 63, 20 (May 2001) 205321, 0163-1829

Nishimura, S.; Abrams, N.; Lewis, B.; Halaoui, L.I.; Mallouk, T.E.; Benkstein, K.D.; Van de Lagemaat, J. & Frank, A.J. (2003). Standing wave enhancement of red absorbance and photocurrent in dye-sensitized titanium dioxide photoelectrodes coupled to photonic crystals. *Journal.of the American Chemical Society* 125, 20, (May 2003) 6306-6310, 0002-7863

Rothenberger, G.; Comte, P. & Graetzel, M. (1999). A contribution to the optical design of dye-sensitized nanocrystalline solar cells. *Solar Energy Materials and Solar Cells* 58, 3, (July 1999) 321-336, 0927-0248

Schwarzburg, K. & Willig, F. (1999). Origin of photovoltage and photocurrent in the nanoporous dye-sensitized electrochemical solar cell. *Journal of Physical Chemistry B* 103, 28, (July 1999) 5743-5746, 1089-5647

Södergren, S.; Hagfeldt, A.; Olsson, J. & Lindquist, E. (1994). Theoretical Models for the Action Spectrum and the Current-Voltage Characteristics of Microporous Semiconductor Films in Photoelectrochemical Cells. *Journal of Physical Chemistry* 98, 21, (May 1994) 5552-5556, 0022-3654

Tachibana, Y.; Hara, K.; Sayama, K. & Arakawa, H. (2002). Quantitative analysis of light-harvesting efficiency and electron-transfer yield in ruthenium-dye-sensitized nanocrystalline TiO_2 solar cells. *Chemistry of Materials* 14, 6, (June 2002) 2527-2537, 0897-4756

Tenakkone, K.; Kumara, G.R.R.A.; Kottegoda, I.R.M. & Perera, V.P.S. (1999). An efficient dye-sensitized photoelectrochemical solar cell made from oxides of tin and zinc. *Chemical Communications*, 1, (January 1999) 15-16, 1359-7345

Usami, A. (1997). Theoretical study of application of multiple scattering of light to a dye-sensitized nanocrystalline photoelectrochemical cell. *Chemical Physics Letters* 277, 1-3, (October 1997) 105-108, 0009-2614

Vlachopoulos, N.; Liska, P.; Augustynski, J. & Gratzel, M. (1988). Very Efficient Visible Light Energy Harvesting and Conversion by Spectral Sensitization of High Surface Area Polycrystalline Titanium Dioxide Films. *Journal of the American Chemical Society* 110, 4, (February 1988) 1216-1220, 0002-7863

Wang, P.; Zakeeruddin, S.M; Moser, J.E.; Humphry-Baker, E. & Graetzel, M. (2004). A solvent-free, SeCN-/(SeCN)(3)(-) based ionic liquid electrolyte for high-efficiency dye-sensitized nanocrystalline solar cells. *Journal of the American Chemical Society, 126, 23* (June 2004) 7164-7165, 0002-7863

Wang, P.; Klein, C.; Humphry-Baker, R.; Zakeeruddin, S.M. & Graetzel, M. (2005). A high molar extinction coefficient sensitizer for stable dye-sensitized solar cells. *Journal of the American Chemical Society,* 127,3, (January 2005) 808-809, 0002-7863

Wu, Z.; Lee, D.; Rubner, M.F. & Cohen, R.E. (2007) Structural color in porous, superhydrophilic, and self-cleaning SiO_2/TiO_2 Bragg stacks. *Small* 3, 8 (August 2007) 1445-1451, 1613-6810

Zaban, A.; Meier, A. & Gregg, B.A. (1997). Electric potential distribution and short-range screening in nanoporous TiO2 electrodes. *Journal of Physical Chemistry B* 101, 40, (October 1997) 7985-7990, 1089-5647

Zeng, L.; Yi, Y.; Hong, C.; Liu, J.; Feng, N.; Duan, X.; Kimerling, L.C. & Alamariu, B.A. (2006). Efficiency enhancement in Si solar cells by textured photonic crystal back reflector. *Applied Physics Letters,* 89, 11, (September 2006) 111111, 0003-6951

Zhang, Z. et al. (2007). The electronic role of the TiO_2 light-scattering layer in dye-sensitized solar cells. *Zeitschrift für physikalische chemie* 221, 3, (March 2007) 319-327, 0942-9352